Analysis and Behavior of Structures

EDWIN C. ROSSOW
Northwestern University
Robert R. McCormick School of Engineering and Applied Science

Prentice Hall, Upper Saddle River, New Jersey 07458

Library of Congress Cataloging-in-Publication Data

Rossow, Edwin C.

    Analysis and behavior of structures / Edwin C. Rossow.

      p. cm.

    Includes bibliographical references and index.

    ISBN 0-02-403913-6

    1. Structural analysis (Engineering)   2. Structural dynamics.

    I. Title.   95-3647

    TA645.R683   1996   CIP

    624.1'71—dc20

Editor: Bill Stenquist
Editorial production service: Electronic Publishing Services Inc.
Manufacturing buyer: Donna Sullivan
Marketing manager: Dan Kaveney
Cover designer: Wendy Alling Judy
Cover photo: Edwin C. Rossow
Editorial assistant: Meg Weist

© 1996 by Prentice-Hall, Inc.
Simon & Schuster/A Viacom Company
Upper Saddle River, NJ 07458

The author and publisher of this book have used their best
efforts in preparing this book. These efforts include the
development, research, and testing of the theories and
programs to determine their effectiveness. The author and
publisher shall not be liable in any event for incidental or
consequential damages in connection with, and arising out of,
the furnishing, performance, or use of these programs.

Printed in the United States of America
10 9 8 7 6 5 4 3 2 1

ISBN 0-02-403913-6

Prentice-Hall International [UK] Limited, *London*
Prentice-Hall of Australia, Pty. Limited, *Sydney*
Prentice-Hall Canada Inc., *Toronto*
Prentice-Hall Hispanoamericana, S.A., *Mexico*
Prentice-Hall of India Private Limited, *New Delhi*
Prentice-Hall of Japan, Inc., *Tokyo*
Simon & Schuster Asia Pte. Ltd,, *Singapore*
Editora Prentice-Hall do Brasil, Ltda., *Rio de Janeiro*

To Janet Vieregg Rossow
1937–1995

Contents

Chapter 4 _____

Chapter 5 _____

Chapter 6 _____

Chapter 7 _____

Contents

Contents

Preface

I had two primary objectives in writing this book. First, I wanted to provide students with a valuable learning tool that they feel comfortable using. Second, I wanted to provide instructors with a text that is easy to use and flexible enough to fit comfortably with the goals and requirements of most any curriculum. This book is intended to be used in a first course in structural analysis in the junior or possibly senior year. It may be used either as a reference or supplemental text in a second-term course in analysis or as the main text for that course, depending on the nature of the curriculum.

The book is written so that both the beginning and serious student with an in-depth interest in structural analysis can benefit from its use. The major concepts of structural analysis are presented succinctly as boxed text or in carefully constructed figures. Example problems are worked in detail with easy-to-follow step-by-step procedures. The final section of each chapter summarizes the important ideas and the limitations of the theory. Additional discussion of the figures and the examples is easily identified in the text for those students who want or need additional exposure to the material. Finally, the body of the text develops the ideas in a logical and straightforward manner.

The book also serves as a reference for the civil engineer and students who work as practicing structural engineers. Not only is the material presented and discussed in detail, but of particular value is the listing at the end of each chapter of the limitations and underlying assumptions of the topics presented in that chapter. Engineers working with computer programs that perform the analysis and design of structures will find these succinct listings a valuable aid to their evaluation of the computer program results. The extension of the classical techniques of analysis in the later sections of Chapter 9 and 10 to nonlinear problems follows as a continuation of linear analysis. This provides students with a valuable introduction to nonlinear analysis, and practicing engineers with a way to check and evaluate the results of computer programs that are capable of performing nonlinear analysis.

The manner in which the book is used in a first course is dependent on educational objectives and the background of the students. However, I would suggest the use of Chapters 1 to 12. This represents a significant amount of material for a single term, so the topics covered in Sections 1.10, 2.5, 4.8, 4.9, 8.7, 10.5 to 10.7, 11.5 and 12.5, and Chapter 9 can be considered as optional or omitted without loss of continuity. In some first courses Chapters 16 and 17 can be included after the material in the first four sections of Chapter 11 has been assigned. The material presented in Chapter 17 presumes that Sections 16.1 to 16.4 have been read and understood.

An unusual opportunity is available to the instructor of a first course in structural analysis. In Sections 3.2 and 5.2, use of the displaced configuration of structures is discussed in writing the equilibrium equations. I have found over the years that students intuitively reason that some changes in geometry must occur in a loaded structure due to member deformations, and they wonder why these changes are not considered in the analysis. The discussion in these sections point out the problems of including the deformations and lead to the conclusion that they can usually be neglected. I believe that these discussions not only heighten the awareness of students to one of the assumptions of the mathematical models for analysis, but also provides some perspective in which they can judge for themselves the reasons for neglecting the effects of structural deformations.

In part, Chapter 6 is a review of material that students have been exposed to in mechanics of materials courses, but the instructor may want to use it to refresh some of the ideas associated with simple bending theory before considering deformation computations using virtual work principles in Chapter 7. A special opportunity exists in Section 6.2, in which the assumptions and limitations of simple bending theory are discussed in some detail. In this section I provide an introduction to the reasons behind a number of the rules and limitations that are set forth in design specifications and codes. Instructors who wish to enhance the background and understanding by students of individual-member bending behavior before undertaking frame analysis will find that the first three or four sections accomplish this end.

The axial shortening of members due to bending curvature is discussed in Section 6.7. Instructors may be tempted to omit this section or make it optional, particularly in a first course. However, my experience has been that even in a first course, students wonder about the magnitude of this effect. I find it useful to refer to this section as a means of establishing a perspective on bending behavior and its relation to deformation behavior of frames.

A second course in analysis is probably populated by students who have a strong interest in structural engineering, for which a more in-depth exposure to analysis is very important. I recommend that the omitted sections and chapters in the first course be assigned. Sections 9.4, 9.5, and 10.8, which deal with nonlinear analysis, provide the instructor with an opportunity to introduce this topic at a level that is appropriate to third- or fourth-year undergraduates.

Sections 1.10, 2.5, 10.5, 11.5, and 12.5 deal specifically with the analysis of symmetric structures. The use of symmetry provides a powerful means of checking the consistency of computer-generated analyses as well as reducing the effort in an analysis, be it a hand or a computer analysis. The uses and limitations of symmetry are set forth carefully in these sections and are discussed in the context of linear or nonlinear material and geometrically nonlinear environments. I feel that these sections provide students of structural engineering with excellent opportunities to learn and understand some very important concepts.

A review of Chapter 6 in a second course will be helpful to structural engineering students. Section 6.2 provides a very compact presentation of the limitations of simple bending theory and anticipates the requirements set forth in design specifications for structural members. In Section 6.4 I review the classical moment-area theorems and generalize them for use in making deformation calculations with nonlinear moment–curvature relations. I, as others before me, have called them the curvature-area theorems to emphasize that generality. Section 6.6 presents a general procedure for the calculation of axial deformations that is usually not presented in mechanics of materials courses. These sections provide the necessary introduction to the methods of analysis for nonlinear materials set forth in Chapter 9 and subsequent chapters.

The classical portal and cantilever methods of approximate analysis of rigid frames are developed in Chapter 14. The portal method is presented in a rather general form that accommodates base fixity conditions that are either pinned or fixed. The cantilever method is derived in a form that is slightly different from that which is commonly presented, but one that emphasizes the slender beam assumption on which the method is based. Finally, in Section 14.6, I present a method of estimating lateral displacements of rigid frames that is not found in other books but is simple to apply for drift calculations.

In Chapter 15 I review and extend the energy methods introduced in mechanics of materials courses and the virtual work principles established in Chapter 7. The relationships between the virtual work principles, Total potential energy principles, and Castigliano's theorems are discussed. Section 15.3 provides an introduction to the concepts of the finite element method. Sections 15.1 and 15.2 also provide the basis for the development of the stiffness matrices presented in Chapters 16 and 17 through the use of energy principles if the instructor wishes to present that approach. In Sections 15.5 and 15.6 I introduce the shear building and the single-bay building as simple models for approximate analysis of frames based on energy principles. These models, which are popular in computer programs for approximate dynamic analysis, are presented in sufficient detail so that students understand how internal moments can be obtained from the deflection solution.

Problems are provided at the end of each chapter and answers to approximately one-third of the problems appear at the end of the book. There are computer problems available at the ends of many chapter which require students to make graphical presentations of the solutions. The intent of these problems is to emphasize some aspect of the behavior of the mathematical model of a structure through a parameter variation. Students do not need special software packages to solve these problems, although commercial software or software that is available on computers that are supplied for student use by the school or department can be used. I encourage the assignment of these problems because their solution will enhance the development of intuition of structural behavior.

Acknowledgments

There is no doubt that this book could not have been completed without the encouragement, patience, unselfish support, and love of my late wife, Janet. While she is not here to help with the finishing touches, her imprint is nevertheless found in many parts of this manuscript.

I wish to thank the many people who have contributed to the completion of this project. I am much indebted to William Stenquist, Executive Editor Prentice Hall, for his faith in this project and his guidance during the tumultuous times of the merger of two publishing companies. The staff of Electronic Publishing Services Inc. was extremely helpful to this first-time author and completed in a very pleasant and cooperative manner the tasks required for publication, including copyediting, proofreading, and production supervision.

The reviews and comments of Dr. Subhash C. Anand of Clemson; Dr. Karen C. Chou of the University of Tennessee, Knoxville; Dr. Ross B. Corotis of the University of Colorado at Boulder; Dr. Phillip L. Gould of Washington University, and Dr. William J. Hall of the University of Illinois, Urbana-Champaign, contributed significantly to the organization and content of the book. I wish to extend special thanks to my colleagues Dr. Chou and Dr. Corotis for generously sharing their time in discussions with me about the book and for the several thoughtful improvements they provided.

I want to thank Sonya Kalfus, Dorothy Carlson, Mary Hill, and Richard Cox, who provided much of the word processing and typing of the equations at various stages in the preparation of the manuscript. The hard work of students Dean Kokubun, Carl W. Kreiter, Matthew C. Olson, and Roman T. Przepiorka, who were of great assistance in the preparation of the document and the creation of the solution manual, is greatly appreciated.

I am grateful for the encouragement of my colleagues and friends at Northwestern and, in particular, the support of Raymond J. Krizek, past Chair of the Civil Engineering Department. Finally, I would like to thank Dean Jerome B. Cohen of the McCormick School of Engineering and Applied Science and Northwestern University for granting me a partial leave of absence so that this project might be completed.

Preface to the Student User

You have in your possession a book that has the potential to be an awesome learning tool. I say *potential,* because just ownership of a saw and a hammer does not make you a carpenter until you have learned how to use them properly. I hope that a few observations and suggestions on my part will enhance the value of this book to your enjoyment and learning of structural engineering.

Be sure to look at both the introductory material at the beginning of each chapter and the final section at the end of the chapter. The introductory material is relatively short and puts into perspective the material that is presented in the chapter. The final section, always entitled "Summary and Limitations," provides a succinct review of the concepts discussed in the chapter and an enumeration of the limitations of these concepts. Knowing and understanding the limitations of your tools gives you peace of mind and contributes to your being a successful practicing professional engineer.

The usual expositions and discussions of structural analysis techniques in the body of the text are enhanced with three major modes of presentation of important ideas and concepts. First, there is boxed text, which sets important concepts apart from the descriptive material. Second, many of the figures have been constructed to convey a single important concept. These figures include a discussion of the concept that is set apart from the general text for easy identification. Third, the examples have been presented in a detailed form that is easy to follow. There is, however, additional discussion of the examples that, as with figures, is set apart from the general text.

When it comes time to review material for an examination, you will find an additional feature in the book that will be useful. Many of the examples have been worked in general form with parameters such as loads, dimensions, and member properties appearing as symbols. Numerical values for these parameters are provided at the end of the discussion of the examples. In reviewing, take the example problem as one having the numerical values listed, solve with these values, and compare your answer with that of the example. Any discrepancies can be resolved by comparing your calculations with those of the example and resolving the differences.

I hope you will find the study of structural analysis both interesting and enlightening. I encourage you to work some of the computer-oriented problems at the end of problem sets. They provide you with some insight into the behavior of structures as parameters that establish the configuration and loading of the structure are varied. I wish you success in the completion of your education and in your future professional practice.

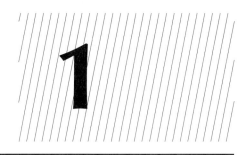

Introduction to Structural Engineering

Engineering and engineering skills are as old as civilization itself. By considering as engineering any modification or use of the natural materials and resources at hand to improve productivity or the environment, it can be argued that everyone has innate engineering skills that are used daily. Whether it be the folding of a piece of paper to create a funnel or the building of a birdhouse for the backyard, everyone practices engineering at some level. What distinguishes those who practice engineering professionally from everyone else is the special training the former receive with their education.

Civil and structural engineering are perhaps the oldest engineering professions. A long and rich history of building extends back more than 5000 years, with structures from that time still in existence today. Architects and engineers were one-in-the-same in those early times and were known as master builders. As time passed, the complexities of architecture and engineering reached levels where single individuals could no longer be competent as master builders. The last of those who might still be classified as master builders are people such as Frank Lloyd Wright, Peter Luigi Nervi, Edwardo Torroja, and Felix Candella (see, e.g., References 28, 29, 36, and 41). Today, architects and structural engineers work together as a team on large and expensive structures.

Structural engineers are employed in many different environments performing work that is widely varied but dependent on the same basic principles. Structural engineers are found in the aerospace industry, in the transportation industry, in government or regulatory agencies, in research laboratories, in universities, in large and specialized consulting firms, in the construction industry, and as individual or small firm consultants. With such a diversity of practice, it is understandable that finding a definition of structural engineering is at best difficult.

Following is a definition of structural engineering that incorporates the important aspects of the profession:

> *Structural engineering* is the art of creatively combining the principles of static and dynamic equilibrium, observations on the deformation behavior of materials and measurements of loads acting on structures with a broad range of mathematical principles and techniques to produce a structure that is safe for its users and the general public and that satisfies a predetermined set of criteria in as economical a way as possible.

The first part of the definition represents the exciting challenge of structural engineering. Truly it is an art, for although the principles of static and dynamic equilibrium and mathematics are well known and agreed upon, the creative execution of the design process for a particular structure by different engineers will produce structures that are different, each structure reflecting the personality and philosophy of its creator. Just like art, structural engineering has evolved from primitive beginnings. Its course has been charted by keen observation of the behavior of building materials, the response of completed structures to loads, and the aesthetics of structural forms, and drastically changed by both the successes of brilliant practitioners and the failures of others.

The ultimate requirement in the definition is that a structure must be safe for the users and the general public. The owner or builder establishes the criteria of performance and acceptance of the structure, which will not be built if its cost is too great. In addition, certain minimum conditions must be met that are set forth in codes and specifications established by the engineering profession and local, state, or national legislative bodies. The responsibility for the safe design of the structure falls directly on the structural engineer, who must bear the consequences of failure.

Everyone has some structural engineering capabilities which have been practiced at various times. In what follows a formalization of these capabilities, together with a presentation and review of well-established engineering, physical, and mathematical principles, will start the process of creating the knowledge and self-confidence that is needed for one to become a successful structural engineer.

1.1   Description of Structural Analysis and Structural Design

The primary role of the structural engineer is design. The design process requires the creation of a structure that fulfills the criteria set by the owner or builder of the structure. In addition, the structure must satisfy all safety requirements and be economical to construct. Often, a preliminary design is carried out on several different structure configurations simultaneously in order that the most economical and aesthetic structure can be built. The materials used in the design, foundation conditions, clearance requirements, and environmental impacts all affect structural configurations.

The designer makes many decisions in execution of the design process. The material must be selected, the layout and shape of the structural members decided, the appropriate loads determined, and a final demonstration or proof of the safety of the final design presented. It is this last task that will occupy the majority of the designer's time and effort in the creation of a structure.

Proof of Performance

The first and early convincing proof of the success and safety of a structure is the placement of the structure into service. An unexpected failure of any kind during the construction of the structure cannot be permitted because it raises questions about its future performance. It is the designer's responsibility to assure that failure of the structure (or any part of it) does not occur during this early phase in the lifetime of the structure. The designer must visualize all aspects of the construction phase and predict the response of the incomplete structure.

Once the structure is in service the designer must predict the response of the structure to the expected functional loadings or actions and environmental hazards. The predictions of structural response or behavior are the proof needed to assure the owner of the structure and the general public that the structure is safe and can be used for its intended purpose. To aid in making predictions of structural response the designer has five important procedures or tools available for use.

First, for some types of structures it is possible to test one or more full-scale or prototype structures for extreme loading or environmental conditions. Electrical transmission towers, transportation vehicles (airplanes, trucks, railroad cars, boats, and automobiles) or other structures that will be mass produced can be tested, with the cost of the tests being recovered or nearly so by the savings in material or construction costs of the delivered structures. This proof of performance is the closest a designer can come to assuring the owner and the general public that the structure will work and is safe. The testing program is, then, a tool of the designer.

Second, for large structures that affect the safety of many people or have unusual features incorporated in their design, small- or medium-scale tests of the structures may be conducted. As with full-scale tests, the designer can vary parameters in the structure to improve the design. While this proof of performance is less convincing than full-scale tests, it provides some physical evidence of the response of the structure to loads. An example of this type of modeling is the tests in wind tunnels of small-scale models of large buildings to determine loads and the effects of the presence of the structure on wind patterns in its local vicinity.

Third, when full- or small-scale model tests of a structure cannot be economically justified, tests of a model or prototype of a single member or a small portion of the structure may be conducted. These tests can be used to assess structural performance under expected loads, to evaluate performance under conditions of fire or special dynamic loads, or to learn more about behavior of the member or assembly of members. Parameters can be varied and designs modified on the basis of the test results. These tests do not necessarily prove the performance of the entire structure but increase the level of confidence the designer has of the success of the structure.

Fourth, the most important and universally used technique is to create a mathematical model of a structure and evaluate structural performance by studying the effects of varying different parameters in the model. Even if it is possible to perform physical tests of a portion or all of the structure at any scale, many computations involving a mathematical model of the structure still are undertaken. The availability of powerful hand-held calculators, personal microcomputers, minicomputers, and high-speed mainframe or supercomputers provides the designer with the capability of constructing sophisticated and accurate mathematical models of any structure. Mathematical modeling of structures is one of the most powerful and versatile tools available to the designer.

Finally, it is impossible to overestimate the value of that tool possessed by the designer known as intuition and experience. The greatest and most successful designers have used this tool extensively in creating the spectacular structures for which they are known. The late Fazlur Kahn, developer of the structural tube and bundled tube concept for high-rise building construction and structural engineer for the world's tallest building, often said that he could "feel" how a structure would respond to a set of loads applied to it. This sense of how a structure will respond is most important when evaluating the results of extensive computations using mathematical models of these structures. The existence of many powerful computer programs that create and analyze structures from input data supplied by structural engineers has made the detailed evaluation of structural behavior a relatively routine task. However, the sheer volume of information supplied by these programs will overwhelm the unsophisticated user. In fact, many structural engineers subscribe to the philosophy that "If you don't know the answer or how your structure will respond, you shouldn't use a computer to analyze it."

Value of Analysis of Simple Structures

The most indispensable tool of the structural engineer, *intuition*, is usually not innate but grows in time with the practice of the profession. The successful engineer adds to his or her personal intuition "database" by drawing on the experience of interactions with other engineers, architects, construction personnel, fabricators, and materials suppliers. A firm foundation for the development of intuition is an in-depth understanding of the behavior of very simple structures. Carrying out hand calculations in the complete analysis of simple structures, studying and observing the effects of variation of parameters on structural behavior, and creating and viewing graphical presentations of structural response establish this early understanding. The examples presented and the problems proposed for solution throughout this book are intended to encourage the development of intuition through a better understanding of structural behavior.

Another major effort in this book is to develop those fundamental techniques of structural analysis that are the basis of the current practice in the profession and enhance the understanding of structural behavior. The presentation of mathematical models provides a means of studying and evaluating the response of both simple and complex structures. But they must be used with caution, because the elegance of mathematical models and simplicity of their use can create in the user a false sense of security about the accuracy or reliability of predicted responses. This is particularly true when computer programs are used to perform extensive computations in an analysis.

It is important to recognize that no matter how many computations have been performed or how much computer time has been expended or how sophisticated these mathematical models may become, the truth of the following statement remains unaltered when evaluating the results of any analysis:

> The errors in an analysis are greater or at best equal to the errors that are created in the mathematical model due to the introduction of simplifying assumptions.

The uncertainties of the analysis because of assumptions about material behavior, loads, and idealization of the members of the structure and their connections will be a major focus in subsequent sections.

1.2   Idealization of a Structure

The word *idealization* suggests immediately that any idealization or modeling of a structure provides an approximation of the behavior of the structure. In the context of structural analysis, idealization means creating a mathematical model of the behavior of structures subjected to loads and other actions that cause deformations. This model must incorporate as accurately as possible a functional representation of the loadings on a structure.

The deformation of the material of the structure must be expressed in a mathematical relation, the stress–strain relation. Finally, the deformation behavior of individual members must be expressed mathematically in terms of parameters related to the member's geometric properties and characteristic dimensions. Neither the magnitude and variation of the loads, the modeling of the stress–strain relation, nor the member deformation behavior can be represented exactly in mathematical form, which means that the mathematical model is an imperfect idealization of the structure.

In the mathematical model described above, the internal variation of axial forces, shears, moments, and stresses and the variation of deflection of the structure can be obtained as mathematical functions based on certain physical laws. Structural behavior is simulated in the solution of the equations established in the mathematical model of the structure. The simulated or predicted behavior of the mathematical model will not match *exactly* the *real* or observed behavior of the actual structure. The challenge of the idealization process is to create a mathematical model of a structure that represents the real or true behavior of the structure to a degree of accuracy that is acceptable from a practical standpoint.

The structural engineer must understand the magnitude of the errors that are created when assumptions are made in forming a mathematical model. The availability of computers and sophisticated mathematical procedures provides opportunities to reduce the errors in idealizations to almost imperceptible levels. However, the cost in time and effort to create the data necessary for these more highly sophisticated models may not justify the subsequent reduction in idealization errors. A balance must be struck between the practical effects of idealization errors and the cost of reducing them.

Types of Structures

The idealization process is dependent on the type of structure being studied. Structures can be classified into three different categories on the basis of how they are constructed and consequently on how they resist applied loads. The first and by far the most common type of structure is one that is assembled from a number of individual elements, such as beams, cables, or other long, slender members. These elements are joined or articulated to form an assembly of members that provide the main support system for a structure that is referred to as an articulated structure. Bridges, office and apartment buildings, industrial buildings, and electric power transmission towers are examples of articulated structures. The mathematical models for these structures require the solution of ordinary differential equations.

The second type of structure is one in which the main load-carrying elements of the structure are made from two- or three-dimensional continuous structural elements. Flat plates, curved shells, and membrane structures derive their load-carrying ability directly from the continuous form that defines the outside boundary of the structure. Storage silos, tanks or bins, arched dams, air-inflated structures, shell or domed structures, and folded plate roofs are examples of these structures. The mathematical models for these structures require the solution of partial differential equations.

The third type of structure is basically a hybrid of the first two types. Long, slender structural elements are joined with two- or three-dimensional continuous forms, and the load-carrying action is shared between the two. Examples of these structures are the hulls of ships, the fuselage and wings of aircraft, spacecraft and rockets, bodies of trucks, automobiles, railroad cars and truck trailers, and pressure vessels. Mathematical models for these structures are more complicated and require the solution of both ordinary and partial differential equations.

The replacement of a real structure with a mathematical model which adequately reflects the actual behavior of the real structure is a task requiring that proper assumptions be made. The focus of this book is on trusses and frames, both of which are articulated structures. The behavior of articulated structures is governed by the behavior of individual members and the manner in which they are connected.

Mathematical Model of a Single Member

The mathematical model for a single member is simply a line which, for linear elastic materials, is the member centroidal axis, as shown in Fig. 1.1.

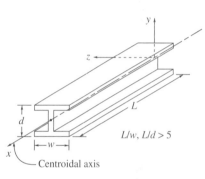

(a) Real three-dimensional beam member

> **Figure 1.1** The structural member shown in Fig. 1.1a has one dimension, its length, that is usually much greater than the other two. Provided that the width and depth/length ratios of the member are 1:5 or less, the member may be idealized with a mathematical model that is simply a line representing an axis of the member, as shown in Fig. 1.1b.

The principles developed in mechanics of materials enables the stresses due to various actions to be computed by mathematical formulas that are derived on the basis of equilibrium, compatibility of deformation, and in some cases assumptions about the manner of deformation of the member. These formulas are referred to the centroidal axis of the member shown in Table 1.1, and represent, for members made of linear elastic materials, the mathematical model of the member stress behavior (see e.g., References 2, 10, and 42). The assumptions that are required to derive the mathematical formulas should be reviewed and evaluated anytime they are used, in particular when the model is used for a situation that is relatively new to the designer. For example, how would the normal stresses be computed using the formulas in the first and third rows of Table 1.1 if an axial force is not applied at the centroid of the cross section?

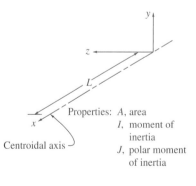

(b) Mathematical model of three-dimensional beam member

Figure 1.1a–b Idealization of a structural member.

Table 1.1   Mathematical Models of Individual Members with Linear Elastic Material That Simulate Stress Behavior

Action	Stress	Cross-Section Properties	Reference Axis	Formula
Axial force, P	Normal, σ	Area, A	Centroid	$\dfrac{P}{A}$
Unconstrained torsion T	Shear, τ	Polar moment of inertia, J, and radius coordinate, ρ	Centroid	$\dfrac{T\rho}{J}$
Bending, M	Normal, σ	Moment of inertia, I and depth coordinate, Y	Centroid	$\dfrac{My}{I}$
Bending shear, V	Shear, τ	Length of cut in cross section, b, and moment of area of cut section, Q	Centroid	$\dfrac{VQ}{bI}$

The manner in which members are supported, or more generally stated, constrained against deformation or motion, is idealized using symbols that connect to the reference axis of the individual members. Figure 1.2 shows several structures with their supports and the idealization of these supports showing the constraints and the corresponding reactive force or moment developed. Table 1.2 shows the one-to-one relation between a constraint against motion and the force or moment reaction necessary to enforce that constraint. Note that the force or moment action developed is always in the direction of the displacement or rotation constraint. For example, in Fig. 1.2e the reactive force is in the direction of the displacement restraint, which is normal to the sliding surface at the end of the member. The guided restraints shown in Fig. 1.2f, which restrain rotation and one displacement, are not idealizations of construction details of real structures but are used as constraints in certain mathematical models of structural behavior.

Table 1.2   Relation Between Deformation Restraints and Reactions

Restraint	Corresponding Reaction
No displacement in specified direction	Force must develop in direction of displacement constraint
No rotation	Moment must develop

(a)

Figure 1.2a   Structural supports and idealized reaction symbols.
(a) Pinned support for tied arch bridge. (Courtesy Alfred Benesch &
Company, Chicago)

Mathematical Model of an Assembly of Members

A structure is simply an assembly of individual members, and the idealization of such an assembly for a planar structure is shown in Fig. 1.3.

> **Figure 1.3** The real three-dimensional structure of Fig. 1.3a is collapsed
> into the plane structure of Fig. 1.3b, which assumes that stresses and
> strains are constant in the direction perpendicular to the plane. In Fig.
> 1.3c the structure is further idealized as a series of lines which are, for
> linear elastic materials, the centroidal axis of the individual members of
> the structure, as shown in Fig. 1.1.
>
> This mathematical model is a reasonably accurate representation of the
> behavior throughout the real structure except for the regions that are cir-
> cled. At joints such as B and C, at points of application of loads and pos-
> sibly at supports such as A and D, the assumptions made in the mechan-
> ics of materials model of a member are not satisfied. In the immediate

9

(b)

(c) Pinned

(d) Fixed

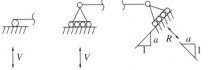

(e) Sticky roller

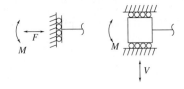

(f) Guided

Figure 1.2b–f   Structural supports and idealized reaction symbols.
(b) Neoprene "roller" support for highway bridge.
(Courtesy Elias M. Gordon, PE)

vicinity of these points, plane sections do not remain plane and there are nonzero vertical normal stresses (σ_y) present (see References 4 and 31). The mathematical model based on these assumptions being satisfied is in error; the results of any analysis using the model are also in error.

Although it is possible to find the exact solution of the equations of the mathematical model of the structure, that analysis is not an "exact" representation of the true physical behavior of the real structure since errors exist in the mathematical model. The concept of exact analysis must be limited to the mathematical model of a structure, which is the reason the admonition stated at the end of Section 1.1 is so important.

(a)

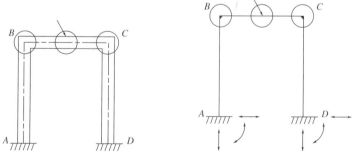

(b) Real structure   (c) Mathematical model

Figure 1.3a–c   Mathematical model of real structure with attention
drawn to regions of incorrect idealization. (a) Rigid bents.
(Courtesy Alfred Benesch & Company, Chicago)

The errors in mathematical models that arise because of a lack of satisfaction of all assumptions about the true behavior of real structures are usually small. Development and solution of more sophisticated mathematical models (with fewer or less limiting assumptions), observations of the behavior of real structures, and laboratory tests of structures show that the errors in the simple mathematical models usually have little or no practical significance. More frequently, the uncertainty about the material properties or occurrence and magnitude of loads overshadows the uncertainty of the results of the analysis due to errors in mathematical models. This does not mean that one can neglect the errors in mathematical models, but that these errors should be viewed within the perspective of other uncertainties.

1.3   Types of Loads on Structures

Several types of loads can act on structures. The most familiar, and the ones that most often control the design of structures, are the static and dynamic loads arising from the functional requirements and the environmental siting of the structure. Other significant loads or structural actions that occur on or within structures arise from the actions caused by thermal changes, erection and fabrication techniques, or differential settlement of supports. All these loads and actions must be considered in design along with an assessment of their importance to the safety of the structure.

Static loads are defined as those loads that cause a response of the structure that yields no significant acceleration of any part of the structure. Static loads are further divided into dead loads and live loads. *Dead load* represents the intrinsic weight of the structure and the weight of any equipment, partitions, or other items in the structure that are firmly attached to and can be expected to remain a permanent part of the structure. *Live loads* are defined as loads that vary with time in location and magnitude but whose time variation is sufficiently slow in relation to the structural response that they are still not dynamic in nature. Examples of live loads are furniture or storage cabinets in rooms, warehouse storage items, people in a building, and wind loads on a structure that do not set the structure into motion.

Dynamic loads cause the structure to respond such that a portion or all of the structure is subject to significant time-varying accelerations. The response of a structure subjected to dynamic loads is measured during periods of time that are of the order of fractions of a second to several seconds. Examples of dynamic loads on structures are impact loads, blast loads, earthquake ground accelerations, loads from moving vehicles or reciprocating machinery, and wind loads that cause significant structural motion. Dynamic loads interact with the mass of the structure to develop inertial forces. The mass is the dead weight of the structure and the weight of any live loads that move with it.

The analysis of structures subjected to dynamic loads requires that the mathematical models be established in the form of partial differential equations in which both time and position or spatial coordinates are independent variables. The solution of these equations by a variety of mathematical techniques is possible for simple situations, but generally becomes complicated and time consuming for most structures. Since the design of a structure requires estimates of maximum stresses and deformations that will occur in a structure under any loading, dynamic loading is often replaced by some form of static loading which causes stresses and deformations in the structure that are comparable in magnitude to those from dynamic loading. While this may be satisfactory for estimating maximum dynamic stresses and deformations, it does not address other features of dynamic structural behavior that can be important, such as the psychological impact of occupant-perceived motion.

Stresses arising from actions caused by temperature differentials, fabrication errors, erection techniques, and differential settlement of the supports of a structure may affect its design. With the exception of erection techniques, loads caused by these actions are known as *self-straining loads* because they involve changes in the geometric relationship between members of the structures. As will be seen later, the primary load-carrying members of the structure are not affected by any of the self-straining types of loads if the structure is statically determinate. However, nonstructural components may suffer undesirable effects, such as cracking of plastered and glass walls or the sticking or binding of doors and windows. In statically indeterminate structures the self-straining loads can cause significant stresses in primary load-carrying members that cannot be ignored in design. In these situations the experience of the designer is needed to evaluate the proper treatment of these stresses. Consider, for example, the effect in a 50-story building if the columns on the outside of the building decrease 54°F (30°C) in temperature while the interior columns remain at constant temperature.

1.4   Specification of Loading

The determination of the appropriate loads for the design of a structure is one of the most uncertain tasks to be undertaken. It is the designer's responsibility to specify loading that reflects the expected maximum loads that the structure will be subjected to in its lifetime. The engineer is assisted in this task by the publication of minimum load standards by local and federal governments or agencies and professional organizations. A significant amount of current research is directed toward a more consistent approach to the specification of loads for structures based on probability theory. For the remainder of this book, the loads used for analysis are assumed to be known and the focus will be on how well the mathematical models of a structure can predict response to a given loading.

It is important to have some sense of the magnitude of loads that act on structures, since the development of intuition requires a sense of what is an appropriate load for a given structural configuration. In Tables 1.3 and 1.4 the values for some common types of loadings are displayed. The loads in the tables can be converted to SI units with the aid of the table on the inside back cover. In Table 1.3 the weights of various building materials are listed. A representative set of minimum live load intensities is given in Table 1.4. These loads are published by the American Society of Civil Engineers (Reference 4) but reflect the type of minimum loads that are specified by other building codes as well (Reference 21). Wind loads for structures are specified according to height, since wind velocity increases with distance above the earth surface. On a flat surface the pressure due to a 90-mph (40.2-m/s) wind is approximately 20 lb/ft² (958 Pa).

Dynamic loads such as those due to ground accelerations caused by earthquake or vibrating machinery are beyond the scope of this text. Books that deal with the topic of dynamics of structures and design for dynamic effects are sources for the interested reader (see, e.g., References 10 and 33).

1.5   Internal Stresses

In Section 1.2 and Table 1.1 computation of the magnitude of certain internal stresses due to the actions of forces, torques, and moments for linear elastic materials is presented. The spatial distribution on the cross section of stress is based on a number of assumptions about deformation behavior and the proportion of the length of the member to its depth and width. For nonlinear materials or different geometric proportions of a member, a different distribution of stress on the cross section can exist. The general state of stress on a small element of a three-dimensional body is shown in Fig. 1.4a. There are three normal stresses (σ_x, σ_y, σ_z) and three shear stresses (τ_{xy}, τ_{xz}, τ_{yz}). The condition from an equilibrium analysis in mechanics of materials of the most general state of three-dimensional stress that the shear stresses τ_{yx}, τ_{zx}, and τ_{zy} are equal to the shear stresses τ_{xy}, τ_{xz}, and τ_{yz}, respectively, has been used.

The existence and magnitude of these internal stresses are due to the presence and magnitude of internal forces, torques, and moments. In Fig. 1.4b a plane cross section of area A taken through a three-dimensional body parallel to the y–z plane shows the positive sense of the three stresses σ_x, τ_{xy}, and τ_{xz} and the six internal forces, torques, and moments that in general

Table 1.3 Weights of Various Building Materials

Material	Weight (lb/ft²)	Material	Weight (lb/ft²)
Ceilings		Partitions	
Channel suspended		Clay tile	
system	1	3 in.	17
Lathing and plastering	See "Partitions"	4 in.	18
Acoustical fiber tile	1	6 in.	28
		8 in.	34
Floors		10 in.	40
Steel deck	See	Gypsum block	
	manufacturer	2 in.	9½
Concrete, reinforced 1 in.		3 in.	10½
Stone	12½	4 in.	12½
Slag	11½	5 in.	14
Lightweight	6 to 10	6 in.	18½
		Wood studs, 2 x 4,	2
Concrete, plain 1 in.		12–16 in. o.c.	4
Stone	12	Steel partitions	
Slag	11	Plaster, 1 in.	
Lightweight	3 to 9	Cement	10
		Gypsum	5
Fills, 1 in.		Lathing	
Gypsum	6	Metal	½
Sand	8	Gypsum board, ½ in.	2
Cinders	4		
		Walls	
Finishes		Brick	
Terrazzo, 1 in.	13	4 in.	40
Ceramic or quarry tile,		8 in.	80
¾ in.	10	12 in.	120
Linoleum, ¼ in.	1	Hollow concrete block	
Mastic, ¾ in.	9	(heavy aggregate)	
Hardwood, ⅞ in.	4	4 in.	30
Softwood, ¾ in.	2½	6 in.	43
		8 in.	55
Roofs		12½ in.	80
Copper or tin	1	Hollow concrete block	
Corrugated steel	See	(light aggregate)	
	manufacturer	4 in.	21
3-ply ready roofing	1	6 in.	30
3-ply felt and gravel	5½	8 in.	38
5-ply felt and gravel	6	12 in.	55
		Clay tile	
Shingles		(load bearing)	
Wood	2	4 in.	25
Asphalt	3	6 in.	30
Clay tile	9 to 14	8 in.	33
Slate, ¼ in.	10	12 in.	45
		Stone, 4 in.	55
Sheathing		Glass block, 4 in.	18
Wood, ¾ in.	3	Windows, glass, frame,	8
Gypsum, 1 in.	4	and sash	
		Curtain walls	See
Insulation, 1 in.			manufacturer
Loose	½	Structural glass, 1 in.	15
Poured-in-place	2	Corrugated cement	3
Rigid	1½	asbestos, ¼ in.	

SOURCE: *Manual of Steel Construction,* 9th ed. American Institute of Steel Construction, Inc., Chicago, 1989.

Table 1.4   Various Live Loads

Occupancy or Use	Live Load (lb/ft²)	Occupancy or Use	Live Load (lb/ft²)
Apartments (see "Residential")		Marquees and canopies	75
Armories and drill rooms	150	Office buildings	
Assembly areas and theaters		File and computer rooms shall	
Fixed seats (fastened to floor)	60	be designed for heavier loads	
Lobbies	100	based on anticipated occupancy	
Movable seats	100	Lobbies	100
Platforms (assembly)	100	Offices	50
Stage floors	150	Penal institutions	
Balconies (exterior)	100	Cell blocks	40
On one- and two-family residences	60	Corridors	100
only, and not exceeding 100 ft²		Residential	
Bowling alleys, poolrooms, and	75	Dwellings (one- and two-family)	
similar recreational areas		Uninhabitable attics	10
Corridors		without storage	
First floor	100	Uninhabitable attics	20
Other floors, same as occupancy		with storage	
served except as indicated		Habitable attics and sleeping	30
Dance halls and ballrooms	100	areas	
Decks (patio and roof)		All other areas	40
Same as area served, or for the		Hotels and multifamily houses	
type of occupancy accommodated		Private rooms and corridors	40
Dining rooms and restaurants	100	serving them	
Dwellings (see "Residential")		Public rooms and corridors	100
Fire escapes	100	serving them	
On single-family dwellings only	40	Schools	
Garages (passenger cars only)	50	Classrooms	40
For trucks and buses use AASHTO[a]		Corridors above first floor	80
lane loads		Sidewalks, vehicular driveways,	
Grandstands (see "Stadium and arena		and yards, subject to trucking[c]	250
bleachers")		Stadium and arena bleachers[d]	100
Gymnasiums, main floors and		Stairs and exitways	100
balconies	100	Storage warehouses	
Hospitals		Light	125
Operating rooms, laboratories	60	Heavy	250
Private rooms	40	Stores	
Wards	40	Retail	
Corridors above first floor	80	First floor	100
Hotels (see "Residential")		Upper floors	75
Libraries		Wholesale, all floors	125
Reading rooms	60	Walkways and elevated platforms	
Stack rooms—not less than[b]	150	(other than exitways)	60
Corridors above first floor	80	Yards and terraces (pedestrians)	100
Manufacturing			
Light	125		
Heavy	250		

SOURCE: Adapted from American Society of Civil Engineers, *Minimum Design Loads for Buildings and Other Structures*, ANSI/ASCE 7–88, 1988.

[a]American Association of State Highway and Transportation Officials.

[b]The weight of books and shelving shall be computed using an assumed density of 65 lb/ft³ (pounds per cubic foot, sometimes abbreviated pcf) and converted to a uniformly distributed load; this load shall be used if it exceeds 150 lb/ft².

[c]AASHTO lane loads shall also be considered where appropriate.

[d]For detailed recommendations, see *American National Standard for Assembly Seating, Tents, and Air-Supported Structures*, ANSI/NFPA 102, 1988.

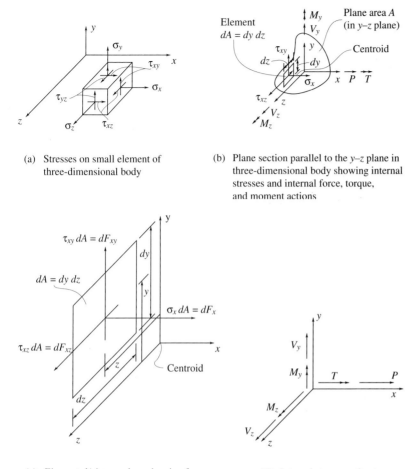

(a) Stresses on small element of three-dimensional body

(b) Plane section parallel to the y–z plane in three-dimensional body showing internal stresses and internal force, torque, and moment actions

(c) Element dA in y–z plane showing forces dF_x, dF_{xy}, and dF_{xz} due to internal stresses

(d) Internal stress resultants

Figure 1.4a–d   Stress and stress resultants and internal actions on a y–z plane section through a three-dimensional body.

are present. The axial force, P, the torque, T, the two shear forces, V_y and V_z, and the two moments, M_y and M_z, are referred to as *internal force or moment actions*. The internal actions are referenced to the centroid of the cross section and are shown in vector form using double-headed arrows to represent the torque and moments. The double-headed arrow vectors are based on the usual right-hand-rule convention.

The internal stresses and internal force actions are related by equilibrium. In Fig. 1.4c the three internal stresses acting on an element of area dA ($= dy\, dz$) give rise to the three differential forces dF_x, dF_{xy}, and dF_{xz}. Summing these forces (which means integrating since the forces are all

17

infinitesimal) over the entire area of the cross section, A, yields the resultant forces

$$P = \int_A dF_x = \int_A \sigma_x \, dA \tag{1.1}$$

$$V_y = \int_A dF_{xy} = \int_A \tau_{xy} \, dA \tag{1.2}$$

$$V_z = \int_A dF_{xz} = \int_A \tau_{xz} \, dA \tag{1.3}$$

These resultant forces, shown in Fig. 1.4d, are the internal forces that are required by equilibrium to act in the body and give rise to internal shear and normal stresses.

As can be seen in Fig. 1.4b and c, the normal differential force dF_x creates moments with respect to the centroid of the cross section of $z \, dF_x$ about the y axis and $y \, dF_x$ about the z axis. Summing (or integrating) these differential moments over the entire cross section produces the two internal moment actions M_y and M_z,

$$M_y = \int_A z \, dF_x = \int_A z\sigma_x \, dA \tag{1.4}$$

$$M_z = -\int_A y \, dF_x = -\int_A y\sigma_x \, dA \tag{1.5}$$

which are shown in Fig. 1.4d. Again equilibrium requires that normal stresses must develop on the cross section when the internal moment actions M_y and M_z shown in Fig. 1.4b are present. The negative sign appears in Eq. (1.5) for M_z since a positive internal force dF_x due to a tensile normal stress σ_x creates a negative moment about the z axis.

Finally, the two differential forces dF_{xy} and dF_{xz} produce an internal torque about the centroid as shown in Fig. 1.4c and d. The torque, T, is obtained by integration over the cross section and is defined as

$$T = \int_A (y \, dF_{xz} - z \, dF_{xy}) = \int_A (y\tau_{xz} - z\tau_{xy}) \, dA \tag{1.6}$$

The internal torque T develops from the internal shear stresses by equilibrium.

Equations (1.1) to (1.6) define the relations between the internal stresses on a plane cross section and the internal actions that exist there. These are completely general relations that are dependent only on equilibrium considerations.

1.6   Free-Body Diagrams

A free-body diagram is a diagram or sketch of a structure, or portion of a structure, which has been set free from any displacement restraints and shows all internal, external, and restraint forces and moments that act. This body, which represents the structure as a diagram, is free to undergo any displacement or rotation that may occur. An important purpose of a free-body diagram is to make use of Newton's second law to obtain equations of equilibrium. When a surface is created by passing a plane through a structure the internal forces and moments that exist on the cut surface are related by the equilibrium equations to the external loadings and accelerations of this free body of the structure. If the free body is in motion, the equations are referred to as *dynamic equilibrium equations* or *equations of motion.* For a free body that is stationary, the equations become the *equations of static equilibrium.* In either case, these relations provide a mathematical model for the internal force and moment actions which can be used to determine the internal stresses required for design.

When a solid structure is "cut" by passing a section through it and one of the two structures created by the cut is taken as a free body, there will be internal normal and shear stresses that act on the surface of the cut. As shown in Section 1.5, summing these yields the internal axial and shear forces, moments, and torques. Figure 1.5 shows the three internal forces and three internal moments defined in Eqs. (1.1) to (1.6) which arise from the summation of the shear and normal stresses that exist on the cut section of a three-dimensional body in static equilibrium. The equations of equilibrium use the centroid of the cross section as a point of reference; hence

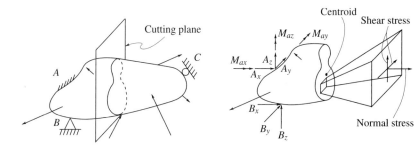
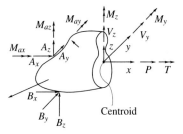

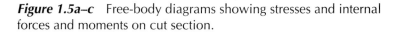

(a) Structure subjected to general loading with cutting plane

(b) Cut section showing stress on small element of area

(c) Cut section showing the internal axial forces, shears, and moments that give rise to the internal shear and normal stresses

Figure 1.5a–c   Free-body diagrams showing stresses and internal forces and moments on cut section.

three-dimensional slender members are mathematically represented as a line that is the centroidal axis and when a cut is made through these members it is perpendicular to this axis. The y and z axes are usually taken to be the principal axes of bending of the cross section.

For each free body there are six independent equations of equilibrium. Figure 1.5c shows the internal unknown forces and moments that act on the surface of the cut and all other external forces and moments. The six equilibrium equations are the summation of forces parallel to each of the three coordinate axes and summation of moments about each of the coordinate axes. If all the external forces and moments that act on a free body are known, the three forces and three moments that exist on any cut surface of the body can be obtained from these six equations. Any attempt to write down more than six equations for the free body does not increase the number of equations for solution since the seventh, eight, and so on, equations can always be obtained as some linear combination of the original six equations. This yields the following conclusion:

In any free-body diagram of a three-dimensional structure there are six and only six independent equations of equilibrium.

1.7   Equilibrium Conditions for Planar Structures

All structures are obviously three-dimensional, and in some cases the only mathematical model that represents the behavior of the structure to a reasonable degree of accuracy is one in which the equations are written in terms of three spatial coordinates. Even with the use of computers, the effort, inconvenience, and in some cases complexity of creating a three-dimensional mathematical model for analysis is time consuming and interpreting the results can also be difficult. In a large number of practical cases, structures are constructed as planes of assembled members, which in turn intersect other planes of assembled members to create a three-dimensional structure. Applied loads acting in three-dimensional space on the structure can be resolved into components that are parallel to the plane assemblies of members.

Each plane of assembled members and the component of the three-dimensional loads in that plane is itself a planar structure. If the effects of the interactions of any one planar structure with another intersecting planar structure are not large, the planar structures can be analyzed and designed separately. Experience has shown that the interaction effects are small in a large number of practical situations, and the analysis and design of each planar component of a structure as an independent plane structure are possible with errors that are within acceptable limits. The analysis of plane structures is a major focus of this book. For convenience, the plane of the structures will be assumed to be the x–y plane, and all loads and actions of the structure will take place in this plane.

The number of equilibrium equations for plane structures with loads acting solely in the x–y plane is reduced from the six for three-dimensional structures to three. The internal actions in long, slender members also reduce to three, an axial force parallel to the x axis, a shear force parallel to the y axis, and a moment about the z axis. The only stresses acting on any section or cut through a slender plane member are the normal stress σ_x and the shear stress τ_{xy}. From Fig. 1.4d and Eqs. (1.1), (1.2), and (1.5), the internal actions are P, V, and M. The subscript y associated with V and z associated with M can be dropped, as there now is only one shear force and one moment that acts internally on the y–z plane, which is the plane of the cross section. Thus if all external moments and forces acting on a structure are known, the three equilibrium equations are sufficient to obtain P, V, and M.

Since every free body of a plane structure must be in equilibrium, there are three independent equilibrium equations for *each* free body. If a structure is divided into two parts, A and B, and three equilibrium equations are written for each (a total of six equations), the six equilibrium equations can be used to obtain unknown forces and moments. However, if parts A and B are joined together to form the complete structure, the three equilibrium equations for the complete structure cannot be combined with the existing six equations because these three equations are no longer independent but are some linear combination of the six equations obtained from the two free bodies A and B. This concept is illustrated in the example shown in Fig. 1.6.

Figure 1.6 The portal frame with the concentrated load has six unknown reactions components, and the three equations of equilibrium for the entire structure can be used to find three relations between those reaction components. Consider that a cut is made in the structure and the free body created to the left of the cut is isolated as shown in Fig. 1.6b. There are six unknown forces acting on the free body, and three equations of equilibrium can be written as indicated. The free body to the right of the cut has six unknown forces acting on it, and again three equations of equilibrium can be written, as shown in Fig. 1.6c. These six equilibrium equations can be solved for six unknowns. For example, three of the reaction forces and the internal axial force, shear force, and moment at the cut can be obtained in terms of the other three reaction components.

At this point all of the independent equations of equilibrium have been established. Any attempt to increase the number of equations of solution is doomed to failure. Consider the three equations of equilibrium for the entire structure written as shown in Fig. 1.6d. These equations are not "new" or independent when they are combined with the six equations established previously. As shown in Fig. 1.6e, Eq. (7) can be

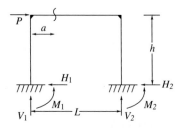

(a) Structure with loading and reaction
components

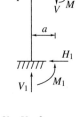

$$\Sigma F\uparrow: \quad V_1 - V = 0 \qquad (1)$$

$$\Sigma F\rightarrow: \quad P - H_1 - F = 0 \qquad (2)$$

$$\Sigma M_{cut}: \quad V_1 a + H_1 h - M_1 - M = 0 \quad (3)$$

(b) Free body to left with equations
of equilibrium

$$\Sigma F\uparrow: \quad V_2 + V = 0 \qquad (4)$$

$$\Sigma F\rightarrow: \quad F - H_2 = 0 \qquad (5)$$

$$\Sigma M_{cut}: \quad M + H_2 h - V_2(L - a) - M_2 = 0 \quad (6)$$

(c) Free body to right with equations
of equilibrium

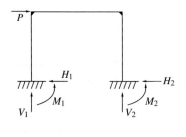

$$\Sigma F\uparrow: \quad V_1 + V_2 = 0 \qquad (7)$$

$$\Sigma F\rightarrow: \quad P - H_1 - H_2 = 0 \qquad (8)$$

$$\Sigma M_1: \quad Ph - V_2 L - M_1 - M_2 = 0 \qquad (9)$$

(d) Free body of entire structure with
equations of equilibrium

Eq. (7) = Eq. (1) + Eq. (4)

Eq. (8) = Eq. (2) + Eq. (5)

Eq. (9) = h [Eq. (2) + Eq. (5)] + Eq. (3)
    + Eq. (6) - a [Eq. (1) + Eq. (4)]

(e) Linear dependence of equilibrium
equations

Figure 1.6a–e Linear independence of equilibrium equations for
free bodies. Note that Eqs. (1) to (6) are independent, but Eqs. (7) to
(9) are dependent.

obtained as the sum of Eqs. (1) and (4), Eq. (8) as the sum of Eqs. (2) and (5), and Eq. (9) by the more complicated linear combination of Eqs. (1) to (6). Thus of the nine equilibrium equations written in the example, only six are independent.

   Since only six independent equilibrium equations exist, there are several combinations of six of the nine equations that can be considered as independent. The original six equations obtained by using the equilibrium equations for the left free body, A, and the right free body, B, have already been shown to be independent. An alternative combination of six independent equilibrium equations is the three equations for the entire structure [Eqs. (7) to (9) in Fig 1.6d] together with the three equations from either free body A [Eqs. (1) to (3)] or free body B [Eqs. (4) to (6)].

As illustrated in the discussion above, the following general statement can be made about using the equations of equilibrium for a plane structure:

> In general, for a planar structure subjected to loadings acting in its plane there are three equations of static equilibrium for the structure as a whole plus three equilibrium equations for each cut or section through the structure that is used to create a free body of a portion of the structure, but no more.

It is extremely important always to keep this concept in mind when analyzing complicated structures. The number of independent equilibrium equations is limited and no "new" equations can be generated from combinations of these equations. As shown in Fig. 1.6, the independent equations of equilibrium available for a structure may not be sufficient in number to solve for all unknown reactions and internal forces and moments. These structures are referred to as *statically indeterminate* and a great amount of grief can be avoided if they are recognized before an analysis using only equilibrium equations is attempted.

1.8   Superposition of Forces in Statically Determinate Structures

The design of a structure involves an evaluation of its response to any applied loading that can be considered to have a reasonable likelihood of acting on the structure during its functional existence. The analysis of the response of the structure to these various loadings can be simplified if the loads are broken down into common configurations that often can be related to one single type of action. When two or more of these actions occur simultaneously, the response of the structure is easily obtained by adding or superimposing the response from each separate action.

The structures in Fig. 1.7 are subjected to two loading configurations, a concentrated load and a uniform load. In both structures the reactions for each loading configuration can be superimposed to obtain the reactions for the two loadings acting simultaneously. Implicit in these calculations are the assumptions that the structure is stable and that the geometry of the structures is *essentially* unchanged by the action of either or both loading configurations. The structures are considered to respond to the action of the loads as though they were undeformed or *rigid*. From a practical standpoint, struc-

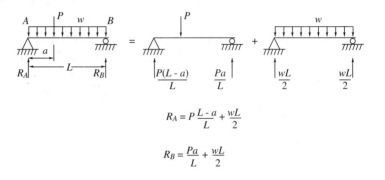

$$R_A = P\frac{L-a}{L} + \frac{wL}{2}$$

$$R_B = \frac{Pa}{L} + \frac{wL}{2}$$

(a) Simply supported beam

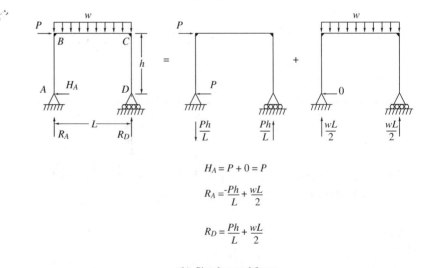

$$H_A = P + 0 = P$$

$$R_A = \frac{-Ph}{L} + \frac{wL}{2}$$

$$R_D = \frac{Ph}{L} + \frac{wL}{2}$$

(b) Simple portal frame

Figure 1.7a–b   Superposition of forces and force responses in "rigid" statically determinate, stable structures.

tures that undergo very small deformations due to applied loadings can still be considered to be *undeformed*. Instability and changes in geometry due to the action of loads are known as *geometric nonlinearity*. When this nonlinearity occurs in any one of the responses of a structure to a loading, the responses cannot be added to obtain the combined response for the combined loadings. Geometric nonlinearity is discussed further in subsequent chapters.

Equilibrium equations are also used to obtain the internal distribution of axial forces, shear, and moments for the structures of Fig. 1.7. Thus the internal force actions from the separate loading configurations can be added to obtain the values for combined loading. The structures are statically determinate because the complete external and internal actions can be obtained from equilibrium considerations alone. The ability to add the external or internal force or moment responses to two or more loading configurations in a statically determinate structure can be stated as the *principle of superposition of forces:*

> The external or internal force actions associated with two or more configurations of loads acting on an undeformed statically determinate, stable structure can be computed separately for each loading configuration and then added or superimposed to obtain the actions for all loadings acting simultaneously.

In making this statement, no restriction has been placed on the type of material in the structure. As long as the structure does not change geometry under the action of the loads, it can be constructed of either linear or nonlinear material and the principle is still valid. For structures made with linear materials the stresses can also be computed separately for each load configuration and then summed to give the stress in the member under the combined loads. In structures made from nonlinear materials, the stresses cannot be so summed. For these structures the stress at any point can be computed from the force and moment at that point obtained as the superposition of the forces and moments due to all loadings.

1.9   Superposition of Deformations

The superposition of deformations of structures due to several load configurations or other structural actions to obtain the deformation for the combined actions can be done in a manner similar to that for forces in Section 1.8. However, to add or superimpose deformations of structures, some additional limitations must be met. The limitations and conditions that are required are illustrated using the simply supported beam of Fig. 1.7a as shown in Fig. 1.8a.

25

Figure 1.8 The deflection of the center of the beam caused by the combination of the concentrated and uniform loading configurations is Δ. The center deflection due to the concentrated load and the uniform load acting separately is shown in Fig. 1.8c and d, respectively. The computation of the deflections shown in Fig. 1.8c and d is based on deformations caused by strains in the material of the member, which in turn are related through the stress–strain relation to the stresses in the member arising from internal force or moment actions. For members that are long in relation to their width and depth, the elastic stress–strain relation reduces to the relation between the normal stress σ_x and the normal strain ε_x as shown in Fig. 1.8b. The nature of this linear elastic stress–strain relation plays an important part in the computation of the deflections of the beam.

Using the linear stress–strain relation shown in Fig. 1.8b for the material of the beam, the strains created by the placement of the concentrated load, P, can be calculated as the stresses from P divided by E, the slope of the stress–strain relation, or $\varepsilon_p = \sigma_p/E$. When the load P is removed from the structure, the stresses and the strains in the beam return to zero

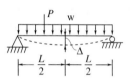

(a) Simple beam with combined loading

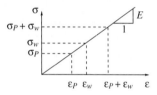

(b) Linear stress–strain relation for beam material

(c) Deflection due to P, Δ_P (d) Deflection due to w, Δ_w

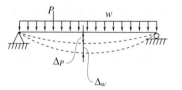

(e) Superposition of deflections from each loading gives deflection for combined loading, $\Delta = \Delta_P + \Delta_w$

Figure 1.8a–e Superposition of deformations in stable structure with linear elastic materials undergoing small deflections.

because the stress–strain relation is elastic. Similarly, strains created when the uniform load, w, is placed on the beam are $\varepsilon_w = \sigma_w/E$, which return to zero when the load w is removed from the structure. Placing P and w on the beam simultaneously creates a total stress $\sigma_p + \sigma_w$ that is obtained by superposition of the stresses due to each load separately. As shown in Fig. 1.8b, the strain ε due to P and w together is $\varepsilon = \sigma/E = (\sigma_p + \sigma_w)/E = \varepsilon_p + \varepsilon_w$, which is the simple addition or superposition of the strains from each separate loading. Only if the stress–strain relation is linear and elastic can the deformations be obtained by superposition. Since the strains ε_P and ε_w yield the displacements Δ_P and Δ_w, the deflection of the center of the beam, Δ, with both loads acting is also obtained by superposition.

The *principle of superposition of deformations* can be stated as follows:

The deformations of a stable structure computed separately for the action of two or more individual load systems can be summed to obtain the deformations of the structure under the combined system of loads provided that the material of the structure is linear and elastic and all deformations are small, so that the geometry of the structure is unchanged.

The principle of superposition of forces stated in Section 1.9 is valid for statically determinate structures having either linear or nonlinear materials. In statically indeterminate structures, the magnitude and distribution of external reactions and internal member actions depend on the deformation behavior and hence the stress–strain relation of the material of the structure. The use of superposition of forces for indeterminate structures is possible only when all the assumptions and limitations of the principle of superposition of deformations are satisfied. The distinction between superposition of forces and superposition of deformations is not presented by authors of most textbooks on structural analysis, who list the limitations of the superposition principle as those of the principle of superposition of deformations.

1.10   Symmetric and Antisymmetric Loading of Symmetric Structures

Even a casual observer has noted the large number of symmetric structures that are built. There are many reasons for structures being symmetric. Symmetric structures have an aesthetic appeal. The forces that control the design of structures are frequently symmetric, and actions of lateral loads due to wind or earthquake action can occur from any direction, which encourages symmetric design of the resisting elements. Fabrication and erection costs of symmetric structures are lower because nearly half of the members of the structure are identical to other members.

The pattern of loading on a symmetric structure can be symmetric, antisymmetric, or of a general nature. A general load pattern is one in which the placement and magnitude of loading have no particular relation to the axis of symmetry of the structure. A symmetric load pattern is one in which the position, magnitude, and direction or sense of all loads is symmetric with respect to the axis of symmetry of the structure. For symmetric loads, the direction of action of loads at symmetric points on opposite sides of the axis of symmetry is the same for loads parallel to the axis and opposite for loads perpendicular to the axis.

An antisymmetric load pattern is one in which the position and magnitude of all loads is symmetric with respect to the axis of symmetry of the structure, but the direction or sense of the loads is antisymmetric. For antisymmetric loads, the direction of action of loads at symmetric points on opposite sides of the axis of symmetry is opposite for loads parallel to the axis and the same for loads perpendicular to the axis. Examples of all of these loading patterns are illustrated in the discussion that follows.

The effort in the analysis and design of symmetric structures can be reduced and the presence of symmetry is an aid to the designer in evaluating computer analyses of structures. Stable, symmetric structures subjected to symmetric loading must resist the loading with complete symmetry of the internal axial forces, shears, and moments as well as the external reactions. In addition, all deflections and rotations or slopes to deflected members must be symmetric. For this situation it is a necessary condition that the results of any computer analysis be symmetric, which serves as an aid in checking the computer simulations for errors.

Symmetrically Loaded Symmetric Structures

A stable, symmetric structure that is symmetrically loaded will always respond in a symmetric manner even if the material of the structure is nonlinear. When the deformations of the structure become relatively large, the deformations still are symmetric and hence symmetry of the deformed structure is maintained. These observations are of great value to the structural engineer who is using computer models or programs that accommodate large displacements and nonlinear materials. They provide a means by which the consistency of these programs can be checked.

There are several other important consequences of symmetry. Figure 1.9 shows two structures that are symmetric and symmetrically loaded. In some cases symmetry can provide a means of calculating the magnitude of selected reaction forces as shown in the portal frame of Fig. 1.9b. A

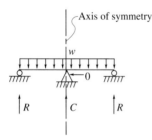

(a) Two-span beam

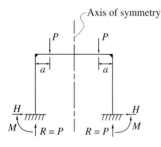

(b) Portal frame

Figure 1.9a–b   Response of symmetric structures to symmetric loading.

more important consequence of symmetry is that the symmetry of deformation provides conditions on deformations of members that cross the axis of symmetry.

The structures of Fig. 1.9 are assumed to be stable and are shown in Fig. 1.10 as being divided in half at the vertical axis of symmetry into symmetric pairs. The conditions of restraint on these symmetric pairs at the axis of symmetry are obtained from the following conditions relating to the symmetric deformations of the structure:

- Horizontal displacements at symmetric points on opposite sides of the axis of symmetry will be equal in magnitude and opposite in direction. *The result is that there is no horizontal displacement of any point on the axis of symmetry.*
- Vertical displacements at symmetric points on opposite sides of the axis will be equal in magnitude and have the same direction. *The result is that there is the possibility of vertical displacement of any point on the axis of symmetry.*
- Rotations or slopes to deflected members at symmetric points on opposite sides of the axis will be equal in magnitude and opposite in direction. *The result is that there is no rotation of any point on the axis of symmetry.*

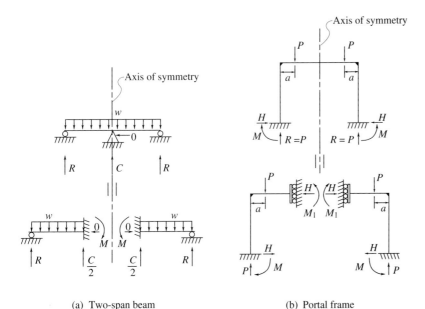

(a) Two-span beam   (b) Portal frame

Figure 1.10a–b Mathematical model created by reduction of a stable, symmetrically loaded structure at axis of symmetry.

29

Because the original structures have been reduced to symmetric pairs, the analysis effort can be cut in half. For example, a mathematical model of the left half of the symmetric structures in Fig. 1.10 can be analyzed and the results used for the right half of the structure because of the symmetry of the structure and loading. Force actions at the axis of symmetry are additive for external restraints, as is shown by comparing results at the center support for the structure in Figs. 1.9a and 1.10a.

The concepts discussed above can be summarized in the following statement:

> A stable, symmetric structure under the action of symmetric loading may be divided in half at the axis of symmetry and one half of the structure analyzed, with the results for the entire structure being obtained by symmetry. Structures made from nonlinear materials or undergoing large displacements causing geometry changes can also be analyzed this way because of the symmetric response of the structure.

Antisymmetrically Loaded Symmetric Structures

In a manner similar to symmetric loading of stable, symmetric structures, antisymmetric loading of symmetric structures such as shown in Fig. 1.11 can, with three additional limitations, be reduced to the analysis of half of the structure. First, the deformations of the structure must be very small, such that the geometry of the deformed structure is for all practical purposes the same as the original (the concept of the "rigid" structure). Second, the stress–strain relation of the material of a structure must be elastic and

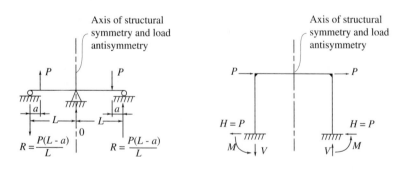

(a) Two-span beam   (b) Portal frame

Figure 1.11a–b   Response of rigid symmetric structures to antisymmetric loading.

symmetric in tension and compression. Third, if the material of the structure has a symmetric nonlinear elastic stress–strain relation, the members of
the structure must also have cross sections that are symmetric with respect
to the axis about which bending takes place (normally, the z axis for members in plane structures as seen in Fig. 1.1). When all of these conditions are
met, the structure will respond in an antisymmetric manner to the antisymmetric loading. All internal and external force actions on the structure will
be antisymmetric and all deformations will be antisymmetric. In some
cases antisymmetry will enable the calculation of certain reaction components as shown for the vertical reactions of the beam of Fig. 1.11a and the
horizontal reactions of the frame in Fig. 1.11b.

The structures of Fig. 1.11 are shown in Fig. 1.12 as being divided at the
vertical axis of symmetry into symmetric pairs but subjected to antisymmetric loading. The conditions of restraint for members crossing the axis of
symmetry of these antisymmetrically loaded structures are obtained from
the antisymmetric deformations of the structure:

- Horizontal displacements at symmetric points on opposite sides of the
  axis will be equal in magnitude and both in the same direction. *The
  result is that there is the possibility of a horizontal displacement of
  any point on the axis of symmetry.*

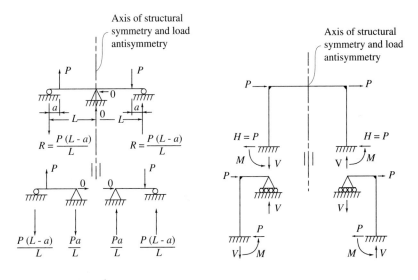

(a) Two-span beam   (b) Portal frame

Figure 1.12a–b Mathematical model created by reduction of
antisymmetrically loaded rigid, stable structures at axis of structural symmetry.

31

- Vertical displacements at symmetric points on opposite sides of the axis will be equal in magnitude but opposite in direction. *The result is that there is no vertical displacement of any point on the axis of symmetry.*

- Rotations or slopes to deflected members at symmetric points on opposite sides of the axis will be equal in magnitude and both in the same direction. *The result is that there is the possibility of a rotation of any point on the axis of symmetry.*

Because the original structures have been reduced to antisymmetrically loaded symmetric pairs, the analysis effort can be cut in half. For example, a mathematical model of the left half of the symmetric structures in Fig. 1.12 can be analyzed and antisymmetry used for the right half of the structure. Force actions at the axis of symmetry are equal and opposite for external restraints, as is shown by comparing results at the center support for the structure in Figs. 1.11a and 1.12a.

The concepts discussed above can be summarized in the following statement:

A symmetric structure under the action of antisymmetric loading may be divided in half at the axis of symmetry and one half of the structure analyzed with the results for the entire structure being obtained by antisymmetry. Antisymmetric analysis is limited to stable structures having members made from linear or nonlinear materials with doubly symmetric cross sections and symmetric stress–strain relations, and undergoing small displacements that cause no significant geometry changes.

General Loading of Symmetric Structures

A stable symmetric structure that is subjected to a general loading can be analyzed under the proper conditions by using half of the structure. Any general loading can be reduced to the sum of two loadings, one symmetric and one antisymmetric, as is shown in Fig. 1.13a for the two structures of Fig. 1.9. In Fig. 1.13b the mathematical models of the left halves of these structures are shown with the loading and constraint conditions at the axis of symmetry that correspond to the symmetric or antisymmetric portion of the general loading. The analyses of the left half of these structures can be summed to obtain the response of the left half of the structure under the general loading. The response of the right half of the structure is obtained by adding the symmetric and antisymmetric responses for the right half that are obtained from the respective left-half analyses.

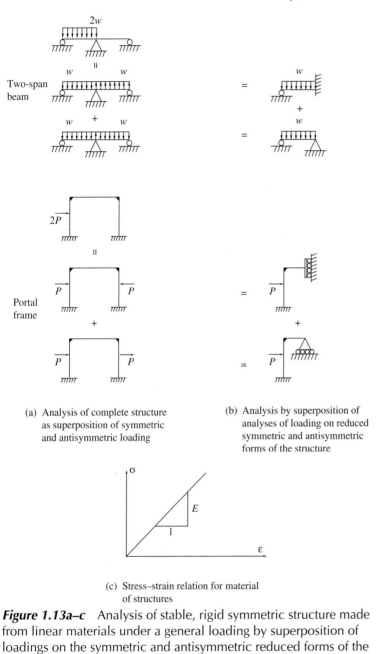

(a) Analysis of complete structure as superposition of symmetric and antisymmetric loading

(b) Analysis by superposition of analyses of loading on reduced symmetric and antisymmetric forms of the structure

(c) Stress–strain relation for material of structures

Figure 1.13a–c   Analysis of stable, rigid symmetric structure made from linear materials under a general loading by superposition of loadings on the symmetric and antisymmetric reduced forms of the structures.

The superposition of the symmetric and antisymmetric analyses of a symmetric structure is possible only if the limitations of the principle of superposition of deformations are met. The analysis of symmetric structures subjected to a general loading can be summarized in the following statement:

> A general loading on a symmetric structure can always be divided into the sum of two loadings, one symmetric and the other antisymmetric with respect to its axis of symmetry. Provided that the limitations of the principle of superposition of deformations are met, the analysis of a stable, symmetric structure can be obtained as the superposition of the analyses of the symmetrically and antisymmetrically loaded halves of the structure.

From the standpoint of effort expended in analysis, it might appear that a single analysis of the whole symmetric structure might not require any more effort than the superposition of the results of the two analyses of the symmetric and antisymmetric halves of the structure. However, the effort required in all cases for the analysis of the symmetric and antisymmetric forms of the structure is always less. This is true not only for hand analyses but for analyses done using computer programs. For a very large majority of symmetric structures designed in engineering practice, the limitations of the principle of superposition of deformations are met and analysis using symmetry of the structure provides significant savings in effort and time.

The existence of structures with more than a single axis of symmetry is not unusual. Provided that the limitations of the principle of superposition of deformations are met, the analyses of these structures can be reduced to ones that involve structural forms that are loaded symmetrically and antisymmetrically about each axis of symmetry of the structure. Figure 1.14 shows a structure that is symmetric about two axes. In Fig. 1.15 a general loading is divided into four combinations of loading patterns that are symmetric and antisymmetric with respect to the two axes of symmetry of the structure. The complete analysis of the structure is obtained as the superposition of the results from each analysis of the four loading patterns which act on a structure that is only one-fourth of the original. The conditions of constraint (or support) of the members crossing the axes of symmetry for each loading pattern are determined in the same manner as those for the single axis of symmetry of the structures in Fig. 1.13.

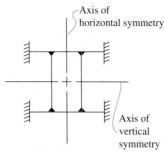

Figure 1.14 Structural symmetric about two axes.

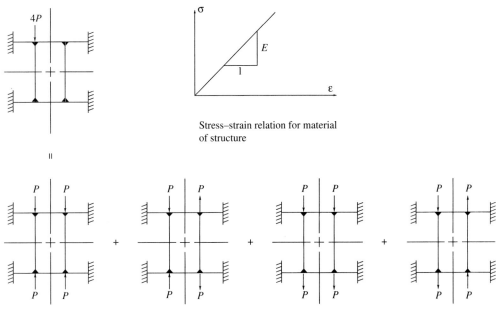

Stress–strain relation for material
of structure

Symmetric vertical–
symmetric horizontal

Symmetric vertical–
antisymmetric horizontal

Antisymmetric vertical–
symmetric horizontal

Antisymmetric vertical–
antisymmetric horizontal

Figure 1.15   Division of general loading on doubly symmetric stable, rigid structure made from linear elastic material into four combinations of symmetric and antisytmmetric loading patterns.

1.11   Summary and Limitations

In the first section of this chapter the role of the structural engineer and the
creative aspects of engineering were discussed. Some of the tools used by
structural engineers to improve designs and to validate the safe performance of structures are the tests on models of structures or, in some cases,
the complete structure, the tests on portions or individual members of structures, and the creation of mathematical models sometimes called idealizations of structures. All of these tools supplement the most important tool
used by the structural engineer, which is intuition and a basic understanding
of structural behavior.

Mathematical models or idealizations of structures are limited and have
innate errors because of the need to make assumptions about the nature and
occurrence of loads, about material behavior, and about geometric configurations of structural members. Equilibrium considerations are used to

establish mathematical relations between internal stresses and internal force and moment actions in Section 1.5. The concept of free-body diagrams is presented in Section 1.6.

The equations of equilibrium are discussed for free-body diagrams and specialized to planar structures. The number of equations of equilibrium for a plane structure is three. For each section or cut that is passed through a plane structure, three equilibrium equations are available to calculate internal force or moment actions that exist on the cut surface.

The principle of superposition of forces and the principle of superposition of deformations are presented separately for structures. Each principle has its own set of limitations, which are discussed in detail. Symmetric structures subjected to symmetric and antisymmetric loads can be analyzed by using a mathematical model of one of the symmetric halves of the structure, with appropriate constraint conditions imposed on members that cross the axis of symmetry. These analyses are possible when the proper limitations are met. Stable, symmetric structures subjected to symmetric loads can be analyzed in this manner for conditions of large displacements and nonlinear materials.

The important limitations addressed in this chapter are highlighted below:

- The mathematical models of structures are limited in accuracy by the assumptions that are required in creating them, and simulations using the mathematical models provide only approximations of the true, observed physical behavior of the real structure.

- The principle of superposition of forces applies to structures made with linear or nonlinear materials, but is limited to stable, statically determinate structures that undergo no significant geometry changes under the action of all loadings.

- The principle of superposition of deformations is limited to stable structures that have linear elastic materials and undergo no significant geometry changes under the action of all loadings.

- The superposition of either force actions or deformations from separate loading systems acting on statically indeterminate structures is limited to stable structures that have linear elastic materials and undergo no significant geometry changes under the action of all loadings.

- The analysis of one half of a symmetric structure subjected to anti-symmetric loading is limited to stable structures that have material which is linear elastic, or if nonlinear elastic, the members of the structure must have doubly symmetric cross sections and the material must have a stress–strain relation that is symmetric in tension and compression.

- The analysis of one half of a symmetric structure subjected to a general loading as the superposition of appropriate symmetric and antisymmetric loading patterns is limited to stable structures that have linear elastic materials and undergo no significant geometry changes under the action of all loadings.

It is important to recognize these limitations because they define the limits of applicability of certain techniques in structural analysis. With the exception of the first, no amount of refinement or increase in sophistication of a mathematical model of a structure will overcome the limitations stated above.

Some questions that should always be asked about the concepts discussed in this chapter are:

- What are the magnitude and pattern of loading that causes the maximum reactions, internal forces, and deformations in a given structure?

- How much error is introduced into a mathematical model of a structure by ignoring the overlapping of members at joints?

- What types of errors occur in the mathematical model of an individual member if the width or height becomes greater than one-fifth of the member length?

- How are loads actually applied to the members of a structure, and how well does the idealization of the application of the loads match the physical reality?

- When should tests be conducted on a structure, a portion of a structure, or a scale model of the structure?

- Should the same level of sophistication be used in a mathematical model of a structure for the preliminary analysis that is used in feasibility studies and the analysis of the final design?

- How does a structural engineer verify the accuracy of a simulation done with a particular mathematical model used in a particular computer program?

- Who is responsible if a failure of a structure occurs because the results of a computer simulation are in error?

- If the results of a computer analysis of a symmetrically loaded symmetric structure are symmetric, does this prove that the computer program is free from errors and the results are correct?

- When dividing a symmetric structure in half for an analysis, how are members that lie on the axis of symmetry treated?

Computation of Reactions for Planar Statically Determinate Structures

In Sections 1.6 and 1.7 it was shown that there are a fixed number of equations of equilibrium available to determine the reactions and the internal forces and moments in the members of a structure. It was assumed implicitly that the structure is supported in a manner such that no rigid-body translation or rotation of the structure can occur. If no significant changes in geometry occur under loading, such a structure is statically determinate when the number of unknown reaction components and internal forces and moments is equal to the number of equations of equilibrium. As will be shown later, when there are fewer unknowns than equations of equilibrium, the structure is geometrically unstable. However, when the number of unknowns is greater than the number of equations of equilibrium, the structure is generally statically indeterminate.

All structures deform under the action of applied loads, but for most practical situations these deformations are small. Calculation of the deflections of structures under load will be discussed in greater detail in subsequent chapters. It must be recognized that deflections or deformations alter the geometry of a structure so that the dimensional relations between loads and reaction components are changed. However, throughout this chapter it is assumed that these dimensional changes are so small that they do not alter the initial or original geometry of a structure before loads are applied. Hence the geometry of structures will be taken as the original undeformed geometry and the structures may be considered to be rigid in the discussions of the concepts presented.

In this chapter the problem of computing the reactions of structures and evaluating whether they are statically determinate with respect to the reactions is addressed. Plane trusses, plane frames, and beams are considered along with the problem of evaluating geometric stability of the structure. The larger problem of obtaining all internal forces and moments in a structure in addition to the reactions is addressed later. Plane trusses are studied in Chapter 3 and plane frames in Chapter 5.

2.1   Static Determinacy of Reactions and Equations of Condition

The number of reaction components is obtained by counting the externally applied reactive forces and moments at each of the idealized reactions of the structure. Figure 1.2 shows the components of force or moment that exist for each reaction type. For example, the reaction in Fig. 1.2d has three components, while Fig. 1.2c and f each have two components. The simply supported beam shown in Fig. 2.1a has three unknown reaction components labeled R_1, R_2, and R_3. Since there are three equilibrium equations for the planar structure, the reactions can be determined from the appropriate use of these equations and the reactions of the structure are statically determinate.

If more than three reaction components are present in a structure, the determination of all components is not possible because there are only three equations of equilibrium for the entire structure. In Fig. 2.1b the simply supported beam in Fig 2.1a is further constrained by fixing the left support against rotation. The rotation constraint requires a moment reaction compo-

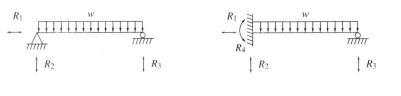

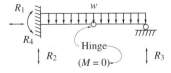

Three unknown reaction components
Three equilibrium equations

Four unknown reaction components
Three equilibrium equations

Four unknown reaction components
Three equilibrium equations
One equation of condition

(a) Simply supported beam

(b) Indeterminate propped cantilever beam

(c) Determinate structure with equation of condition

Figure 2.1a–c   Simple determinate and indeterminate structures.

40

nent to enforce the no-rotation condition. There are four unknown reaction components and only three equilibrium equations, so the structure is now statically indeterminate.

The construction of structures may include some sort of internal release of continuity of deformation. Concurrent with the release of deformation continuity at some point in the structure is a release at the same point of an internal force action. Consider the structure of Fig. 2.1c, which is similar to the one of Fig. 2.1b but with a hinge near the center of the member. A hinge releases the continuity of slope of the deformed member between its right and left sides, which also releases the internal moment at that point. Consequently, a condition that the moment at this point be zero is introduced at the hinge.

The structure of Fig 2.1c has four unknown reaction components, but there are three equations of equilibrium and one known condition of zero internal moment at the hinge. If a section is passed through the hinge to create a free body of either the left or right half of the structure, the summation of moments about the cut must be zero. This creates a new and independent equilibrium equation for the structure. The three equilibrium equations and the zero internal moment condition provide a total of four equations. These four equations can be solved for the four unknown reactions and the structure is statically determinate.

Equations of Condition for Plane Frames

For an individual plane structural member there are three deformation continuities that can be released. The release of continuity of a particular deformation means that the force action that normally ensures continuity of that deformation is also released, or set to zero. If at some point in a member the continuity of axial displacement is released, as shown in Fig. 2.2d, the axial force at the same point is zero. Similarly, as shown in Fig. 2.2c, the release of continuity of vertical or transverse displacement makes the shear at that point zero.

Finally, as already shown in Figs. 2.1c and 2.2b, the release of continuity of slope makes the moment at that point zero. The internal release of a deformation continuity reduces force action to zero, which creates a condition that can be used as an equilibrium equation. This equation is called an *equation of condition* for the structure.

The equation of condition is obtained by establishing a free body created with a section through the point in the structure where the release of continuity occurs. One force or moment action is zero because of the release and a summation of forces in the direction of released force action or a summation of moments for a released moment action in the free body provides an independent equilibrium equation. This equilibrium equation is the equation of condition for the released deformation continuity in the structure. These ideas are illustrated in Fig. 2.2b to 2.2d.

(a)

Figure 2.2a   Equations of condition for beams and frames with dashed lines showing a potential unrestricted relative movement created by the indicated release of continuity of deformation. (a) Construction detail in steel member for a hinge. (Courtesy Bruce W. Abbott, PE)

At any internal point in a structure not only one but two force actions can be set to zero to provide release of the corresponding two continuities of deformation. When two releases occur, a free body established by passing a section through the point of the releases provides two zero-force actions. Two independent equations of equilibrium are available in the free body which are obtained in the same manner as the single equation described in the preceding paragraph. Two equations of condition are obtained for two releases as shown, for example, in Fig. 2.2e, where the moment and axial force are zero.

When investigating the number of equations available to compute the reaction components for a structure, each zero-force action condition due to a release of a continuity of deformation provides an independent equilibrium equation, or equation of condition. The sum of the three equilibrium equa-

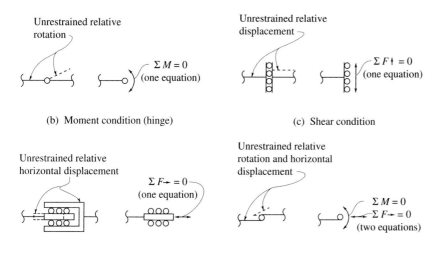

(b) Moment condition (hinge)   (c) Shear condition

(d) Axial force condition   (e) Axial force and moment condition

Figure 2.2b–e   Equations of condition for beams and frames with dashed lines showing a potential unrestricted relative movement created by the indicated release of continuity of deformation.

tions for the whole structure and all the equations of condition yields the total number of equations available to compute reaction components. When the number of equations of equilibrium and condition equals the number of reaction components, the structure is statically determinate. This is a necessary condition for static determinacy, but not a sufficient one, as will be shown later.

Equations of Condition for Plane Trusses

In trusses, equations of condition evolve in a manner similar to that described above for beams and frames. Each member of a truss is capable of carrying only a single axial force. A free body created by passing a section through a truss will establish one unknown member force for each member cut by the section. Any section through the structure yields a free body for which three equations of equilibrium can be written. When a section passing through a truss cuts two rather than three members, there are only two unknown member forces but three equations of equilibrium, and one of the equilibrium equations becomes an equation of condition.

Figure 2.3 Special construction details for plane trusses are shown in Fig. 2.3 with an enlargement of the deformation release created. In Fig. 2.3a a relative rotation can occur in the structure at joint h, shown by the dashed lines. A free body created by passing a section through the two

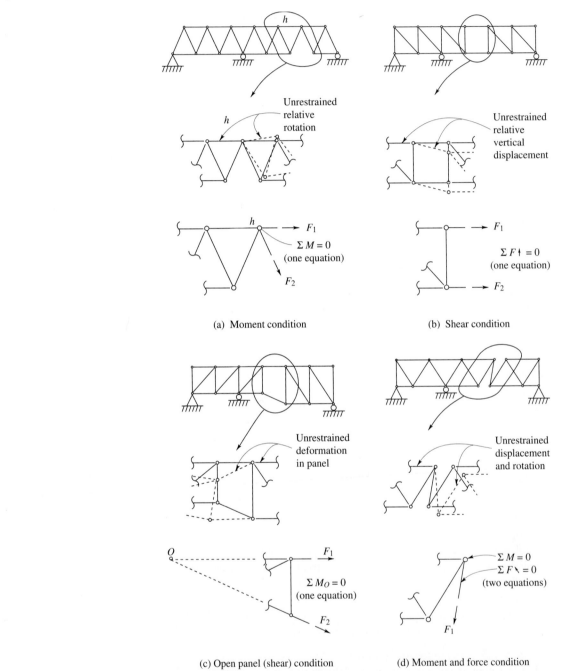

(a) Moment condition

(b) Shear condition

(c) Open panel (shear) condition

(d) Moment and force condition

Figure 2.3a–d   Equations of condition for trusses with dashed lines showing a potential unrestricted relative movement created by the indicated release of continuity of deformation.

members provides an equation of equilibrium that does not involve either force. Summing moments at joint h yields an equation of condition since the moment is zero at the joint. This equation of condition is analogous to the equation of condition associated with a hinge in a beam or frame member.

In Fig. 2.3b the two parallel links in the structure release the relative displacement normal to the links of the right side of the structure with respect to the left as indicated by the dashed lines. No force normal to the links can be transmitted, a condition that is analogous to the shear condition in Fig. 2.2c for beam and frame members. The equation of condition is obtained in the free body created by passing a section through the two links, taking the left or right portion of the structure, and summing forces normal to the links. This equilibrium equation is the equation of condition for the release of continuity of vertical displacement of the truss. Other equations of conditions are shown for plane trusses in Fig. 2.3c and d. Note that two equations of condition are obtained from the two released deformation continuities in Fig. 2.3d.

In summary , the equations of condition for both frames and trusses are internal equilibrium equations and are derived from the release of a continuity in the deformation behavior at an internal point in a structure. This release establishes a zero magnitude for an internal moment or force action at the same point within the structure.

Examples of Statically Determinate Reactions

Once a structure has been identified as statically determinate with respect to the reactions, the reactions can be computed using the equilibrium equations and, if present, any equations of condition. Examples 2.1 and 2.2 illustrate this process.

Example 2.1 The beam shown has four reaction components, but the hinge at C, as indicated in Fig. 2.2b, provides the needed fourth equation for solution of the reaction components. It is usual to start the analysis for reaction components by immediately employing the equation of condition, the zero moment at the hinge at C in this example, which gives R_D. Since there are no horizontal applied loads, the horizontal reaction H_A is zero. The remaining reactions are obtained from the equations of equilibrium for the entire structure. (For numerical computations with U.S. units, take $L = 30$ ft and $P = 8$ kips. For SI units, take $L = 10$ m and $P = 36$ kN.)

Example 2.2 The plane truss frame shown has four reaction components. The construction of the truss at joint 3 corresponds to that shown in Fig. 2.3a, which creates an equation of condition that is added to the three equilibrium equations for the entire structure for a total of four

Example 2.1

Obtain the reactions of the beam shown.

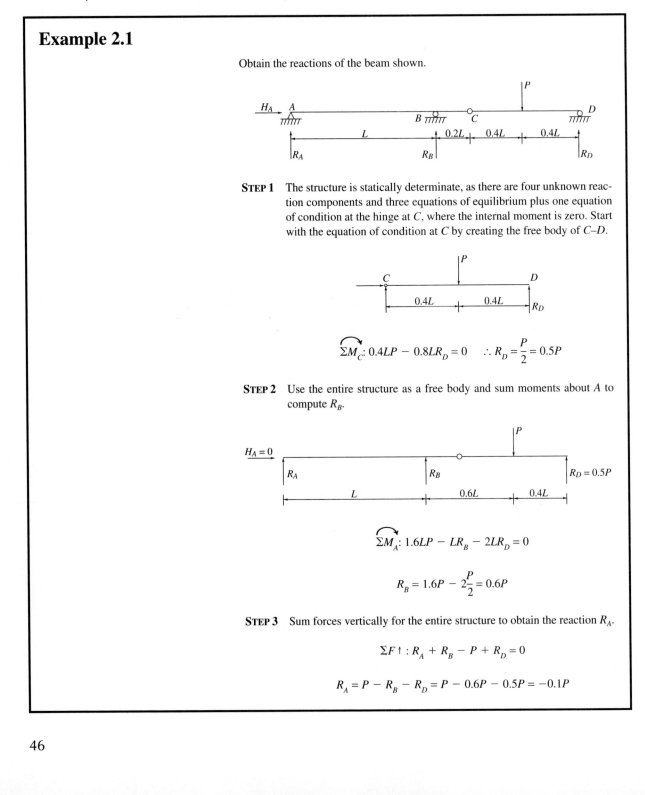

STEP 1   The structure is statically determinate, as there are four unknown reaction components and three equations of equilibrium plus one equation of condition at the hinge at C, where the internal moment is zero. Start with the equation of condition at C by creating the free body of C–D.

$$\Sigma M_C: 0.4LP - 0.8LR_D = 0 \quad \therefore R_D = \frac{P}{2} = 0.5P$$

STEP 2   Use the entire structure as a free body and sum moments about A to compute R_B.

$$\Sigma M_A: 1.6LP - LR_B - 2LR_D = 0$$

$$R_B = 1.6P - 2\frac{P}{2} = 0.6P$$

STEP 3   Sum forces vertically for the entire structure to obtain the reaction R_A.

$$\Sigma F\uparrow : R_A + R_B - P + R_D = 0$$

$$R_A = P - R_B - R_D = P - 0.6P - 0.5P = -0.1P$$

Example 2.2

Obtain the reactions of the truss structure shown.

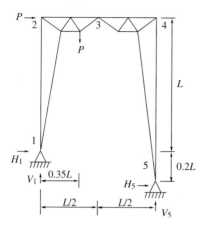

STEP 1   The structure is statically determinate. There are four reaction components, but the equation of condition from Fig. 2.3a at 3 combines with
the three equilibrium equations for the entire structure to create four
equations.

STEP 2   Use the equation of condition at 3 by taking the Free body 3–4–5 and
summing moments about 3.

$$\overset{\curvearrowright}{\Sigma M_3}: \frac{V_5 L}{2} + H_5(1.2L) = 0 \quad \therefore V_5 = -2.4 H_5$$

STEP 3   Take a free body of the entire structure and sum moments about 1 to
obtain H_5 and hence V_5. From the same free-body, sum forces horizontally to obtain H_1 and vertically to obtain V_1.

$$\overset{\curvearrowright}{\Sigma M_1}: LP + 0.35LP - LV_5 - 0.2LH_5 = 0 \quad \therefore H_5(2.4 - 0.2) + 1.35 = 0$$

$$H_5 = \frac{-1.35}{2.2} = -0.6136P \quad \therefore V_5 = -2.4(-0.6136P) = 1.4727P$$

$$\Sigma F \uparrow : V_1 + V_5 - P = 0 \quad \therefore V_1 = P - V_5 = P - 1.4727P = -0.4727P$$

$$\Sigma F \rightarrow : H_1 + P + H_5 = 0$$
$$\therefore H_1 = -P - H_5 = -P - (-0.6136P) = -0.3864P$$

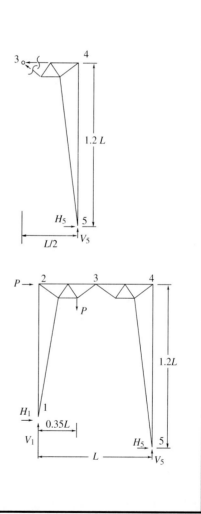

equations. Again, the analysis is started with the equation of condition at 3 by passing a section through the structure cutting the members just to the right of 3. The axial forces in these members intersect at 3, so that summation of moments in step 2 about 3 gives a relation between the reaction components V_5 and H_5. The use of the three equations of equilibrium for the entire structure in step 3 completes the reaction computation. (For numerical computations with U.S. units, take $L = 40$ ft and $P = 4$ kips. For SI units, take $L = 12$ m and $P = 15$ kN.)

2.2   Geometric Stability

The geometric configuration of a structure significantly affects its performance and the level of stresses that develop. The stability of a structure as determined by its geometry alone is called *geometric stability*. If a structure can undergo any displacement or rotation, even a small one, without causing any significant stress in the members of the structure, the structure is said to be *geometrically unstable*. An investigation of the geometric stability of a structure requires a study of the kinematics of all possible deformation characteristics of a structure. The considerations in this treatment will be limited to simple cases.

Figure 2.4 Some examples of geometrically unstable structures along with the reaction components that can develop under a general loading are presented in this figure. No loads are shown to act on the structures, because geometric instability is a characteristic of the structure and is independent of its loading. The dashed arrows indicate the potential form of unrestrained motion. In Fig. 2.4a and c the instability occurs due to the potential of unrestrained horizontal rigid-body displacement of the entire structure in Fig. 2.4a or the right portion (*B–C*) of the structure in Fig. 2.4c. In Fig. 2.4b, e, and f the reaction components or the resultant reaction, *R*, all pass through a common point, *I*, called a *center of instantaneous motion*, which permits possible rigid-body rotation of the structure about *I*.

Figure 2.4d causes confusion, because, intuitively, one feels that the structure can carry any load that is applied to it. However, the load-carrying capability of the structure is dependent on the development of a large vertical displacement of the hinge at *C*. It can be seen that a small rotation of *A–C* about *A* and *B–C* about *B* causes a correspondingly small vertical displacement to occur at *C*. Consequently, there is little (or practically speaking, no) change in length of *A–C* or *B–C*, which means there is little (or no) axial force developed in these members. This type of instability is referred to as *instantaneous geometric instability*. It is just as important in design to avoid this type of instability in the structural configuration as it is the other types of instability depicted in Fig. 2.4, because this type of instability produces extremely large internal forces or moments in a structure.

Associated with each instability shown in Fig. 2.4 is an inadequacy in the equilibrium equations. It is possible in all of these structures to write at least one equilibrium equation in a form that will contain no reaction components. For example, the summation of horizontal forces in Fig. 2.4a for the entire structure will yield an inconsistent result when horizontal loads are applied since there is no horizontal reactive force in the equation. The same is true for the B–C portion of the structure in Fig 2.4c. The summation of moments about point I in Fig. 2.4b, e, and f yields an equation without a reactive component. Thus, under a general system of loading, these structures will be unable to carry these loads in a satisfactory manner, if at all!

A modified form of the structure of Fig. 2.4d is examined in detail in Fig. 2.5.

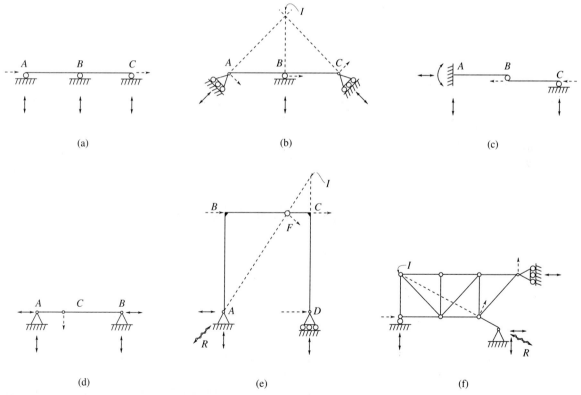

Figure 2.4a–f Examples of geometrically unstable structures. Dashed arrows indicate the direction of possible rigid-body motion.

49

Figure 2.5 The geometric parameter, d, defines the inclination of members A–C and B–C and greatly affects the magnitude of the forces in these members due to the applied load P. The two equilibrium equations at joint C are established in Fig. 2.5b and then solved for the forces in members A–C and B–C, F_{AC} and F_{BC}. These expressions are shown as a

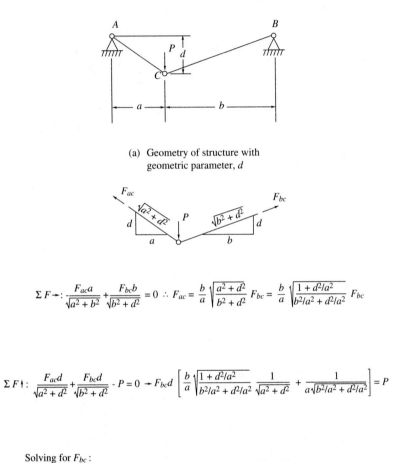

(a) Geometry of structure with geometric parameter, d

(b) Joint equilibrium at C

$$\Sigma F \rightarrow: \, \frac{F_{ac}a}{\sqrt{a^2 + b^2}} + \frac{F_{bc}b}{\sqrt{b^2 + d^2}} = 0 \quad \therefore \, F_{ac} = \frac{b}{a} \sqrt{\frac{a^2 + d^2}{b^2 + d^2}} \, F_{bc} = \frac{b}{a} \sqrt{\frac{1 + d^2/a^2}{b^2/a^2 + d^2/a^2}} \, F_{bc}$$

$$\Sigma F \uparrow: \, \frac{F_{ac}d}{\sqrt{a^2 + d^2}} + \frac{F_{bc}d}{\sqrt{b^2 + d^2}} - P = 0 \rightarrow F_{bc} \left[\frac{b}{a} \sqrt{\frac{1 + d^2/a^2}{b^2/a^2 + d^2/a^2}} \, \frac{1}{\sqrt{a^2 + d^2}} + \frac{1}{a\sqrt{b^2/a^2 + d^2/a^2}} \right] = P$$

Solving for F_{bc}:

$$F_{bc} = \frac{d/a}{\sqrt{b^2/a^2 + d^2/a^2}} \, \left(\frac{b}{a} + 1\right) = P \quad \therefore \quad F_{bc} = \frac{P\sqrt{b^2/a^2 + d^2/a^2}}{(1 + b/a)\,(d/a)}$$

Backsubstitute for F_{ac}:

$$F_{ac} = P\frac{(b/a)\sqrt{1 + d^2/a^2}}{(1 + b/a)\,(d/a)}$$

Figure 2.5a–b   Relation between geometry and member forces for various configurations of a two-member truss.

function of P, b/a, and the ration d/a, which is a measure of the vertical position of joint C in relation to its distance from A.

In Fig. 2.5c the forces in the two members obtained from the joint force equilibrium equations in Fig 2.5b are divided by applied load P and entered into a spreadsheet as a function of d/a for the particular case where the ratio b/a is taken as 2. In Fig. 2.5d a plot is generated by the spreadsheet analysis of the member forces divided by P as a function of d/a. As d, or d/a approaches zero, the form of the structure becomes identical to that of the structure in Fig. 2.4d. As d/a approaches zero, the member forces of the structure in Fig. 2.5 become increasingly large,

d/a	F_{AC}/P	F_{BC}/P
0.02	33.340	33.335
0.04	16.680	16.670
0.06	11.131	11.116
0.08	8.360	8.340
0.1	6.700	6.675
0.2	3.399	3.350
0.3	2.320	2.247
0.4	1.795	1.700
0.5	1.491	1.374
1	0.943	0.745
1.5	0.801	0.556
2	0.745	0.471
2.5	0.718	0.427
3	0.703	0.401
3.5	0.693	0.384
4	0.687	0.373
4.5	0.683	0.365
5	0.680	0.359

Spreadsheet for calculation of the ratio of the member forces to the load, P, in the two-member truss for the special case of $b = 2a$ and various values of the parameter d/a. Values in columns 2 and 3 are obtained in the following manner:

$$\text{Col. 2} = \frac{2\sqrt{1 + (\text{col. 1})^2}}{3 \,(\text{col. 1})}$$

$$\text{Col. 3} = \frac{\sqrt{4 + (\text{col. 1})^2}}{3 \,(\text{col. 1})}$$

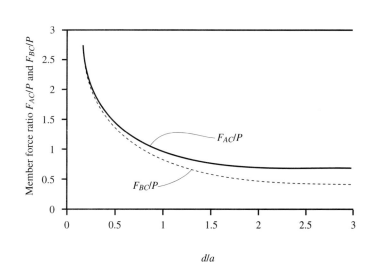

(c) Spreadsheet calculation

(d) Relation of member forces to geometric parameter, d

Figure 2.5c–d   Relation between geometry and member forces for various configurations of a two-member truss.

and in the limit, infinite. Another way of stating this is that no capability exists in the structure to resist vertical loads as the structure approaches the geometric configuration of Fig. 2.4d.

The instantaneous character of geometric instability is illustrated by the results presented in Fig. 2.5. For even a very, very small value of d, it is possible to satisfy vertical equilibrium at C, but it is not possible to satisfy vertical equilibrium when d is zero. Therefore, d being zero in the structure of Fig. 2.5a creates the structure of Fig. 2.4d, which is geometrically unstable. A vertical load applied at joint C in Fig. 2.4d will cause a deflection downward because the members of the structure are in tension and will elongate. The small, downward deflection of joining C "creates" a new, stable geometry for the structure. However, the magnitudes of the forces in the members in this circumstance are extremely large, perhaps 100 or 1000 times the magnitude of P. This condition is undesirable in a structure and must be avoided in design.

Geometric instability is only a function of the geometry of the structure, but it can be identified in some cases from a count of reaction components and equations of equilibrium and condition. In Fig. 2.4a, b, d, and f the number of reaction components is equal to the number of equations of condition plus the three equilibrium equations, although each of those structures is unstable. However, in Fig. 2.4c and e the number of equations of equilibrium and condition is one greater than the number of unknown reaction components. For example, in Fig. 2.4c there are four unknown reaction components, but there are two equations of condition and three equations of overall equilibrium, for a total of five equations. In this case the structure is unconditionally unstable since there are too many deformation releases in the structure.

A principle relating to geometric instability can be stated based on the reaction component and equation count:

> When the number of unknown reaction components is less than the sum of the number of equations of equilibrium and condition, the structure is geometrically unstable.

Based on the discussion of this and the preceding section, evaluation of the static determinancy of reactions of a plane structure becomes simply a counting game. The number of unknowns is the number of reaction components and the number of equations is three (for the equilibrium equations of the entire structure) plus the number of equations of condition, if any. In any structure that is geometrically stable, the following statement can be made:

> In a plane structure that is geometrically stable and undergoes no significant changes in geometry due to loading, the reactions are statically determinate if the sum of the three equations of equilibrium and the number of equations of condition is equal to the number of reaction components and statically indeterminate if the number of reaction components exceeds the total number of equations.

In some cases geometric instability may be overlooked or is too subtle to recognize. When an analysis is attempted in these cases, the equilibrium equations of solution will not be linearly independent and inconsistencies will arise. The form of the inconsistencies depends on how the analysis is executed. If the equilibrium equations and the equations of condition are written down as a system of simultaneous linear equations, the determinant of the coefficient matrix for the equations will be zero. If the equations are formed and at the same time solved for some unknown reaction component in terms of applied loads and other reaction components, a point will be reached where an inconsistent result is obtained. These ideas are shown in Example 2.3.

Example 2.3 This structure is geometrically unstable because the portion D–B–E can rotate as a rigid body about joint I. It would appear that the five unknown reaction components can be obtained from the three equations of equilibrium for the whole structure and the two equations of condition, one at hinge D and a second at hinge E. By count the structure is statically determinate, but the count does not give any indication of the potential geometric instability.

In the first three steps the complete set of five equations involving the five unknown reaction components is established. The determinant of the coefficient matrix of these equations is shown to be zero in step 4. This means that the five equations are not linearly independent and that some combination of four of the equations will give the fifth. If the first equation, (1), is subtracted from the second, (2), and the result multiplied by 2, the left-hand side of the fifth equation, (5), is obtained. This shows the lack of linear independence of the five equations.

An alternative approach to the calculation is shown in step 5 when the equations are solved in the order in which they are established. Eventually, the last equation can be written in terms of the single unknown, H_C, but this fifth equation becomes an inconsistent expression, with H_C disappearing from it completely. In a geometrically unstable structure the equations of equilibrium and condition are not independent, and trying to solve them leads to inconsistent or meaningless results. (For numerical computations with U.S. units, take $h = 15$ ft, $a = 20$ ft, and $P = 2$ kips. For SI units, take $h = 5$ m, $a = 6$ m, and $P = 9$ kN.)

53

Example 2.3

Check the geometric stability of the frame shown.

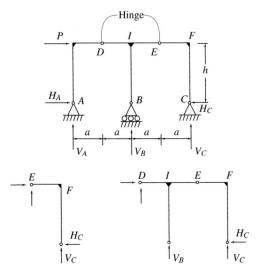

Free body 1   Free body 2

STEP 1   In free body 1 use the first equation of condition and sum moments about E.

$$\overset{\curvearrowleft}{\Sigma M_E}: aV_C - hH_C = 0 \tag{1}$$

STEP 2   In free body 2 use the second equation of condition and sum moments about D.

$$\overset{\curvearrowleft}{\Sigma M_D}: aV_B + 3aV_C - hH_C = 0 \tag{2}$$

STEP 3   Establish the three remaining equilibrium equations using a free-body diagram of the complete structure.

$$\Sigma F \rightarrow : H_A + P - H_C = 0 \rightarrow \quad \therefore H_A - H_C = -P \tag{3}$$

$$\Sigma F \uparrow : V_A + V_B + V_C = 0 \tag{4}$$

Example 2.3 (continued)

$\overset{\curvearrowright}{\Sigma M_A}: -hP + 2aV_B + 4aV_C = 0 \;\rightarrow\; \therefore 2aV_B + 4aV_C = Ph$ (5)

STEP 4   Collect the five equations into matrix form for solution.

$$\begin{bmatrix} 0 & 0 & 0 & -h & a \\ 0 & 0 & a & -h & 3a \\ 1 & 0 & 0 & -1 & 0 \\ 0 & 1 & 1 & 0 & 1 \\ 0 & 0 & 2a & 0 & 4a \end{bmatrix} \begin{Bmatrix} H_A \\ V_A \\ V_B \\ H_C \\ V_C \end{Bmatrix} = \begin{Bmatrix} 0 \\ 0 \\ -P \\ 0 \\ Ph \end{Bmatrix}$$

STEP 5   Compute the determinant of the coefficient matrix, D.

$$D = \begin{vmatrix} 0 & 0 & -h & a \\ 0 & a & -h & 3a \\ 1 & 1 & 0 & 1 \\ 0 & 2a & 0 & 4a \end{vmatrix} = \begin{vmatrix} 0 & -h & a \\ a & -h & 3a \\ 2a & 0 & 4a \end{vmatrix}$$

or

$$D = -a \begin{vmatrix} -h & a \\ 0 & 4a \end{vmatrix} + 2a \begin{vmatrix} -h & a \\ -h & 3a \end{vmatrix} = 4ha^2 - 4ha^2 = 0$$

The determinant being zero indicates that the structure is geometrically unstable because the equations are not independent. Now solve the equations in the order in which they are established. From (1) $V_C = (h/a)H_C$; from (2) and (1) $V_B = -2H_C(h/a)$; from (3) $H_A = H_C - P$; and from (4) $V_A = (h/a)H_C$.
With everything in terms of H_C, substitute (1) and (2) into (5):

$$2a\left(-2\frac{h}{a}H_C\right) + 4a\left(\frac{h}{a}H_C\right) = Ph \;\rightarrow\; 0 = Ph \;?$$

This result is inconsistent which again indicates the geometric instability of section D–B–E which is free to rotate about I.

2.3 Computation of Reactions for Structures with Concentrated Loads

Calculation of the reactions of a structure should not be started until it has been determined that the reactions are statically determinate and the structure is geometrically stable. In Examples 2.1 and 2.2 it was assumed (correctly) that the structures in those examples were geometrically stable, and hence the reactions could be computed successfully. In Sections 2.1 and 2.2 it was also shown that a count of the number of reaction components and the number of equations of solution can establish that the structure is either geometrically unstable or possible statically determinate or indeterminate. Even if the count indicates that the reactions are statically determinate or indeterminate, an evaluation of geometric stability is still required. For example, by count the structures in Fig. 2.1a and c and Fig. 2.4a, b, d, and f are all statically determinate, but only the structures of Fig. 2.1a and c are geometrically stable. Geometric stability is determined from a careful inspection of the structure for possible modes of rigid-body displacement or rotation.

The equations of equilibrium that are used in all free bodies and for the structure as a whole are a summation of forces in two separate (usually orthogonal) directions and summation of moments about some point. Applied concentrated loads and reaction components at roller supports have known directions of action, while the two force reaction components at pinned or fixed supports should be taken as two orthogonal components. The calculations are less troublesome and more error free if all forces, known and unknown, are resolved into a convenient set of orthogonal directions, usually, horizontal and vertical.

The summation of moments about a point can be done using vectors and the cross-product definition for a moment. For two-dimensional structures it is easier and more convenient to compute the moment of a force about a point as the product of the magnitude of the force and the perpendicular distance from that point to the line of action of the force.

In summing moments about some point, two important observations can simplify the resulting equilibrium equation. First, if the line of action of a force passes through the point of summation of moments, the moment of that force or the components of that force about the point are zero. The second observation is that when the selection of a point for summation of moments is taken at the point of intersection of the lines of action of two or more forces, those forces will not enter into the moment equilibrium equation. This establishes the following helpful concept:

> Equilibrium equations derived from the summation of moments about a point are most effective if the point used for the summation is the intersection of the lines of action of two or more forces.

Making an ingenious or careful selection of points for moment summation will reduce the computational effort required for the analysis.

There are some simple techniques that can make the equilibrium calculations less prone to errors. Since the moment of a force is equal to the sum of the moment of the components of the force, forces that are inclined to the two orthogonal directions selected for the analysis should be resolved into the components parallel to those directions for the calculation. A force can act anywhere along its line of action without changing the magnitude of its moment about a given point. Sometimes it is possible to move or slide a force along its line action to a point where one of its components will pass through the point of summation of moments. Fewer mistakes are made if the positive direction of action for the summation forces and moment summation about a point is written down next to the actual equation. These techniques are used throughout this book in all examples and discussions of equilibrium.

Example 2.4 In this example the equation of condition for a hinge is used to compute the reaction for a simple portal frame. The reactions of the frame are statically determinate by count, and the geometric stability of the frame is obtained by inspection. Member A–B–C is considered to be a rigid body that can only rotate about A. Similarly, a member C–D–E is a rigid body that can only rotate about E. Those two rigid bodies are joined at hinge C, which creates an assembled and geometrically stable structure. How would the analysis change if it were started using the equations of equilibrium for the entire structure? (For numerical computations with U.S. units,, take $h = 10$ ft, $L = 20$ ft, and $P = 3$ kips. For SI units, take $h = 4$ m, $L = 6$ m, and $P = 12$ kN.)

Example 2.5 In this example a truss with a release of two continuities of deformation at joints 4 and 5 is analyzed. This is the same release as the one shown in Fig. 2.3d. The reactions of the structure are statically determinate by count, and an investigation of geometric stability is required. The truss is developed from a series of triangular forms that are rigid. The assembled triangular forms for the 1–2–4 portion of the structure create a rigid body that is fixed against any kind of motion by the reaction constraints at 1 and 2. The assembled triangular forms for the 5–3 portion of the structure create a rigid body that can only rotate about 3.

Example 2.4

Obtain the reactions of the frame shown.

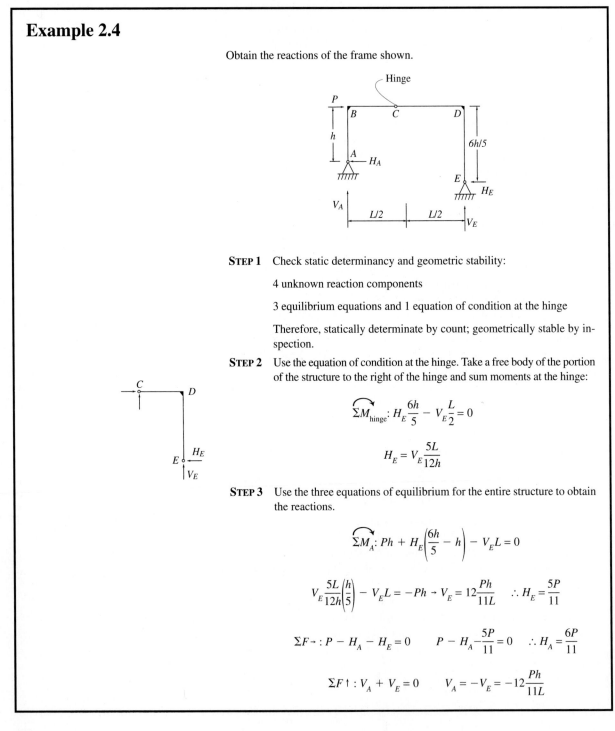

STEP 1   Check static determinancy and geometric stability:

4 unknown reaction components

3 equilibrium equations and 1 equation of condition at the hinge

Therefore, statically determinate by count; geometrically stable by inspection.

STEP 2   Use the equation of condition at the hinge. Take a free body of the portion of the structure to the right of the hinge and sum moments at the hinge:

$$\Sigma M_{\text{hinge}}: H_E \frac{6h}{5} - V_E \frac{L}{2} = 0$$

$$H_E = V_E \frac{5L}{12h}$$

STEP 3   Use the three equations of equilibrium for the entire structure to obtain the reactions.

$$\Sigma M_A: Ph + H_E \left(\frac{6h}{5} - h \right) - V_E L = 0$$

$$V_E \frac{5L}{12h}\left(\frac{h}{5}\right) - V_E L = -Ph \rightarrow V_E = 12\frac{Ph}{11L} \quad \therefore H_E = \frac{5P}{11}$$

$$\Sigma F \rightarrow : P - H_A - H_E = 0 \qquad P - H_A - \frac{5P}{11} = 0 \quad \therefore H_A = \frac{6P}{11}$$

$$\Sigma F \uparrow : V_A + V_E = 0 \qquad V_A = -V_E = -12\frac{Ph}{11L}$$

Example 2.5

Obtain the reactions for the plane truss.

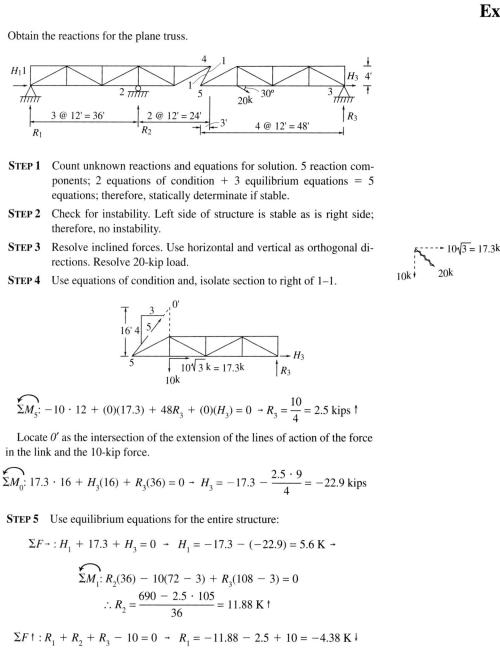

STEP 1   Count unknown reactions and equations for solution. 5 reaction components; 2 equations of condition + 3 equilibrium equations = 5 equations; therefore, statically determinate if stable.

STEP 2   Check for instability. Left side of structure is stable as is right side; therefore, no instability.

STEP 3   Resolve inclined forces. Use horizontal and vertical as orthogonal directions. Resolve 20-kip load.

STEP 4   Use equations of condition and, isolate section to right of 1–1.

$$\Sigma M_5: \; -10 \cdot 12 + (0)(17.3) + 48R_3 + (0)(H_3) = 0 \; \rightarrow \; R_3 = \frac{10}{4} = 2.5 \text{ kips} \uparrow$$

Locate $0'$ as the intersection of the extension of the lines of action of the force in the link and the 10-kip force.

$$\Sigma M_0: \; 17.3 \cdot 16 + H_3(16) + R_3(36) = 0 \; \rightarrow \; H_3 = -17.3 - \frac{2.5 \cdot 9}{4} = -22.9 \text{ kips}$$

STEP 5   Use equilibrium equations for the entire structure:

$$\Sigma F \rightarrow : \; H_1 + 17.3 + H_3 = 0 \; \rightarrow \; H_1 = -17.3 - (-22.9) = 5.6 \text{ K} \rightarrow$$

$$\Sigma M_1: \; R_2(36) - 10(72 - 3) + R_3(108 - 3) = 0$$

$$\therefore R_2 = \frac{690 - 2.5 \cdot 105}{36} = 11.88 \text{ K} \uparrow$$

$$\Sigma F \uparrow : \; R_1 + R_2 + R_3 - 10 = 0 \; \rightarrow \; R_1 = -11.88 - 2.5 + 10 = -4.38 \text{ K} \downarrow$$

Note: Analysis is started at section where the equations of condition exist. This is the usual case.

Because the 1–2–4 rigid body is fixed in position, joint 4 is also fixed in position. Link 4–5 is free to rotate about 4, but when it is joined at joint 5 with the rigid body 5–3 that can only rotate about 3, the resulting structure is fixed in position so that the entire structure is geometrically stable.

In both examples, calculation of the reactions starts with a free body that utilizes the equation(s) of condition for the structure. It is not always necessary to start the analysis with the equations of condition, but they must be used at some point. Frequently, starting with these equations will shorten the computations. In Example 2.5 there are two equations of condition for the deformation releases at joints 4 and 5. Rather than use a summation of forces perpendicular to link 4–5 as shown in Fig. 2.3d for the second equation of condition, it is more convenient in this case to sum moments about the second point, o'. The equations obtained by summation of moments about two points along the line of action of the link is statically equivalent to equations obtained by summing moments about one point and summing forces perpendicular to the link.

2.4   Computation of Reactions for Structures with Distributed Loads

There are no conceptual differences between the analysis of structures for reactions when they are subjected to distributed rather than concentrated loads. It is, however, useful to discuss some techniques for the treatment of distributed loads in forming the equilibrium equations. These techniques replace distributed loads in free-body diagrams with statically equivalent concentrated loads.

A general distributed load can be regarded as a plane geometric shape defined by a known function $q(x)$ between points $x = a$ and $x = b$ and the horizontal x axis.

Figure 2.6 In Fig. 2.6a, the computation of a statically equivalent concentrated load reduces to the use of the properties of a geometric shape. The functional variation of the applied distributed loading, or its shape, is also called the loading diagram, as indicated in the figure. In Fig. 2.6b, summation of forces parallel to the action of the distributed load requires that the infinitesimal concentrated forces, $q(x) \cdot dx$, be summed or integrated over the length of the loading from a to b. Evaluation of the integral indicated is defined as the concentrated load, Q, and is simply the area of the plane geometric body defined by the loading function, $q(x)$, between $x = a$ and $x = b$ and the x axis.

The summation of moments of the distributed load in the loading diagram, $q(x)$, which is defined with respect to the origin of the coordinate system, about an arbitrary point, O, is shown in Fig. 2.6c. The integration

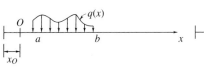

(a) Loading diagram for general distributed load

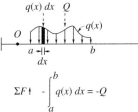

$$\Sigma F \uparrow \quad - \int_a^b q(x)\, dx = -Q$$

Q = area under loading diagram

(b) Sum of forces with distributed load

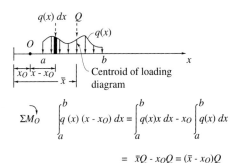

$$\Sigma M_O \curvearrowright \int_a^b q(x)(x - x_O)\, dx = \int_a^b q(x)x\, dx - x_O \int_a^b q(x)\, dx$$

$$= \bar{x}Q - x_O Q = (\bar{x} - x_O)Q$$

$\bar{x}$ = location of centroid of loading diagram with respect to origin of coordinate system ($x = 0$)

(c) Sum of moments about an arbitrary point with distributed load

Figure 2.6a–c   Reduction of distributed load diagram to a statically equivalent concentrated load.

of the moment of the infinitesimal concentrated forces about O, $(x - X_0)q(x)\, dx$, breaks into two parts. One part is the product of the distance from the origin to point O, X_O, and the area, Q, of the loading diagram. The other is the moment of the area of the loading diagram defined by $q(x)$ about the origin ($x = 0$).

The moment of the area of a plane body about a point can be calculated as the product of the total area of the body and the distance between the centroid of the plane body and the point of the summation of moments. The area of the loading diagram is the concentrated force of magnitude, Q, and the distance from the origin to the centroid of the loading diagram is $\overline{X}$. Thus the moment of the distributed load about point O is the product of the equivalent concentrated force for the distributed load, Q, and the distance from point O to the centroid of the loading diagram, $(\overline{X} - X_0)$, as is shown in Fig. 2.6c.

The concepts presented in Fig. 2.6 provide a simple means of treating any distributed load in a free body:

> In each free body the effect in equilibrium considerations of a distributed load can be calculated by replacing it with a concentrated force equal in magnitude to the area of the loading diagram of the distributed load and acting at its centroid.

Several common areas and their centroids are shown in Fig. 2.7. The calculation of the reactions of structures with distributed loads is illustrated in Examples 2.6 and 2.7

2.5 Reactions for Symmetric Structures

Symmetric structures that undergo little or no deformation and have symmetric or antisymmetric applied loads correspondingly will have reactions that are symmetric or antisymmetric. The effort required to obtain the value of reactions in these structures can be significantly reduced. In Section 1.10 the analysis of symmetric structures is discussed and the limitations on the use of symmetry and antisymmetry of loading in analysis are specifically stated. It is assumed in the discussion in this section that all limitations on the analysis of symmetric structures are satisfied.

Before any attempt is made to obtain the reactions of a symmetric structure, it must be checked for static determinacy and geometric stability. For statically determinate and geometrically stable structures, the analysis proceeds as shown in the previous examples, except that the reaction of the structure can be assumed to be symmetric under a symmetric loading or antisymmetric under an antisymmetric loading.

Example 2.8 The symmetrically loaded, symmetric portal frame is statically determinate and geometrically stable. Geometric stability is established by treating all members of the frame as rigid. This makes the hinges at 2 and 6 fixed in position and the 2–3–4 and 4–5–6 members can only rotate about 2 and 6, respectively. When joined by the hinge at 4, these members form a stable structural configuration.

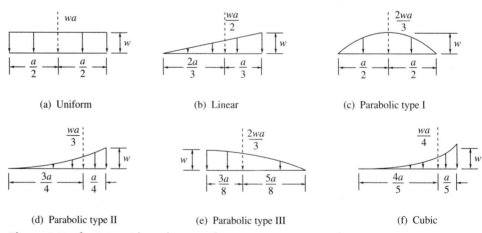

(a) Uniform (b) Linear (c) Parabolic type I

(d) Parabolic type II (e) Parabolic type III (f) Cubic

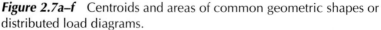

Figure 2.7a–f Centroids and areas of common geometric shapes or distributed load diagrams.

Example 2.6

For the structure shown, compute the reactions.

STEP 1 Check the static determinacy and geometric stability. Four reaction components and four equations—three equilibrium and one equation of condition at hinge D. Structure is stable and statically determinate.

STEP 2 Isolate the free body of A–D and use the equation of condition.

$$\Sigma M_D: 50 \cdot 3 - R_A \cdot 5 = 0$$

$$R_A = 30 \text{ kN} \uparrow$$

STEP 3 Isolate the free body of the entire structure. Replace the distributed loads with statically equivalent concentrated loads. Use the equilibrium equations for the entire structure.

$$\Sigma F \rightarrow : H_A = 0$$

$$\Sigma M_B: -30 \cdot 10 + 50 \cdot 8 + 25 \cdot \frac{5}{3} - 100 \cdot 5 + R_C \cdot 10 = 0$$

$$R_C = \frac{215}{6} = 35.83 \text{ kN} \uparrow$$

$$\Sigma F \uparrow : 30 - 50 - 25 + R_B - 100 + \frac{215}{6} = 0$$

$$R_B = \frac{655}{6} = 109.17 \text{ kN} \uparrow$$

Note: Analysis should always start at the point where equation(s) of condition are available.

Example 2.8

Obtain the reactions for the symmetrically loaded symmetric portal frame.

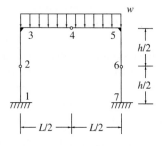

STEP 1   Check the static determinacy and geometric stability. There are six unknown reaction components, three equations of equilibrium, and three
equations of condition. The frame is statically determinate and is geometrically stable by inspection.

STEP 2   Because of symmetry the reaction components are also symmetric.
Sum moments about joint 1 to obtain the vertical reactions.

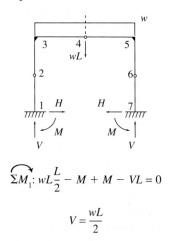

$$\overset{\curvearrowright}{\Sigma M_1}: wL\frac{L}{2} - M + M - VL = 0$$

$$V = \frac{wL}{2}$$

STEP 3   Take the free body of member 6–7. Use the equation of condition at
hinge 6 to obtain M in terms of H.

$$\overset{\curvearrowright}{\Sigma M_6}: H\frac{h}{2} - M = 0 \therefore M = H\frac{h}{2}$$

Example 2.8 (continued)

STEP 4   Take the free body of member 4–5–6–7. Use the equation of condition at hinge 4 to obtain values of H and then M.

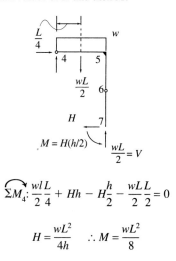

$$\Sigma M_4: \frac{wl}{2}\frac{L}{4} + Hh - H\frac{h}{2} - \frac{wL}{2}\frac{L}{2} = 0$$

$$H = \frac{wL^2}{4h} \quad \therefore M = \frac{wL^2}{8}$$

An alternative analysis is to use symmetry of structure and loading to reduce the structure to the right half. Then:

ΣM_6 in free body 4–5–6 gives H_4.

$\Sigma F \uparrow$ gives vertical reaction, V.

$\Sigma F \rightarrow$ gives horizontal reaction, H.

ΣM_7 gives moment reaction, M.

Example 2.9

Obtain the reactions for the antisymmetrically loaded symmetric portal frame of Example 2.8.

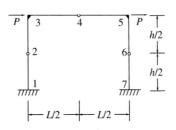

STEP 1   Check the static determinacy and geometric stability. As in Example 2.8, the frame is statically determinate and geometrically stable.

STEP 2   Because of antisymmetry of the loading, the reaction components are also antisymmetric. Sum forces horizontally to obtain the horizontal reactions.

$$\Sigma F \rightarrow : P + P - H - H = 0$$

$$H = P$$

STEP 3   Take the free body of member 6–7. Use the equation of condition at hinge 6 to obtain M in terms of P.

$$\Sigma M_6: P\frac{h}{2} - M = 0 \quad \therefore M = P\frac{h}{2}$$

STEP 4   Sum moments about joint 1 to obtain the value of V.

$$\Sigma M_1: Ph + Ph - P\frac{h}{2} - P\frac{h}{2} - VL = 0$$

$$V = P\frac{h}{L}$$

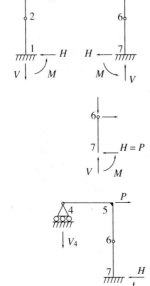

An alternative analysis is to use the symmetry of structure and the antisymmetry of the loading to reduce the structure to the right half. Then:

ΣM_6 in free body 4–5–6 gives V_4.

$\Sigma F \uparrow$ gives vertical reaction, V.

$\Sigma F \rightarrow$ gives horizontal reaction, H.

ΣM_7 gives moment reaction, M.

Example 2.10

Obtain the reactions for the uniformly loaded bridge.

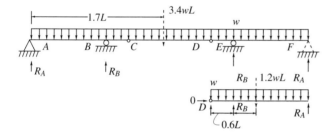

STEP 1   Check the static determinacy and geometric stability. There are five reaction components, three equilibrium equations, and two equations of condition at the hinges at C and D. The structure is statically determinate and geometrically stable by inspection.

STEP 2   The structure can be considered to be symmetric and symmetrically loaded because there are no horizontally applied loads. The vertical reactions at E and F are therefore equal to the vertical reactions at B and A, respectively.

STEP 3   Take a free body of D–E–F and use the equation of condition at the hinge at D to obtain a relation between R_B and R_A.

$$\overset{\curvearrowright}{\Sigma M_D}: \; 1.2wL \cdot 0.6L - 0.2LR_B - 1.2LR_A = 0$$

$$\therefore R_B = 3.6wL - 6R_A$$

STEP 4   Sum moments about A in the entire structure to obtain the final value of R_A and hence R_B.

$$\overset{\curvearrowright}{\Sigma M_A}: \; 3.4wL \cdot 1.7L - R_B L - R_B \cdot 2.4L - R_A \cdot 3.4L = 0$$

$$5.78wL - 3.4(3.6wL - 6R_A) - 3.4R_A = 0$$

$$\therefore R_A = 0.38wL \quad \text{and} \quad R_B = 1.32wL$$

components, but the antisymmetry of the vertical reactions serves as a condition that has the same effect as an equation of condition. The same structure under the action of a symmetric load remains statically indeterminate even though the reactions are symmetric. The condition of symmetry does not provide an independent condition that can be added to the equations of equilibrium, as shown in Figs. 1.9a and 1.10a.

2.6   Summary and Limitations

The computation from equilibrium considerations of reactions for statically determinate structures is presented. Each structure has three overall equilibrium equations and possibly one or more equations of condition. Each equation of condition is derived from the release of some type of continuity of deformation in the structure, with a corresponding release of a force action.

Considerations of geometric stability are discussed with some general principles established to help identify its existence. When the sum of the number of equations of equilibrium and condition exceeds the number of reaction components, the structure is geometrically unstable. A structure is also geometrically unstable if it, or any portion of it, can translate or rotate about one or more points as a rigid body. An attempt to compute the reactions of a geometrically unstable structure will lead to inconsistent results and if the equations are solved simultaneously, the determinant of the coefficient matrix will be zero.

Techniques are illustrated for computation of the reaction components of statically determinate and geometrically stable structures. The treatment of distributed loads in equilibrium calculations is simplified by replacing them with statically equivalent concentrated loads. The distributed load in any free body is considered to be a plane geometric shape, defined by the nature of the distributed load. The statically equivalent concentrated load has a magnitude equal to the area of the geometric shape and acts at the centroid of the geometric shape. When equations of condition are used in the analysis, the analysis nearly always starts with the establishment of these equations.

The effort required in computing reactions of symmetric structures under the action of symmetric or antisymmetric loading is shown to be reduced significantly. The reaction components in these structures will correspond in pattern to the symmetry or antisymmetry of the loading which reduces the number that must be computed. Alternatively, the structure can be divided at its axis of symmetry, with support conditions corresponding to the symmetric or antisymmetric nature of the loading and the reaction of this reduced structure computed.

The concepts of analysis presented and discussed in this chapter are based on mathematical models or idealizations of the structure and are subject to certain assumptions and limitations. The important ones are listed below and should be considered in evaluating the results of any analysis:

- The idealized representation of the reaction components of a structure are a close approximation of the true conditions of support of the real structure.
- The deformations of the structure are very small, so that the equilibrium equations can be established using the dimensions of the original undeformed structural geometry.
- The magnitude and distribution of applied loads are not affected by the deformations of the structure.
- The idealized representation of the internal releases of continuity of deformation in the mathematical model are a close approximation of the true conditions of release in the real structure.
- The calculation of the reaction components considered in this chapter is for statically determinate structures and does not consider any deformation, so the results are valid for structures fabricated from either linear or nonlinear materials.

In addition to being aware of the limitations listed above, some questions that should be uppermost in the minds of structural engineers when an analysis for reactions of a structure is completed, whether the analysis is performed by hand techniques or by a computer analysis, are the following:

- Are the magnitude, distribution, and location of the applied loads on the structure represented accurately in the analysis?
- Is each release of continuity of deformation in the real structure considered appropriately in the mathematical model of the structure?
- If a structure is geometrically unstable, will the results of a computer analysis indicate this instability?

PROBLEMS

2.1   Indicate if the structures shown are statically determinate or indeterminate and geometrically stable or unstable. Indicate the reasons for your answers.

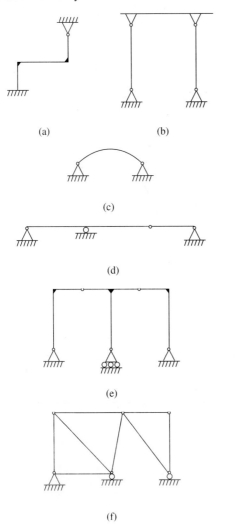

(a)

(b)

(c)

(d)

(e)

(f)

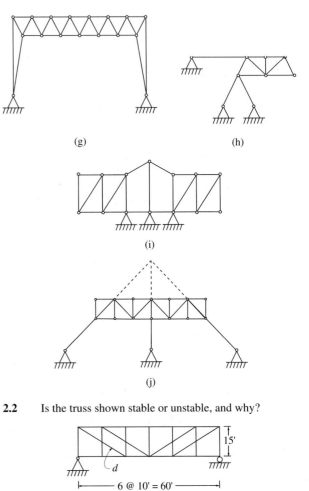

(g)

(h)

(i)

(j)

2.2   Is the truss shown stable or unstable, and why?

15'

d

6 @ 10' = 60'

2.3   Compute the reactions at A and B.

A

B

C

P

L

$0.3L$

2.4   Compute all reactions. Note the hinge at 3.

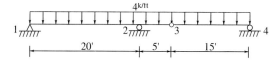

2.5   Compute all reactions. Note the hinge at 3.

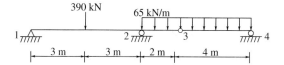

2.6   Compute all reactions.

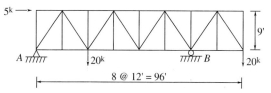

2.7   Compute all reactions for the plane truss shown.

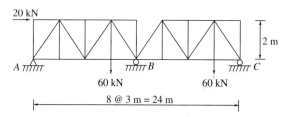

2.8   Compute all reactions for the plane truss shown.

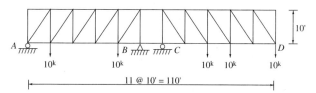

2.9   Compute all reactions.

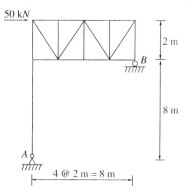

2.10   Compute all reactions. Note the hinge at 3.

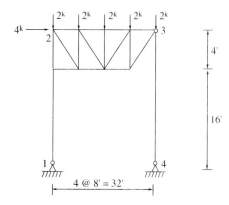

2.11   Compute all reactions. Note the hinge at 3.

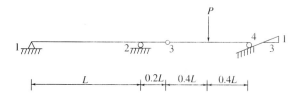

2.12 Compute the reactions of the frame due to the uniform loading and horizontal load. Note the hinges at C and F.

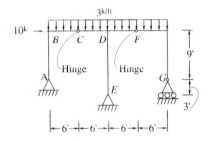

2.13 For the truss loaded as shown, compute reaction forces at A and B.

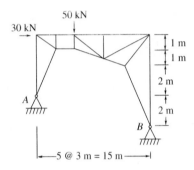

2.14 Compute reaction forces at A and B. Note the hinge at C.

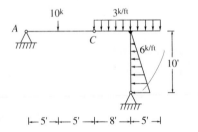

2.15 Compute the reactions for the cantilever bridge. Note the hinges at C and D.

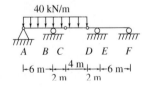

2.16 Obtain the reactions for the structure with hinges at B and C.

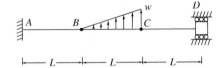

2.17 Find the reactions of the truss.

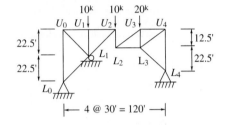

2.18 Obtain the reactions for the truss shown and verify that $R_1 = 18$ kN, $R_2 = 288$ kN, $R_3 = 102$ kN, and $H = 0$.

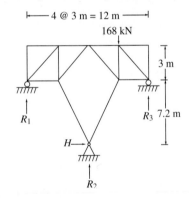

2.19 The plane truss is loaded with the single 20-kip load at the end. Compute all reactions.

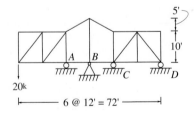

2.20   Compute the reactions of the truss.

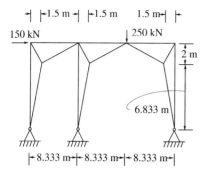

2.21   Compute the reactions of the frame. Note the hinge at 3.

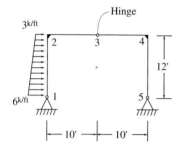

2.22   Compute the reactions of the frame. Note the hinges at *C* and *E*.

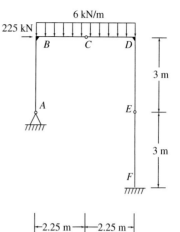

2.23   Compute all reactions for the plane truss shown.

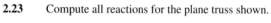
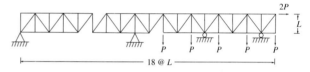

75

COMPUTER PROBLEMS

For the problems that follow, obtain the graphical results requested using one of the following methods:

(a) Create a spreadsheet and use the results to obtain the graphs.

(b) Write a small computer program that obtains the results requested and create a file that can be plotted to obtain the graphs.

(c) Use available software programs that will do the analysis requested. Run the programs as many times as required to obtain the results requested and create a file that can be plotted to obtain the graphs.

C2.1   The location of support B is a distance h below the top of the truss structure, as shown. Find the horizontal reaction at A as a function of h, for $10 \leq h \leq 15$. Plot the results.

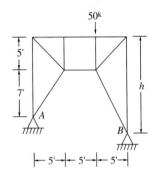

C2.2   The location of support B is a distance a to the right of the support at A of the truss structure, as shown. Find the vertical reaction at A as a function of a, $6 \text{ m} \leq a \leq 15 \text{ m}$. Plot the results.

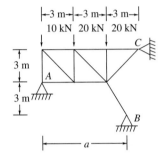

C2.3   Find the vertical reaction at R as a function of a and b. Take values of b at $0, 0.1, 0.5$, and 1. Use $0 < a \leq 1.0$. Plot the results.

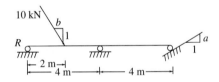

C2.4   Obtain the reaction at R as a function of a, where a varies from $1 \text{ ft} \leq a \leq 29 \text{ ft}$. Plot the results.

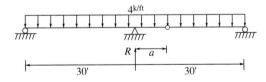

C2.5   Obtain the horizontal reaction of the uniformly loaded portal frame as a function of position of the hinge for $0 \leq a \leq 7$ m. Plot the results.

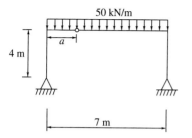

C2.6   Obtain the moment reaction at 1 as a function of the location of the hinge at 2. Take $0 \leq a \leq 16$ ft and plot the results. Note the hinge at 4.

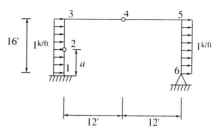

Analysis of Plane Trusses

Trusses are used to carry loads over relatively long spans in situations where there are few or no limitations on the depth of the structure. They are frequently used for roofs over large open areas, such as in warehouses, sports arenas, or industrial buildings. They also find extensive use in highway and railroad bridges. A great many trusses were constructed in this country during the nineteenth and early twentieth centuries because of the need for bridges with the expansion of the railroads. Many individuals and small companies developed special forms and member configurations for trusses which they patented. Trusses became known by their names, such as Howe, Pratt, Fink, Bowman, and Warren, to name a few. Examples of the configuration of these trusses are shown in References 11 and 15.

Later in the twentieth century the development of the automobile led to a bridge-building bonanza, with many of them over rivers or gorges that required long spans. Trusses were again a popular choice, particularly because of the strong growth of the steel industry, which made it a readily available and economical material. Not only were trusses used for bridges in which they were the primary load-carrying elements, they were also used as the stiffening elements of long-span suspension bridges.

Trusses are a very popular structural form since they provide for an economical use of material. The individual member stresses are uniform across the area of the cross section and throughout the length because the member is subjected to axial loads with little or no bending moments. Simple cross sections can be used for the members, and the joining of members at joints is not difficult. Different materials can also be mixed so that each material is used efficiently. For example, wood and steel can be used together in trusses where the wood is for those members that are in compression and steel for those in tension.

In Chapter 2, reactions for statically determinate trusses were obtained as well as an evaluation of the geometric stability of the structures. In this chapter the internal distribution of member forces and the behavior of trusses are studied. Important assumptions about the analysis of trusses are established and determinations made of their static determinacy and geometric stability.

3.1   Truss Action

A truss is constructed with long, slender members which are designed on the assumption that axial forces are the primary stress-inducing quantity. These structures are analyzed on the basis of the members being idealized as pinned or pin connected at joints. This creates a mathematical model which is a collection of lines representing the centroidal axes of the members ending in pins at the joints. By convention, the sign of the axial force is taken as positive when the member is in tension and negative when in compression. Figure 3.1 shows a mathematical model of a truss member and Fig. 3.2 shows the sign convention for member forces.

$$A \qquad\qquad B$$
$$V_A \; \} = 0 \qquad\qquad V_B \; \} = 0$$

$$\Sigma M_A = 0 \qquad\qquad \Sigma M_B = 0$$

$$\therefore V_B = 0 \qquad\qquad \therefore V_A = 0$$

Figure 3.1 Mathematical model of plane truss member action.

Figure 3.1 An individual truss member is isolated as a free body with the axial and shear forces acting at the ends. There are no moments on the ends of the member because the truss members are assumed to be pin connected, so the four end forces are the only unknowns that appear in the free body. No lateral loads are applied on the member, as loads in a truss are always applied at the joints of the structure. There are three equations of equilibrium for the free body. Summing moments in the free body about each end of the member yields the conditions that the end shears are zero because no loads act between the member ends. The last equilibrium equation, the sum of forces parallel to the axis of the member, yields the result that the end axial forces are equal and opposite. This establishes that truss members are two force members and have only one internal action of unknown magnitude, which is an axial force.

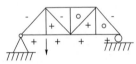

Truss with load and sign of force in members

$$F \xleftarrow{\qquad + F \qquad} F$$

Tension is positive

$$F \xrightarrow{\qquad - F \qquad} F$$

Compression is negative

Figure 3.2 Sign convention for truss member forces.

Figure 3.3 The parts of a truss and other notation are defined in this figure. The notation for a general truss is shown in Fig. 3.3a. The top and bottom members of the truss are called *chords,* and the members in between are called *verticals* and *diagonals.* The panels of the truss are delineated by the verticals, and a typical notation for joints is U_i for joints located on the upper chord and L_i for joints located on the lower chord, as indicated.

Figure 3.3b shows a common configuration for a truss highway bridge with the deck of the bridge at the level of the bottom chord of the truss. The bridge is supported by the two main trusses on either side, and the portal frames at each end of the bridge help provide lateral stability to the structure. Loads applied to the deck of the bridge in Fig. 3.3b are carried by means of the stringers to the floor beams, which, in turn, carry them to the joints or panel points at the bottom chord of the trusses. The details of the floor system are shown in the elevation and plan views in Fig. 3.3c. As can be seen, the major loads are applied to the truss at the joints by the secondary structural system made up of the stringers and floor beams, which ensures that the primary axial force action in the members of the truss is preserved.

A truss can be thought of as a very deep, wide flange (a W section) or an I beam. The cross-section configurations of an I beam or W section and a truss are compared in Fig. 3.4.

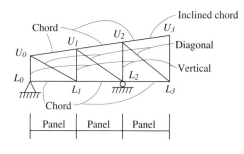

The U_i and L_i
are panel points

(a) Truss member definition
and terminology

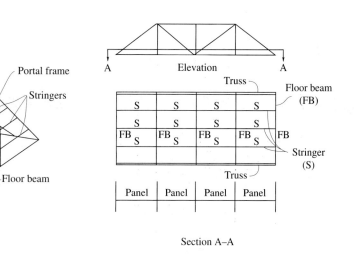

(b) Schematic of highway truss bridge
showing the deck and floor system

(c) Floor system of a highway truss bridge

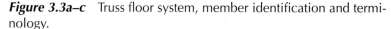

Figure 3.3a–c   Truss floor system, member identification and terminology.

Figure 3.4 In I beams and W sections internal moments are developed by the equal and opposite forces that develop in the flanges (F_f) of the beam with little internal moment developed by the forces F_w in the web. The internal shear force V_w is developed from shear stresses in the web, and no significant shear is developed in the flanges. In a similar manner, the chord members of the truss which appear at the top and bottom correspond to the flanges of the beam as they develop large forces (F_t and F_b) to resist the internal moment. These members can develop large axial forces at the top and bottom extremities of the structure, which provides a

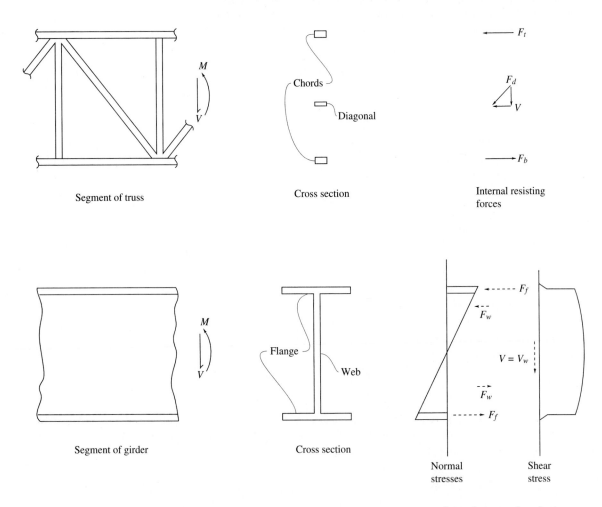

Segment of truss

Cross section

Internal resisting forces

Segment of girder

Cross section

Normal stresses

Shear stress

Internal stress and resultant resisting forces

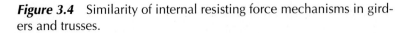

Figure 3.4 Similarity of internal resisting force mechanisms in girders and trusses.

large resisting couple or moment. The verticals and diagonals develop the shear force as shown in Fig. 3.4, and consequently are often referred to as *web members* because their role is very similar to that of the web of a wide-flange beam. The chord members of trusses have large cross-sectional areas, while web members have smaller cross-sectional areas. This configuration of members and distribution of cross-sectional areas make the truss a very efficient structure for carrying large loads over long spans.

In summary, a truss is divided into panels, with all loads being applied at either top or bottom panel points. In roof trusses the loads are applied at the top panel points by a system of purlins and in bridges at the top or bottom panel points by a system of stringers and floor beams. Trusses may have either parallel or inclined chords and may have a large variety of configurations for the web members. No significant bending action occurs in the members of a truss structure. However, there are some manufactured or prefabricated components that are referred to as trusses. For example, lightweight steel joists are considered to be trusses even though there is bending action in the chords of the joist. These members have a shallow depth, usually have no vertical members, and have a relatively short spacing of panel pints, defined as those points where diagonal members intersect top or bottom chords. The structural action of these members is more "trusslike" than "beamlike" and they are thus called trusses.

3.2   Unknowns and Assumptions in Truss Analysis

Each member of a truss, such as those shown in Fig. 3.2, has an unknown force associated with it. In addition, there are one or two unknown reaction components at each support of the structure. In a statically determinate structure there are a sufficient number of equilibrium equations available to determine all unknown reactions and internal member forces. Consider the simple two-member truss shown in Fig. 3.5, with the horizontal and vertical loads H and V acting at A. Under the action of these loads the structure deforms because the axial forces in the members cause a change in member length, and a horizontal and vertical displacement develops at A which is exaggerated in Fig. 3.5b for clarity. The structure must be in equilibrium in the displaced position shown, or the two equilibrium equations involving the two unknown internal member forces F_1 and F_2 and the known applied loads H and V at joint A become, by reference to Fig. 3.5c,

$$F_1 \cos(\theta_1 + \Delta\theta_1) - F_2 \cos(\theta_2 + \Delta\theta_2) = H$$
$$F_1 \sin(\theta_1 + \Delta\theta_1) + F_2 \sin(\theta_2 + \Delta\theta_2) = V$$

(3.1)

In Eqs. (3.1), $\Delta\theta_1$ and $\Delta\theta_2$ represent the change in angles θ_1 and θ_2 caused by the change in length of members 1 and 2. Angles θ_1 and θ_2 define the

83

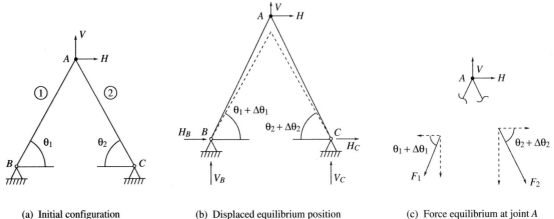

(a) Initial configuration   (b) Displaced equilibrium position   (c) Force equilibrium at joint A

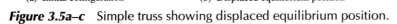

Figure 3.5a–c   Simple truss showing displaced equilibrium position.

initial (or unloaded) positions of the members of the truss. There are four additional unknowns associated with this truss, the reaction components at joints B and C. However, the two equilibrium equations at each reaction joint can be used to obtain these four unknown reaction components in terms of F_1 and F_2 and the two angle changes, $\Delta\theta$.

Equations (3.1) contain four unknowns, the two member forces, F_i, and the two angle changes, $\Delta\theta_i$. Thus these equilibrium equations are insufficient in number to solve for both the unknown member forces and unknown angle changes. Since the two equations of equilibrium (or equations of statics) are insufficient to determine the unknown member forces, the structure is statically indeterminate. This means that the structure cannot be analyzed unless two additional equations or conditions can be found relating to the behavior of the structure or involving the member forces and the angle changes. Several possibilities will be considered in the next few paragraphs that will resolve the difficulty.

If the two members of the truss are assumed to be perfectly rigid, two conditions are provided which are that each member of the truss does not change length. These conditions will not permit any displacement of joint A, and as a result the $\Delta\theta$'s would be zero. Now Eqs. (3.1) are sufficient for the analysis of the truss because it has rigid members, and under these conditions the truss of Fig. 3.5a can be defined as statically determinate.

All real structures undergo deformations when loaded, these deformations being related directly to the forces that develop in the members. A more appropriate approach to the analysis of this truss is to develop some evaluation of its deformation behavior. This approach provides two additional relations between the member forces F_1 and F_2, the change in length of these members and the angle changes $\Delta\theta_1$, and $\Delta\theta_2$. These relations can be used to eliminate the dependence of Eqs. (3.1) on the two angle changes, leaving only the member forces as unknowns. If this analysis is carried out, the resulting two equations involving F_1 and F_2 will have an extremely complex, nonlinear form which makes their solution difficult at best.

For most materials the elongation of a member is very small in relation to the load that it sustains. This implies that the angle changes will also be very small. So a third approach to the problem is to explore the possibility that some simplification of the equations can be obtained based on the assumption that the angle changes are very small. First, expand the sine and cosine terms in Eqs. (3.1) into single-angle forms, which changes Eqs. (3.1) to

$$F_1 \cos \theta_1 \cos_1 \Delta\theta_1 - F_1 \sin \theta_1 \sin \Delta\theta_1$$
$$- F_2 \cos \theta_2 \cos \Delta\theta_2 + F_2 \sin \theta_2 \sin \Delta\theta_2 = H$$

(3.2)

$$F_1 \sin \theta_1 \cos \Delta\theta_1 + F_1 \cos \theta_2 \sin \Delta\theta_1$$
$$+ F_2 \sin \theta_2 \cos \Delta\theta_2 + F_2 \cos \theta_2 \sin \Delta\theta_2 = V$$

For very small angle changes, the sine and cosine of the angle changes can be replaced with the first term of a Taylor series expansion of these functions, with the error being at most less than the square of the angle change. Thus, replacing $\sin \Delta\theta$ with $\Delta\theta$ and $\cos \Delta\theta$ with 1 changes Eqs. (3.2) to the form

$$F_1 \cos \theta_1 - F_1 \Delta\theta_1 \sin \theta_1 - F_2 \cos \theta_2 + F_2 \Delta\theta_2 \sin \theta_2 = H$$

(3.3)

$$F_1 \sin \theta_1 + F_1 \Delta\theta_1 \cos \theta_1 + F_2 \sin \theta_2 + F_2 \Delta\theta_2 \cos \theta_2 = V$$

Unfortunately, Eqs. (3.3) still contain four unknown quantities, and although the form of these equations is simplified, it must be remembered that these equations are no longer identical with the equilibrium equations, Eqs. (3.1) or (3.2), but are now approximations of them. If the approximation is taken one step further by assuming that $\Delta\theta_1$ and $\Delta\theta_2$ are very, very small, to the point where they may be taken to be approximately zero, the equations can be further simplified by deleting these terms. In this case Eqs. (3.3) reduce to

$$F_1 \cos \theta_1 - F_2 \cos \theta_2 = H$$

(3.4)

$$F_1 \sin \theta_1 + F_2 \sin \theta_2 = V$$

These equations are not exact, but they contain only the two unknown member forces, which may be calculated by direct solution. This solution can be considered the first approximation of the exact solution to Eqs. (3.1), based on the assumption of very, very small angle changes.

Comparing Eqs. (3.1) and (3.4) shows that Eqs. (3.4) can be obtained by letting $\Delta\theta_1$ and $\Delta\theta_2$ go to zero in Eqs. (3.1). Thus Eqs. (3.4) are nothing more than the rigid-body equilibrium equations for this truss. The solution for F_1 and F_2 based on Eqs. (3.4) assumes that the geometry of the structure does not change due to the application of the loads. Since structures are generally very stiff (i.e., they deform very little under the action of normal loads), the assumption of unchanged geometry is usually a good one. This assumption is incorporated into the basic analysis procedure detailed for trusses below.

Once the member forces are obtained from Eq. (3.4), why can't these forces be used in Eq. (3.3) to provide a first approximation to the solution for the angles $\Delta\theta_1$ and $\Delta\theta_2$? [*Hint:* Reread the third and fifth paragraphs of this section or try to substitute the solutions of Eq. (3.4) into Eq. (3.3) and solve for the $\Delta\theta$'s.]

3.3 Determinacy and Geometric Stability

It has been established in previous sections that under the action of a general loading, each member of a truss has one unknown axial force, and an additional unknown is associated with each reaction component. If the assumption is made that the geometry of the truss is essentially unchanged, these represent the only unknown force actions, and the following statement can be made:

> If it is assumed that the deformations of a truss are small so that its geometry remains essentially unchanged under the action of the applied loads, the number of unknowns involved in the analysis of any truss is equal to the sum of the number of reaction components and the number of members in the truss.

There are two equilibrium equations at each joint of a plane truss. If the structure undergoes significant deformations so that its geometry is changed, these equations will be similar in form to Eqs. (3.1) and will involve the unknown angle changes for each member at the joint. For a structure undergoing small deformations so that the geometry is essentially unchanged, the form of these equations will be similar to Eqs. (3.4). In either case:

> There are two equilibrium equations at each joint of a plane truss, which makes the total number of equilibrium equations for the structure twice the number of joints.

In the simple structure shown in Fig. 3.5, there are six unknowns (two member forces and four reaction components) if the deformations of the structure are sufficiently small so that the angle changes are zero. There are six equations of equilibrium (two at each of the three joints). Hence the structure is statically determinate and stable since the geometry of the structure remains unchanged under the applied loading. The analyses that follow in this chapter are all based on the following:

> It is assumed that under the action of applied loading, the deformations of a truss are so small that the geometry of the truss does not change and the initial geometry of the structure can be used for reference in writing the equations of equilibrium.

Using the structure of Fig. 3.5 as a base or point of beginning, the structure can be enlarged sequentially by adding one new joint at a time as illustrated in Fig. 3.6.

Figure 3.6 A new joint can be added to the structure of Fig. 3.6a by the introduction of two new members, as shown in Fig. 3.6b and c. Alternatively, a new joint can be created by the addition of one new member and one new reaction constraint (or support), as shown in Fig. 3.6d. Each new joint introduces two new equilibrium equations, but two new unknown are added at the same time. The two new unknowns can

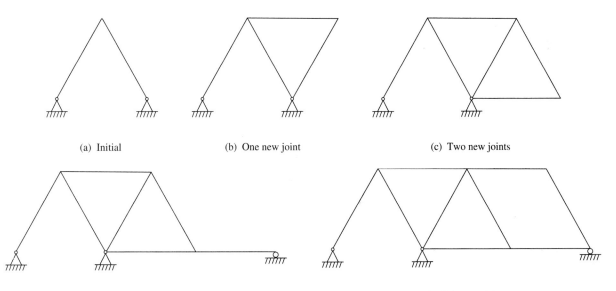

(a) Initial   (b) One new joint   (c) Two new joints

(d) Three new joints   (e) Four new joints

Figure 3.6a–e Construction of a statically determinate, geometrically stable truss.

be either two new member forces or one new member force and one new reaction component. As long as the enlarging of the truss is carried on in this manner the structure will remain statically determinate and stable.

Figure 3.7 If a new joint is added to a geometrically stable truss by using only one new member, as shown in Fig. 3.7a, there will be two new equations of equilibrium but only one new unknown member force. For the structure of Fig. 3.7a there are now more equations of equilibrium than there are unknowns and it is also clear that the new joint is unstable. This idea can be generalized to conclude that if there are fewer unknowns in a structure than there are equations of equilibrium, the structure must be geometrically unstable.

A structure that appears to be statically determinate can be geometrically unstable. For example, if a new joint is added to a structure using two collinear members, as shown in Fig. 3.7b, the structure is locally unstable. Another example of this instantaneous geometric instability is shown in Fig. 2.3d.

Figure 3.8 Certain arrangements of the members or the reactions of a structure that is apparently statically determinate may make the entire structure or some local part geometrically unstable. This is illustrated by

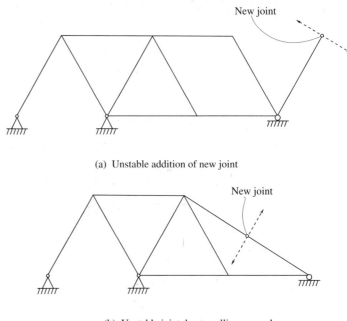

(a) Unstable addition of new joint

(b) Unstable joint due to collinear members

Figure 3.7a–b Truss constructions giving local geometrical instabilities.

the trusses shown in Figs. 2.3f and 3.8. In all of these structures, the instability is due to some type of instantaneous rigid-body displacement or rotation. For all of the structures shown in these figures, the dashed arrows indicate the sense of a possible rigid-body translation or rotation of a part or all of the structure.

A statically indeterminate structure is created when, in a statically determinate structure, members are added between existing joints or more reaction components are introduced as shown in Fig. 3.9.

Figure 3.9 Adding a new member, as in Fig. 3.9b, or a new reaction component, as in Fig. 3.9c, adds an unknown force quantity, but no new equations of equilibrium are introduced because no new joints are added to the structure. The degree of indeterminacy is defined as the number of unknowns in excess of the number of equilibrium equations. Hence each new member or reaction component introduced into a statically determi-

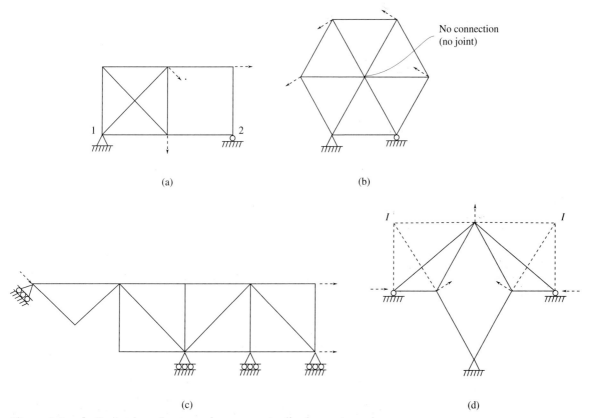

(a)

(b)

(c)

(d)

Figure 3.8a–d Examples of trusses that are statically determinate by count, but are geometrically unstable.

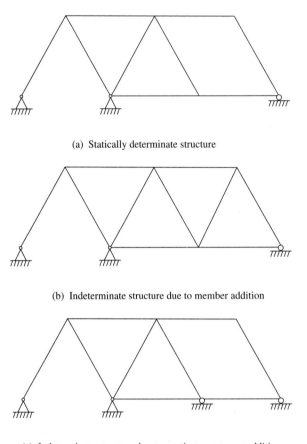

(a) Statically determinate structure

(b) Indeterminate structure due to member addition

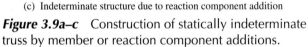

(c) Indeterminate structure due to reaction component addition

Figure 3.9a–c Construction of statically indeterminate truss by member or reaction component additions.

nate, geometrically stable structure without creating a new joint will increase the degree of indeterminacy by one.

The geometric configuration of the members and the reaction components, not their number, determine whether or not a structure is unstable. Consequently, the introduction of new members or reaction components to a structure that is geometrically unstable may not take away the instability. For example, adding a reaction component to the roller support at 2 of the structure in Fig. 3.8a (and changing it to a pin support) does not remove the geometric instability even though a new unknown has been added with no new joints. Similarly, adding one or more new members to the rectangular panels of the structure of Fig. 3.8c adds unknown(s) but does not remove the geometric instability, which is a rigid-body horizontal translation of the structure.

The ideas discussed above may all be summarized in the following statement:

A structure is geometrically unstable if the number of unknowns is less than the number of equations of equilibrium. If, in a geometrically stable structure, the number of unknowns is equal to the number of equations of equilibrium, the structure is statically determinate, but if the number of unknowns exceeds the number of equations, the structure is indeterminate.

A quantitative or more mathematical statement of these concepts is also useful. In a plane truss, let b represent the number of members, r the number of reaction components, n the number of joints, and p the index of static indeterminacy, then p can be determined from the expression

$$p = b + r - 2n \qquad (3.5)$$

where the following conditions on p define the structure to be

Geometrically unstable	when $p < 0$
Statically determinate if not geometrically unstable	when $p = 0$
Statically indeterminate to degree p if not geometrically unstable	when $p > 0$

The requirement that p be equal to or greater than zero is a necessary but not sufficient condition for the geometric stability of a truss. Similarly, p being equal to zero is a necessary but not sufficient condition for the structure to be statically determinate. Can you sketch some examples of trusses that validate the correctness of these two statements? (*Hint:* Reexamine Figs. 3.7b and 3.8.)

3.4   Plane Truss Analysis by the Method of Joints

Every statically determinate and geometrically stable plane truss can *always* be analyzed successfully by the method of joints. One may start at any joint and from a free-body diagram of that joint write two equilibrium equations in terms of the unknown member forces, any unknown reaction components, and any applied loads. Proceeding systematically through all n joints of the structure establishes a total of $2n$ equations. For statically determinate structures these equations are sufficient to obtain the $2n$ unknowns by solving them simultaneously. The method is tedious, but it always works and can be programmed for solution on a computer.

Example 3.1 The simple three-joint truss is analyzed by starting at joint 3, where there are only two unknowns, the forces in the two members. The resolution of the inclined member force, F_1, into horizontal and vertical components permits the direct calculation of the force in this member from the horizontal equilibrium equation. The calculation of the reaction components at 1 and 2 is again aided by using the two components of the force F_1. For numerical calculation using U.S. units, take $a = 10$ ft, $b = 5$ ft, and $p = 10$ kips; for SI units, take $a = 3$ m, $b = 1.5$ m, and $P = 50$ kN.

In Example 3.1, calculation of all the unknowns was possible simply by starting at one joint in the structure, solving for the unknowns there with the two equations of equilibrium, and proceeding to other joints to complete the analysis. It is not always possible in the method of joints to solve for the unknowns joint by joint in this manner. The analysis may require the solution of a system of simultaneous equations without a modification in the procedure, which is illustrated next in a more comprehensive example.

Example 3.2 For hand computation the solution of simultaneous equations can almost always be avoided by using the three-step procedure process illustrated. The first step of the process is to compute the reaction forces on the structure. To do this a determination of whether or not the structure is statically determinate and geometrically stable must be made. A count of the number of members (13), the number of reaction components (3), and the number of joints (8) gives from Eq. (3.5) that $p = 13 + 3 - 2 \cdot 8 = 0$. A careful inspection of the member and support configuration confirms that the structure is geometrically stable. The reaction components can then be determined as shown here and in most other trusses simply by writing and solving the three equations of equilibrium for the entire structure.

In some more complicated situations where there are more than three reaction components, the equation(s) of condition that accompany internal releases of continuity of deformation as shown in Fig. 2.3 are required to complete the computation of reaction components. Example 2.3 is an illustration of the latter situation.

Because loads are applied at panel points on the truss which often are evenly spaced, computation of the reactions can be reduced to using proportions. As illustrated in the example, the reaction R_{L0} for the three-panel truss can be computed as the sum of two-thirds of the 100-kip load at panel point L_1 and one-third of the 120-kip load at panel point L_2.

The second step is to resolve all member forces, applied loads, and reaction forces into a convenient set of orthogonal components, usually horizontal and vertical. This is accomplished most conveniently by creating a large sketch of the truss, with all reactions and applied loads shown horizontally and vertically. Forces in inclined members are

Example 3.1

Use the method of joints to obtain the member forces and reaction of the truss.

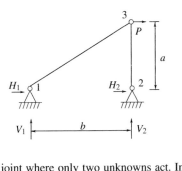

STEP 1   Start at a joint where only two unknowns act. In this case, joint 3 has only the two unknown member forces acting. The two equations of equilibrium can be solved for the member forces. Use the orientation of the member 1–3 to resolve its force into horizontal and vertical components.

$$L = \sqrt{a^2 + b^2}$$

$$\Sigma F \rightarrow : P - F_1 \frac{b}{L} = 0$$

$$F_1 = \frac{PL}{b}$$

$$\Sigma F \uparrow : -F_1 \frac{a}{L} - F_2 = 0$$

$$F_2 = -F_1 \frac{a}{L} = \frac{-Pa}{b}$$

Joint 3

STEP 2   Proceed to joints 1 and 2 and use the two equations of equilibrium at each joint to compute the reaction components.

$$\Sigma F \rightarrow : H_1 + F_1 \frac{b}{L} = 0$$

$$H_1 = -F_1 \frac{b}{L} = -P$$

$$\Sigma F \uparrow : V_1 + F \frac{a}{L} = 0$$

$$V_1 = -F_1 \frac{a}{L} = \frac{-Pa}{b}$$

Joint 1

$$\Sigma F \rightarrow : H_2 = 0$$

$$\Sigma F \uparrow : V_2 + F_2 = 0$$

$$V_2 = -F_2 = \frac{Pa}{b}$$

Joint 2

Example 3.2

Obtain all member forces for the truss loaded as shown.

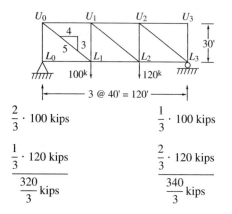

$$\frac{2}{3} \cdot 100 \text{ kips} \qquad \frac{1}{3} \cdot 100 \text{ kips}$$

$$\frac{1}{3} \cdot 120 \text{ kips} \qquad \frac{2}{3} \cdot 120 \text{ kips}$$

$$\frac{320}{3} \text{ kips} \qquad \frac{340}{3} \text{ kips}$$

STEP 1 Compute the reactions. (*Note*: The reactions can be obtained by proportion using the number of panels.)

$$\Sigma M_{L0} : -100 \cdot 40 - 120 \cdot 2 \cdot 40 + R_{L3} \cdot 3 \cdot 40 = 0$$

$$R_{L3} = \frac{40}{3 \cdot 40} \cdot 100 + \frac{2 \cdot 40}{3 \cdot 40} \cdot 120 = \frac{1}{3} \cdot 100 + \frac{2}{3} \cdot 120 = \frac{340}{3} \text{ kips}$$

$$\Sigma M_{L3} : -R_{L0} \cdot 3 \cdot 40 + 100 \cdot 2 \cdot 40 + 120 \cdot 40 = 0$$

$$R_{L0} = \frac{2 \cdot 40}{3 \cdot 40} \cdot 100 + \frac{40}{3 \cdot 40} \cdot 120 = \frac{2}{3} \cdot 100 + \frac{1}{3} \cdot 120 = \frac{320}{3} \text{ kips}$$

STEP 2 Resolve member forces horizontally and vertically at each joint. Create diagram below for all member forces. Enter forces as they are computed at each joint by equilibrium.

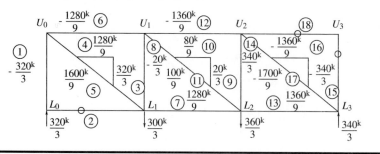

Example 3.2 (continued)

STEP 3 Starting at joint L_0, proceed joint by joint in the sequence U_0, L_1, U_1, L_2, U_2, L_3, U_3, until all member forces have been computed. Note that the last three equilibrium equations provide a check on the computations. The circled numbers in the summary diagram correspond with the numbers in the calculations below.

Joint L_0:

$$\Sigma F \uparrow \quad \frac{320}{3} + F_1 = 0$$

$$F_1 = \frac{-320}{3} \text{ kips} \tag{1}$$

$$\Sigma F \rightarrow F_2 = 0 \tag{2}$$

Joint U_0:

$$\Sigma F \uparrow \quad \frac{320}{3} - F_3 = 0$$

$$F_3 = \frac{320}{3} \text{ kips} \tag{3}$$

$$F_4 = \frac{4}{3} F_3 = \frac{1280}{9} \text{ kips} \tag{4}$$

$$F_5 = \frac{5}{3} F_3 = \frac{1600}{9} \text{ kips} \tag{5}$$

$$\Sigma F \rightarrow \quad \frac{1280}{9} + F_6 = 0$$

$$F_6 = \frac{1280}{9} \text{ kips} \tag{6}$$

Joint L_1:

$$\Sigma F \rightarrow \quad \frac{-1280}{9} + F_1 = 0$$

$$F_1 = \frac{1280}{9} \text{ kips} \tag{7}$$

$$\Sigma F \uparrow$$
$$\frac{-300}{3} + \frac{320}{3} + F_8 = 0$$

$$F_8 = \frac{-20}{3} \text{ kips} \tag{8}$$

Joint U_1:

$$\Sigma F \uparrow \quad \frac{20}{3} - F_9 = 0$$

$$F_9 = \frac{20}{3} \text{ kips} \tag{9}$$

$$F_{10} = \frac{4}{3} F_8 = \frac{80}{9} \text{ kips} \tag{10}$$

$$F_{11} = \frac{5}{3} F_8 = \frac{100}{9} \text{ kips} \tag{11}$$

$$\Sigma F \rightarrow$$
$$\frac{1280}{9} F_{12} + \frac{80}{9} = 0$$

$$F_{12} = \frac{-1360}{9} \text{ kips} \tag{12}$$

Example 3.2 (continued)

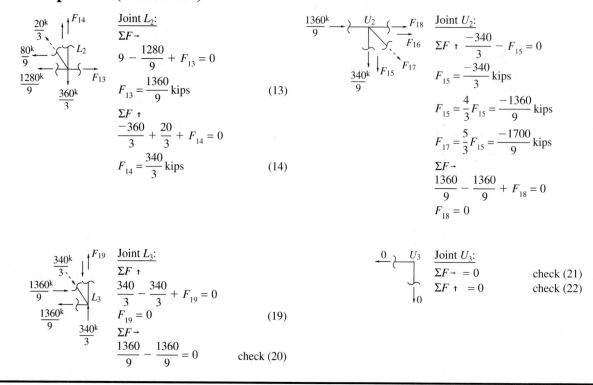

Joint L_2:

$\Sigma F \rightarrow$

$$9 - \frac{1280}{9} + F_{13} = 0$$

$$F_{13} = \frac{1360}{9} \text{ kips} \tag{13}$$

$\Sigma F \uparrow$

$$\frac{-360}{3} + \frac{20}{3} + F_{14} = 0$$

$$F_{14} = \frac{340}{3} \text{ kips} \tag{14}$$

Joint U_2:

$\Sigma F \uparrow$ $\dfrac{-340}{3} - F_{15} = 0$

$$F_{15} = \frac{-340}{3} \text{ kips}$$

$$F_{15} = \frac{4}{3} F_{15} = \frac{-1360}{9} \text{ kips}$$

$$F_{17} = \frac{5}{3} F_{15} = \frac{-1700}{9} \text{ kips}$$

$\Sigma F \rightarrow$

$$\frac{1360}{9} - \frac{1360}{9} + F_{18} = 0$$

$$F_{18} = 0$$

Joint L_3:

$\Sigma F \uparrow$

$$\frac{340}{3} - \frac{340}{3} + F_{19} = 0$$

$$F_{19} = 0 \tag{19}$$

$\Sigma F \rightarrow$

$$\frac{1360}{9} - \frac{1360}{9} = 0 \qquad \text{check (20)}$$

Joint U_3:

$\Sigma F \rightarrow\ = 0$ check (21)

$\Sigma F \uparrow\ = 0$ check (22)

shown along with their horizontal and vertical components as illustrated in Example 3.2. The resolution of forces horizontally and vertically expedites the writing and solving of the joint equilibrium equations.

The final step is to start with a free body of a joint at one end of the structure such as L_0, where there are only two unknown member forces. Using the two equilibrium equations for the free body of the joint, the two unknown member forces can be obtained and entered on the large sketch of the truss. This procedure is repeated at a second joint, such as U_0. Continuing from joint to joint and using the two equations of equilibrium in the free body at each joint, all member forces can be obtained. The order in which the joints are considered for solution in this process is determined by the sequential selection of joints where only two unknowns exist. At the completion of the process, all member forces will appear on the enlarged sketch of the structure, as shown.

At some point near the end of the joint-by-joint analysis the free body at a joint may not have two unknown member forces acting. This first occurs at joint L_3, where the only unknown member force is that for member L_3–U_3. The reason for this occurrence is the fact that three equilibrium equations were used to compute the unknown reaction components at L_0 and L_3 from overall equilibrium considerations of the structure in step 1. When the analysis reaches L_3, all but one of the $2 \cdot 8 = 16$ independent equations of equilibrium have been used, and all but one of the unknown member forces and reaction components have been obtained. The last independent equilibrium equation is summation of forces vertically in the free body of joint L_3 which gives the value of the unknown force in member L_3–U_3.

The analysis is now complete because all unknowns have been determined. However, since equilibrium must be satisfied both horizontally and vertically at every joint, writing the horizontal equilibrium equation at joint L_3 and the two equilibrium equations at the last joint, U_3, provides a check of the consistency of the numerical calculations. As can be seen, these three equilibrium equations at joints L_3 and U_3 are satisfied identically, which verifies the consistency of the analysis.

The joint-by-joint analysis in Example 3.2 was started at joint L_0. What other joint could have been used as a starting point of the analysis, and what sequence of joints would have been followed to obtain all the member forces? (*Hint:* Reread the sixth paragraph of the discussion of Example 3.2.)

Example 3.3 Example 3.3 is a presentation of a spreadsheet computation of the member forces in the truss of Example 3.2. The spreadsheet is a rather simple one to construct and provides the capability of varying the panel lengths, truss height, and the pattern of vertical loads on the truss. The calculation of member forces is through the use of the equilibrium equations at each joint, as was illustrated in Example 3.2. Note that

Example 3.3

Create a spreadsheet for the analysis of the truss problem of Example 3.2. The analysis is to be carried out by the method of joints as in Example 3.2. Input to the spreadsheet is the following:

Panel length:	cell B1
Height:	cell D1
Loads: L_0 to L_3:	cells B3 to E3

Vertical positive loads act upward on the truss. The possibility of applied loads at L_0 and L_3 is included. Horizontal loads and loads acting at top chord joints are not considered.

The vertical reactions at L_0 and L_3 are computed from the moment equilibrium equations and the results stored in cells B4 and E4. The horizontal reaction at L_0 is assumed to be zero. The length of diagonal members is computed by the Pythagorean theorem and stored in cell F1. The reactions are computed as in Example 3.2:

$$\text{At } L_0 \text{ (cell B4)} = -\frac{3 \cdot B3 + 2 \cdot C3 + D3}{3}$$

$$\text{At } L_3 \text{ (cell E4)} = -\frac{C3 + 2 \cdot D3 + 3 \cdot E3}{3}$$

The forces in the members are computed from joint equilibrium. The calculations follow from the free bodies of Example 3.2.

$$L_0\text{–}U_0 \text{ (cell B6)} = -B4 - B3 \qquad\qquad L_0\text{–}L_1 \text{ (cell B7)} = 0$$

$$U_0\text{–}L_1 \text{ (cell B8)} = -\frac{F1}{D1} \cdot B6 \qquad\qquad U_0\text{–}U_1 \text{ (cell B9)} = -\frac{B1}{F1} \cdot B8$$

$$L_1\text{–}U_1 \text{ (cell B10)} = -C3 - \frac{D1}{F1} \cdot B8 \qquad L_1\text{–}L_2 \text{ (cell B11)} = B7 - \frac{B1}{F1} \cdot B8$$

The other member forces are computed in a similar manner.

	A	B	C	D	E	F
1	PANEL:	40	HEIGHT:	30	DIAGONAL:	50
2	PANEL POINT	L0	L1	L2	L3	
3	LOAD	0	-100	-120	0	
4	REACTIONS	106.67	–	–	113.33	
5	MEMBER	FORCE				
6	L0-U0	-106.67				
7	L0-L1	0.00				
8	U0-L1	177.78				
9	U0-U1	142.22				
10	L1-U1	-6.67		Note: All loads and reactions		
11	L1-L2	142.22		in kips; lengths in feet		
12	U1-L2	11.11				
13	U1-U2	-151.11				
14	L2-U2	113.33				
15	L2-L3	151.11				
16	U2-L3	-188.89				
17	U2-U3	0.00				
18	L3-U3	0.00				

the member forces are listed in the spreadsheet in the same order as they are calculated in Example 3.2.

The member forces are displayed in cells B6 to B18 and are calculated from mathematical expressions using the contents of the cells of the spreadsheet. For example, calculation of the force in member L_1–U_1 which appears in cell B10 comes from the vertical equilibrium equation at joint L_1:

$$F_{L1U1} + \frac{30}{50} \cdot F_{U0L1} + (-100) - 0$$

$$\therefore\ F_{L1U1} = -(-100) - \frac{30}{50} \cdot F_{U0L1}$$

Since the applied load of (-100) is in cell C3, the truss height in cell D1, the diagonal length in cell F1, and the force in member U_0–L_1 in cell B8, the equilibrium equation above for the force in member L_1–U_1 in cell B10 is

$$\text{cell B10} = -(\text{cell C3}) - \frac{\text{cell D1}}{\text{cell F1}} \cdot \text{cell B8}$$

as indicated in the example. The other member forces are calculated in a similar manner. Although the member forces in the spreadsheet results are listed with as many as five digits, the results in reality are of no better accuracy than three digits because of uncertainties regarding the constructed geometry of the truss and the small, but not zero, displacements of the joints of the truss.

The spreadsheet was created to do an analysis, but by adding columns for member area, length, and stress, it can become part of a design procedure for the truss. The spreadsheet is specific to the configuration of the truss of Example 3.2, but it can serve as a template that is easily modified for any other statically determinate truss. With a little imagination, the form of the spreadsheet can be modified to serve a more general configuration of truss and to accommodate loads applied vertically or horizontally at top or bottom panel points.

3.5   Special Considerations at Truss Joints

The joint-by-joint analysis described in Section 3.4 is a long and tedious process for obtaining the member forces in a truss. There are a number of special circumstances related to the configuration of the members of a truss that can simplify the analysis process. The simplification may allow the evaluation of the magnitude of member forces by inspection or with the aid of a very short computation.

Figure 3.10 Consider the situation where only two members connect at a joint at an arbitrary angle. In Fig. 3.10a the joint is unloaded, so that using the two joint equilibrium equations to solve for the member forces leads to the result that the member forces must be zero. This condition occurs at joint U_3 of the truss in Example 3.2.

In Fig. 3.10b there is a load acting at the joint where the members join. The member forces can easily be obtained if the applied load, P, is resolved into components parallel to the orientation of the members. The two member forces will be equal and opposite to these components of P. This condition also occurs in Example 3.2 at joint L_0. In that particular case the applied load at L_0 was the vertical reaction parallel to the vertical member L_0–U_0, which made the force in that member equal to the reaction force (in compression), and the force in the other member (L_0–L_1) equal to zero.

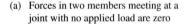

(a) Forces in two members meeting at a joint with no applied load are zero

(b) Forces in two members meeting at a joint with an applied load are obtained from vector resolution

Figure 3.10a–b   Member forces at joints having only two members.

The observations summarized in Fig. 3.10 can speed the analysis of the truss. They can be summarized in the following statement:

> If two members that are not collinear meet at a single joint in a truss and no load is applied to that joint, the member forces are zero. If a load is applied to the joint, the member forces are equal in magnitude and opposite in sense to the components of the applied force that are parallel to the lines of action of the two members.

A second special situation is illustrated in Fig. 3.11, where two members that are collinear at a joint are joined by a third member at some angle to the line of action of the first two.

Figure 3.11 In Fig. 3.11a, the force in the third member joining the collinear pair is always zero since no load is applied at the joint. Resolving the force in that member into components that are parallel

(F_p) and perpendicular (F_c) to the line of action of the collinear members and summing forces in the direction perpendicular to the line of collinear action makes the perpendicular component F_c and hence the force in the third member zero. If there is a load applied to the joint as shown in Fig. 3.11b, the force in the third member can be obtained by summing forces perpendicular to the line of action of the collinear members using the components of the applied load, P_c, and member force, F_c.

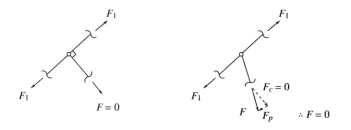

(a) Zero member force at unloaded three-member joints

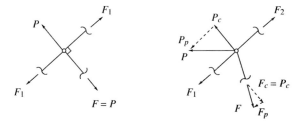

(b) Member forces at loaded three-member joints obtained by resolution

Figure 3.11a–b Member force at three member joints having two collinear members.

The situation depicted in Fig. 3.11 will often provide the third member force from simple inspection. This can be summarized in the following:

If two collinear members at a joint in a truss are joined by a third member at some angle with respect to the first two, the force in the third member is zero if no load is applied at the joint. If a load acts at the joint, the force in the third member can be obtained from the condition that its component perpendicular to the line of action of the collinear members is equal in magnitude and opposite in direction to the corresponding component of the applied load.

A situation that is similar to that in Fig. 3.11 is presented in Fig. 3.12, where two collinear members at a joint are joined by two additional members at arbitrary angles. This is a common situation, and the use of the force equilibrium equation perpendicular to the line of action of the collinear members is extremely useful.

Figure 3.12 Summation of forces perpendicular to the collinear members yields the relation that the components, F_{c1} and F_{c2} of the forces F_1 and F_2 perpendicular to this line of action must be equal and opposite. A similar relation is obtained when a load is applied to the joint, as shown in Fig. 3.12b, but with the load component, P_c, also appearing in the equilibrium equation.

The conditions shown in Fig. 3.12 provide a useful tool for analysis and can be summarized as follows:

It two members are collinear at a joint in a plane truss and are joined by two additional members, 1 and 2, at arbitrary angles to the collinear members, the force components perpendicular to the collinear members of members 1 and 2 are equal and opposite if no applied load acts at the joint. If an applied load acts at the joint, the perpendicular force components of the load and members 1 and 2 are related through equilibrium.

The condition shown in Fig. 3.12a occurs in Example 3.2 at the top chord joints U_1 and U_2. Note in that example that the vertical components of the forces in the diagonal members are equal and opposite to the forces in the verticals at the same joint. The condition shown in Fig. 3.12b occurs at the lower chord joints L_1 and L_2. Again note that the vertical components of the forces in the diagonal members can be obtained from a simple vertical equilibrium condition.

A review of the analysis presented in Example 3.2 in light of the concepts presented in Figs. 3.10 and 3.12 provides some interesting observations. Once the reactions are computed at L_0 and L_3, the forces in the members that join at joints L_0 and U_3 can be obtained by inspection using the concepts presented in Fig. 3.10. The forces in the vertical members of the truss and the *vertical component* of the forces in the diagonal members can be obtained by inspection or by very simple computation using the concepts presented in Fig. 3.12 by proceeding from joints U_0 to L_1 to U_1 to L_2 to U_2. This leaves a few chord member forces that can be computed from joint equilibrium equations that have been simplified by the process described above.

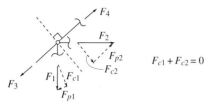

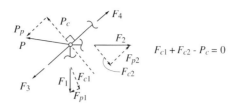

(a) Relation between member forces at unloaded joint   (b) Relation between member forces at loaded joint

Figure 3.12a–b   Member forces at four member joints having two collinear members.

Trusses are constructed of long, slender members that are nearly uniformly stressed by axial loads. This works well for members in tension, but slender members in compression are subject to instability considerations which may limit their effectiveness. To improve the efficiency of compression members designers add bracing (or secondary) members to the truss which have little effect on the distribution of axial forces in the structure. The presence of the bracing members seems to complicate the analysis of the structure, but with some of the observations presented in Figs. 3.11 and 3.12 the complications are in fact small.

Figure 3.13 The truss shown has a bottom chord that is clearly in compression. To improve the effectiveness of the compression member, it is braced with the diagonal members shown at points 3, 5, and 7. These extra members seem to complicate the analysis, but a few observations from Fig. 3.11 will simplify the situation. At joint 7, using the concept presented in Fig. 3.11a, it can be concluded that the force in member 6–7 must be zero. At joint 6, using the concept presented in Fig. 3.12a and the fact that the force in member 6–7 is zero, it can be concluded that the force in member 5–6 is also zero. Proceeding in like manner to joints 5, 4, and 3 leads to the conclusion that all the diagonal bracing member forces are zero, which greatly simplifies the analysis of the truss.

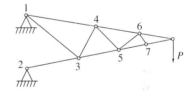

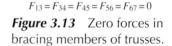

Figure 3.13   Zero forces in bracing members of trusses.

Figure 3.14 A panel of a truss with a top chord and diagonal member that carry compressive axial forces is shown in Fig. 3.14. Again bracing members are added to the panel in Fig. 3.14b to brace the compression members. The analysis of the panel of Fig. 3.14b seems to be more formidable than it actually is. The concept presented in Fig. 3.11a can be used at joints 1 and 2 to establish that members m–1 and m–2 have zero force. The concept in Fig. 3.12a now establishes that member m–3 also has a zero force. The analysis of the forces in the members of the panel of the Fig. 3.14b reduces to the same analysis that is needed for the panel in Fig. 3.14a.

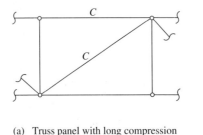

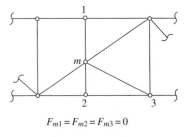

$$F_{m1} = F_{m2} = F_{m3} = 0$$

(a) Truss panel with long compression members

(b) Truss with bracing members that have zero force

Figure 3.14a–b   Bracing or secondary members in truss panel with zero force.

3.6   Truss Analysis by the Method of Sections

Often, only a few member forces may be needed in a truss analysis. Determining the member forces by the method of joints can be long and tedious unless the members are very close to one end of the truss. Another approach, called the *method of sections*, can be used to isolate any member of interest in carefully selected free-body diagrams and obtain the forces in the members with an appropriate application of equilibrium. The method of sections is also a three-step process, and its application is shown in the next four examples.

Example 3.4 The first step is to determine that the structure is statically determinate and geometrically stable and to compute all unknown reactions. The structure is geometrically stable because it is composed of triangles that make it a rigid body. It is supported against all rigid-body displacements and rotations with the reactions at L_0 and L_6. The structure is statically determinate by count of 25 members and three reaction components, for a total of 28 unknowns. There are 14 joints giving 28 equations; thus by Eq. (3.5) it is statically determinate. In step 1 the reactions are computed by proportion in the same manner as in Example 3.2.

In step 2 the free body labeled "Free-body section 1–1" created by cutting members L_2–L_3 and U_3–U_4 and a has three member forces shown in tension acting on it. By finding the point of intersection of the two chord member forces at point O and summing moments about it, a single equation for the member force F_a is obtained. The free body labeled "Free-body section 2–2" created in step 3 provides the section for the computation of forces F_b and F_c. Summation of moments about point O in this free body gives force F_b directly. Note that by moving F_b along its line of action and resolving it into components at L_2 simplifies the moment equilibrium equation. The same technique is used with F_c by mov-

Example 3.4

Obtain the forces in the members designated.

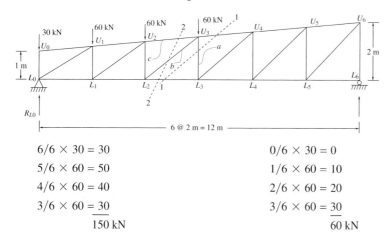

$$6/6 \times 30 = 30 \qquad\qquad 0/6 \times 30 = 0$$
$$5/6 \times 60 = 50 \qquad\qquad 1/6 \times 60 = 10$$
$$4/6 \times 60 = 40 \qquad\qquad 2/6 \times 60 = 20$$
$$\underline{3/6 \times 60 = 30} \qquad\qquad \underline{3/6 \times 60 = 30}$$
$$150 \text{ kN} \qquad\qquad\qquad 60 \text{ kN}$$

STEP 1   Note that the truss is statically determinate and geometrically stable. Calculate the reactions by using proportions. The horizontal reaction is zero.

STEP 2   To obtain the force in member a, create a free body by passing the section 1–1 through member a, member U_3–U_4, member L_2–L_3, and taking the portion to the right of the cuts. Forces in the top and bottom chord members have lines of action that intersect at point O which is 12 m to the left of L_0 since the 1-in-12 slope of the top chord means that the top chord drops 1 m vertically for each 12 m horizontally. Summing moments about O gives the force in member a directly.

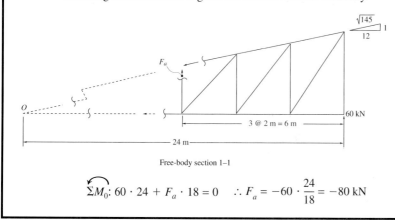

Free-body section 1–1

$$\overset{\frown}{\Sigma M_0}\text{: } 60 \cdot 24 + F_a \cdot 18 = 0 \quad \therefore F_a = -60 \cdot \frac{24}{18} = -80 \text{ kN}$$

105

Example 3.4 (continued)

STEP 3 The force in members b and c is from the free body obtained by passing section 2–2 through members b, c, and L_2–L_3 and taking the portion to the right. The vertical distance L_3–U_3 is 1.5 m and the panel width is 2 m, making the slope of member b 3–4–5 as shown in the free body. The line of action of F_b passes through L_2, so it can be slid along that line until it acts at L_2. Resolve F_b into horizontal and vertical components at L_2 and sum moments about O to obtain the vertical component of F_b acting at L_2. The chord force, F_c, is slid along its line of action until it acts at U_2, a distance of $\frac{4}{3}$ m above L_2. Resolve F_c into components at U_2, and sum moments about L_2 to obtain the horizontal component of F_c.

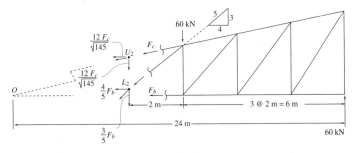

Free-body section 2–2

$$\sum M_O: \frac{3}{5} \cdot F_b \cdot 16 + 60 \cdot 18 - 60 \cdot 24 = 0$$

$$\therefore F_b = \frac{5}{3}(24 - 18) \cdot \frac{60}{16} = \frac{75}{2} = 37.5 \text{ kN}$$

$$\sum M_{L2}: 60 \cdot 2 - 60 \cdot 8 - F_c \cdot \frac{12}{\sqrt{145}} \frac{4}{3} = 0$$

$$\therefore F_c = \frac{\sqrt{145}}{16}(2 - 8)\,60 = -\frac{45}{2\sqrt{145}} = 270.90 \text{ kN}$$

ing it along its line of action to U_2, so that summing moments about L_2 gives F_c directly. In computing force F_c, how could it be obtained by summing forces vertically once F_b is known?

The truss shown in Fig. 3.15 is used to span over a large column-free room in an office building. The forces in selected members of the truss are computed in the next example.

Example 3.5 The weight of the truss is assumed to be applied uniformly along its length and is applied arbitrarily to the top panel points.

Figure 3.15 Placing of 150-foot truss for Chicago Mercantile Exchange. (Courtesy Alfred Benesch & Company)

Example 3.5

The truss shown in Fig. 3.15 weighs 65 tons (130 kips). Compute the forces in the indicated members of the truss due to its own weight.

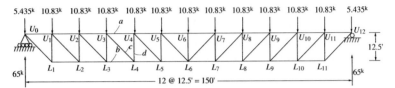

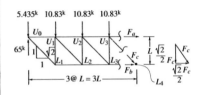

Free-body section 1–1

STEP 1 Take the weight of the truss to be uniformly distributed. There are 12 panels in the truss, so the weight per panel is $130/12 = 10.83$ kips. The truss is obviously stable and statically determinate, and the symmetry of the loading makes the two end vertical reactions $130/2 = 65$ kips.

STEP 2 Pass a section through members a, b, and c to create the free-body section 1–1. From the slope of 1 on 1, resolve F_c into vertical and horizontal components of magnitude $(\sqrt{2}/2) F_c$. Since panel lengths and heights are equal, it is convenient to call them length L. Obtain F_a by summing moments about L_4, F_b, by summing moments about U_3, and F_c by summing forces vertically.

$$\overset{\frown}{\Sigma M_{L4}}: (65 - 5.435)4L - 10.83(3L + 2L + L) + F_a L = 0$$
$$\therefore F_a = 6 \cdot 10.83 - 4 \cdot 59.565 = -173 \text{ kips}$$

$$\overset{\frown}{\Sigma M_{U3}}: (65 - 5.435)3L - 10.83(2L + L) - F_b L = 0$$
$$\therefore F_b = 3(49.735) = 146 \text{ kips}$$

$$\Sigma F \uparrow: 65 - 5.435 - 3 \cdot 10.83 - \frac{\sqrt{2}}{2} F_2 = 0 \quad \therefore F_C = \sqrt{2} \cdot 27.1 = 38.3 \text{ kips}$$

STEP 3 Create free-body section 2–2 by cutting through members a, d, and L_4–L_5 and taking the portion to the left. Sum forces vertically to obtain F_d.

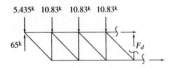

Free-body section 2–2

$$\Sigma F \uparrow: 65 - 5.435 - 3 \cdot 10.83 + F_d = 0$$
$$F_d = -27.1 \text{ kips}$$

The weight could be divided equally between the top and bottom panel points, but this would have a very small effect on some of the member forces. (Which member forces would be affected?) By symmetry the two reactions are 65 kips and the horizontal reaction is zero. The free bodies obtained with sections 1–1 and 2–2 are both taken from the left portion of the structure because there are fewer forces in the free body.

In step 2 the member forces F_a and F_b are computed by summing moments about the bottom and top chord points L_4 and U_3, respectively, since only the unknown member force appears in that equilibrium equation. Because the chords of the truss are parallel, unlike the previous example, forces F_c and F_d can be obtained simply by summing forces vertically in steps 2 and 3. Why is the vertical component of force F_c equal and opposite to force F_d? (*Hint:* Review Fig. 3.12a.)

Example 3.6 This example is an analysis by the method of sections to determine the force in three selected members of a K-truss. This truss presents some unusual problems in the application of the method. Step 1 of the process is unchanged, and the structure is seen to be statically determinate and geometrically stable. A new problem arises when sections through the structure are considered to create free bodies for any of the three member forces, because every section cuts at least four members. Since there are only three equations of equilibrium for a free body, it appears that it is not possible in general to obtain the desired member force.

For chord member a, consider the serpentine section through the structure labeled 1–1 cutting through the four members shown in the free body. Force F_a can be obtained by summation of moments about the point of intersection of the lines of action of the forces in the bottom chord and the vertical members. Because the two cut vertical members are collinear, point L_2 for summing moments is a point of intersection for three of the four unknown member forces. This type of section will work for any chord member of a K-truss, even if one or both of the chords are inclined.

The calculation of the force in a diagonal or vertical member of the truss can only be done with an additional consideration. Any section passing through the structure cuts four members, so the analysis for vertical or diagonal member forces can be continued only by using the concepts presented in Fig. 3.12 at the center joints, m_i, of the truss. A free body of joint m_2 is used to obtain a relation between the forces in the two diagonal members, m_2–L_3 and m_2–U_3 (or b), of the panel by summing forces normal to the vertical members at this joint. The relation between diagonal member forces enables the diagonal force in member b to be obtained by summation of forces vertically in the free body labeled section 2–2 in step 4 of the example. With the force in member b known, why can't the force in the vertical member U_2–m_2 be obtained from joint equilibrium at joint m_2?

Example 3.6

Compute the forces in the indicated members of the K-truss.

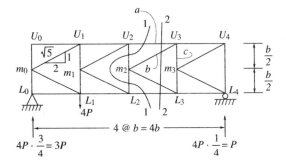

STEP 1   Structure is statically determinate and geometrically stable by inspection. Obtain reactions by proportion.

STEP 2   Isolate the free-body cut by section 1–1.

$$\overset{\curvearrowleft}{\Sigma M_{L2}}: P \cdot 2b + F_a b = 0$$

$$F_a = -2P$$

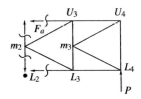

Free-body section 1–1

STEP 3   For the force in b, first isolate the joint at m_2. Resolve the forces into horizontal and vertical components.

$$\Sigma F \rightarrow : 2 \cdot \frac{F_b}{\sqrt{5}} + 2 \cdot \frac{F_1}{\sqrt{5}} = 0$$

$$\therefore F_1 = -F_b$$

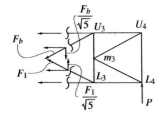

Joint m_2

STEP 4   Use section 2–2 to obtain F_b. Resolve into vertical components.

$$\Sigma F \uparrow : P + \frac{F_1}{\sqrt{5}} - \frac{F_b}{\sqrt{5}} = 0$$

$$F_b = P \cdot \frac{\sqrt{5}}{2}$$

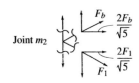

Free-body section 2–2

STEP 5   For F_c, isolate joint U_3 and use components of F_b.

$$\Sigma F \uparrow : -F_c - \frac{F_b}{\sqrt{5}} = 0$$

$$F_c = \frac{-P}{2}$$

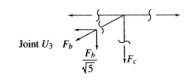

Joint U_3

When the chord(s) of the truss are inclined, the diagonal member forces can still be obtained by summing moments in the free body of the structure about the point where the line of action of the inclined chord member forces intersect. This is similar to the computation of the force F_b illustrated in Example 3.4.

To obtain the force in a vertical such as member c, the force in the diagonal that meets the vertical at the top or bottom chord must first be determined. In the example, F_c cannot be obtained until the diagonal force F_b is known. With the diagonal force known, the concept presented in Fig. 3.12a can be employed to obtain the vertical member force from a free body of the joint at the chord where the vertical and diagonal member join. This is illustrated in step 5 at joint U_3. For numerical computation is U.S. units, take $b = 30$ ft and $P = 5$ kips; for SI units, take $b = 10$ m and $P = 100$ kN.

When the top chord members are inclined and collinear the procedure for calculating the force in vertical members such as U_3–M_3 is slightly more complicated. Again the force in the diagonal that meets the vertical at the top chord must be obtained. Then one of the two following approaches can be used: The first is to employ the concept of Fig. 3.12a and resolve the known diagonal and unknown vertical member force into components that are perpendicular to the inclined chord and sum forces in that direction to determine the vertical member force. The second is to obtain the forces in the two inclined chords at the joint, find the vertical components of the chord and diagonal member forces, and sum forces vertically at the top chord joint to obtain the vertical member force.

When the top chord members are not collinear, the computation of the force in a vertical member is limited to the second procedure described above. In this case the diagonal member force and both inclined chord member forces that intersect with the vertical at the top chord joint must be found. Resolving all forces into horizontal and vertical components and summing forces vertically yields an equilibrium equation that can be used to obtain the vertical member force.

The truss shown in Fig. 13.16 is a conveyer bridge used to move material between two buildings of a manufacturing plant. The truss has two simple spans of 34 and 120 ft and the larger span has an additional 6-ft section cantilevered beyond the support on the right. The larger span is idealized in Example 3.7 and the force computed in three different members by the method of sections. Note that the structure is statically determinate and geometrically stable.

Example 3.7 The force in the three members is obtained by the method of sections, but the manner in which the uniform load is transmitted to the truss must first be reconciled. The secondary structural system brings the loads to the bottom panel joints of the truss. The uniform load is multiplied by the panel length (10 ft) to obtain the panel load ($10w$),

Figure 3.16 Conveyer bridge at industrial plant. (Courtesy Bruce W. Abbott, PE)

which represents the load in a single panel on the secondary structural system. This load is divided equally between the joints at each end of the panel, which reduces the effect of the uniform load to a series of concentrated loads at the lower chord panel joints. The remainder of the solution for member forces is a direct application of moment or force equilibrium equations in the appropriate free bodies.

3.7   Summary and Limitations

The discussions in this chapter have focused on the analysis of plane, statically determinate trusses by methods that are conducive to hand computation. The results from these analyses are not dependent on the material

Example 3.7

Obtain the force in members U_4–U_5, U_4–L_5, and U_4–L_4 for a uniform load, w, extending from L_1 to L_5 on the conveyer bridge of Fig. 3.16.

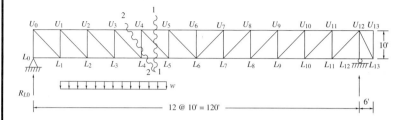

STEP 1   Compute the reactions due to the uniform load from L_1 to L_5. Total load = $40w$ and resultant is 30 ft from L_0.

$$\stackrel{\curvearrowright}{\Sigma M_{L12}}:\ R_{L0} \cdot 120 - 40w(120 - 30) = 0\ \therefore\ R_{L0} = 30w$$

STEP 2   Compute panel load of $10w$ and divide it between ends of panels as $5w$. The loads on the bottom joints are $5w$ at L_1 and L_5 and $10w$ at L_2 to L_4.

STEP 3   Take section 1–1 through structure and isolate left portion as a free body. Resolve F_2 into horizontal and vertical components. Sum moments about L_5 to obtain force in U_4–U_5 (F_1) and forces vertically to obtain force in U_4–L_5 (F_2).

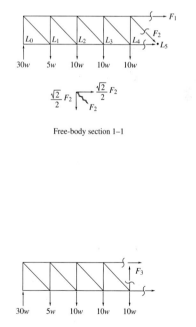

Free-body section 1–1

$$\stackrel{\curvearrowright}{\Sigma M_{L5}}:\ 30w \cdot 50 - 5w \cdot 40$$

$$-10w(30 + 20 + 10) + 10F_1 = 0$$

$$F_1 = \frac{w(200 + 600 - 1500)}{10} = -70w$$

$$\Sigma F \uparrow:\ 30w - 5w - 3 \cdot 10w - F_2 \frac{\sqrt{2}}{2} = 0$$

$$F_2 = -5w \frac{2}{\sqrt{2}} = -5w\sqrt{2}$$

STEP 4   Take section 2–2 through structure and isolate left portion as a free body. Sum forces vertically for force in member U_4–L_4(F_3).

$$\Sigma F \uparrow:\ 30w - 5w - 3 \cdot 10w + F_3 = 0$$

$$F_3 = 5w$$

Free-body section 2–2

properties of the truss members. These techniques can equally well be used for spot checks of computer-generated analyses of plane trusses. In particular, the concepts presented in Section 3.5 are extremely useful in reviewing computer analyses.

There are important limitations that accompany these analyses that must be considered when using the results for design. The limitations are a result of the assumptions that have gone into the creation of the mathematical model that simulates the behavior of the structure under loading.

- The members of the truss are connected together at joints in such a way that the centroidal axes of all members meet at a single point.
- The members of the truss carry axial loads with no significant bending actions being present so that they can all be considered two-force members.
- When subjected to applied loads, the deformations of the structure are so small that the structure can be considered to be rigid and the undeformed geometry can be used in writing all equilibrium equations.
- The individual member of the structure can sustain without material failure the axial loads that are obtained in the analysis.
- Individual members of the structure subjected to compression do not become unstable or buckle.
- The structure as a whole or a local group of members of the structure do not become unstable under the action of the applied loads.

All the limitations listed above can be removed with a significantly more sophisticated mathematical model of the truss behavior. However, most computer programs used for analysis and design of trusses do not have a level of sophistication that overcomes these limitations. The computer results and often not even the documentation of the computer programs give an indication of what limitations should accompany the analysis that has been performed. The structural engineer is the one who bears the responsibility for recognizing the effect of all of these limitations on the final design of the structure.

There are some questions about the analysis of plane trusses that are worthy of thoughtful consideration. Some of these questions will be answered in future chapters, but others are raised for the purpose of establishing a perspective for judging truss behavior.

- What is the effect of not having the centroidal axes of all members meet at a single point?
- The details of construction of a truss often create a joint that is rigid rather than pinned. For long, slender members the effects of bending

are usually negligible, but at what point is the truss more appropriately considered to be a rigid frame?

- How large can the deformations of a truss be before it is necessary to consider its change in geometry in the equations of equilibrium?

- How important in design is a truss member's own weight or the presence of a load on a member between points of attachment?

PROBLEMS

3.1   Obtain all member forces.

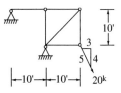

3.2   Obtain all member forces.

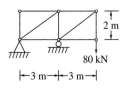

3.3   Obtain all member forces.

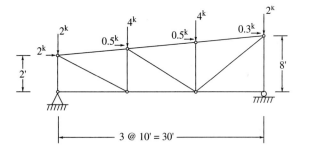

3.4   Obtain all member forces.

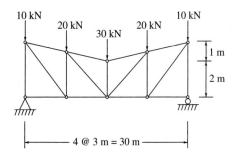

3.5   Find all member forces in the truss loaded as shown.

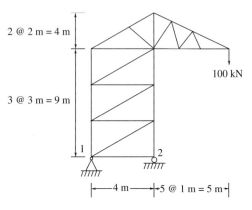

3.6   Find all member forces in the truss shown.

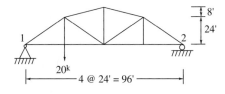

3.7   Obtain all member forces in the truss.

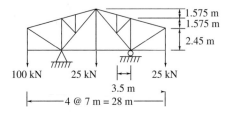

3.8   For the truss shown, obtain all member forces.

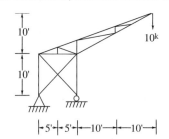

Problems

3.9 The plane truss is loaded as shown. Calculate the forces in all members.

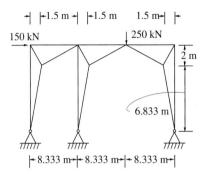

3.10 Compute the force in the members indicated.

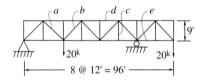

3.11 Compute the force in the members indicated.

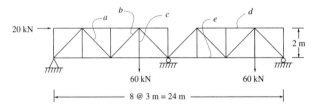
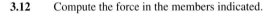

3.12 Compute the force in the members indicated.

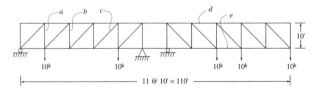

3.13 Compute the force in the members indicated.

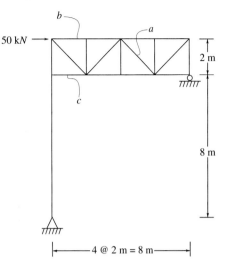

3.14 Determine the forces in members a through d of the truss.

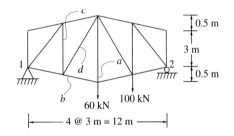
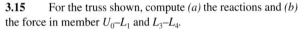

3.15 For the truss shown, compute (a) the reactions and (b) the force in member U_0–L_1 and L_3–L_4.

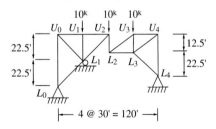

117

3.16 Find the forces in the members indicated. Note the symmetry of the truss.

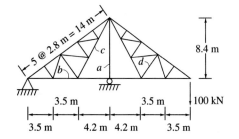

8.4 m

100 kN

3.5 m   3.5 m

3.5 m   4.2 m 4.2 m   3.5 m

3.17 Obtain the forces in the members indicated in the plane truss shown.

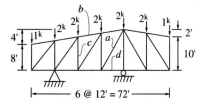

6 @ 12' = 72'

3.18 Obtain the final forces in member a, b, and c in the truss loaded as shown.

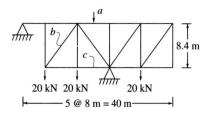

8.4 m

20 kN 20 kN   20 kN

5 @ 8 m = 40 m

3.19 Obtain the forces in the members indicated.

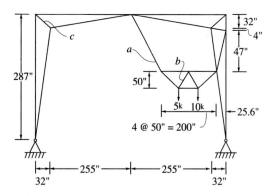

3.20 Obtain the forces in the members indicated in the plane truss shown.

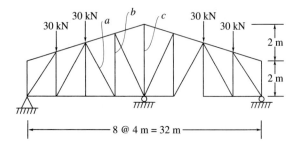

30 kN 30 kN a b c 30 kN   30 kN

2 m

2 m

8 @ 4 m = 32 m

3.21 The plane truss is loaded as shown. Compute the force in the three members indicated.

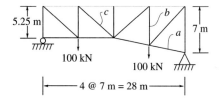

5.25 m

c

b

a

7 m

100 kN

100 kN

4 @ 7 m = 28 m

3.22 The plane truss is loaded with the single 20-kip load at the end.

(a) Compute all reactions.

(b) Show or list all members that have zero member force.

(c) Determine the force in the members indicated.

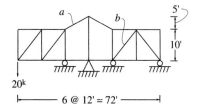

a b 5'

10'

20^k

6 @ 12' = 72'

3.23 Find the force in members *a* through *d* in the truss shown.

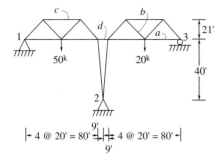

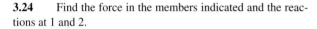

3.24 Find the force in the members indicated and the reactions at 1 and 2.

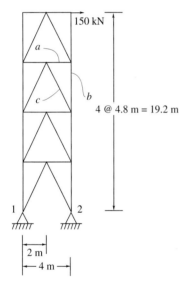

3.25 Find the force in the members indicated and the reactions at 1 and 2.

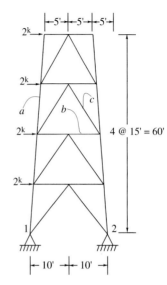

3.26 Obtain the reactions for the truss and verify that $R_1 = -102$ kN, $R_2 = 288$ kN, $R_3 = -18$ kN, and $H = 0$. Using these reactions, obtain the forces for the members indicated.

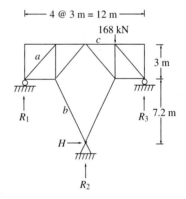

119

3.27   For the truss loaded as shown, compute the force in the members labeled.

3.28   Compute the force in the members indicated.

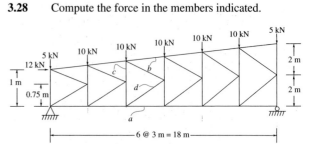

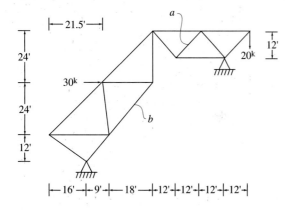

COMPUTER PROBLEMS

For the problems that follow, obtain the graphical results requested, using one of the following methods:

(a) Create a spreadsheet and use the results to obtain the graphs.

(b) Write a small computer program that obtains the results requested and create a file that can be plotted to obtain the graphs.

(c) Use available software programs that will do the analysis requested. Run the program as many times as required to obtain the results requested and create a file that can be plotted to obtain the graphs.

C3.1 Plot the force in members $L_0–U_1$ and $U_1–U_2$ for $0.5 \le b/h \le 2.0$. Which member force is affected the most by b/h?

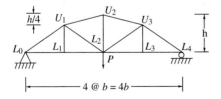

C3.2 Plot separately the force in members $L_0–U_1$ and $U_1–U_2$ for $0.5 \le b/h \le 2.0$, where $P = wb$ (w is a constant). Which member force is affected the most by b/h?

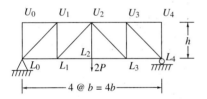

C3.3 Plot the force in members $L_0–U_1$, $U_1–U_2$ and $U_1–L_2$ for $0.5 \le b/h \le 2.0$.

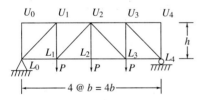

C3.4 Plot the force in members $U_1–U_2$ and $U_1–L_2$ for $0.5 \le b/h \le 2.0$, where $P = wb$ (w is a constant). Which member force is affected the most by b/h?

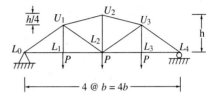

C3.5 Plot the force in members $M_1–L_2$ and $M_1–U_2$ as a function of a/h for $b/h = 1,2$ ($0 \le a/h \le 1$).

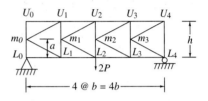

C3.6 Plot the force in member A as a function of a/h for $b/h = 0.5, 1$, and $1.5(0 \le a/h \le 1)$.

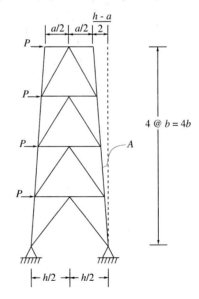

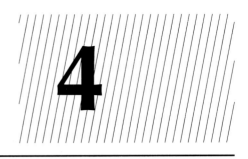

Analysis of Beams: Shear, Moment, and Axial Force Diagrams

The structural engineer is interested in the maximum magnitudes of the shears, axial forces, and moments that exist in each member of a structure. The variation of these actions along a member is important, because the designer often varies cross-sectional proportions or material properties to achieve an economical design. For example, a member may be fabricated from steel of higher strength in a part of the member where the actions have greater magnitude than in other parts. Reinforced concrete members have additional reinforcing steel placed in regions where the magnitude of moment or shear is particularly high.

A means of visualizing the variation of moment, shear, and axial force along a member is through the construction of diagrams. Computer programs capable of displaying graphical output and sketches based on design computations will present the final results in the visual form of these diagrams. In this chapter the sign convention most frequently used in creating these diagrams for single members is established. The techniques for drawing the diagrams and the relationships between applied loads, internal axial loads, shears, and moments required by equilibrium in beams are presented.

123

4.1   Sign Convention

The use of computers for the analysis of structures makes the selection of coordinate systems and sign conventions important for proper interpretation of graphical input and output. Figure 4.1a shows the coordinate system used for a member, which is usually referred to as a *local* or *member coordinate system*. The most commonly used reference axis parallel to the member is the centroidal axis. For a plane member the y and z axes are taken as the principal axes of the cross section, and the z axis is normal and out of the plane of the paper, making the coordinate system in Fig. 4.1b right-handed.

Positive applied loads are taken to act in the positive y-coordinate direction for transverse loads and the positive x-coordinate direction for axial loads, as indicated in Fig. 4.1b. It is assumed that the loads are all applied at or along the reference local or x axis, which is usually the centroidal axis of the member. This assumption enables continuation of the representation of structures by their centroidal axis line.

Moments acting on a short segment of beam are considered to be positive when the segment is bent concave upward. The center of curvature is then located on the positive y side of the member when it undergoes positive bending. This definition for positive moments implies that moments are plotted on the compression side of the member as summarized in Fig. 4.1c.

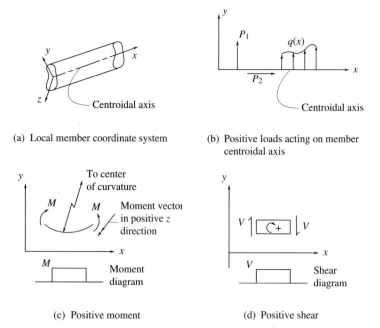

(a) Local member coordinate system

(b) Positive loads acting on member centroidal axis

(c) Positive moment

(d) Positive shear

Figure 4.1a–d   Convention for loading, moment, and shear in beams.

The positive sense of shear is shown in Fig. 4.1d and follows the convention generally used in most textbooks on mechanics of materials. It is most easily remembered by looking at a short element of beam along the axis of the member. When the element is subjected to positive shear, the element will rotate in a clockwise sense in the absence of an equilibrating moment. As with moment, positive shear is plotted in the positive y direction. The sign convention for axial forces is discussed in a later section.

This sign convention for shear and moment is in common usage and will be followed throughout the remainder of this book. It should be pointed out that there are some different sign conventions used for moments, axial forces, shears, and applied loads by other authors. It is therefore very important when using computer programs to determine what sign convention is used when the program output is graphical. The convention will have a significant impact on the form of the diagrams for the variation of shear, moment, or axial force. The user manual for the program must be consulted for the sign convention before the program is used.

4.2   Differential Equations of Equilibrium for a Transversely Loaded Beam

The differential equations of equilibrium for a transversely loaded beam are important because they define the relation between moment, shear, and applied load. It is assumed in developing these equations that there is no axial force acting, hence there is no coupling in this formulation between axial force and bending. It is also assumed that the deformations of the member are small, so that the slopes do not affect the direction of action of shear forces. The assumption is the same as saying that the equilibrium equations are written in the original geometry of the undeformed member.

An element of infinitesimal length taken from the beam in Fig. 4.2a is shown with positive moments and shears acting on it in Fig. 4.2b. Only the centroidal axis is shown since all shears and moments are referred to it. The element is of infinitesimal length, dx, so the distributed load, $q(x)$, may be taken as constant over the length with no significant error. Summation of forces acting on the element in the y direction yields

$$V + q(x)\,dx - (v + dV) = 0$$

Upon dividing by dx, the relation between shear and applied load becomes

$$\frac{dV}{dx} = q(x) \tag{4.1}$$

Summation of moments about the left end of the differential element yields

$$-M + q(x)\,dx\,\frac{dx}{2} + (M + dM) - (V + dV)\,dx = 0$$

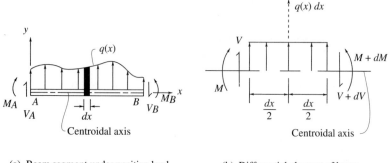

(a) Beam segment under positive load

(b) Differential element of beam
referred to centroidal axis

Figure 4.2a–b   Equilibrium considerations for beams.

Simplifying, dividing by dx, and recognizing that the remaining terms, $q(x)$ $dx/2 + dV$, are a differential order smaller than the others gives

$$\frac{dM}{dx} = V \tag{4.2}$$

Equation (4.2) expresses the relation between moment and shear in a beam. If Eq. (4.2) is differentiated with respect to x and substitution made from Eq. (4.1), the relation between moment and applied load can be expressed as

$$\frac{d^2M}{dx^2} = q(x) \tag{4.3}$$

Equations (4.1) and (4.2) are equilibrium equations for the beam segment, representing the summation of forces and summation of moments, respectively. Equation (4.3) incorporates both equilibrium equations and is referred to as the *differential equation of equilibrium for a beam*. It may be used along with appropriate boundary (or end) conditions to obtain the variation of moment in a beam for any given loading $q(x)$.

4.3   Drawing Shear and Moment Diagrams Using Direct Integration

Direct integration of Eq. (4.3) can be used to obtain the shear and moment diagrams for a member as illustrated in Fig. 4.3.

Figure 4.3 The shear and moment diagrams are obtained through integration of Eq. (4.3) using the condition at each end that the moment is zero. The beam is geometrically stable and there is no horizontal reaction at the left end. The load intensity is w (in units of force/length) at the

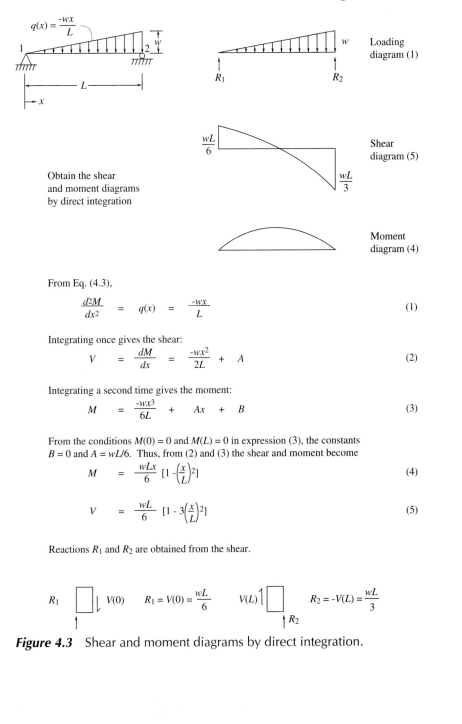

Loading diagram (1)

Obtain the shear and moment diagrams by direct integration

Shear diagram (5)

Moment diagram (4)

From Eq. (4.3),

$$\frac{d^2M}{dx^2} = q(x) = \frac{-wx}{L} \qquad (1)$$

Integrating once gives the shear:

$$V = \frac{dM}{dx} = \frac{-wx^2}{2L} + A \qquad (2)$$

Integrating a second time gives the moment:

$$M = \frac{-wx^3}{6L} + Ax + B \qquad (3)$$

From the conditions $M(0) = 0$ and $M(L) = 0$ in expression (3), the constants $B = 0$ and $A = wL/6$. Thus, from (2) and (3) the shear and moment become

$$M = \frac{wLx}{6}\left[1 - \left(\frac{x}{L}\right)^2\right] \qquad (4)$$

$$V = \frac{wL}{6}\left[1 - 3\left(\frac{x}{L}\right)^2\right] \qquad (5)$$

Reactions R_1 and R_2 are obtained from the shear.

R_1 $\qquad V(0)$ $\qquad R_1 = V(0) = \frac{wL}{6}$ $\qquad V(L)$ $\qquad R_2 = -V(L) = \frac{wL}{3}$

R_2

Figure 4.3   Shear and moment diagrams by direct integration.

right end of the beam, so that $q(x)$ is defined as $q(x) = -wx/L$ and is negative because the load acts downward. The vertical reactions at the left and right ends of the beam are R_1 and R_2, respectively. Why is it not necessary to compute reactions R_1 and R_2?

All of the externally applied actions on the member are shown in the loading diagram for the beam. The loading diagram is a graphical presentation of the location and variation of applied concentrated or distributed loads and the support reactions. Not only does the diagram provide a convenient way of showing all load actions that occur on a member, but it represents, in a mathematical form, the right-hand side of Eqs. (4.1) and (4.3).

Using the loading diagram and Eq. (4.3) yields expression (1). Integrating this once gives the shear, V, in expression (2), and integrating a second time gives the moment, M, in expression (3), where the constants A and B can be evaluated from the moment conditions at each end of the beam. The condition $M = 0$ at the left and $(x = 0)$ requires that $B = 0$ while the condition $M = 0$ at the right and $(x = L)$ requires that $A = wL/6$. Substituting these values into the moment and shear expressions, (3) and (2), yields the final moment and shear expressions (4) and (5). The moment and shear diagrams are plotted directly from these expressions as shown in the figure. The reactions can be obtained from the shear expression. At $x = 0$, the shear is $wL/6$, which is equal to reaction R_1, as shown in the free-body diagram for the left end. At $x = L$ the shear is $-wL/3$, which is equal to the negative of reaction R_2 as shown in the free-body diagram for the right end.

The results presented in Fig. 4.3 take the usual graphical form, where the shear diagram is drawn directly below the loading diagram and the moment diagram directly below the shear diagram. Computation of the reactions at the ends of the beam from the mathematical expression for shear requires careful attention to the sign convention that is employed. The reactive forces are shown as positive loads on the beam, hence upward. The sign convention for positive shear in Fig. 4.1d requires that in Fig. 4.3 the shear act down in the free body of the left end of the beam and act up in the free body for the right end of the beam.

Direct integration of Eq. (4.3) is a straightforward method of obtaining the shear and moment diagrams as illustrated in Fig. 4.3. When a single member has a loading that extends over its entire length, no matter how complicated it may be, this approach is useful and may even be the simplest to employ. But when loadings are not continuous over the entire length, or when concentrated loads or moments are present, it is awkward to use this approach.

Example 4.1 A simply supported beam has a uniform load over half of its length. Because of the discontinuity in the uniform loading at the center of the member, the loading diagram is divided into two regions labeled I and

Example 4.1

Obtain the shear and moment diagrams by direct integration.

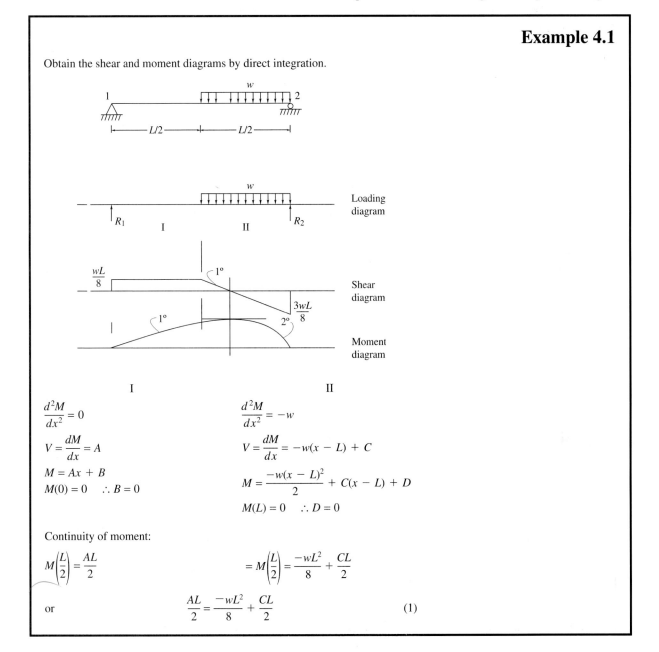

$$\text{I}$$

$$\frac{d^2M}{dx^2} = 0$$

$$V = \frac{dM}{dx} = A$$

$$M = Ax + B$$

$$M(0) = 0 \quad \therefore B = 0$$

$$\text{II}$$

$$\frac{d^2M}{dx^2} = -w$$

$$V = \frac{dM}{dx} = -w(x - L) + C$$

$$M = \frac{-w(x - L)^2}{2} + C(x - L) + D$$

$$M(L) = 0 \quad \therefore D = 0$$

Continuity of moment:

$$M\left(\frac{L}{2}\right) = \frac{AL}{2} \qquad\qquad = M\left(\frac{L}{2}\right) = \frac{-wL^2}{8} + \frac{CL}{2}$$

or $$\frac{AL}{2} = \frac{-wL^2}{8} + \frac{CL}{2} \tag{1}$$

Example 4.1 (continued)

I   II

Continuity of shear:

$$V\left(\frac{L}{2}\right) = A \qquad\qquad = V\left(\frac{L}{2}\right) = \frac{wL}{2} + C$$

or $$A = \frac{-wL}{2} + C \qquad\qquad (2)$$

Solving (1) and (2) for A and C yields $C = -3wL/8$ and $A = wL/8$. This yields the final shears and moments in each region. The reactions are obtained from the shear.

$$V = \frac{wL}{8}$$

$$M = \frac{wLx}{8}$$

$$R_1 = V_{x=0} = \frac{wL}{8}$$

V

R_1

$$V = \frac{wL(5 - 8x/L)}{8}$$

$$M = \frac{wL^2[5x/L - 4(x/L)^2 - 1]}{8}$$

$$R_2 = -V_{x=L} = -\left(-3\frac{wL}{8}\right) = 3\frac{wL}{8}$$

V

R_2

II, and Eq. (4.3) must be applied in each region independently. The integrations of Eq. (4.3) yield four undetermined constants, two in each region. Two of the four constants are evaluated from the zero-moment condition at each end of the beam. At the center of the beam ($x = L/2$) there are no concentrated loads or moments applied to the beam. Consequently, both the shear and moment must vary continuously. This requires that the shear and moment expressions for the left and right regions be continuous and yield identical values at the center, where the uniform loading starts.

The values of reactions R_1 and R_2 are obtained from the expressions developed for the internal shear, V, in the same manner as in Fig. 4.3. At the left end of the beam, the internal shear just to the right of the end is positive downward and R_1 is positive upward, so R_1 is equal to the shear. At the right end of the beam, the positive sense of the internal shear just to the left of the end and R_2 are in the same direction, so R_2 is equal to the negative of the internal shear. For numerical computations in U.S. units, take $L = 40$ ft and $w = 2$ kips/ft; for SI units, take $L = 16$ m and $w = 50$ kN/m.

As illustrated in Example 4.1, direct integration of the loading functions for shear and moment becomes unwieldy if the loading is discontinuous. When concentrated loads or moments are present, evaluation of the constants of integration from requirements on the relationships of shear and moment between regions is more complicated. The function representing shear is discontinuous at concentrated loads and the function representing moment is discontinuous at concentrated moments, and these conditions must be accounted for properly when the constants of integration are being evaluated. There are alternative methods of using direct integration in these situations as illustrated in References 7 and 17.

4.4   Drawing Shear and Moment Diagrams Using Equilibrium Directly

The drawing of shear and moment diagrams is a simple three-step process. After ascertaining that the structure is statically determinate and geometrically stable, the reactions are computed. As shown in Chapter 2, this computation starts with any equations of condition that may exist within the structure.

The second step in the process is to draw the loading diagram for the structure and establish the regions where the shear and moment will be continuous functions of the applied loading. The first region in the beam is between its left end and the first abrupt change in the external loading. An abrupt change of loading occurs at points where a distributed loading begins or ends, where a sudden change in magnitude or slope occurs in a distributed loading, or where a concentrated force, moment, or reaction force acts. The division process is continued by proceeding from left to right along the axis of the member until the right end is reached.

The third step is to create a series of free-body diagrams by successively passing a section through the beam in each defined region. The use of the equilibrium equations in each free body provides mathematical expressions for variation of the shear and moment. These expressions are then plotted, creating the shear and moment diagrams. Because the structure has been divided into regions defined by the loading, the mathematical expressions obtained for each region are valid for that region only. The following example illustrates the three-step process.

Example 4.2 Example 4.2 is a simple structure that is loaded by two concentrated loads. There are three regions for this structure, defined by the left and right ends of the structure, the concentrated load at B, and the reaction at C. The reactions are computed in step 1 using the entire structure as a free body. Note that no axial forces are shown in the free bodies since no horizontal forces or reactions exist in the structure. The free bodies for each region are taken in as simple a form as possible. A convenient position coordinate is selected independently in each free body to define the variation of shear and moment.

In region I, the interval $A–B$ in the structure, the position coordinate t is used to define the location of the cut through the beam, which creates free body I of the left portion of the beam. At the cut, the unknown magnitude of the internal shear is V and the unknown magnitude of the internal moment is M. Because the cut can be made anywhere in the region, the coordinate t varies from zero to L and defines the variation of the shear and the moment in the region. Summing forces vertically defines the shear, V, to be of constant magnitude $P/2$ and summing moments about the cut yields a linear variation of the moment, M. These expressions for shear and moment, labeled (3) and (4), respectively, are plotted in region I of the shear and moment diagrams.

The process just described for region I is repeated for region II, interval $B–C$ in the beam, using the position coordinate r to define the location of the cut that creates free body II. The expressions labeled (5) and (6) define the variation of the shear and moment, which is plotted in the respective diagrams.

The cut used to create free body III in region III is located by the coordinate u, which, for convenience, is measured from the right end of the beam. The expressions labeled (7) and (8) define the variation of the shear and moment in this last region, and the shear and moment diagrams are plotted accordingly. How would calculation of the shear and moment in region III change if the free body used were that of the structure to the left of the cut? For numerical computations in U.S. units, take $L = 10$ ft and $P = 5$ kips; for SI units, take $L = 3$ m and $P = 30$ kN.

132

Example 4.2

Obtain the shear and moment diagrams for the structure shown.

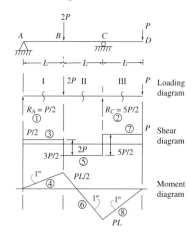

STEP 1   The structure is determinate and geometrically stable by inspection. From the free-body diagram of the whole structure, sum the moments about A and sum the forces vertically to obtain R_A (1) and R_C (2). There are no axial forces in structure, by inspection.

$$\overset{\curvearrowleft}{\Sigma M_A} = 0 : R_C \cdot 2L - 2P \cdot L - P \cdot 3L = 0 \quad \therefore R_C = \frac{5P}{2} \quad (1)$$

$$\Sigma F \uparrow = 0 : R_A - 2P + \frac{5P}{2} - P = 0 \quad \therefore R_A = \frac{P}{2} \quad (2)$$

STEP 2   Based on the loading diagram, divide structure into regions I (A–B), II (B–C), and III (C–D).

STEP 3   Use free bodies I, II, and III to obtain and plot variation of shear (V) and moment (M) in each region. The equation numbers for shear or moment variation obtained from the equilibrium calculation in each region correspond to those in the diagrams.

Free body I	Free body II	Free body III

$\Sigma F \uparrow -V + P/2 = 0 \quad \therefore V = P/2$ | $\Sigma F \uparrow : P/2 - 2P - V = 0$ | $\Sigma F \uparrow : V - P = 0 \quad \therefore V = P$

V constant (3)

$\overset{\curvearrowleft}{\Sigma M_{\text{cut}}} : M - P \cdot \dfrac{t}{2} = 0$

$M = Pt/2 \qquad M \text{ linear (4)}$

$V = -3P/2 \qquad V \text{ constant (5)}$

$\overset{\curvearrowleft}{\Sigma M_{\text{cut}}} : M + 2P \cdot r - \dfrac{P}{2} \cdot (L + r) = 0$

$M = PL/2 - 3Pr/2 \qquad M \text{ linear (6)}$

V constant (7)

$\overset{\curvearrowleft}{\Sigma M_{\text{cut}}} : -M - Pu = 0$

$M = -Pu \qquad M \text{ linear (8)}$

Several techniques illustrated in Example 4.2 have been employed to reduce the chance for error. First, the internal shear and moment that act at the cut in each free body have been shown to act in a positive sense. The expressions derived from the equilibrium equations for the variation of shear and moment will then have the correct sign for these quantities, which facilitates plotting their variation. Second, the two equilibrium equations used to obtain the shear and moment are written so that only one of the two quantities appears in each equation. Third, the use of different position coordinates in each free body results in the shear and moment diagrams being defined by simple mathematical expressions. These expressions contrast with those obtained in Example 4.1, which are defined by a single coordinate and as a consequence have more complicated forms. How is the procure changed if the same coordinate is used in all three free bodies?

4.5   Relationships Between Loading, Shear, and Moment Diagrams

Equation (4.1) states that the derivative of the shear is equal to the sign and magnitude of the load intensity. Stated another way, the slope of the shear diagram at any point is equal to the load intensity. Similarly, Eq. (4.2) embodies the concept that the slope of the moment diagram at any point is equal to the sign and magnitude of the shear. The full understanding of these concepts is only important but also very useful. By inspection of the relations between loading, shear, and moment the variation of shear and moment can be checked qualitatively. These concepts can be stated succinctly as follows:

At each point along a member, the slope of the shear diagram is equal to the intensity of the applied load, and the slope of the moment diagram is equal to the shear.

Figure 4.4 In Fig. 4.4a the variation of the shear in a member is easily sketched by observing the sense and the variation of the load intensity. Four possibilities are shown for the variation of the load intensity and the corresponding variation of the shear. The same relations are shown between shear and moment in Fig. 4.4b. Once these relations are understood, the appearance of the loading, shear, and moment diagrams can quickly be checked for consistency.

These concepts are illustrated in Fig. 4.3, where the load intensity is negative and varies linearly from zero at the left end of the beam to $-w$ at the right end. Thus, the slope of the shear diagram starts at zero at the left end

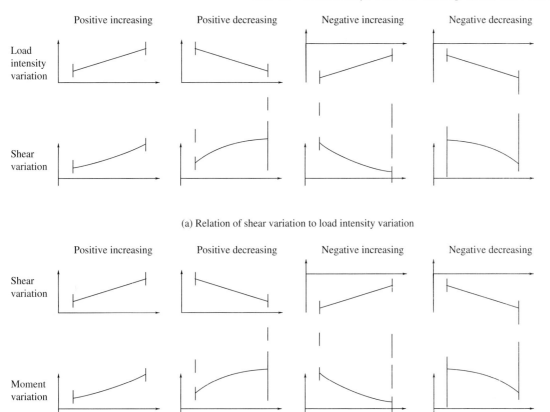

Positive increasing   Positive decreasing   Negative increasing   Negative decreasing

Load intensity variation

Shear variation

(a) Relation of shear variation to load intensity variation

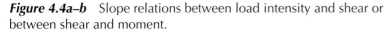

Positive increasing   Positive decreasing   Negative increasing   Negative decreasing

Shear variation

Moment variation

(b) Relation of moment variation to shear variation

Figure 4.4a–b Slope relations between load intensity and shear or between shear and moment.

of the beam and becomes increasingly negative as the right end is approached. The slope of the moment diagram follows the variation of the magnitude of the shear. It starts as positive at the left end where the shear is positive, and decreases to zero and becomes negative as the right end is approached. The maximum moment occurs near the center of the beam, where the shear and hence the slope to the moment diagram are zero.

The same relations are also seen in Examples 4.1 and 4.2. In Example 4.1 the load intensity is zero in the left half of the beam and $-w$ in the right half. In Example 4.2 the load intensity in the regions between the concentrated loads is zero, hence the slope in each region of the shear diagram is also zero. The slope of the moment diagrams in these two examples is again seen to be the magnitude of the shear.

These concepts can be used to advantage in plotting or drawing the shear and moment diagrams in problems where the loading variation is more complicated, as shown in Example 4.3.

Example 4.3   The concepts embodied in Eqs. (4.1) and (4.2) are used as an aid in obtaining the final shear and moment diagrams. Note that the analysis begins at the hinge at B, which enables the calculation of the reaction at D. There are three regions for equilibrium considerations in the structure based on the loading diagram. In the third step, the variation of the shear and moment from A to B is second and third degree, respectively. The diagrams can be drawn correctly if the relationships defining the slopes between moment and shear and shear and applied load are used.

In region A–B the slope in the shear diagram starts at zero at A and decreases, since the loading is negative, as one approaches B. This defines uniquely the curve for the shear diagram. The variation of the moment can be obtained in A–B by noting that the shear is positive at A and becomes less positive as one moves toward B. Thus the slope to the moment diagram is positive at A and decreases as one moves to B.

Example 4.4   By using the relation between the slope of the shear diagram and the magnitude of the load intensity and the slope of the moment diagram and the magnitude of the shear, the shape of shear and moment diagrams may be sketched directly from the loading diagram. Although relative values of quantities are not known without computation, the variations of shear and moment can be obtained quite easily by inspection. How would the shear and moment diagrams change if w were applied upward instead of downward?

4.6   Incremental Techniques for Obtaining Shear and Moment Diagrams

Equations (4.1) and (4.2) define the relation between shear and applied loading and moment and shear, respectively. The relations were used in Section 4.5 as an aid in drawing the shear and moment diagrams for members. There relations can be further developed to aid in the drawing of shear and moment diagrams by what can be called *incremental techniques*.

If the distributed loading is continuous and there are no applied concentrated loads between the two points $x = a$ and $x = b$, both sides of Eq. (4.1) can be integrated between a and b, with the result

$$\int_a^b \frac{dV}{dx}\,dx = V(b) - V(a) = \int_a^b q(x)\,dx = \begin{array}{l}\text{area under loading dia-}\\ \text{gram between } a \text{ and } b\end{array} \quad (4.4)$$

Example 4.3

Obtain the shear and moment diagrams.

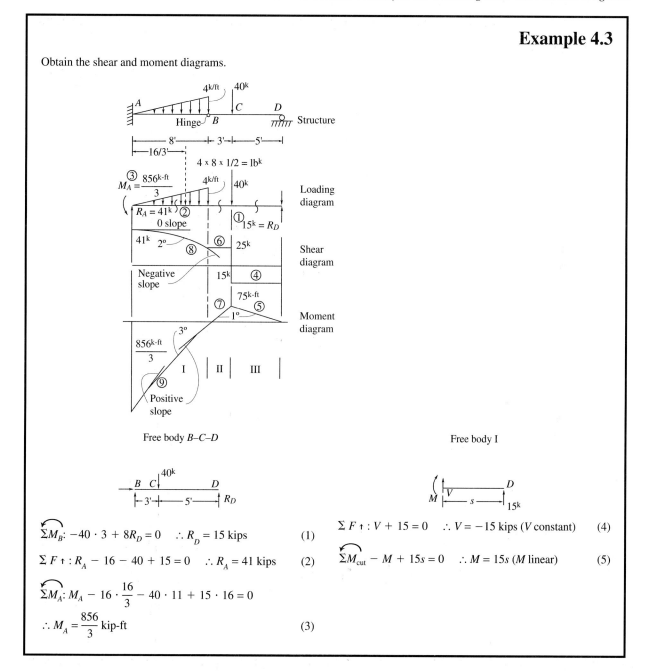

Structure

Loading diagram

Shear diagram

Moment diagram

Free body B–C–D

Free body I

$\overset{\curvearrowleft}{\Sigma M_B}: -40 \cdot 3 + 8R_D = 0 \quad \therefore R_D = 15 \text{ kips}$ (1)

$\Sigma F \uparrow: R_A - 16 - 40 + 15 = 0 \quad \therefore R_A = 41 \text{ kips}$ (2)

$\overset{\curvearrowleft}{\Sigma M_A}: M_A - 16 \cdot \dfrac{16}{3} - 40 \cdot 11 + 15 \cdot 16 = 0$

$\therefore M_A = \dfrac{856}{3} \text{ kip-ft}$ (3)

$\Sigma F \uparrow: V + 15 = 0 \quad \therefore V = -15 \text{ kips } (V \text{ constant})$ (4)

$\overset{\curvearrowleft}{\Sigma M_{\text{cut}}} - M + 15s = 0 \quad \therefore M = 15s \ (M \text{ linear})$ (5)

Example 4.3 (continued)

Free body II

Free body III

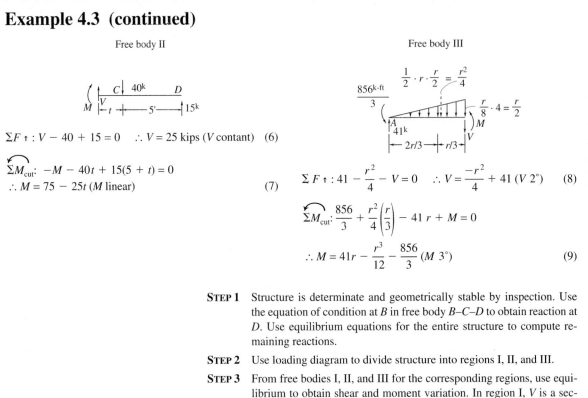

$$\Sigma F \uparrow : V - 40 + 15 = 0 \quad \therefore V = 25 \text{ kips } (V \text{ contant}) \quad (6)$$

$$\overset{\frown}{\Sigma M_{\text{cut}}}: -M - 40t + 15(5 + t) = 0$$
$$\therefore M = 75 - 25t \ (M \text{ linear}) \tag{7}$$

$$\Sigma F \uparrow : 41 - \frac{r^2}{4} - V = 0 \quad \therefore V = \frac{-r^2}{4} + 41 \ (V \ 2°) \tag{8}$$

$$\overset{\frown}{\Sigma M_{\text{cut}}}: \frac{856}{3} + \frac{r^2}{4}\left(\frac{r}{3}\right) - 41 \, r + M = 0$$

$$\therefore M = 41r - \frac{r^3}{12} - \frac{856}{3} \ (M \ 3°) \tag{9}$$

STEP 1   Structure is determinate and geometrically stable by inspection. Use the equation of condition at *B* in free body *B–C–D* to obtain reaction at *D*. Use equilibrium equations for the entire structure to compute remaining reactions.

STEP 2   Use loading diagram to divide structure into regions I, II, and III.

STEP 3   From free bodies I, II, and III for the corresponding regions, use equilibrium to obtain shear and moment variation. In region I, *V* is a second- and *M* a third-degree curve. The slope of *V* is zero at *A* and negative to left of *B* since the load intensity is zero at *A* and −4 kips/ft to left of *B*. The slope of *M* is positive at *A* and *B* since *V* is 41 kips at *A* and 25 at *B*. These slopes define the shapes of the shear and moment diagrams. Numbers in the diagrams correspond to equation numbers in the free-body diagrams.

Example 4.4

Sketch the loading, shear, and moment diagrams for the structure loaded as shown.

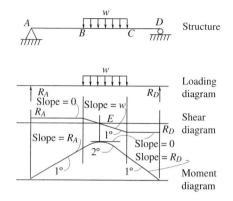

STEP 1   Sketch the loading diagram starting with R_A on the left and ending with R_D. For the structure and loading, the sense of R_A and R_D is correct, and since the centroid of the load is to the right of center of the structure, the magnitude of R_D will be greater than the magnitude of R_A.

STEP 2   Sketch the shear diagram. The slope of the shear diagram in each region is equal to the load intensity, which is zero in regions A–B and C–D and negative in B–C since w is acting downward. The slope of the shear diagram coupled with the estimated magnitude of the shear at the right and left ends defines its shape uniquely.

STEP 3   Sketch the moment diagram. The moment is zero at A and D because there are no concentrated moments acting on the ends of the beam. The slope of the moment diagram is constant in A–B and C–D with the magnitude and sense being defined by the ordinates (R_A and $-R_D$) of the shear diagram. In B–C the moment is second degree and is uniquely defined by the slopes at B and C and the zero slope at E, the point where the shear is zero.

When the value of the shear is known at the point $x = a$, the value of the shear at the point $x = b$ can be obtained by adding to the shear at a the change in shear between the two points. The change in shear is simply the area under the loading diagram between the two points. If a concentrated load is present, the concept expressed in Eq. (4.4) can be used in the regions on either side of the load. Then, in passing from the left to the right side of the concentrated load, the shear will change by an amount that is equal to the magnitude of the concentrated load. These concepts can be summarized as follows:

> The change in shear between two points in a member is equal to the area under the loading diagram between those points, provided that there are no concentrated loads or abrupt changes in distributed loads between the points. The change in shear between the left and right sides of a concentrated load is equal to the magnitude of the concentrated load.

These concepts are clearly shown in the previous examples. Consider their use in Example 4.3. Once the reaction at the left end of the member (41 kips) is known, the shear diagram can be obtained using the incremental technique. The shear at B in the member will be the shear at A plus the area under the loading diagram between A and B. Since the loading is downward, and hence negative, the shear at B will be less than at A and is given as $V_A - 1/2 \cdot 4 \cdot 8 = 41 - 16 = 25$ kips as shown. By knowing the relation between the slope of the shear diagram and the applied loading, the shear diagram for region A–B can be drawn immediately.

For regions B–C and C–D the distributed load is zero and no change in shear occurs. In passing from the left to the right side of the concentrated load at C, the shear changes abruptly by $25 + 15 = 40$ kips. The fact that the shear decreases by 40 kips can be seen if a free body is created by making cuts through the beam immediately to the left and right sides of the load and summing forces vertically, or by recognizing that the 40-kip load is negative.

An examination of Eq. (4.2) shows that the change in moment between two points can be obtained in a manner similar to the change in shear discussed above. If the shear is continuous and there are no applied concentrated moments between the two points $x = a$ and $x = b$, both sides of Eq. (4.2) can be integrated between a and b, with the result that

$$\int_a^b \frac{dM}{dx}\, dx = M(b) - M(a) = \int_a^b V(x)\, dx = \begin{array}{l}\text{area under shear dia-}\\\text{gram between } a \text{ and } b\end{array} \quad (4.5)$$

When the value of the moment is known at the point $x = a$, the value of the moment at the point $x = b$ can be obtained by adding to the moment at a the change in moment, which is simply the area under the shear diagram between the two points. If an applied concentrated moment is present, the concept expressed in Eq. (4.5) can be used in the regions on either side of the moment. The moment in passing from the left to the right side of the concentrated moment applied will change by an amount that is equal to the magnitude of the concentrated moment. These concepts can be summarized as follows:

The change in moment between two points in a member is equal to the area under the shear diagram between those points, provided that there are no applied concentrated moments between the points. The change in moment between the left and right sides of an applied concentrated moment is equal to the magnitude of that concentrated moment.

The concepts that apply to moments can also be seen in Example 4.3. The moment at A is known to be $-856/3$ kip-ft and the change in moment between A and B can be obtained as the area under the shear diagram. This area can be divided into the sum of a rectangular area of height 25 kips and an area of the same form as that in Fig. 2.7e of height 16 kips. The moment, M_B, is then given as $M_A + 25 \cdot 8 + 2/3 \cdot 16 \cdot 8 = -856/3 + 200 + 256/3 = 0$, as it should be since there is a hinge at B. The change in moment from B to C is $25 \cdot 3 = 75$ kip-ft and from C to D is $-15 \cdot 5 = -75$ kip-ft, as shown in the diagram. Using this incremental technique for moments and knowing the relation between slopes to the moment diagram and the shear, the moment diagram can be constructed with relative ease.

The incremental techniques can be used to check the consistency of shear and moment diagrams that have been constructed by computer programs or other engineers. The reader is encouraged to review the shear and moment diagrams developed in the other examples presented in the preceding sections. Another example of the use of these techniques to create a complete shear and moment diagram is also presented in Example 4.5.

Example 4.5 The uniformly loaded simply supported beam also has a concentrated load and applied moment at point B, a situation that can arise in a structure when a column is rigidly attached to a beam. The reactions of the beam are calculated in the usual way, which establishes the values of the internal shear and moment at point A. The shear diagram between A and B is obtained by using Eq. (4.4) to calculate the shear to the left of B and recognizing that the slope to the shear diagram

Example 4.5

Draw the loading, shear, and moment diagrams for the structure shown.

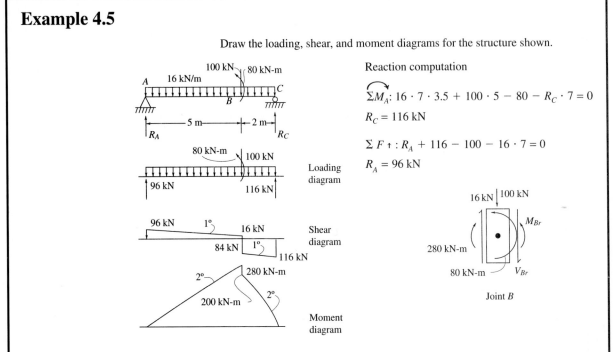

Loading diagram

Shear diagram

Moment diagram

Joint B

Reaction computation

ΣM_A: $16 \cdot 7 \cdot 3.5 + 100 \cdot 5 - 80 - R_C \cdot 7 = 0$

$R_C = 116$ kN

$\Sigma F \uparrow : R_A + 116 - 100 - 16 \cdot 7 = 0$

$R_A = 96$ kN

STEP 1   The structure is statically determinate and geometrically stable by inspection. Obtain the reactions at A and C from equilibrium of moments about A and vertical forces. No horizontal forces are applied, so the horizontal reaction at A is zero.

STEP 2   Draw the loading diagram using the reactions and applied forces and moments. The shear and moment at A are 96 kN and 0, respectively. Calculate the shear at B using the area of the loading diagram.

$$V_B - V_A = -16 \cdot 5 \therefore V_B = V_A - 80 = 96 - 80 = 16 \text{ kN}$$

Draw the shear diagram from A to B. Calculate the moment at B using the area of the shear diagram and draw the moment diagram from A to B.

$$M_B - M_A = \left(\frac{1}{2}\right)(96 + 16) \cdot 5 \therefore M_B = M_A + 56 \cdot 5 = 0 + 280 \text{ kN-m}$$

STEP 3   Get the shear and moment to the right of B from the free body above.

$$\Sigma F \uparrow : 16 - 100 - V_{Br} = 0 \quad \therefore V_{Br} - 84 \text{ kN}$$

$$\Sigma M: 280 - 80 - M_{Br} = 0 \quad \therefore M_{Br} = 200 \text{ kN-m}$$

STEP 4   Check the change in shear and moment from B to C.

$$V_C - V_{Br} = -16 \cdot 2 \quad \therefore V_C = -84 - 32 = -116 \text{ kN}   \text{OK}$$

$$M_C - M_{Br} = \frac{1}{2} \cdot (-84 - 116) \cdot 2 \quad \therefore M_C = 200 - (84 + 116) = 0   \text{OK}$$

is constant and negative. Having the shear diagram between A and B, the moment diagram is obtained by using Eq. (4.5) to calculate the moment to the left of B and recognizing that the slope of the moment diagram is positive at A and becomes less positive as point B is approached. This creates the second-degree curve shown.

The concentrated load and applied moment at B requires the use of a free body at that point formed by making a cut through the beam on each side of B. Equilibrium in that free body provides the values of the shear and moment on the right side of B, as indicated in step 3. The remainder of the shear and moment diagrams can now be sketched in since the values of shear and moment are known at C. However, as shown in step 4, Eqs. (4.4) and (4.5) are used again to obtain the values of shear and moment at C as a check on the analysis process.

It is interesting to note that the shear and moment diagrams in Example 4.5 were obtained with the use of only two free-body diagrams. They were required for equilibrium of the structure as a whole and at B to establish the sudden change in shear and moment due to the concentrated load and applied moment. As one becomes more experienced in drawing shear and moment diagrams, the need to form a free body at the point of a concentrated load or applied moment will decrease (or perhaps vanish). Once the value of shear and moment is known at a single point, these diagrams can be created rather quickly by using the concepts presented in Eqs. (4.1), (4.2), (4.4), and (4.5).

4.7 Shear and Moment Diagrams in Girders of Beam-and-Girder Systems

The construction of a number of highway bridges is similar to the form shown in Fig. 4.5.

Figure 4.5 The major load-carrying members that span between the supports at each end of the bridge are called *girders*. As shown in Fig. 4.5a, the deck of the bridge, which carries the roadway and its traffic, is supported on a series of relatively small beams called *stringers*, which parallel the girders but which span a much shorter distance. The stringers are supported by the floor beams, which, in turn, are perpendicular to and span between the girders.

The sole function of the deck–stringer–floor beam structural system is to convey any applied load on the roadway or deck to the girders, which then carry the load to the supports and either end of the bridge. The design of the girders of the bridge requires the construction of the loading, shear, and moment diagrams for these members. Figure 4.5b shows the manner in which the system is idealized. The points of attachment of the floor beams to the girder, called the *panel points* of the girder, are represented as pin or roller supports on top of the girder. The stringers are shown as simply supported beams that span between the floor beams.

The loads on the bridge are applied to the deck, which in turn conveys the loads to the stringers. As shown in Fig. 4.5c, the stingers are simply supported beams which convey the load through their end reactions to the floor beams. Finally, the load enters the girder as concentrated loads at the panel points where the floor beams join to the girder. The loading, shear, and moment diagrams for the girder are obtained from the magnitudes of the concentrated loads at the panel points. No matter what form of loading is applied to the deck, it ends up as a series of concentrated loads on the girder.

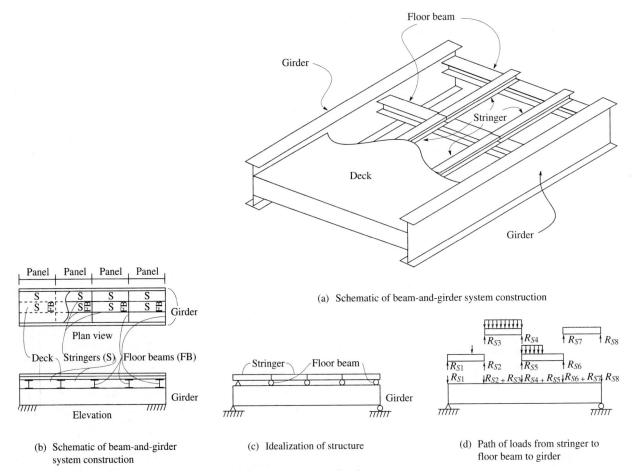

(a) Schematic of beam-and-girder system construction

(b) Schematic of beam-and-girder system construction

(c) Idealization of structure

(d) Path of loads from stringer to floor beam to girder

Figure 4.5a–d Construction details, idealization, and path of applied loads in beam-and-girder highway bridge.

The structural system shown in Fig. 4.5 is called a *beam-and-girder system*. It has been presented in the context of a highway bridge, but the same type of system is also found in some buildings. Example 4.6 shows how the loading, shear, and moment diagrams for a girder are obtained for a particular loading on the structural system.

> **Example 4.6** The beam-and-girder system shown has a uniform load over the left half of the structure. The key to the analysis of these structures is to obtain the loads that appear at the panel points of the girder. As shown in step 1 of the example, the floor system is taken apart, with each individual stringer represented as a simply supported beam. The stringer reactions are computed by equilibrium of the simply supported beam and combined at each floor beam to obtain the panel point load on the girder. In this example, as in the previous one, the incremental technique for drawing shear and moment diagrams is used. In step 4, the shear and moment are computed for the *C–D–E* portion of the girder using Eqs. (4.4) and (4.5) as a check on the analysis. What would the shear moment diagrams for the girder look like if the uniform load in *A–B* and *B–C* were replaced with 60-kip concentrated loads in the middle of stringers *A–B* and *B–C*? (The same?)

4.8 Differential Equation of Equilibrium for Axially Loaded Member

Axially loaded members are found in every articulated structure and are important elements of some foundations. In articulated structures the major axial loading of members is due to loads applied to their ends and the self-weight of the member if it is inclined from the horizontal. In foundations, axially loaded elements called piles are sometimes used. These elements are subjected to axial loads at their ends and to distributed applied loads along the member if they are friction piles.

The determination of the internal axial forces from equilibrium is a relatively easy task but one that is frequently overlooked in treatments of structural analysis. The construction of loading diagrams for the members, as is the case for the transversely loaded beams discussed in the preceding sections, simplifies the determination of the internal axial force variation and enables the presentation of the variation in graphical form.

An axially loaded member is shown in Fig. 4.6a. The member is constrained against axial displacement at its left end by the reactive force, R. A distributed axial loading of intensity $p(x)$ in units of force per unit length, and an axial end force P are applied to the member, all in the positive sense of applied axial loads as defined in Fig. 4.1b. It is assumed that the end load and the distributed load are applied in a manner such that they may be considered to act at the centroid of the cross section.

Example 4.6

Obtain the loading, shear, and moment diagrams for the girder AE.

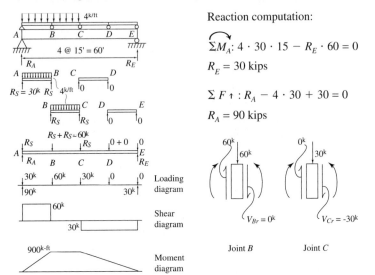

Reaction computation:

$$\Sigma M_A: 4 \cdot 30 \cdot 15 - R_E \cdot 60 = 0$$

$$R_E = 30 \text{ kips}$$

$$\Sigma F \uparrow : R_A - 4 \cdot 30 + 30 = 0$$

$$R_A = 90 \text{ kips}$$

Loading diagram

Shear diagram

Moment diagram

Joint B Joint C

$V_{Br} = 0^k$ $V_{Cr} = -30^k$

STEP 1 The structure is statically determinate and geometrically stable by inspection. Obtain the reactions at A and E using the complete structure as a free body. Separate the floor beam–stringer system from the girder A–E. Compute each individual stringer reaction, sum these at the floor beams to get the floor beam reaction, and apply this reaction to girder A–E. All stringer reactions are equal and are obtained from summation of moments about one end. The resultant reactions are:

$$\Sigma M : R_S \cdot 15 - 4 \cdot 15 \cdot 15/2 = 0 \quad \therefore R_S = 30 \text{ kips}$$

STEP 2 Draw the girder loading diagram using floor beam reactions.

STEP 3 Obtain the shear diagram using an incremental technique. Use free bodies at B and C to account for the change in shear due to concentrated loads. Draw the shear diagram by moving from left to right. Note that the area of the loading diagram is zero between all loads.

$$V_B - V_A = 0 \quad \therefore V_B = V_A = 60 \text{ kips} \qquad V_{Br} = 0 \quad V_C - V_{Br} = 0$$
$$\therefore V_C = V_{Br} = 0 \quad V_{Cr} = -30 \text{ kips}$$
$$V_E - V_{Cr} = 0 \quad \therefore V_E = V_{Cr} = -30 \text{ kips}$$

STEP 4 Use the incremental technique left to right for the moment diagram.

$$M_B - M_A = 60 \cdot 15 \quad \therefore M_B = 0 + 900 = 900 \text{ kip-ft}$$
$$M_C - M_B = 0 \quad \therefore M_C = 900 \text{ kip-ft}$$
$$M_E - M_C = -30 \cdot 30 \quad \therefore M_E = 900 - 900 = 0$$

A free body of an infinitesimal length of the member, dx, is shown in Fig. 4.6d, in which it is assumed that over the length dx the distributed load, $p(x)$, can be taken as constant. The internal axial force, F, is assumed to be in tension and to increase with an increase in x. Summing forces horizontally in the free body gives the relation

$$\Sigma F \rightarrow : -F + p(x) \cdot dx + F + dF = 0$$

Simplifying and dividing by dx yields

$$\frac{dF}{dx} = -p(x)$$

Equation (4.6) is the differential equation of equilibrium for the member and provides the relation between the variation of the internal axial force and the applied axial load. Once the loading diagram has been constructed, variation of the internal axial force can be obtained by direct integration of Eq. (4.6).

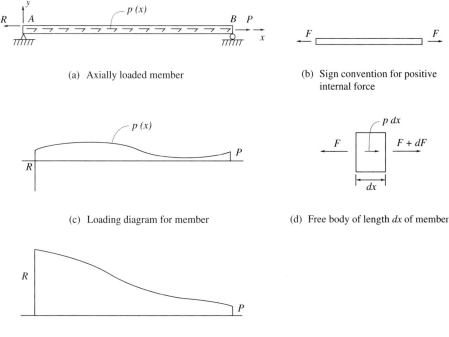

(a) Axially loaded member

(b) Sign convention for positive internal force

(c) Loading diagram for member

(d) Free body of length dx of member

(e) Internal axial force diagram

Figure 4.6a–e Axially loaded member showing sign convention for internal force, loading diagram, and axial force diagram.

4.9   Loading and Internal Axial Force Diagrams _____

The loading diagram for the member in Fig. 4.6a is shown in Fig. 4.6c where positive loads and load intensities are plotted above the horizontal axis. Concentrated loads are indicated by a vertical line in the diagram. The reaction force, R, is shown below the line because it acts in the negative x direction on the end of the member. Note the similarity of the loading diagram for a transversely loaded member, shown in Fig. 4.3, to the loading diagram of Fig. 4.6c.

Internal axial forces are assumed to be positive if they create tension in the member. The sign convention is presented in Fig. 4.6b. Note that the free body of length dx in Fig. 4.6d follows this sign convention, so that the internal axial force expressed in Eq. (4.6) is positive in tension.

Equation (4.6) has the same form as Eq. (4.1), which relates the internal variation of shear to applied loading in transversely loaded beams. As a consequence of this observation, the same concepts that are associated with Eqs. (4.1) and (4.4) for shear in transversely loaded beams can be used to relate internal axial forces to applied loads in axially loaded members. The presence of the minus sign in Eq. (4.6) reverses the relation of the slope of the internal axial force diagram to the loading diagram, but the concept can be stated as follows:

> The slope at any point to the internal axial force diagram for an axially loaded beam is the negative of the intensity of the applied load.

Recognizing that the internal axial force at the left end of the member in Fig. 4.6 is R and at the right end is P, the internal axial force diagram can be sketched using the concept of the relation of its slope to the intensity of the applied load. Figure 4.6e shows the internal axial force diagram resulting from the application of these ideas.

Following the concept in Eq. (4.4), both sides of Eq. (4.6) can be integrated between the points $x = a$ and $x = b$, which yields the result

$$\int_a^b \frac{dF}{dx}\,dx = F(b) - F(a) = -\int_a^b p(x)\,dx = -\left(\begin{array}{l}\text{area under loading dia-}\\ \text{gram between } a \text{ and } b\end{array}\right) \quad (4.7)$$

It is assumed in Eq. (4.7) that no concentrated axial loads are applied to the member between the points $x = a$ and $x = b$. The concept expressed in Eq. (4.7) can be stated as follows:

The change in internal axial force between two points of an axially loaded member is equal to the negative of the area under the loading diagram between the two points, provided that there are no applied concentrated axial loads between the points.

By using the concepts established for axially loaded members, it becomes an easy task to draw the internal axial force diagrams for any applied loading, as illustrated in Example 4.7.

Example 4.7 The member shown is loaded with a distributed axial load. Once the reactive force, R, is obtained by equilibrium in step 1, the loading diagram is easily obtained using the convention that positive applied loads act in the positive x direction. With the loading diagram being established, the use of the concepts presented in Eqs. (4.6) and (4.7) makes drawing of the axial force diagram a very simple task. For numerical computations using U.S. units, take $L = 40$ ft and $p = 100$ lb/ft; for SI units, take $L = 15$ m and $p = 1200$ N/m.

4.10   Summary and Limitations

Loading, shear, and moment diagrams present graphically the variation along a member of a particular pattern of transverse loading, the corresponding internal shear, and internal moment, respectively. Using a convenient sign convention, the shear and moment diagrams can be constructed in three different ways. Direct integration of the differential equation of equilibrium, Eq. (4.3), is useful when loadings are continuous on simply supported or cantilever beams. Direct calculation of the shear and moment from a series of free-body diagrams is the most general approach and can be used for any sort of member geometry. The use of incremental techniques giving the change in shear and moment between two points is useful for members with numerous discontinuities in the loading pattern.

The drawing and checking of the form of these diagrams is enhanced by using the relationships between shear and applied load and moment and shear. The relations are derived from the requirements of force equilibrium in Eq. (4.1) and moment equilibrium in Eq. (4.2). From Eq. (4.1) the slope of the shear diagram is equal to the applied load intensity, and the integration of this equation to obtain Eq. (4.4) provides a basis for calculating the change in shear between two points as the area under the loading diagram. From Eq. (4.2) the slope of the moment diagram is equal to the shear, and

Example 4.7

Obtain the loading and internal force diagrams for the axially loaded member.

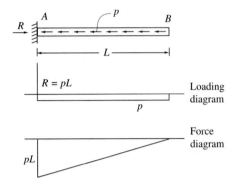

STEP 1   The member is restrained at the left end against horizontal displacement by the unknown force R. The member is stable and statically determinate. Obtain the reaction force R by summing forces horizontally.

$$\Sigma\, F\rightarrow\, :\, R - pL = 0 \quad \therefore R = pL$$

STEP 2   Draw the loading diagram for the member. The applied load p in units of force per length is negative because it is directed to the left.

STEP 3   Draw the internal axial force diagram. Since tension is positive, the internal force at the left end is $-R$, which is $-pL$. The internal force at the right end is computed using Eq. (4.7), and the slope of the diagram is positive because of the negative applied load, p.

$$F_B - F_A = -(-p)L \quad \therefore F_B = F_A + pL = -pL + pL = 0$$

PROBI

4.1 Draw
shown.

4.2 Draw
shown. Note

4.3 Draw
shown. Note

4.4 Draw
shown. Note

4.5 Draw
shown.

the integration of this equation to obtain Eq. (4.5) provides a basis for calculating the change of moment between two points as the area under the shear diagram.

A structural support system known as a beam-and-girder system is used in many highway bridges and some buildings as a means of supporting loads over relatively long distances. The loading, shear, and moment diagrams for the girders, which provide the major long-span support of the system, are similar to those of other members, but the loads are applied to these at specific panel points. An analysis of the manner in which loads applied to the system eventually enter the girder is required before the loading, shear, and moment diagrams can be drawn.

The analysis of axially loaded members is important in the design of piles in foundations and in some specialized structural configurations. Only one equation of equilibrium, which is summation of forces parallel to the axis of the member, is available to relate applied loads to the internal axial force, and its derivation is presented as Eq. (4.6). The techniques for the drawing of the loading and internal axial force diagrams follow closely those for transversely loaded beams. The only difference is the negative sign in Eq. (4.6), which makes the slope of the axial force diagram the negative of the load intensity and, in Eq. (4.7), the change in internal axial force between two points is the negative of the area under the loading diagram between the points.

The concepts of analysis presented and discussed in this chapter are based on mathematical models or idealizations of the structure and are subject to certain assumptions and limitations. The important ones are listed below and should be considered in evaluating the results of any analysis:

- All of the assumptions and limitations listed at the end of Chapter 2 are equally applicable here, as the mathematical models or idealizations used in this chapter are extensions of those in Chapter 2.

- The deformations of the structure are not considered, so it is assumed that any moment or axial force distribution in a member can be sustained without the member becoming unstable.

- The moments and shears in a transversely loaded beam are obtained with the assumption that no axial loads are present, or if they are present, they do not affect the magnitude of distribution of the moments and shears.

- The internal axial loads are obtained with the assumption that there are no transversely applied loads, or if there are, they have no effect on the magnitude or distribution of the axial forces.

In addition to being aware of the limitations listed above, some questions about the accuracy of an analysis for internal actions in a member, whether

4.11 Draw the shear and moment diagrams for the beam shown. Note that *C* and *D* are hinges.

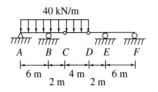

4.12 The shear diagram for a simply supported member is shown. Draw the moment and loading diagrams. That is, show the loads acting and the moment diagram.

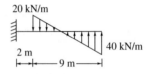

4.13 Draw the shear and moment diagrams for the linearly varying applied load on the cantilever beam.

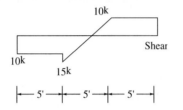

4.14 The structure shown has hinges at *B* and *C*. Draw the shear and moment diagrams for the loading shown.

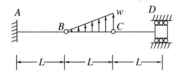

4.15 Draw the shear and moment diagrams for the structure shown. The diagrams for the inclined right portion of the structure should be drawn perpendicular to the member axis.

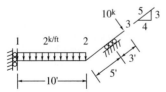

4.16 The shear diagram for a simply supported member is shown. Draw the moment and loading diagram. That is, show the loads acting and the moment diagram.

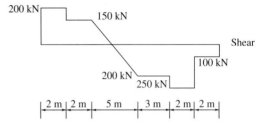

4.17 The beam with the hinge at *B* is loaded with the upward-distributed load shown. Compute and draw the shear and moment diagrams.

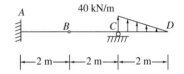

4.18 Draw shear and moment diagrams for the beam *A–B–C–D–E–F*. Note the hinges at *B* and *D*.

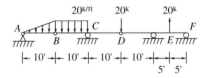

4.19 For the beam shown, draw the shear and moment diagrams. Note the hinge at *C*.

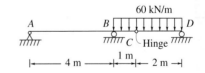

4.20 Draw the complete shear and moment diagrams for member *A–B* of the structure shown. The stringers of the super-structure are a series of simply supported beams with the support at 1 separate for girder *A–B*.

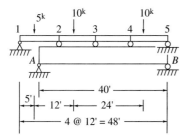

4.21 Draw the shear and moment diagrams for girder *A–E* due to a uniform load from *C* to *F* of 30 kN/m.

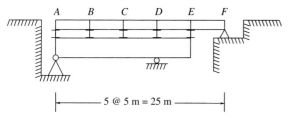

4.22 Draw the shear and moment diagrams for girder *A–F* for a uniform load of 2 kips/ft acting from *B* to *E*.

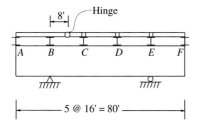

4.23 Draw the loading and internal force diagrams for the axially loaded member shown.

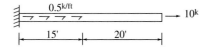

4.24 Draw the loading and internal force diagrams for the axially loaded member shown.

4.25 Draw the loading and internal force diagrams for the axially loaded member shown.

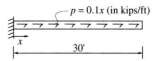

4.26 Draw the loading and internal force diagrams for the axially loaded member shown.

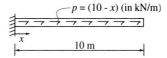

4.27 Draw a loading diagram for the arbitrarily varying axial load $p(x)$.

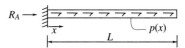

Show that the reaction R_A, to the distributed loading $p(x)$ along the member is given by

$$R_A = -\int_0^L p(x)\, dx$$

Also show that at some point x_r, the internal axial force is given by

$$F_r = \int_{x_r}^L p(x)\, dx$$

155

Analysis of Statically Determinate Plane Frames

Frames differ from trusses in that the members of the structure are connected in a rigid manner at joints. That is, the angle between the members that connect at a joint remains unchanged during all loading and deformation action of the structure. The internal actions in frame members consist of shear force and moment in addition to axial force. Frames most commonly are the major load-carrying elements in buildings and are constructed with a variety of materials, including steel, wood, reinforced concrete, aluminum, and certain plastics. The horizontal members of frames typically have relatively short spans lengths because the depth of these members is generally limited to 3 ft (1 m) or less.

A one-bay, one-story rectangular frame such as that shown in Fig. 5.1 is called a *portal frame*. Stories are defined by the spacing in the vertical direction of horizontal members called *beams* or *girders*. Bays are defined by the horizontal spacing of vertical members called *columns*. A one-story, one-bay frame in which the horizontal beam is replaced with two upwardly inclined members that meet in a peak is called a *gable frame*. Portal and gable frames are used in small residential and industrial structures where the width of the structure (bay width) is typically 20 to 60 ft (7 to 20 m) and story heights typically are 10 to 30 ft (3 to 10 m).

Structures having multiple bays and stories are generally called *multistory structures*. They are classified as rigid frames or braced frames, depending on

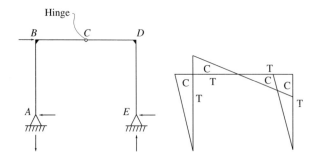

(a) Loaded frame   (b) Moment diagram plotted on
                             compression side of member

Figure 5.1a–b   Sign convention for presentation of
frame moment variation.

their construction. A rigid frame is one in which the members are horizontal
and vertical and joined together with rigid connections at joints. A braced
frame is a structure in which diagonal members are present and provide trusslike action in resisting horizontal loads caused by wind or the action of an
earthquake. These diagonals, referred to as *bracing members*, can have several different geometric configurations. Structures that have continuous vertical walls, called *shear walls*, typically constructed in reinforced concrete, are
also referred to as *braced frames*, the bracing being provided by the shear
walls. The connection of the horizontal members in braced frames to columns
may be rigid, semirigid (some flexibility in the connection), or pinned.

By the nature of their construction, nearly all frames are statically indeterminate, usually highly indeterminate. Single-bay frames of one or two
stories or frames of one story and several bays may be statically determinate. However, indeterminate frames are far more common because they
are better able to resist applied loading with the least amount of deformation. Some of the essential features of frame behavior will be established in
this chapter, but only statically determinate frames will be considered. The
analysis of frames produces moments, shears, and axial forces in the members, but the graphical display of moments obtained from a frame analysis
is emphasized. The determination of static determinacy and geometric stability is also studied.

5.1   Frame Action

Frames are loaded and respond to these external actions in a manner significantly different than trusses. In addition to carrying axial load, frame members develop an internal shear force and bending moment. Frames are much

more flexible than trusses since the bending action in members of the frame is not as effective in resisting deformation as is the axial force action in trusses. There is also a difference in the manner of loading of frames, because the loads are often applied transverse to members between joints as well as at the joints of the frame. The design of the members of a frame is controlled primarily by the bending moments that develop, with axial loads being important in columns of high-rise structures and shear in the design of reinforced concrete members.

The deformation behavior of frame members is exactly the same as that in beams. There is axial deformation due to internal axial forces, and transverse deformation or deflection perpendicular to the member centroidal axis due to internal moments and shears. The deformations of frame members are continuous along the member axis unless there is some release of continuity due to special construction features. As shown in Fig. 2.1, a release of axial continuity renders the axial force zero at the point of release. Similarly, a release of relative vertical displacement continuity renders the shear zero and a release of slope continuity renders the moment zero at release points.

Since bending action is the most important feature of frame design, the primary goal of frame analysis is to obtain the bending moment diagram for all members. Because members are oriented vertically as well as horizontally, and even inclined in some cases, it is necessary to establish a convention for the plotting of moments that is unambiguous. Figure 5.1b shows the distribution of moments in a frame that follows the convention that the moment is plotted on the compression side of the member. This convention is consistent with the convention shown in Fig. 4.1c for plotting positive moments in beams. The completed analysis for a frame, which is presented in the form of a moment diagram for each member, immediately indicates the magnitude and direction of action of the moments at any point in the structure. The "C" and "T" in Fig. 5.1b indicate which side of the member is in compression or tension with the moment diagram drawn as shown. The designation is redundant since the moment diagram itself is on the compression side of the member, so "C" and "T" will not be used in future moment diagrams.

5.2   Equations of Equilibrium _____

When an individual frame member is isolated as a free body, there are end forces and moments as well as applied transverse loads that must be in equilibrium.

Figure 5.2 As shown, there are six unknown forces and moments associated with the three internal actions at each end of a frame member as well as the distributed load acting on the member. However, the member as a free body must be in equilibrium, which means that three equilibrium

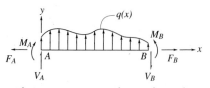

Figure 5.2 Typical member of a frame.

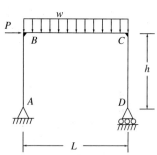

(a) Initial configuration

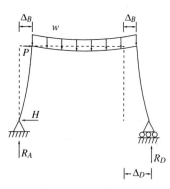

(b) Displaced position

Figure 5.3a–b Equilibrium considerations for a simple frame.

equations are available to relate those end forces and moments. There are then only three independent forces and moments that can be developed in any member of a frame due to applied loading. For example, in Fig. 5.2 one might consider the axial force, F_B, the shear force, V_B, and the moment, M_B, as the independent forces and moment for member $A–B$. The three equations of equilibrium would provide the values of F_A, M_A, and V_A in terms of the applied loading $q(x)$ and the end actions at B, F_B, V_B, and M_B.

Similar to trusses discussed in Chapter 3, the equilibrium equations of a frame structure must be written using the geometry of its final displaced position. The simple frame shown in its displaced equilibrium position in Fig. 5.3 must satisfy the three equilibrium equations:

$$\Sigma F\rightarrow:\ P - H = 0$$

$$\Sigma F\uparrow:\ R_A + R_D - wL = 0$$

$$\Sigma M_A:\ Ph + wL\left(\frac{L}{2} + \Delta_B\right) - R_D(L + \Delta_D) = 0 \qquad (5.1)$$

Displacements Δ_B and Δ_D have been exaggerated in Fig. 5.3b for clarity, but for normal structures under design service loads these displacements would nearly always be very small.

The unknown reaction components in Eqs. (5.1) are easily obtained, provided that displacements Δ_B and Δ_D are known. With the reaction components and the joint displacements known, free-body diagrams for members $A–B$, $B–C$, and $C–D$ can be used to obtain the three internal member forces and moment using the equilibrium equations for each free body. If Δ_B and Δ_D are known, the frame is statically determinate since the three equilibrium equations for each free body and Eqs. (5.1) are sufficient in number to solve for all unknown external and internal actions.

The joint displacements Δ_B and Δ_D are, however, not known. As will be seen in later chapters, they can be obtained from an analysis of the deformations of the members of the structure due to the internal bending and axial force actions caused by the applied loading. But the internal member forces and moments, and hence deformations, depend on the values of the unknown reaction forces in Eqs. (5.1). The dependence of the unknown values of Δ_B and Δ_D on the unknown reaction forces makes Eqs. (5.1) highly nonlinear. It is possible to establish and solve the nonlinear forms of Eqs. (5.1), but this is a lengthy and very complicated procedure.

The dilemma in the solution process is removed as it was for trusses by assuming that the deflections Δ_B and Δ_D are negligibly small in relation to the length proportions of the structure. They may, for practical purposes, be taken as zero in Eqs. (5.1). This means that the reaction components and the internal forces and moments are computed as though the structure were composed of rigid members and that the equilibrium equations for the structure are based on its initial or undeformed geometry.

In applying the equations of equilibrium to the analysis of frames, the release of internal force actions in members of a frame should be recalled. Figure 2.1 shows the types of releases that can occur in a frame member and the resulting equilibrium equations of condition. It is assumed that each release of a deformation continuity does not cause any significant change in the geometry of the structural member so that the assumption of small (or zero) changes in geometry is still valid for the analysis of frames. The analysis of statically determinate frames requires that these equations of condition be used at some point in the analysis for the distribution of moments and forces. Many times no reaction can be computed without the immediate use of one or more of these equations. A simple frame is analyzed in Example 5.1 that does not require the use of equations of condition.

> **Example 5.1** The simple frame is loaded with the single concentrated load, P. Once the reactions are computed in step 1, the individual member moment diagrams are obtained by using free-body diagrams. Moments are plotted on the left side of member A–B and the top of member B–C which are the compression side of the members. For numerical computations in U.S. units, take $L = 12$ ft and $P = 2$ kips; for SI units, take $L = 4$ m and $P = 10$ kN.

5.3 Determinacy and Geometric Stability

Number of Unknowns and Number of Equations of Solution

Before any frame is analyzed, a determination of whether the structure is geometrically stable and statically determinate must be made. As discussed in Section 5.2, the deformations of the frame must be taken to be very small or zero if joint deflections or member deformations are not to be considered as unknowns. Under these conditions, there are three unknown forces or moments associated with each member of a frame. Added to these are the unknown reaction components, which leads to the following:

> If it is assumed that the deformations of a frame are small so that its geometry remains essentially unchanged under the action of applied loads, the total number of unknowns involved in the analysis of any frame is equal to the sum of the number of reaction components and three times the number of members.

Example 5.1

Obtain the final moment diagram for all members of the frame.

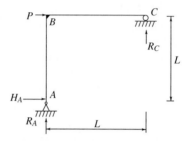

STEP 1   Calculate the reactions of the frame. Sum moments about A to obtain R_C, sum forces horizontally for H_A and vertically for R_A.

$$\overset{\curvearrowright}{\Sigma\, M_A}: PL - R_C L = 0 \therefore R_C = P$$

$$\Sigma\, F\uparrow: R_A + R_C = 0 \therefore R_A = -R_C = -P$$

$$\Sigma\, F\rightarrow: H_A + P = 0 \therefore H_A = -P$$

STEP 2   Take free bodies of each individual member of the frame and use moment equilibrium to calculate the moment variation in the member. For convenience, use the coordinates r for member A–B and t for member B–C.

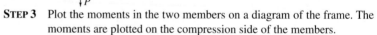

$$\Sigma\, M_{cut}: M - Pt = 0 \qquad \therefore M = Pt \quad \text{Linear}$$

$$\Sigma\, M_{cut}: M - Pr = 0 \qquad \therefore M = Pr \quad \text{Linear}$$

STEP 3   Plot the moments in the two members on a diagram of the frame. The moments are plotted on the compression side of the members.

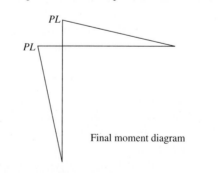

Final moment diagram

When a free body is taken at a joint in a frame, all of the end forces and moments for the members that connect to that joint must satisfy equilibrium. Thus at each joint of a frame there are three equilibrium equations, two force and one moment.

Within the individual members of a frame, special construction details can affect continuity of deformation. As discussed in Section 2.1 and shown in Fig. 2.1, it is possible to release any continuity of deformation in a member of a structure. The consequence of the release of a force action that ensures the continuity of deformation in a member is the creation of an equation that is independent of the equilibrium equations at the joints of a frame. It is important to recognize that the release of a continuity of deformation condition always occurs in a member not at a joint. This can cause confusion unless the definition of a joint of a frame is made clear.

A joint of a frame is defined as a point where one or more reaction components act or where the centroidal axes of two or more noncollinear members intersect. The hinge at C in Fig. 5.1 is not counted as a joint because members B–C and C–D are collinear at C. The hinge at C is simply a condition of release of slope continuity within member B–D. If, in that figure, the hinge were moved from the middle of B–D toward joint D, the release in slope continuity clearly remains unchanged. Even in the limit when the hinge is moved to D, the slope continuity release is still considered to be in member B–D, but located at the right end of the member.

With these ideas in mind, a count of the number of equations of equilibrium and condition for a frame structure becomes an easy task. This count can be summarized as follows:

> The number of equations of equilibrium available for the analysis of a plane frame structure is the sum of the number of equations of condition, if any, and three times the number of joints.

The simple frame shown in Fig. 5.1 has three members and four joints. The number of unknowns for this structure is 13 because there are $3 \cdot 3 = 9$ unknown member actions and 4 unknown reaction components. As discussed in Section 5.2, the number of unknowns is increased if deflections of the frame, or geometry changes, are considered in the equilibrium equations. However, if the assumption is made that the structure is rigid or the geometry unchanged, the number of unknowns remains as 13. The assumption of small deformations is basic to the analyses illustrated in the remainder of this chapter and can be stated as follows:

It is assumed that under the action of applied loading the deformations of a frame are so small that the geometry of the frame does not change and the initial geometry of the structure can be used for reference in writing the equations of equilibrium.

Geometric Instability

The problem of geometric instability of frames is similar to that for trusses. As discussed and shown by examples in Chapters 2 and 3, the problem of geometric instability requires careful attention to the manner in which the structure can displace or rotate as a rigid body. The local instability of a portion of the structure must also be investigated. Fortunately, frames usually are not geometrically unstable because they tend by their construction to be highly restrained. An example of a geometrically unstable frame is shown in Example 2.3.

A type of local instability that is not likely to occur in a real structure because of the manner of its construction can occur in the idealization or mathematical model of a frame. In the frame of Fig. 5.4, a joint is intended to be idealized as a pin joint, that is, a joint at which the moment acting on

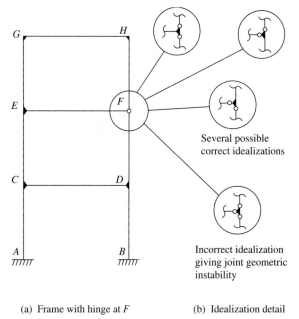

Several possible correct idealizations

Incorrect idealization giving joint geometric instability

(a) Frame with hinge at *F*　　　(b) Idealization detail

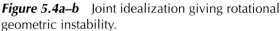

Figure 5.4a–b Joint idealization giving rotational geometric instability.

the ends of all members connecting to the joint is zero. If the mathematical model for this condition is created by releasing the continuity of slope of all members at the joint, the joint becomes unstable. It is possible for the joint to undergo a rigid-body rotation, which creates problems with the solution of the equations established in the mathematical model of the frame.

As with trusses, geometric instability is based on the undeformed geometry of the structure and hence may be only instantaneous in nature. That is, a small but measurable change in geometry may stabilize the structure. Nevertheless, this type of situation must be avoided in design because if it is not, forces and moments develop in the members of the structure that are very large in proportion to the loads applied.

Knowing the number of equations and the number of unknowns provides some important information about the frame structure. Similar to trusses, the following concept applies to frames:

> If, in a frame, the number of unknowns is less than the sum of the number of equations of equilibrium and condition, the frame is geometrically unstable. If, in a geometrically stable frame, the number of unknowns equals the number of equations, the frame is statically determinate, but if the number of unknowns exceeds the number of equations, the frame is statically indeterminate.

This concept can be expressed quantitatively in a mathematical form. In a plane frame having n joints, b members, r reaction components, and s releases of deformation continuity, the index of static indeterminacy, p, is given by

$$p = 3b + r - 3n - s \qquad (5.2)$$

where the following conditions on p define the structure as:

$p < 0$ geometrically unstable
$p = 0$ statically determinate if not geometrically unstable
$p > 0$ statically indeterminate to degree p if not geometrically unstable

The requirement that p be equal or greater than zero is a necessary but not sufficient condition for a frame to be geometrically stable. In Example 2.3, there are five reaction components and $5 \times 3 = 15$ unknown member actions, for a total of 20 unknowns. There are six joints, giving $3 \times 6 = 18$ equilibrium equations, and there are two equations of condition, for a total of 20 equations of solution. By Eq. (5.2) the frame is statically determinate, but as shown in Example 2.3 it is geometrically unstable. If the hinge were moved to joint F, would the frame be stable?

5.4 Analysis of Plane Frames for the Distribution of Internal Moments by Method of Sections _____

Statically determinate frames can be analyzed by the method of joints, but more frequently are analyzed in hand computations by the method of sections described for trusses. Again, the first step in the analysis is to check whether or not the frame is statically determinate and geometrically stable. The usual next step is to compute the reactions of the structure, although in multistory frames the analysis is best started with the top story. Statically determinate frames usually have several internal releases creating equations of condition, so calculation of the reactions or internal forces always starts with free bodies taken through the releases. The final step in the analysis is to proceed member by member through the frame to obtain the variation of moments and draw the moment diagrams. Examples 5.2 to 5.4 show how the procedure is implemented.

Example 5.2 The method of sections is illustrated with the simple frame shown. The geometric stability of the structure is checked by considering the members of the frame as being perfectly rigid. The A–B–C portion of the structure can be considered as a single rigid link, as can the C–D–E portion. The A–B–C link can rotate about A and the C–D–E link about E. However, when the two are joined together with the hinge at C, neither link can undergo any rotation, thus making the structure geometrically stable. The reactions are computed by starting with the condition of zero moment at the hinge at C. A free body is taken of the C–D–E portion of the structure by making a cut through the hinge at C. Summing moments about C gives a relation between H_E and V_E. With this relation established, the reactions can be calculated from a free body of the entire structure using the three equilibrium equations.

With all reaction components known, the variation of the moment in each member of the frame is obtained from a free body of that member. Note that the variable used to compute the moments in each of the members is arbitrary and has been taken differently in each free body. At each joint of the frame, moment equilibrium must be satisfied. This condition can be used as an aid and a check on the computations. At joints where only two members connect, the moments on the ends of these members must be equal in magnitude, as is seen in the final moment diagram for the entire structure. For numerical calculations in U.S. units, take $L = 10$ ft and $w = 4$ kips/ft; for SI units, take $L = 3$ m and $w = 60$ kN/m.

Example 5.3 The location of the lower support of the simple frame can be anywhere from a distance of $0.5L$ to $1.5L$ from the left side of the structure. The value of the absolute maximum moment at any point in the frame for the varying positions of the support is required in the design decision process. The reactions of the structure in terms of a are calculated in step 1. Note that the horizontal reaction at 3 is zero for all positions of the support.

The maximum moment can occur at one of two points in the frame. The vertical reaction R_3 will cause a moment at 2 that may be a maximum, and the uniform load causes a moment that may be the maximum somewhere between the ends of member 1–2. In step 2 these moments are calculated from the two free-body diagrams. The point of maximum moment in member 1–2 can be found by differentiating the moment expression with respect to r and setting the result to zero. This is equivalent to finding the point where the shear is zero, which is accomplished in step 3.

The two potential maximum moments are given by expressions (2) and (3). In step 4 these two moment expressions are evaluated through the use of a spreadsheet and the resulting absolute maximum moment is plotted. The notation ( )^2 indicates that the expression in parentheses is squared. As shown by the graph, the maximum moment occurs at joint 2 for values of a/L less than about 0.85 and in member 1–2 for greater values of a/L. As seen in the graph, the location of the support has a significant effect on the maximum moment in the frame. How can it be shown that the exact value of a/L that gives the minimum moment is $\left(2 + \sqrt{2}\right)/4$?

Example 5.4 A more complicated structure is analyzed in this example. The geometric stability is obtained as in Example 5.2 by considering as rigid links the A–B–G, G–H, B–C–D, and D–E–F portions of the structure. These rigid links are seen to be connected together in a stable configuration by the hinges at G, B, and D. With geometric stability established, the static determinacy of the structure is obtained from an application of Eq. (5.2). Immediate use of the three equations of condition reduces the number of unknown reaction components to three that can then be obtained using the equilibrium equations for the entire structure. As in Example 5.2, it is easier to obtain the moment diagram for each member using an appropriate free body of a portion of the structure rather than simply a free body of the individual member. This approach avoids the necessity of computing axial and shear forces in all members of the frame. Note again in the final moment diagram of the frame that moment equilibrium is satisfied at each joint.

5.5   Moment Variation by Superposition

The drawing of moment diagrams of individual members of frames has been done using free-body diagrams of a portion of the member in Examples 5.1 and 5.3 or a portion of the member combined with a portion

Example 5.2

Obtain the final moment diagram for all members of the structure.

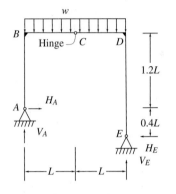

STEP 1 Structure is statically determinate. From Eq. (5.2), $n = 4$, $b = 3$, $s = 1$, $r = 4$, which makes $p = 3 \cdot 3 + 4 - (3 \cdot 4 + 1) = 0$. Structure is geometrically stable by inspection.

STEP 2 Isolate free body by taking section through hinge at C.

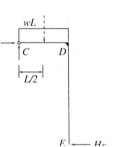

$$\Sigma M_C:\ -wL \cdot \frac{L}{2} - H_E \cdot 1.6L + V_E L = 0$$

$$H_E = \frac{V_E - wL/2}{1.6}$$

With H_E known in terms of V_E, consider the free body of the entire structure.

$$\Sigma M_A:\ -2wLL - H_E \cdot 0.4L + V_E \cdot 2L = 0$$

Substituting for H_E allows solving for V_E and then in turn for H_E and the other reactions.

$$V_E = \frac{15wL}{14} \quad \therefore H_E = \frac{5wL}{14}$$

$$\Sigma F\rightarrow:\ H_A - H_E = 0 \rightarrow H_A = H_E = \frac{5wL}{14}$$

$$\Sigma F\uparrow:\ V_A - 2wL + V_E = 0 \rightarrow V_A = 2wL - V_E = \frac{13wL}{14}$$

Example 5.2 (continued)

STEP 3 Isolate the free body of each member and use moment equilibrium.
Plot the final result.

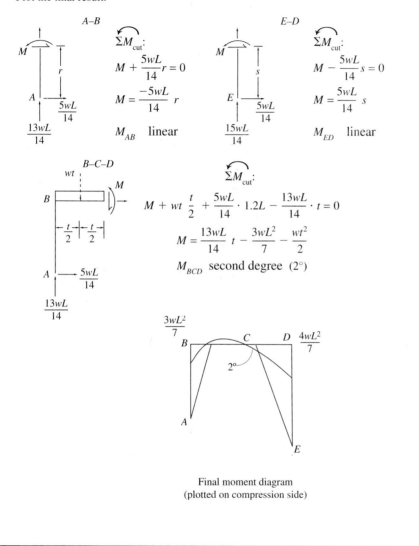

A–B

ΣM_{cut}:

$$M + \frac{5wL}{14}r = 0$$

$$M = \frac{-5wL}{14}\, r$$

M_{AB} linear

E–D

ΣM_{cut}:

$$M - \frac{5wL}{14}s = 0$$

$$M = \frac{5wL}{14}\, s$$

M_{ED} linear

B–C–D

ΣM_{cut}:

$$M + wt\,\frac{t}{2} + \frac{5wL}{14}\cdot 1.2L - \frac{13wL}{14}\cdot t = 0$$

$$M = \frac{13wL}{14}\, t - \frac{3wL^2}{7} - \frac{wt^2}{2}$$

M_{BCD} second degree $(2°)$

Final moment diagram
(plotted on compression side)

169

Example 5.3

Obtain a plot of the variation of the maximum moment in the frame loaded with the uniform load as a function of the location of the pin support, a/L. Use a spreadsheet to create the plot, and vary a/L from 0.5 to 1.5.

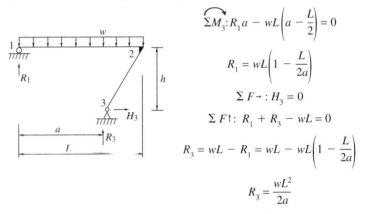

$$\overset{\frown}{\Sigma M_3}: R_1 a - wL\left(a - \frac{L}{2}\right) = 0$$

$$R_1 = wL\left(1 - \frac{L}{2a}\right)$$

$$\Sigma F\rightarrow: H_3 = 0$$

$$\Sigma F\uparrow: R_1 + R_3 - wL = 0$$

$$R_3 = wL - R_1 = wL - wL\left(1 - \frac{L}{2a}\right)$$

$$R_3 = \frac{wL^2}{2a}$$

STEP 1 The frame is geometrically stable for the range of values of a. Use a free body of the entire structure to compute the reactions.

STEP 2 The maximum moment will occur either at joint 2 or somewhere in member 1–2. Take free bodies to obtain the moment at 2 and the moment and shear in member 1–2.

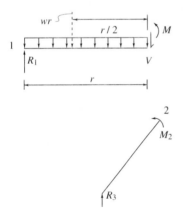

$$\overset{\frown}{\Sigma M_{cut}}: M = wr\left(\frac{r}{2}\right) - R_1 r = 0$$

$$M = R_1 r - \frac{wr^2}{2} \qquad (1)$$

$$\Sigma F\uparrow: R_1 - wr - V = 0 \therefore V = R_1 - wr$$

$$\overset{\frown}{\Sigma M_2}: M_2 - R_3 \cdot (L - a) = 0$$

$$M_2 = R_3(L - a)$$

$$M_2 = \frac{wL^2(L/a - 1)}{2} \qquad (2)$$

STEP 3 The maximum moment occurs in member 1–2 at the point where the shear is zero. Solving for the r which makes the shear zero and substituting into (1) gives the final expression for the maximum moment in member 1–2.

$$V = 0 \therefore r_{max} = \frac{R_1}{w} = L\left(1 - \frac{L}{2a}\right)$$

From (1),

$$M_{max} = r_{max}\left(R_1 - \frac{wr_{max}}{2}\right)$$

Example 5.3 (continued)

$$= L\left(1 - \frac{L}{2a}\right)\left[wL\left(1 - \frac{L}{2a}\right) - \frac{wL(1 - L/2a)}{2}\right]$$

$$= \frac{wL^2(1 - L/2a)^2}{2} \tag{3}$$

STEP 4 Create the spreadsheet based on the moment expressions (2) and (3). The columns of the spreadsheet are created as indicated and the variation of the maximum moment with a/L is plotted. Note that the maximum moment in columns 2, 3, and 4 are multiplied by wL^2.

Spreadsheet Construction

Column 1: Enter a/L range.

Column 2: Enter expression (3)

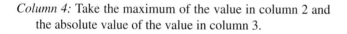

$$\text{Col. 2} = [(1 - 1/\text{col. } 1/2)^2]/2$$

Column 3: Enter expression (2)

$$\text{Col. 3} = (1/\text{col. } 1 - 1)/2$$

Column 4: Take the maximum of the value in column 2 and the absolute value of the value in column 3.

The values in column 4 are plotted against the values in column 1.

a/L	Maximum Moment		
	1–2	At 2	In Frame
0.5	0.000	0.500	0.500
0.6	0.014	0.333	0.333
0.7	0.041	0.214	0.214
0.8	0.070	0.125	0.125
0.9	0.099	0.056	0.099
1	0.125	0.000	0.125
1.1	0.149	-0.045	0.149
1.2	0.170	-0.083	0.170
1.3	0.189	-0.115	0.189
1.4	0.207	-0.143	0.207
1.5	0.222	-0.167	0.222

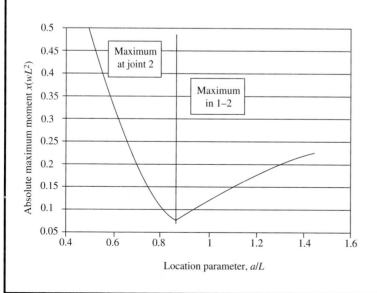

Location parameter, a/L

Example 5.4

Obtain the final moment diagram for all members of the structure.

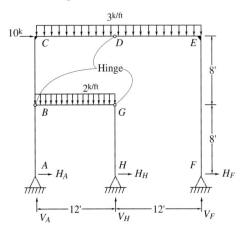

STEP 1   The structure is statically determinate. From Eq. (5.2), $b = 6$, $n = 7$, $s = 3$, $r = 6$, which makes $p = 3 \cdot 6 + 6 - (3 \cdot 7 + 3) = 0$. The structure is also geometrically stable by inspection.

STEP 2   Isolate the free bodies by cutting the structure at the hinge points.

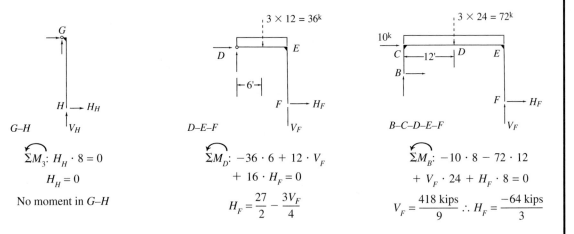

G–H

$$\Sigma M_3: H_H \cdot 8 = 0$$

$$H_H = 0$$

No moment in G–H

D–E–F

$$\Sigma M_D: -36 \cdot 6 + 12 \cdot V_F$$

$$+ \; 16 \cdot H_F = 0$$

$$H_F = \frac{27}{2} - \frac{3V_F}{4}$$

B–C–D–E–F

$$\Sigma M_B: -10 \cdot 8 - 72 \cdot 12$$

$$+ \; V_F \cdot 24 + H_F \cdot 8 = 0$$

$$V_F = \frac{418 \text{ kips}}{9} \quad \therefore \; H_F = \frac{-64 \text{ kips}}{3}$$

Example 5.4 (continued)

STEP 3   Return to the entire structure for the remaining reaction forces.

$$\widehat{\Sigma M}_A: \ -10 \cdot 16 - 72 \cdot 12 - 24 \cdot 6$$

$$+ \frac{418}{9} \cdot 24 + 12 \cdot V_H = 0 \rightarrow V_H = \frac{40}{9} \text{ kips}$$

$$\Sigma F \rightarrow : \ H_A + 10 - \frac{64}{3} = 0 \rightarrow H_A = \frac{34}{3} \text{ kips}$$

$$\Sigma F \uparrow : \ V_A + \frac{40}{9} + \frac{418}{9} - 24 - 72 = 0$$

$$V_A = \frac{406}{9} \text{ kips}$$

STEP 4   Isolate a free body for each member, develop the appropriate moment expression, and plot final results.

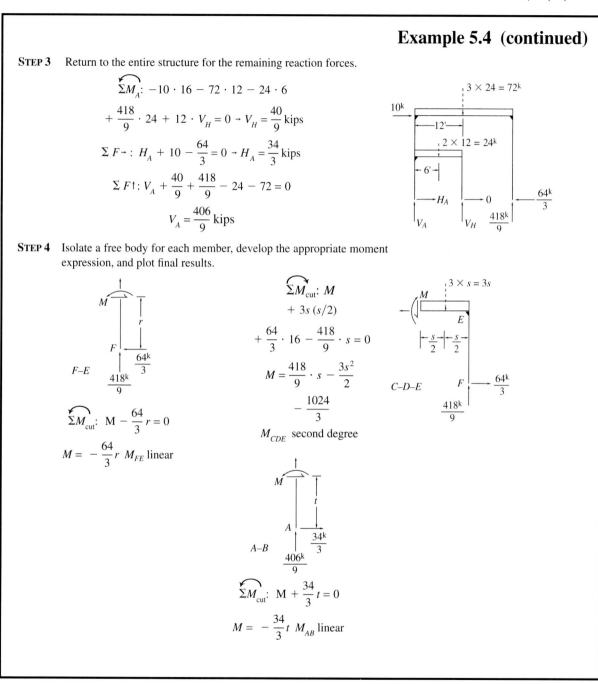

$$\widehat{\Sigma M}_{\text{cut}}: \ M - \frac{64}{3} r = 0$$

$$M = -\frac{64}{3} r \ M_{FE} \text{ linear}$$

$$\widehat{\Sigma M}_{\text{cut}}: \ M + 3s \ (s/2)$$

$$+ \frac{64}{3} \cdot 16 - \frac{418}{9} \cdot s = 0$$

$$M = \frac{418}{9} \cdot s - \frac{3s^2}{2}$$

$$- \frac{1024}{3}$$

M_{CDE} second degree

$$\widehat{\Sigma M}_{\text{cut}}: \ M + \frac{34}{3} t = 0$$

$$M = -\frac{34}{3} t \ M_{AB} \text{ linear}$$

Example 5.4 (continued)

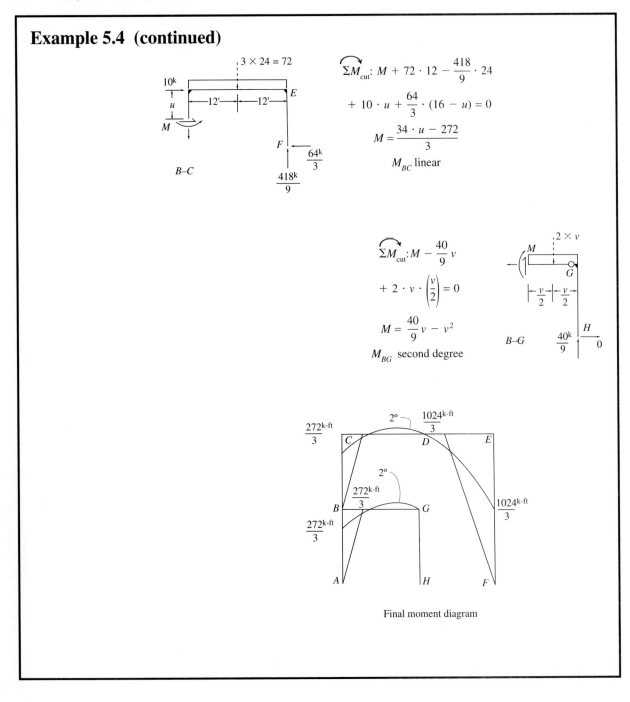

$3 \times 24 = 72$

$$\overset{\curvearrowright}{\Sigma M}_{cut}: M + 72 \cdot 12 - \frac{418}{9} \cdot 24$$

$$+\ 10 \cdot u + \frac{64}{3} \cdot (16 - u) = 0$$

$$M = \frac{34 \cdot u - 272}{3}$$

$$M_{BC} \text{ linear}$$

$B{-}C$

$\dfrac{64^k}{3}$

$\dfrac{418^k}{9}$

$$\overset{\curvearrowright}{\Sigma M}_{cut}: M - \frac{40}{9} v$$

$$+\ 2 \cdot v \cdot \left(\frac{v}{2}\right) = 0$$

$$M = \frac{40}{9} v - v^2$$

$$M_{BG} \text{ second degree}$$

$2 \times v$

$B{-}G$

$\dfrac{40^k}{3}$

$\dfrac{272^{k\text{-ft}}}{3}$

$\dfrac{1024^{k\text{-ft}}}{3}$

$2°$

$\dfrac{272^{k\text{-ft}}}{3}$

$\dfrac{272^{k\text{-ft}}}{3}$

$\dfrac{1024^{k\text{-ft}}}{3}$

Final moment diagram

of the frame in Examples 5.2 and 5.4. Alternatively, once the end moments have been obtained for an individual member, the use of superposition can aid the drawing of the moment diagrams. The requirements for superposition of force actions presented in Section 1.8 are satisfied by the limitations that must be met in performing a statically determinate frame analysis.

The concept of superposition of the moment diagrams obtained from several loading systems is derived from the behavior of a simple beam subjected to these same loading systems.

Figure 5.5 The equivalence of applied and reactive forces in a simple beam with end moments, shears, and applied loads on a frame member by virtue of equilibrium considerations is illustrated. In Fig. 5.5a the equilibrium equations for a frame member acted upon by two end moments, M_A and M_B, and an arbitrary distributed load, $q(x)$, are shown. Since there is no interaction between moment and axial forces, the axial forces and the equilibrium equation associated with them have not been included. The shear forces at each end of the frame member are obtained from the two equilibrium equations. As shown, V_B is obtained directly from the moment equilibrium equation and V_A from the force equilibrium equation.

In Fig 5.5b a simply supported beam is shown with the same end moments and distributed load acting on it as on the frame member in Fig. 5.5a. The two equilibrium equations that are used to obtain the end reactions R_A and R_B are also shown. Again the lack of any axial forces has obviated the need for the third equilibrium equation.

The equivalence of the force distributions on the beam and the frame member is established in Fig. 5.5c. The equilibrium considerations for the two structural systems are identical, which makes the reactions of the beam equal to the shear forces on the ends of the frame member. Thus loading a simply supported beam with the same distributed load and end moments that act on a frame member leads to the same distribution of external, and hence also of internal forces and moments in the beam as in the frame member.

The superposition of the moment diagrams for the loading of a simply supported beam with a distributed $q(x)$ and two end moments to obtain the moment diagram for all three actions occurring simultaneously is shown in Fig. 5.6a and b. By virtue of the equivalence established in Fig. 5.5 between the actions in a simply supported beam and a frame member, the loading on the frame member can be broken down in the manner shown in Fig. 5.6c. The moment diagrams for each separate loading on the frame member are the same as those of the beam, and hence the superposition process for the beam also applies to the frame member.

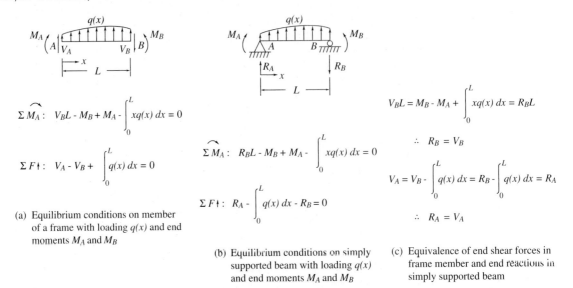

(a) Equilibrium conditions on member of a frame with loading $q(x)$ and end moments M_A and M_B

(b) Equilibrium conditions on simply supported beam with loading $q(x)$ and end moments M_A and M_B

(c) Equivalence of end shear forces in frame member and end reactions in simply supported beam

Figure 5.5a–c   Equivalence between force systems acting on a member of a frame and a simply supported beam.

In practice it is not necessary to create a simply supported beam to carry out the superposition of the moment diagrams for multiple loadings on a frame member. It is only necessary to imagine from the standpoint of equilibrium considerations that the frame member develops the same force system as that of a simply supported beam. These ideas are used in Example 5.5 to develop the moment diagram for selected members from a previous example.

Example 5.5 The moment diagram for member B–C–D of Example 5.2 is obtained by the superposition process.

Member B–C–D of Example 5.2 is a member that has a hinge in the center. The presence of the hinge does not change the superposition process, because the process is based on force systems that are always in equilibrium and on the requirement that they act on a structure that, as a whole, is geometrically stable. Note that the moment at the hinge at C in the final moment diagram is zero from the superposition process. This must be the case and, in fact, is a check of the consistency of the analysis that produced the end moments at B and D in Example 5.2. That analysis used the zero moment condition at C to obtain the reactions of the frame, which in turn led to the moments at B and D for member B–C–D.

The use of superposition of moment diagrams of several loadings acting on a simply supported beam often simplifies the drawing of moment

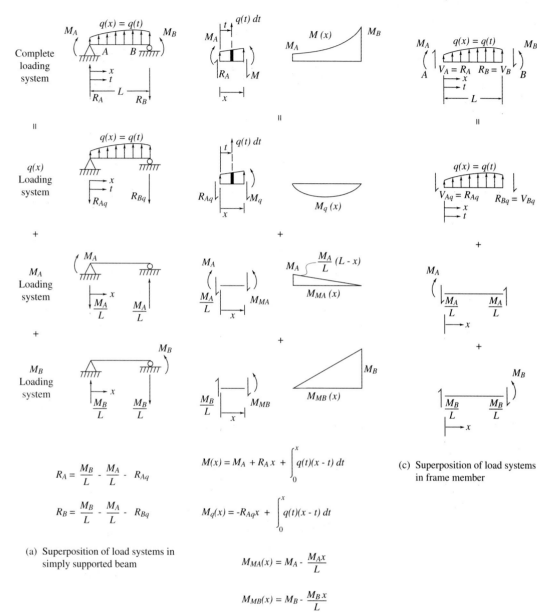

$$R_A = \frac{M_B}{L} - \frac{M_A}{L} - R_{Aq}$$

$$R_B = \frac{M_B}{L} - \frac{M_A}{L} - R_{Bq}$$

(a) Superposition of load systems in simply supported beam

$$M(x) = M_A + R_A x + \int_0^x q(t)(x - t)\, dt$$

$$M_q(x) = -R_{Aq}x + \int_0^x q(t)(x - t)\, dt$$

$$M_{MA}(x) = M_A - \frac{M_A x}{L}$$

$$M_{MB}(x) = M_B - \frac{M_B x}{L}$$

(b) Superposition of moment diagrams for load systems in simply supported beam and frame member

(c) Superposition of load systems in frame member

Figure 5.6a–c Moment diagrams for frame member and equivalent simply supported beam by superposition of loading systems and corresponding moment diagrams.

177

Example 5.5

Use the concept of superposition of loading systems and corresponding moment diagrams to obtain the final moment diagram for member B–C–D of Example 5.2.

STEP 1   For the member, separate the applied distributed loading and end moments into the sum of several single loading systems.

STEP 2   For each loading system, draw the moment diagram that the loading system would create if it acted on a simply supported beam. Compute the moment at the center of the member for each loading system for reference.

STEP 3   Draw the final moment diagram for the member by superimposing the individual loading moment diagrams.

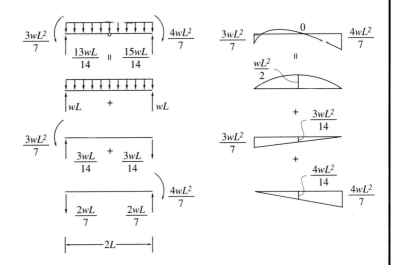

Member B–C–D from Example 5.2

diagrams for a frame member subjected to the same loadings. The form of moment diagrams for simply supported beams under a single particular loading is easily recognized. End moments have triangular diagrams, uniform loads have parabolic diagrams with a maximum midpoint ordinate of $wL^2/8$, and concentrated loads produce triangular diagrams with a maximum ordinate at the point of application of the load. This technique is used extensively throughout the book.

5.6   Summary and Limitations

Frames are distinguished from trusses in that frame members have different geometric proportions and are rigidly attached at joints. The members of frames carry axial forces as well as shear forces and moments. The six member end actions shown in Fig. 5.2 are related by the three equations of equilibrium for the member, so that there are only three unknown end force actions associated with each member of a frame.

The design and selection of the members of a frame are largely dependent on the bending moments in the member. An important part of frame analysis is the presentation of the moment variation in all members of the frame. The moments are plotted using the centroidal axes of the members of the frame as reference axes. The direction or sense of the internal member moments is established by using the convention that the moments are plotted on the compression side of the member. This convention is consistent with the sign convention established and used for beams in Chapter 4.

Like trusses, statically determinate frames must be geometrically stable before they can be analyzed. Once the frame is determined to be stable, either the method of joints or the method of sections can be used to obtain member shears, moments, and axial forces. The method of joints emphasizes the requirements of equilibrium at joints for all member end forces and moments and provides a means of making a check of the consistency of the results of any analysis. The method of sections is better suited for analysis by hand computations.

The drawing of the moment diagrams for members of frames is simplified if the end moments and applied loading (if any) are imagined to be applied to a simply supported beam. The moment diagrams for the simply supported beam under the action of each end moment or applied loading acting alone can be superimposed to obtain the final moment diagram for the member.

There are important limitations and assumptions that accompany these frame analyses and that must be considered when using the results for design. The limitations are a result of the assumptions that have gone into the creation of the mathematical model that simulates the behavior of the frame under loading.

- The members of the frame are connected together rigidly at joints in such a way that the centroidal axes of all members meet at a single point.
- When subjected to applied loads the deformations of the structure are so small the structure can be considered to be rigid and the undeformed geometry can be used in writing all equilibrium equations.
- The individual members of the structure can sustain without material failure the axial loads, shear forces, and moments that are obtained in the analysis.
- Individual members of the structure subjected to axial compression do not become unstable or buckle.
- Individual members of the structure subjected to applied loads and bending do not become unstable.
- The structure as a whole or a local group of members of the structure do not become unstable under the action of the applied loads.

Some of the limitations above can be removed with significantly more sophisticated mathematical models of frame analysis and behavior. However, most computer programs used for analysis and design of both statically determinate and indeterminate frames do not have a level of sophistication that overcomes these limitations. The computer results and often even the documentation of the computer programs do not give an indication of what limitations should accompany the analysis that has been performed. The structural engineer is the one who bears the responsibility of recognizing the effect of all of these limitations on the final design of the structure.

There are some questions about the analysis of plane frames that are worthy of thoughtful consideration. Some of these questions will be answered in future chapters, but others are raised for the purpose of establishing a perspective for judging frame behavior.

- What is the effect of not having the centroidal axes of all members at a joint meet at a single point?
- The details of construction of a frame create a joint of finite size. What effect does this have on the internal forces and moments in a frame?

- How large can the deformations of a frame be before it is necessary to consider its change in geometry in the equations of equilibrium?

- Concentrated loads are never applied as "point" loads on a frame member. What is the effect on the shear and moment variation in a member of the load being applied as a distributed load over a short distance?

- What should be the form of the distributed load representing a "concentrated" load?

PROBLEMS

5.1   Draw the complete moment diagram for the frame shown. Note the hinge at C.

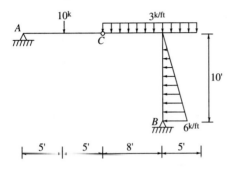

5.2   Draw the complete moment diagram for the frame shown.

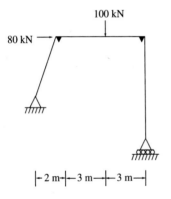

5.3   Members A–B, B–C, and A–C are rigidly connected together at A, B, and C, although joints A and B are externally supported as shown. There is a hinge in the center of each member at D, E, and F. Draw the complete moment diagram for each member of the structure.

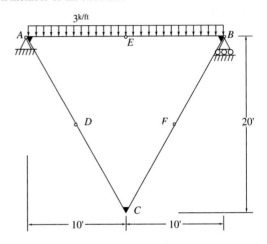

5.4   Draw the complete moment diagram for the frame shown.

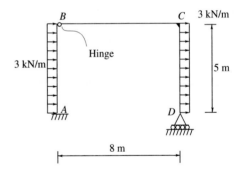

5.5   Draw the complete moment diagram for the frame shown.

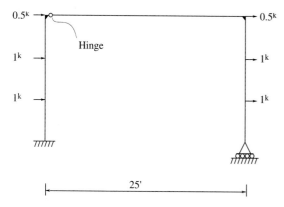

5.6   Draw the complete diagram for the frame loaded as shown.

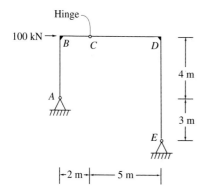

5.7   Draw the shear and moment diagrams for the frame shown.

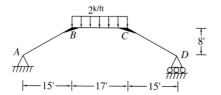

5.8   Draw the final moment diagram for the plane frame loaded as shown. Note the hinge at 3.

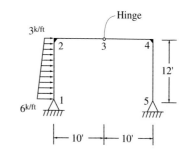

5.9   The plane frame loaded as shown has hinges at *C* and *E*. Draw the final moment diagram for the structure.

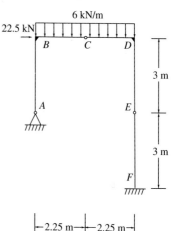

5.10   For the frame shown, draw the complete moment diagram.

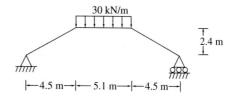

5.11   For the frame shown, draw the complete moment diagram.

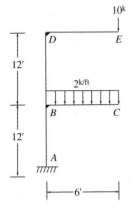

5.12   Obtain the final moment diagram for all members of the frame shown. Note the hinges at 3 and 6.

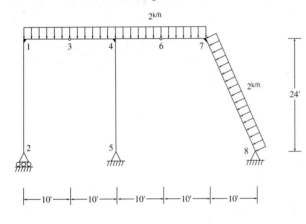

5.13   Draw the complete moment diagram for the final frame loaded as shown. Note that the hinge at C is in the girder and to the left of the joint.

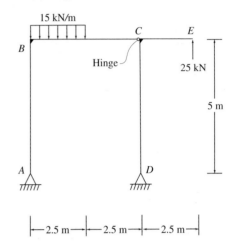

5.14   For the frame loaded as shown, obtain the final moment diagram and axial forces in the members.

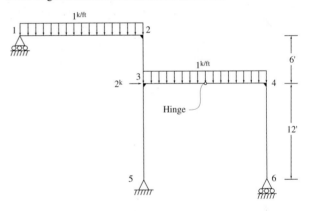

5.15 Draw the complete moment diagram for the frame shown.

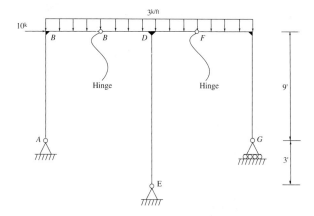

5.17 For the plane frame shown, draw the final moment diagrams for all members. Note the hinges at 2, 4, 5, and 7 and the double hinge at 6.

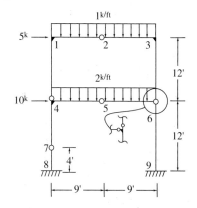

5.16 For the frame shown, draw the moment diagram. Note the hinges in the center of C–D and B–E and that hinges at B and E are in the column just above the joint.

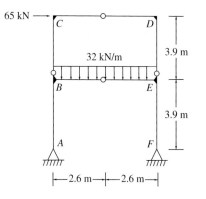

5.18 Obtain the final moment diagram for the frame loaded as shown. Note the hinges at B and D.

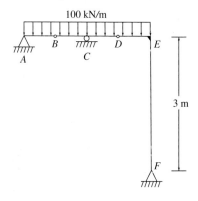

COMPUTER PROBLEMS

For the problems that follow, obtain the graphical results requested, using one of the following methods:

(a) Create a spreadsheet and use the results to obtain the graphs.

(b) Write a small computer program that obtains the results requested, and create a file that can be plotted to obtain the graphs.

(c) Use available software programs that will do the analysis requested. Run the program as many times as required to obtain the results requested and create a file that can be plotted to obtain the graphs.

C5.1 Obtain a plot of M_{max}/wh^2 versus x/b, where M_{max} is the maximum moment anywhere in the frame. Note the hinge at 3 and vary x/b over the range $0 \le x/b \le 1$.

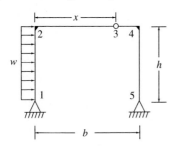

C5.2 Obtain a plot of M_{max}/wh^2 versus α, where M_{max} is the maximum moment anywhere in the frame. Note the hinge at 3 and vary α over the range $0 \le \alpha \le 1$.

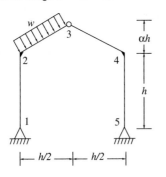

C5.3 Obtain a plot of of M_{max}/wb^2 versus x/b, where M_{max} is the maximum moment anywhere in the frame. Note the hinge at 3 and vary x/b over the range $0 \le x/b \le 1$ for values of h/b of 1, 2, and 3.

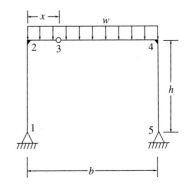

C5.4 Obtain a plot of the horizontal reaction component (acting to the left) at 7 versus a/L, for $0 \le a/L \le 1$. Note the hinges at 3 and 6.

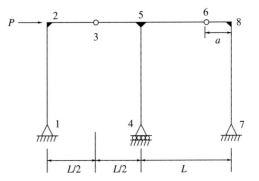

C5.5   The hinges at 2 and 6 are both located the same distance *a* from the base. Create plots of M_5/PL and M_7/PL versus *a/L*, where M_5 is the moment at joint 5 and M_7 the moment at joint 7. Vary *a/L* over the range $0 \leq a/L < 1$.

C5.6   Obtain a plot of M_{max}/wh^2 versus *a/b*, where M_{max} is the maximum moment anywhere in the frame. Note the hinge at 3 and vary *a/b* over the range $0 \leq a/b \leq 1$ for a value of *b/h* of 1.5.

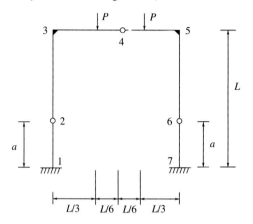

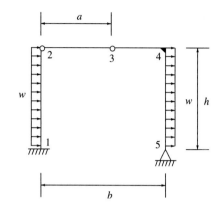

6

Simple Bending Theory and Deformation Analysis of Beams

In Chapter 4 the internal distribution of moments, shears, and axial forces for arbitrary loading of individual members was presented. In Chapters 3 and 5 the analysis of statically determinate trusses and frames for internal member forces, shears, and moments was studied. The force actions in structures are an important design consideration since the safety of the structure is related directly to the capability of structural materials to resist the stresses due to applied loads.

The design of structures is not limited to considerations of force action alone. The deformation of structures is also important in design and commonly will control the selection or configuration of its members. Beam deformations may have to be limited for aesthetic or functional reasons. The drift of buildings, the deflections of bridges, or the vibration of structures which is dependent on member deformations are important design criteria. The analysis of statically indeterminate structures, addressed in Chapter 10, requires an analysis of deformation behavior. This chapter is limited to the analysis of deformation of individual prismatic members. The calculation of displacements and rotations of multiple-member structures is taken up in Chapter 7 using techniques more suited to the analysis of general structural deformation.

The deformations of beams is introduced in the study of mechanics of materials or strength of materials. The *simple bending theory* for beams relates the internal strains in beams due to bending deformations or curvatures to the moment action that causes these deformations. The curvatures are further related to the transverse or vertical displacements of the beams by means of a differential equation. These concepts are reviewed in some detail in the first two sections of this chapter, so that the assumptions and limitations underlying simple bending theory can be listed and discussed. The implications of the assumptions are also presented as further explicit limitations that are not always stated or perhaps are presented without explanation in analyses of beam deformation behavior.

The computation of the deflections and displaced shapes of beams through applications of the differential equation of bending is then presented for prismatic members. Direct integration as well as techniques that permit the computation of displacements or rotations at specific points in beams are developed and discussed. Finally, the axial deformation of beams is presented together with techniques for obtaining the variation of displacements along a member or at specific points in a member.

6.1 Relationship Between Bending and Curvature

Consider a member under the action of applied loads and bending moments and the deformed shape that it takes, as shown in Fig. 6.1a. The deflected shape of this member is exaggerated and is represented by the displaced reference axis, which is also called the *elastic curve*. For linear elastic materials and special cases of nonlinear elastic materials, the reference axis is the familiar centroidal axis of the member. The deflection of the reference axis of the beam occurs in the x–y plane because of limiting assumptions and is defined mathematically by the function $v(x)$, where $v(x)$ is positive in the positive y-coordinate direction. All geometric consideration of deformation behavior is assumed to be adequately described by actions that occur in that plane.

In Fig. 6.1b a segment at point A of the deformed reference axis of the member is shown. The distance O–A is the radius, R, of a circle tangent to the curve at A and having the same curvature as the curve, $v(x)$, at that point. At point B, a distance Δs along the curve $v(x)$, a line perpendicular to the tangent to the curve at B intersects O–A at point O', and makes the small angle $\Delta\theta$ which is also the angle between the tangents at A and B. For a small angle $\Delta\theta$, the arc length Δs is very nearly given by the expression $\Delta s \approx (O'A)\,\Delta\theta$, where $O'A$ is the length of the line O'–A. If point B is taken closer and closer to A, or as Δs goes to zero, the ratio $\Delta\theta/\Delta s$ becomes, in the limit,

$$\lim_{\Delta s \to 0} \frac{\Delta\theta}{\Delta s} = \frac{d\theta}{ds} = \lim_{\Delta s \to 0} \frac{1}{O'A} = \frac{1}{O'A} = \frac{1}{R} = \phi \qquad (6.1)$$

Equation (6.1) relates the curvature ϕ $(-1/R)$ of the deflected reference axis of the member to the rate of change of the angle that tangents to the curve make with the x axis. This is purely a geometric relation and is not dependent on the material properties of the beam. It does assume that the angle θ of the curve varies continuously along the member.

The relation between the displacement function $v(x)$ and the slope of the tangent to $v(x)$ at a typical point A is shown in Fig. 6.1c. From the geometry, the tangent at point A is

$$\frac{dv}{dx} = \tan\theta \tag{6.2}$$

Differentiating both sides of Eq. (6.2) with respect to s and using the chain rule gives

$$\frac{d^2v}{dx^2}\frac{dx}{ds} = \sec^2\theta\,\frac{d\theta}{ds} \tag{6.3}$$

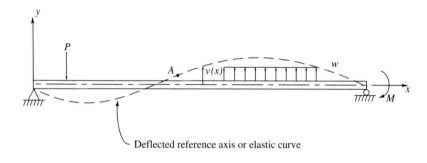

Deflected reference axis or elastic curve

(a) Loaded member showing deflected shape

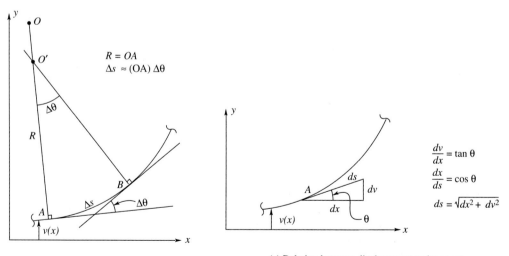

$$\frac{dv}{dx} = \tan\theta$$

$$\frac{dx}{ds} = \cos\theta$$

$$ds = \sqrt{dx^2 + dv^2}$$

(b) Segment of deformed reference axis

(c) Relation between displacement and tangent to deformed reference axis

Figure 6.1a–c   Geometric relations for deformed beam.

191

Solving Eq. (6.3) for the curvature, $d\theta/ds$, substituting into Eq. (6.1), and recognizing the geometric relation between dv, dx, and ds shown in Fig. 6.1c yields

$$\frac{d\theta}{ds} = \frac{1}{R} = \phi = \frac{d^2v}{dx^2}\frac{dx}{ds}\frac{1}{\sec^2\theta} = \frac{d^2v}{dx^2}\frac{dx}{ds}\cos^2\theta = \frac{d^2v}{dx^2}\left(\frac{dx}{ds}\right)^3$$

Substituting for dx/ds from Fig. 6.1c and factoring dx out of the square root in the denominator yields

$$\frac{d^2v}{dx^2}\left(\frac{dx}{\sqrt{dx^2 + dv^2}}\right)^3 = \frac{d^2v/dx^2}{[1 + (dv/dx)^2]^{3/2}} = \frac{1}{R} = \phi \qquad (6.4)$$

Both Eqs. (6.1) and (6.4) are based on the geometry of deformation of the beam and not on its material properties. It simply relates the displacement function $v(x)$ to the curvature $\phi\ (= 1/R)$. Equation (6.4) is a nonlinear differential equation that is difficult to solve except in very special cases. The beams used in structures nearly always undergo very small displacements, which means that the slope, dv/dx, of the tangents to the deflected shape of these beams is also very small. Consequently, the term (dv/dx^2) in the denominator of Eq. (6.4) is essentially zero and can be dropped in comparison to one. This reduces the equation to the more tractable form

$$\frac{d^2v}{dx^2} \approx \frac{1}{R} = \phi \qquad (6.5)$$

Equation (6.5) can be solved for $v(x)$ once the variation of the curvature $\phi\ (= 1/R)$ is known.

The curvature ϕ in Eq. (6.5) can be related to the deformation of the material of the beam with some further geometric considerations. In Fig. 6.2 a segment of beam that is deformed through the action of a pure moment is shown along with the strains that take place in the x–y plane. The undeformed segment of length $\Delta s\ (= \Delta x)$ is shown in Fig. 6.2b with the line segment AB at the reference axis and a second line segment A_1B_1 of the same length a distance y above the x axis. The distances from AB to the top and bottom of the cross section are y_t and y_b, respectively, and the depth of the cross section in the x–y plane is $h = y_t - y_b$.

In Fig. 6.2c the deformed member segment in the x–y plane is shown and, similar to Fig. 6.1b, the distance from point A on the undeformed reference axis to the center of curvature OA is R. Tests of beams made with materials that have either linear or nonlinear stress–strain relations and analytical studies of slender beams subjected to pure moment show that the variation of the axial strain or strain normal to the plane of the cross section

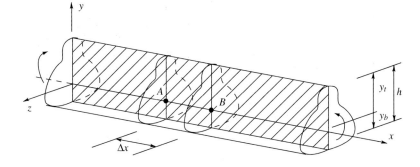

(a) Member showing trace of x–y plane

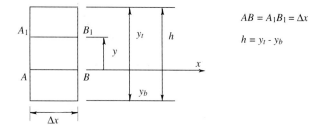

$AB = A_1B_1 = \Delta x$

$h = y_t - y_b$

(b) Member segment in x–y plane before deformation

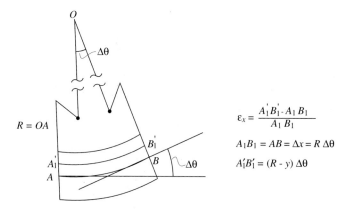

$$\varepsilon_x = \frac{A_1'B_1' - A_1 B_1}{A_1 B_1}$$

$$A_1B_1 = AB = \Delta x = R\,\Delta\theta$$

$$A_1'B_1' = (R - y)\,\Delta\theta$$

(c) Member segment in x–y plane after deformation

Figure 6.2a–c Deformation in the x–y plane of beam segment due to pure bending.

of the member can be assumed to be linear through the depth (i.e., it varies linearly in the y-coordinate direction). This gives rise to the assumption that "plane sections remain plane," or the vertical straight lines at the two sides of the cross section before deformation in Fig. 6.2b remain straight and perpendicular to the x axis after deformation in Fig. 6.2c. J. Bernoulli was the first to propose this type of deformation, and bending action based on this assumption is sometimes referred to as the *Bernoulli theory of bending*.

Along some line in the x–y plane of the beam, which for linear materials is the centroidal axis of the member, no deformation occurs due to the action of the pure moment as is shown for the line A–B in Fig. 6.2c. Lines such as A_1B_1 above the centroidal axis are shortened to the length $A_1'B_1'$ and those below are lengthened. The strain a distance y above the centroidal axis is, by definition,

$$\varepsilon_x = \frac{A_1'B_1' - A_1B_1}{A_1B_1}$$

From Fig. 6.2c it can be seen that for the small angle $\Delta\theta$, $AB = R\,\Delta\theta$ and $A_1'B_1' = (R - y)\,\Delta\theta$. Recalling that $A_1B_1 = AB$, the strain at any point a distance y from the centroidal axis becomes

$$\varepsilon_x = \frac{(R - y)\,\Delta\theta - R\,\Delta\theta}{R\,\Delta\theta} = \frac{-y}{R}$$

or

$$\frac{-\varepsilon_x}{y} = \frac{1}{R} = \phi \tag{6.6}$$

which is an expression for the curvature ϕ $(= 1/R)$. The strains at the top and bottom of the cross section ε_t and ε_b, respectively, are given by

$$\varepsilon_t = \frac{-y_t}{R}$$

$$\varepsilon_b = \frac{-y_b}{R}$$

Subtracting these two expressions and rearranging yields

$$\varepsilon_t - \varepsilon_b = \frac{-y_t}{R} - \frac{-y_b}{R} = \frac{-(y_t - y_b)}{R} = \frac{-h}{R}$$

or

$$\frac{1}{R} = \phi = -\frac{\varepsilon_t - \varepsilon_b}{h} = \frac{-\Delta\varepsilon_{tb}}{h} \tag{6.7}$$

which is an expression for the curvature in terms of the depth of the section and the difference, $\Delta\varepsilon_{tb}$, of the strains at the top and bottom. Combining Eqs.

(6.5) and (6.7) yields the relation between the displacement function, $v(x)$, and deformations or curvature of a beam under the action of pure moment

$$\frac{d^2v}{dx^2} = \frac{1}{R} = \phi = \frac{-\Delta\varepsilon_{tb}}{h} \tag{6.8}$$

Equation (6.8) is strictly a geometric relation based on the assumption of plane sections remaining plane during deformation. It is valid for both linear and nonlinear materials. The most common application of the equation is for linear materials. In linear materials the strain, ε, is defined as $\varepsilon = \sigma/E$, in which E is the modulus of elasticity of the material, and the stress is given, from Table 1.2 or any text on the mechanics of materials, as $\sigma = -My/I$ for a moment M. Recalling that $h = y_t - y_b$, Eq. (6.7) becomes

$$\phi = -\frac{\varepsilon_t - \varepsilon_b}{h} = -\frac{1}{h}\left(\frac{-My_t}{EI} - \frac{-My_b}{EI}\right) = \frac{1}{h}\left(\frac{M}{EI}\right)(y_t - y_b) = \frac{M}{EI}$$

Substituting into Eq. (6.8) gives the moment–curvature relation for linear materials

$$\frac{d^2v}{dx^2} = \phi = \frac{M}{EI} \tag{6.9}$$

This equation is used extensively throughout the remainder of the book.

6.2   Assumptions and Limitations of Simple Bending Theory ____

Several explicit assumptions have been made in establishing Eq. (6.9), which is the basic mathematical model used to simulate the deformation behavior of beams made from linear and elastic materials. It is known as the moment–curvature relation and is the basis of the simple bending theory. The equation also is based on and limited by several implied assumptions.

The entire development of the geometric relations embedded in Eq. (6.9) is based on deformation action of the member that is defined only in the x–y plane and that all displacements are assumed to be small. It is also assumed that the material deformation behavior can be characterized by a linear stress–strain relation. To create mathematical models of structural behavior that are valid for analysis, the structural engineer must recognize and understand these and a number of additional implied assumptions and potential limitations.

Strains and deformations are small, so that geometry changes are negligible and slopes are very small. This assumption permitted dropping dv/dx in the denominator of the curvature relation in Eq. (6.4). For nearly all practical situations the slopes in a deformed beam will not exceed 1/10, or approximately 6°. For this slope the error in neglecting it in Eq. (6.4) is slightly less than 1.5%.

Plane sections before bending remain plane after bending. This assumption is based on observations of material behavior and is exactly satisfied for linear materials under the action of a constant moment. When constant shear is present with moment, the deformed cross section is no longer plane, but the normal strains still vary linearly through the depth. When variable shear acts with moment, the normal strains no longer vary linearly with depth, but the assumption of linear distribution of strains with depth has only a small error if the depth/span ratio of the beam is small (on the order of one-fifth or less), which it is in nearly all practical situations. See Reference 40 for more exact solutions of bending in the presence of variable shear.

The material is homogeneous and has a linear stress–strain relation. It is this assumption that enables the curvature for linear materials to be expressed as M/EI. The assumption also ensures that the point or axis of zero strain in the cross section will be at the *centroid*. For nonhomogeneous beams such as those encountered in reinforced concrete, the zero strain or neutral axis will not coincide with the centroid. For nonlinear materials the neutral axis will still coincide with the centroidal axis when the cross section is symmetric with respect to both the y and z axes and if the stress–strain relation itself is symmetric in tension and compression. The curvature in this case will have to be obtained through use of Eq. (6.7) and the nonlinear stress–strain relation.

The modulus of elasticity of the material is the same in both tension and compression. This assumption is nearly always satisfied, or the difference in the moduli is not large, so that the error in ignoring it is negligible. The effect of different moduli is to shift the zero strain or neutral axis away from the centroidal axis of the cross section and to modify the relation of curvature to moment. In cases where this effect is not small, a reevaluation of the equilibrium conditions for the cross section with the stress distribution arising from the assumed linear strain distribution is required to locate the neutral axis and establish the relation between curvature and moment.

The cross section is symmetric with respect to the y axis. It is assumed in the mathematical model that the deflected centroidal axis remains in the x–y plane and that no displacement of the axis can occur in the z direction. This implies that strains at corresponding points on opposite sides of the y axis in the cross section must be equal, that the correspondence must be complete, and hence that the cross section must be symmetric. This assumption also ensures that the stress distribution is symmetric, that the y axis is a principal axis, and that the shear center will lie on the y axis. Loads will pass through the shear center and bending without twisting will occur. This limitation can be relaxed without significant error if the y axis is a principal axis, but then the possibility of bending and twisting of the member can occur as well as deflection of the member out of the plane (x–y) of loading. See References 9 and 42 for a more complete presentation.

Strains and hence stresses are constant across the width (in the z-axis direction) of the cross section. This assumption is made when the distribution of stress in the cross section is established and equilibrium used to locate the neutral (centroidal) axis. For solid narrow cross sections the assumption is very good. However, cross sections which have portions that are very wide in relation to other portions do not satisfy this assumption.

Figure 6.3 A beam with a cross section shaped like a "T," known as a *T beam*, under the action of a moment has a very wide flange. In Fig. 6.3b the distribution of the strain, and hence the stress, in the flanges of the member as assumed in simple bending theory and as actually occurs is shown. For very wide flanges the error can be substantial. The introduction of an *effective width* of flange (Reference 40), less than the true width, is used to correct for this error in design applications. A similar situation is shown in Fig. 6.3c for a wide box beam where the nonuniform strain distribution is referred to as the *shear lag* effect (Reference 40). Again modifications of the simple bending theory must be introduced to correct the mathematical model.

The beam is initially straight. Equation (6.9) is valid for initially straight beams only, but it can be extended to beams that are initially curved. If the ratio of the depth of a slender beam to its undeformed radius of curvature is sufficiently small, the error in using Eq. (6.9) and simple bending theory is acceptably small. For a ratio of depth to radius of curvature of $1/5$, the error in the maximum stress calculated using simple bending theory is about 7 and decreases rapidly as the ratio decreases. Curved beams are discussed in more detail in References 9 and 40.

The cross section does not vary along the member x axis. The member being prismatic is implied in the simple bending theory. However, it can be shown that if the member cross section varies in a continuous and gradual manner, Eq. (6.9) can be used with no significant error in the results (Reference 40). The theory does not apply for sudden changes of cross section.

No axial load is present. The development of Eq. (6.9) is based on the action of bending moments only. For nonlinear materials the presence of axial force requires a modification of the equation because curvature is

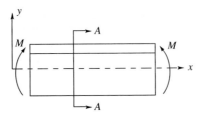
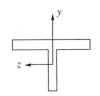

Section *A–A*

(a) T beam under constant moment

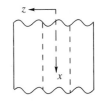

Assumed

Actual

(b) Strain distribution in flanges of T section

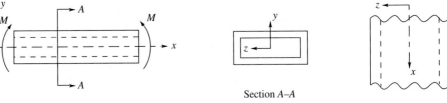

Section *A–A*

Strain

(c) Wide box beam showing nonuniform strain distribution

Figure 6.3a–c   Examples of nonuniform strain distributions in very wide beams.

affected by the magnitude of the axial force. The interaction of axial force and bending for linear materials can be neglected and Eq. (6.9) used for bending as long as axial forces, displacements, and slopes remain small.

No instability of the beam occurs under the action of bending.   In the development of the moment–curvature relation the deformations are all assumed to occur in the x–y plane. The mathematical model does not include displacements of points in the cross section or along the member in the z direction or rotation of the cross section about the x axis. As a consequence, the theory cannot recognize lateral, torsional buckling behavior of the beam. The provision of adequate bracing of the compression region of the member to prevent this instability is required if simple bending theory is to be used (References 6, 8, 15, and 39). This type of instability can be a serious problem for cross sections having thin-walled elements, such as in a W section.

No local instability of elements of the cross section occurs.   The integrity of the cross section is assumed to be maintained under the bending action. If the cross section has thin or slender elements such as the flanges of a W section in a compression region, there is danger of instability or local buckling of one or more of those elements. Under these circumstances the deflections and stresses predicted by simple bending theory are seriously in error. Limits on width/thickness ratios for compression elements in such cross sections are set in design codes and specifications to ensure the performance of the cross section as assumed in simple bending theory (References 3, 6, 8, 15, and 39).

The cross section of the beam does not change shape during bending.   The cross section is assumed to be solid in the development of Eq. (6.9) and there is no significant change in its shape during bending action. When the cross section is made of thin-walled elements such as members used in long spans, the cross section may change shape during bending.

Figure 6.4 The cross sections are made from thin-walled elements which may change shape when subjected to the action of bending. A simple example that is easy to illustrate is the circular cross-sectional shape of a soda straw, shown in Fig. 6.4a, which, under the action of a moment, will change from circular to the elliptical shape of Fig. 6.4b. The change of cross-sectional shape is one that can produce significant increases in stress above that predicted by simple bending theory. The use of diaphragms or internal bracing at spaced intervals along the member axis is necessary to maintain the integrity of the cross-sectional shape (References 6, 27, and 31).

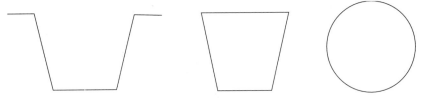

(a) Possible forms of some thin-walled cross sections

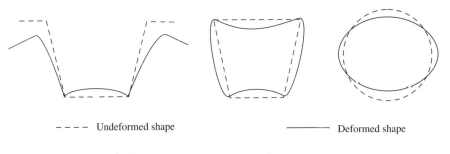

- - - - Undeformed shape ————— Deformed shape

(b) Potential deformed cross section shapes due to bending

Figure 6.4a–b Potential deformations of unconstrained thin-walled cross sections under an applied moment.

The deflections calculated from Eq. (6.9) do not include deformations due to shear. This assumption is related to the "plane sections remain plane" assumption, in that additional deformation takes place when the latter assumption is not met due to shear deformation of the cross section. For the usual geometries of cross sections and span lengths, additional deflection due to shear, which is directly proportional to the square of the ratio of the depth of the beam to its length, $(d/L)^2$, is small. For a d/L ratio of 1/8, the additional deflection due to shear is approximately 5%. If simple bending theory is used with composite members with cross sections constructed from metallic materials at the top and bottom with a different core material having a significantly lower shear modulus, shear deflections can become the dominant part of the total deflection, regardless of the depth/span ratio, and must not be neglected (References 9 and 40).

As the discussion above illustrates, there are many implied assumptions and limitations to the simple bending theory. The computation of deflections through the use of Eq. (6.9) and the assumption of the linear distribution of normal or axial strains and stresses in the cross section are not valid unless all of the conditions listed above are satisfied. Design codes and specifications have limitations that aid the structural engineer in maintaining the applicability of the simple bending theory in analysis (see References 1 to 3). Prefabricated or manufactured beams having a standard cross section also generally satisfy simple bending theory assumptions. Of course, any or all of these assumptions and limitations can be removed with more advanced treatments that create more sophisticated mathematical models of structural behavior. However, these modifications to the theory will remain beyond the scope of this book.

6.3 Determination of Beam Displacements by Direct Integration

Once the variation of the curvature, or for linear materials the moment, along the axis of a member is known, the displacements of the centroidal axis of the member can be obtained from Eq. (6.9). Direct integration with the introduction of the appropriate support restraint conditions will yield the function $v(x)$, which describes the shape of the deformed member.

Example 6.1 Example 6.1 shows the calculation of the displaced shape of a uniformly loaded cantilever beam. The double integration of the curvature function derived from the moment divided by EI is straightforward and yields a fourth-order polynomial with two undetermined constants. These two constants are evaluated from the constraints placed on the displacement function $v(x)$ by the fixed support at the left end (at $x = 0$). For this problem, direct integration is fast and gives the complete displaced shape of the member. For numerical computations using U.S. units, take $w = 1$ kip/ft, $L = 15$ ft, $E = 29{,}000$ ksi, and $I = 500$ in^4; for SI units, take $w = 15$ kN/m, $L = 5$ m, $E = 200$ GPa, and $I = 200 \times 10^6$ mm^4.

Example 6.2 Example 6.2 is considerably more complicated, but it emphasizes the concept of continuity of the displacement function $v(x)$. The application of Eq. (6.9) in this case must be for two regions of the beam because of the piecewise linear character of the curvature function, which comes directly from the moment diagram. Separate displacement functions must be obtained for regions to the left and right of the concentrated load. The double integration in these two regions yields four constants. The two supports at A and C provide a zero-displacement restraint, which eliminates two of the unknown constants.

The remaining two constants are evaluated from the conditions of continuity of the displacement functions at their common point B at $x = L/3$. There can be no sudden jump in displacement of the beam or no sudden

Example 6.1

Obtain an expression for the displaced shape $v(x)$ for the beam. Take EI to be constant.

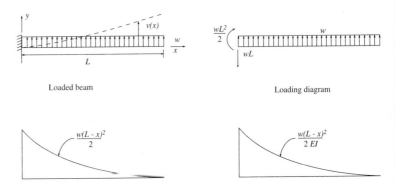

Loaded beam

Loading diagram

Moment diagram

Curvature diagram

STEP 1 Compute the reactions and draw the loading and then the moment diagram.

STEP 2 Obtain the curvature diagram by dividing the moment diagram by EI.

STEP 3 Use Eq. (6.9) and integrate twice to obtain the displaced centroidal axis.

$$\frac{d^2v}{dx^2} = \phi = \frac{M}{EI} = \frac{w}{2EI}(L - x)^2$$

$$\frac{dv}{dx} = -\frac{w}{6EI}(L - x)^3 + A$$

$$v(x) = \frac{w}{24EI}(L - x)^4 + Ax + B \tag{1}$$

STEP 4 Evaluate the integration constants by imposing the zero displacement and zero-slope restraint at $x = 0$.

$$v(0) = 0 = \frac{wL^4}{24EI} + B \rightarrow B = -\frac{wL^4}{24EI} \tag{2}$$

$$\frac{dy}{dv}(0) = 0 = -\frac{wL^3}{6EI} + A \rightarrow A = \frac{wL^3}{6EI} \tag{3}$$

Substituting (2) and (3) into (1) gives the final $v(x)$ expression.

$$v(x) = \frac{wL^4}{24EI}\left[\left(1 - \frac{x}{L}\right)^4 + \left(4\frac{x}{L} - 1\right)\right]$$

change of slope or "kink" of the beam at point B. Thus there must be continuity of displacement and slope at point B, as shown in step 4. The resulting two displacement expressions provide the complete deflected shape of the beam. At the applied load point, B, both expressions yield the displacement $v(L/3) = -4PL^3/243EI$. The negative sign in the displacement indicates that the beam displaces downward or in the negative y-coordinate direction. Would you expect the slope at B to be positive or negative? For numerical computation in U.S. units, take $L = 20$ ft, $P = 20$ kips, $E = 3000$ ksi, and $I = 1500$ in⁴; for SI units, take $L = 7$ m, $P = 100$ kN, $E = 20$ GPa, and $I = 800 \times 10^6$ mm⁴.

Examples of 6.1 and 6.2 show how the complete displaced shape of a member can be obtained from direct application of Eq. (6.9). In most practical applications of displacement analysis the complete displacement function is not required, only the displacement or rotation at a single point. There are other computational techniques that are more efficient for finding a single displacement or rotation quantity. Nevertheless, the ability to obtain the fully deflected shape of a member is a valuable tool of analysis and is used from time to time for special problems.

Direct integration of Eq. (6.9) becomes awkward when the curvature function has several piecewise continuous segments. Modifications to the technique that generally overcome these problems are covered in nearly all strength of materials textbooks (see, e.g., Reference 7).

6.4   Curvature-Area or Moment-Area Theorem _____

In Sections 6.1 and 6.2 the presentation of simple bending theory established that the curvature of the deflected centroidal axis of a beam subjected to bending can be approximated by the second derivative of the function, $v(x)$, which defines that deflected shape. From considerations of geometry, the curvature of a member subjected to an arbitrary bending action is ϕ, as defined in Eq. (6.5). When the material of the member is linear and elastic, ϕ is simply M/EI, where M is the moment, E the modulus of elasticity of the material of the member, and I the moment of inertia of the member cross section. This yields a simple bending expression relating moments to the deflection function $v(x)$ in Eq. (6.9). For nonlinear materials, Eq. (6.8) gives the relation between v and curvature with the curvature being evaluated from Eq. (6.7) and the stress–strain relation.

As shown in Fig. 6.1 and repeated in Fig. 6.5, the assumed sign convention is that the deflection v is positive in the positive y-coordinate direction, M is positive when it causes compression in the top (or upper) portion of the beam cross section, and the curvature is positive when the deflected

Example 6.2

Obtain an expression for the displaced shape $v(x)$ for the prismatic beam in which EI is constant.

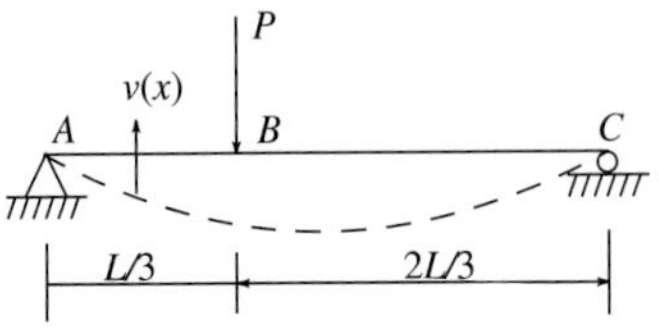

Loaded beam

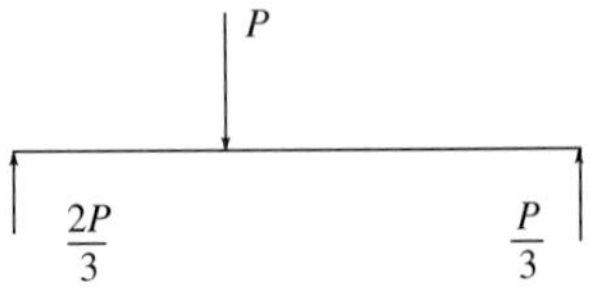

Loading diagram

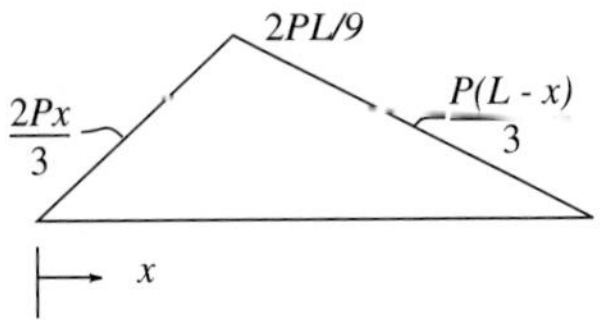

Moment diagram

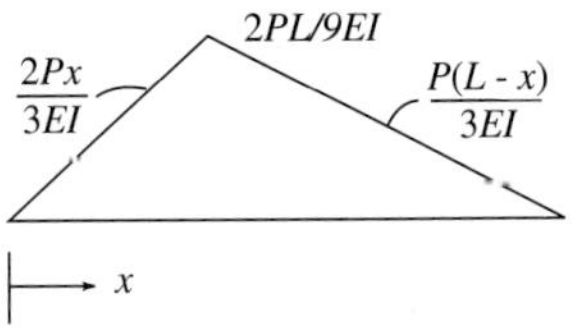

Curvature diagram

STEP 1   Compute the reactions; draw the loading and moment diagrams.

STEP 2   Divide the moment diagram by EI to obtain the curvature diagram.

STEP 3   Use Eq. (6.9) separately in region A–B ($0 \leq x \leq L/3$) and region B–C ($L/3 \leq x \leq L$); integrate twice to obtain $v(x)$'s.

$0 \leq x \leq L/3$	$L/3 \leq x \leq L$
$\dfrac{d^2v}{dx^2} = \dfrac{M}{EI} = \dfrac{2Px}{3EI}$	$\dfrac{d^2v}{dx^2} = \dfrac{M}{EI} = \dfrac{P(L-x)}{3EI}$
$\dfrac{dv}{dx} = \dfrac{Px^2}{3EI} + A$	$\dfrac{dv}{dx} = \dfrac{-P(L-x)^3}{6EI} + C$
$v(x) = \dfrac{Px^3}{9EI} + Ax + B$	$v(x) = \dfrac{P(L-x)^3}{18EI} - C(L-x) + D$

204

Example 6.2 (continued)

STEP 4 Evaluate constants of integration from conditions of zero displacement at beam ends and continuity at $x = L/3$.

$$v(0) = 0 \rightarrow B = 0 \qquad\qquad v(L) = 0 \rightarrow D = 0$$

Continuity of displacement at $x = L/3$:

$$v\left(\frac{L}{3}\right) = \frac{PL^3}{243EI} + A\frac{L}{3} = v\left(\frac{L}{3}\right) = \frac{4PL^3}{243EI} - \frac{2CL}{3} \qquad (1)$$

Continuity of slope at $x = L/3$:

$$\frac{dv}{dx}\left(\frac{L}{3}\right) = \frac{PL^3}{27EI} + A = \frac{dv}{dx}\left(\frac{L}{3}\right) = -\frac{2PL^2}{27EI} + C \qquad (2)$$

Solving (1) and (2) gives $A = -\dfrac{5PL^2}{81EI}$ and $C = \dfrac{4PL^2}{81EI}$

The final displacements are

$0 \leq x \leq L/3$

$$v = \frac{PL^2x}{81EI}\left(9\frac{x^2}{L^2} - 5\right)$$

$L/3 \leq x \leq L$

$$v = \frac{PL^2(L - x)}{162EI}\left[9\left(1 - \frac{x}{L}\right)^2 - 8\right]$$

shape is convex downward. Why is the moment variation in Fig. 6.5b convex downward? [Hint: See Eq. (4.3).] Note that for linear materials the x axis is the undeformed centroidal axis of the beam cross section and that the deflected centroidal axis of the member is called the elastic curve. For nonlinear materials the x axis is simply the reference axis of zero strain in the cross section.

If both sides of Eq. (6.8) or (6.9) are integrated between two points, say a and b, the left side integrates to the first derivative, which is the slope of a tangent to the elastic curve of the member. This slope is also the rotation, θ, with respect to the horizontal at a point on the elastic curve of the member. Thus, integrating both sides of Eq. (6.8) or (6.9) yields

$$\left.\frac{dv}{dx}\right|_b - \left.\frac{dv}{dx}\right|_a = \int_a^b \phi\, dx = \int_a^b \frac{M}{EI}\, dx$$

or

$$\theta_b - \theta_a = \Delta\theta_{ab} = \int_a^b \phi\, dx = \int_a^b \frac{M}{EI}\, dx \qquad (6.10)$$

In words, this can be stated as

$$(\text{slope at } b) - (\text{slope at } a) = \frac{\text{area under curvature function}}{\text{between } a \text{ and } b}$$

As can be seen in Fig. 6.5c, the shaded area under the curvature function or diagram of the curvature function between the two points $x = a$ and $x = b$ is, by Eq. (6.10), the change in slope between a and b. The integration in Eq. (6.10) assumes that there is no sudden change in the slope, dv/dx, between a and b (i.e., dv/dx is continuous between a and b). Equation (6.10) is called the *first curvature-area theorem*, or, more commonly, the *first moment-area theorem*:

If the slope of the elastic curve varies continuously between two points, the area under the curvature function (or M/EI diagram) between these two points is equal to the change in slope between tangents to the elastic curve at the two points.

The calculation of relative displacements between two points in a deformed beam can be obtained from further consideration of the curvature function. The tangent to the elastic curve at $x = a$ has a vertical deviation from the elastic curve at $x = b$ defined as d_{ba}, as shown in Fig. 6.5e. Using the assumption of very small deflections and slopes, this deviation can be calculated from simple geometric relations. As seen in Fig. 6.5e, at a

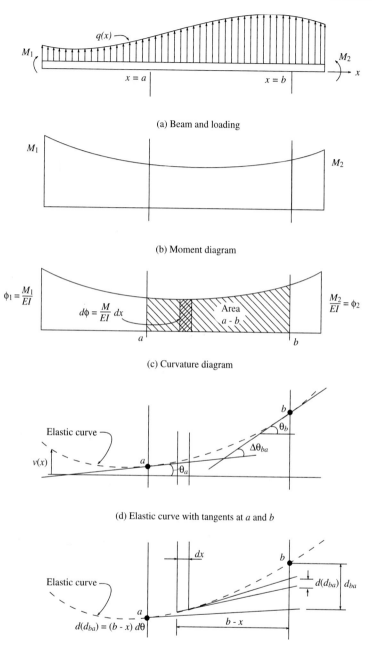

(a) Beam and loading

(b) Moment diagram

(c) Curvature diagram

(d) Elastic curve with tangents at *a* and *b*

(e) Deviation of tangent at *a* from point *b* on elastic curve

Figure 6.5a–e   Deformation and geometric relations for curvature area (or moment-area) theorems.

207

distance of $(b-x)$ from point b, two tangents to the elastic curve, an infinitesimal distance apart, dx, make a contribution to the vertical deviation, d_{ba}, of

$$d(d_{ba}) = (b - x)\, d\theta \qquad (6.11)$$

where $d\theta$ is the change in slope between the tangents. From the first curvature-area (first moment-area) theorem, $d\theta$ is simply the area under the curvature function, which for the infinitesimal distance dx is simply $\phi\, dx = (M/EI)\, dx$. If all of the infinitesimal vertical increments $d(d_{ba})$ are summed, the result is the vertical distance d_{ba}. Thus integrating both sides of Eq. (6.11) yields

$$d_{ba} = \int_a^b (b - x)\, d\theta = \int_a^b (b - x)\, dx = \int_a^b (b - x) \frac{M}{EI}\, dx \qquad (6.12)$$

As illustrated in Fig. 6.5e, the left-hand side of Eq. (6.12) is the vertical deviation of a tangent to the elastic curve at point a from point b on the elastic curve. The right-hand side of Eq. (6.12) represents the sum between a and b of the areas under the curvature function $\phi \cdot dx = (M/EI)\, dx$ times the distance to those areas from point b, $(b-x)$, which, when stated in words, is the moment of the area under the curvature function between a and b about point b. Again, the slope dv/dx to the elastic curve must vary continuously between two points. These observations lead directly to the second curvature-area (second moment-area) theorem:

> If the slope to the elastic curve varies continuously between two points a and b, the vertical deviation at point b on the elastic curve of a tangent to the elastic curve at point a is equal to the moment of the area under the curvature function (or M/EI diagram) between the two points about point b.

The two curvature-area theorems enable the computation of relative deformation behavior of beams subjected to bending. Since curvature functions frequently have a variation that can be represented in the form of a rectangle, triangle, or parabola, the computation of area and moment of area can be accomplished using simple geometric concepts. The areas and locations of centroids of a few common geometric shapes are shown in Fig. 6.6.

Example 6.3 The curvature-area theorems are most useful in computing the deformations of cantilever beams. Example 6.3 shows the computation of the slope and deflection of the end of a cantilever beam. For this beam, the slope or rotation of the right end is obtained by using the first curvature-area theorem, Eq. (6.10). The slope at the right end is

STEP 1   Obtai

STEP 2   Divid
figure

STEP 3   Obtai
vatur

STEP 4   Obtai
of the

Th
diagra
pler ii
the su
show
from
areas,
A_i, an

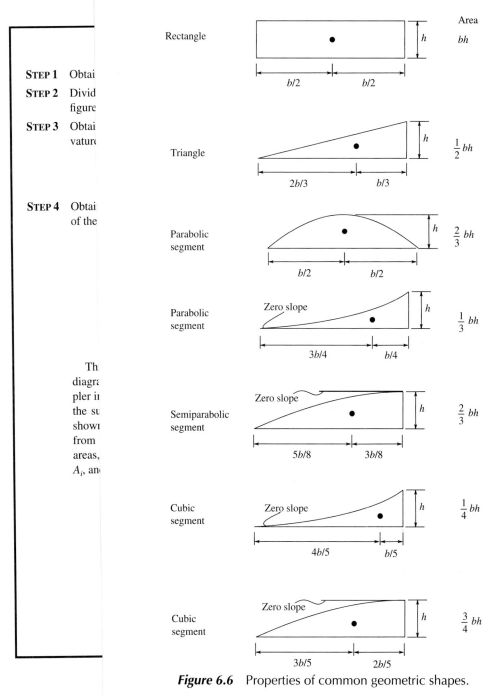

Figure 6.6   Properties of common geometric shapes.

Exan

Example 6.3 (continued)

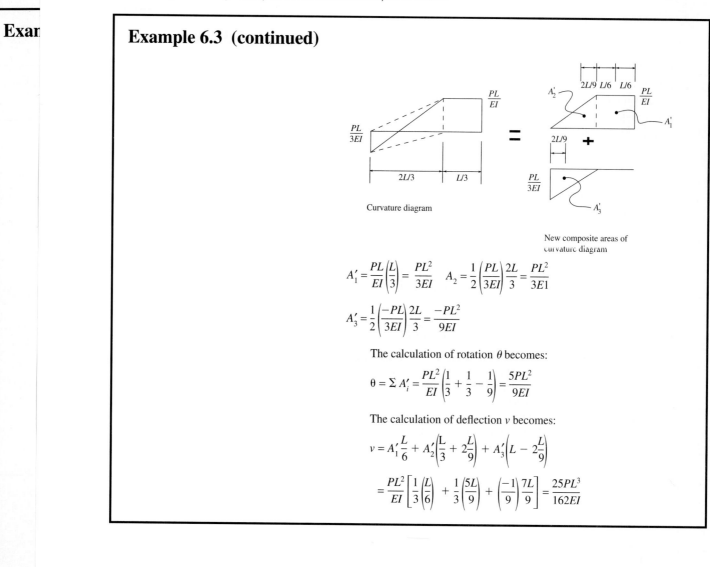

Curvature diagram

New composite areas of
curvature diagram

$$A'_1 = \frac{PL}{EI}\left(\frac{L}{3}\right) = \frac{PL^2}{3EI} \quad A_2 = \frac{1}{2}\left(\frac{PL}{3EI}\right)\frac{2L}{3} = \frac{PL^2}{3EI}$$

$$A'_3 = \frac{1}{2}\left(\frac{-PL}{3EI}\right)\frac{2L}{3} = \frac{-PL^2}{9EI}$$

The calculation of rotation θ becomes:

$$\theta = \Sigma\, A'_i = \frac{PL^2}{EI}\left(\frac{1}{3} + \frac{1}{3} - \frac{1}{9}\right) = \frac{5PL^2}{9EI}$$

The calculation of deflection v becomes:

$$v = A'_1\frac{L}{6} + A'_2\left(\frac{L}{3} + 2\frac{L}{9}\right) + A'_3\left(L - 2\frac{L}{9}\right)$$

$$= \frac{PL^2}{EI}\left[\frac{1}{3}\left(\frac{L}{6}\right) + \frac{1}{3}\left(\frac{5L}{9}\right) + \left(\frac{-1}{9}\right)\frac{7L}{9}\right] = \frac{25PL^3}{162EI}$$

simply the total area of the curvature diagram since the tangent to the elastic curve at the left support is horizontal. Note how simple the computation becomes if the curvature diagram is divided into areas that are rectangles and triangles as shown in the example. The calculated slope is positive, indicating that the right-end rotation is counterclockwise, or the tangent to the elastic curve at the end slopes upward to the right.

The computation of the deflection of the right end of the beam is obtained by using the second curvature-area theorem, Eq. (6.12). The deflection simply becomes the vertical deviation from the tangent to the beam at the fixed end because the tangent and vertical displacement at that point are both zero. The moment of the area under the curvature diagram about the right end gives the desired deflection directly. Note how the use of the simple areas with their known centroids makes the computation straightforward. The positive result indicates that the displacement of the beam is upward.

An alternative computation is also presented in step 4. The leftmost portion of the curvature diagram can be reconfigured so that it is the sum of two triangular diagrams in which the geometry of the areas and location of the centroids can be obtained without having to find the point where the curvature function changes from positive to negative. This calculation using the primed areas gives the same results as those obtained previously. For numerical computations in U.S. units, take $L = 12$ ft, $P = 10$ kips, $E = 29,000$ ksi, and $I = 600$ in^4; for SI units, take $L = 4$ m, $P = 50$ kN, $E = 200$ GPa, and $I = 250 \times 10^6$ mm^4.

Although integration can always be used to obtain areas or moments of areas, the example illustrates that breaking the curvature diagrams into simple geometric figures enables a simpler and less error-prone computation. It is possible to use other geometric configurations of the curvature diagram. Verify that the results are the same if a rectangular diagram of magnitude PL/EI over the full length of the beam is combined with a triangular diagram over the left $2L/3$ of the beam with an ordinate at the left end of $-4PL/3EI$.

Example 6.4 Example 6.4 shows the application of the curvature-area theorems to the deformation analysis of the simple beam in Example 6.2. The vertical displacements of the ends of the beam are known to be zero, but the point in the beam where the slope of a tangent to the elastic curve is zero is unknown. Because of this, additional geometric considerations are required to obtain the desired results.

The geometry of the deformed beam in an exaggerated form is sketched in the example. As developed in step 3, the vertical deflection, v_B, can be obtained from simple geometric considerations with the assumption that the displacements and slope of the tangents to the deformed beam are very small. This permits the tangent of θ_A, which is

213

Example 6.4

Compute the deflection under the concentrated load of the beam of Example 6.2 using the curvature-area theorems. Take EI constant.

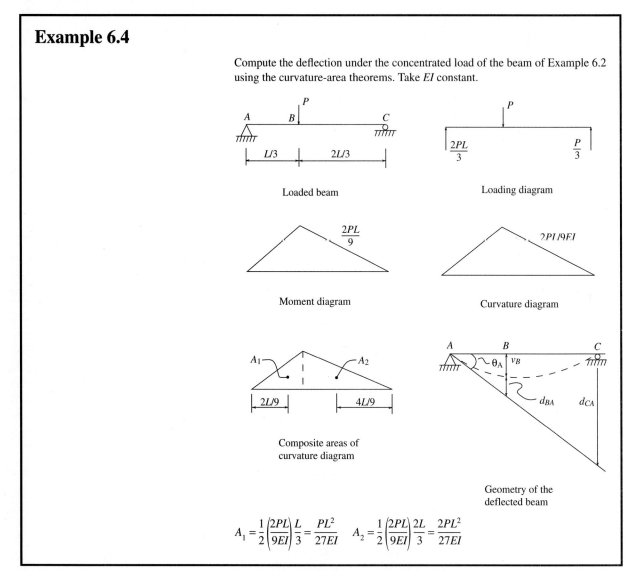

Loaded beam

Loading diagram

Moment diagram

Curvature diagram

Composite areas of
curvature diagram

Geometry of the
deflected beam

$$A_1 = \frac{1}{2}\left(\frac{2PL}{9EI}\right)\frac{L}{3} = \frac{PL^2}{27EI} \qquad A_2 = \frac{1}{2}\left(\frac{2PL}{9EI}\right)\frac{2L}{3} = \frac{2PL^2}{27EI}$$

Example 6.4 (continued)

STEP 1   Draw a curvature diagram from the loading and moment diagram.

STEP 2   Divide the curvature diagram into a composite of two triangles.

STEP 3   Establish equations for the geometry of the deflected beam.

$$\theta_A \frac{L}{3} = d_{BA} + v_B \quad \therefore v_B = \theta_A \frac{L}{3} - d_{BA}$$

$$\theta_A L = d_{CA} \quad \therefore \theta_A = \frac{d_{CA}}{L}$$

STEP 4   Compute d_{CA} and d_{BA} using Eq. (6.12), then compute θ_A and v_B.

$$d_{CA} = A_1\left(L - \frac{2L}{9}\right) + A_2 \frac{4L}{9} = \frac{PL^2}{27EI}\left[\frac{7L}{9} + 2\left(\frac{4L}{9}\right)\right] = \frac{5PL^3}{81EI}$$

$$d_{BA} = A_1\left(\frac{L}{3} - \frac{2L}{9}\right) = \frac{PL^2}{27EI}\left(\frac{L}{9}\right) = \frac{PL^3}{243EI}$$

$$\theta_A = d_{CA} = \frac{1}{L}\left(\frac{5PL^3}{81EI}\right) = \frac{5PL^2}{81EI} \quad \therefore v_B = \frac{5PL^2}{81EI}\left(\frac{L}{3}\right) - \frac{PL^3}{243EI} = \frac{4PL^3}{243EI}$$

This is the same result as was obtained in Example 6.2.

equal to d_{CA}/L, to be taken as simply θ_A and the vertical distance $(v_B + d_{BA})$ to be computed as $\theta_A(L/3)$.

The slope or rotation, θ_A, of the right end of the beam is first established through application of the second curvature-area theorem to compute d_{CA}. Again using the second curvature-area theorem, Eq. (6.12), the vertical deviation, d_{BA}, of the tangent to the elastic curve at A from the elastic curve at B provides the final quantity needed to compete the computation of v_B. As illustrated in the example, the result has the same magnitude as that obtained in Example 6.2, but with a positive sign. The reason for this is that the geometric considerations established in step 3 of the current example envisioned the displacement v_B as being positive downward. For numerical computations, see the values in Example 6.2.

While the procedure for the computation of the vertical deflection of the beam at the applied load in Example 6.4 using the curvature-area theorems is straightforward and emphasizes the physical behavior of the loaded beam, it becomes increasingly tedious if the loading or the geometry of the beam is more complicated. In addition, if there is a break in slope of the elastic curve in a beam due to the presence of a hinge, the curvature-area theorems can only be applied on each side of the hinge, which further complicates their use for these problems. Extensions of the curvature-area theorems are available which eliminate the geometric considerations and make applications to more complicated problems more straightforward. The conjugate beam method and the elastic load method are examples of such extensions and are presented in Section 6.5.

It is possible to use the curvature-area theorems to perform a bending deformation analysis of frames. The computation of displacements and rotations of specific points in frames requires careful geometric analyses of the effects of individual member deformations and is generally awkward to carry out with a direct application of theorems. There are extensions known as the conjugate frame method and the column analogy method which simplify these applications (see References 22, 32, and 43), but they are somewhat specialized and are difficult to generalize to more complicated structures. Usually, a more flexible technique, such as the virtual work approach, which is presented in the next chapter, is easier to use.

Finally, it should be noted that the curvature-area theorems can provide only relative bending deformations. In structures where axial deformations, support movements, temperature changes, or shrinkage can occur, a special treatment of these effects has to be added to the analysis. In these cases there are alternative deformation analysis techniques which are much easier to use. However, the curvature-area theorems do show, in a straightforward manner, the relation between bending moments and deformations in simple structures which aids the understanding and visualization of structural behavior. In Chapter 9 these concepts are extended to the computation of deformations of nonprismatic members and members that have nonlinear stress–strain (or moment–curvature) relations.

6.5   Elastic Load and Conjugate Beam Analysis _____

Computation of beam deformations by the curvature-area theorems is simple for cantilever beams and only slightly more complicated for simply supported beams. Their use has the advantage of emphasizing the geometry of deformation and the assumption of small displacements and rotations. For more complicated problems of beam deformations an alternative approach based on the parallel form of the mathematical model for the computation of slopes and deflections of a beam and the mathematical model for computation of shears and moments in a beam organizes these computations in a familiar and easy-to-use format.

The direct integration of Eq. (6.9) to obtain the slope dv/dx or θ and the displacement v provides one means of computing beam deformations. Turning the process around gives the relations that the derivative of the displacement is the slope and the derivative of the slope is the curvature.

The form of the differential equation (4.3) relating the internal moment in a beam to the applied loading in Section 4.2 is identical to that of Eq. (6.9). Equation (4.2) indicates that the derivative of the moment is the shear and Eq. (4.1) that the derivative of the shear is the applied load. The similarity between the mathematical model presented in Eqs. (6.9) and (4.3) suggests that there can be a correspondence between displacements and slopes of a real beam and moments and shears in some properly constructed beam. The key element in establishing this correspondence is that the curvature on the right-hand side of Eq. (6.9) and the load on the right-hand side of Eq. (4.3) must be identical.

The variation of the curvature or the curvature diagram in a beam, called the *real beam*, deformed by some applied loading is taken to be a load on a second beam, which will be called a *conjugate beam*. The curvature diagram taken as a load on the conjugate beam is called an *elastic load* to reflect the fact that this "load" represents the curvature that defines the elastic curve or deflected shape of the real beam. The beam loaded with the elastic load is called the conjugate beam because *conjugate* means united or tied, and this beam is united with or tied to the deformation behavior of the real beam.

The principle behind the conjugate beam method is summarized in Fig. 6.7.

Figure 6.7 The relations among load, shear, moment, and curvature in the real beam are shown in Fig. 6.7a, while the same relations among elastic load $\phi(x)$, shear V_c, and moment M_c in the conjugate beam are shown in Fig. 6.7b. Figure 6.7c and d illustrate the similarity between

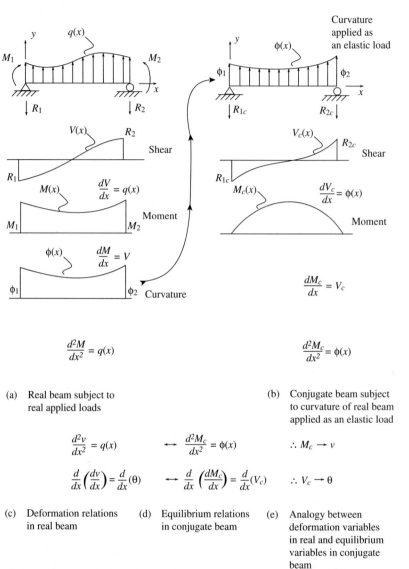

$$\frac{d^2M}{dx^2} = q(x)$$

$$\frac{d^2M_c}{dx^2} = \phi(x)$$

(a) Real beam subject to
real applied loads

(b) Conjugate beam subject
to curvature of real beam
applied as an elastic load

$$\frac{d^2v}{dx^2} = q(x) \qquad \longleftrightarrow \qquad \frac{d^2M_c}{dx^2} = \phi(x) \qquad \therefore M_c \to v$$

$$\frac{d}{dx}\left(\frac{dv}{dx}\right) = \frac{d}{dx}(\theta) \qquad \longleftrightarrow \qquad \frac{d}{dx}\left(\frac{dM_c}{dx}\right) = \frac{d}{dx}(V_c) \qquad \therefore V_c \to \theta$$

(c) Deformation relations
in real beam

(d) Equilibrium relations
in conjugate beam

(e) Analogy between
deformation variables
in real and equilibrium
variables in conjugate
beam

Figure 6.7a–e   Correspondence or analogy between deformation behavior of a real beam and internal equilibrium in the corresponding conjugate beam.

the differential equations that define real beam deformation and conjugate beam internal actions. Finally in Fig. 6.7e the correspondence or analogy between the slope, θ (or dv/dx), and displacement, v, of the real beam and the shear, V_c, and moment, M_c, in the conjugate beam is established.

The computation of slopes and deflections of a beam subject to applied loads can be done by analyzing a conjugate beam having an applied elastic load that varies as the curvature of the real beam. The correspondence in behavior is that shear and moment at any point in the conjugate beam is equal to the slope and vertical displacement of the elastic curve of the real beam at the same point. The computation of slopes and deflections of a beam is reduced to obtaining the variation of the shear and moment in a conjugate beam.

The principal ideas in the conjugate beam method can be summarized as follows:

The deformations of any real beam can be computed by applying the curvature variation in that beam as an elastic load on a conjugate beam. The shear and moment at any point in the conjugate beam due to the elastic load is identically equal to the slope and vertical deflection of the corresponding point in the real beam.

The assumptions and limitations on the use of the conjugate beam method are exactly the same as those associated with the simple bending theory discussed in Section 6.2.

The first step in the conjugate beam method is to establish a conjugate beam that reflects the deformation behavior of the real beam as it is constrained by support or construction characteristics.

Figure 6.8 The relation between a variety of support and construction configurations in real beams and the corresponding conjugate beams is shown. The left side of Fig. 6.8 shows the displacement and rotation conditions for the left, middle, and right points in the real beam. The corresponding conditions on shear and moment in the conjugate beam is shown on the right side of Fig. 6.8. Note that when a hinge appears in a real beam, the slope of the elastic curve is not continuous (i.e., a sudden change, $\Delta\theta$, in slope occurs at the hinge). The corresponding condition in the conjugate beam must be a sudden change in shear which is accomplished with the placement of a support at that point.

The exact interpretation of the correspondence of shear and moment in the conjugate beam with slope and deflection in the real beam is facilitated with the use of a familiar sign convention.

Figure 6.9 The convention presented on the left side of Fig. 6.9 reaffirms the direction of positive slope and deflection in the real beam that was used in developing Eqs. (6.8) and (6.9). On the right side of Fig. 6.9 the sign convention shown for positive elastic loading, shear, and moment in the conjugate beam is identical to the one already

Real beam			Conjugate beam		
Left end support	Middle continuous	Right end support	Left end support	Middle continuous	Right end support

$v_L = 0$	v	$v_R = 0$	$M_C = 0 \rightarrow v_L$	$M_C \rightarrow v$	$M_C = 0 \rightarrow v_R$
θ_L	θ	θ_R	$V_C = R_L \rightarrow \theta_L$	$V_C \rightarrow \theta$	$V_C = R_R \rightarrow \theta_R$
Fixed support	Continuous	Free	Free	Continuous	Fixed support

$v_L = 0$	v	v_R	$M_C = 0 \rightarrow v$	$M_C \rightarrow v$	$M_C = M_R \rightarrow v_R$
$\theta_L = 0$	θ	θ_R	$V_C = 0 \rightarrow \theta$	$V_C \rightarrow \theta$	$V_C = R_R \rightarrow \theta_R$
Fixed support	Hinge	Support	Free	Support	Support

$v_L = 0$	v	$v_R = 0$	$M_C = 0 \rightarrow v_L$	$M_C \rightarrow v$	$M_C = 0 \rightarrow v_R$
$\theta_L = 0$	$\Delta\theta$	θ_R	$V_C = 0 \rightarrow \theta_L$	$\Delta V_C = R \rightarrow \Delta\theta$	$V_C = R_R \rightarrow \theta_R$
Support	Support	Free	Support	Hinge	Fixed support

$v_L = 0$	v	v_R	$M_C = 0 \rightarrow v_L$	$M_C \rightarrow v$	$M_C = M_R \rightarrow v_R$
θ_L	θ	θ_R	$V_C = R_R \rightarrow \theta_L$	$V_C \rightarrow \theta$	$V_C = R_R \rightarrow \theta_R$
Guided Support	Continuous	Support	Guided support	Continuous	Support

v_L	v	$v_R = 0$	$M_C = M_L \rightarrow v_L$	$M_C \rightarrow v$	$M_C = 0 \rightarrow v_R$
$\theta_L = 0$	θ	θ_R	$V_C = 0 \rightarrow \theta_L$	$V_C \rightarrow \theta$	$V_C = R_R \rightarrow \theta_R$

Figure 6.8 Relation between support and constructed configuration of real beam and corresponding conjugate beam.

established in Fig. 4.1 for loading, shear, and moment for real beams. Thus Fig. 6.9 illustrates that positive moment and shear in the conjugate beam correspond exactly to positive displacement and slope in the real beam.

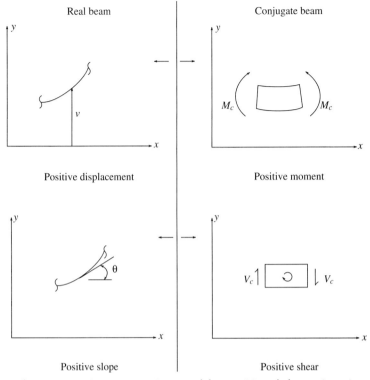

Figure 6.9 Sign convention used for positive deformations in real beam and corresponding positive internal force actions in conjugate beam.

Example 6.5 The application of the conjugate beam method to the analysis of the deformation behavior of the beam considered in Example 6.4 is illustrated. In Example 6.4 the vertical deflection under the applied load was computed using the curvature-area theorems. In step 1 a conjugate beam is created to reflect the deformation behavior of the corresponding real beam. In this case the conjugate beam is simply supported. Once the curvature diagram has been obtained for the real beam, it is applied as an elastic load on the conjugate beam. Because the curvature is all positive, it is applied as an upward triangularly varying load. The reaction of the left end of the conjugate beam, R_{CA}, is computed from equilibrium in step 2. Note that the value of the reaction is also equal to the shear in the conjugate beam at the left end and, consequently, is the slope to the elastic curve of the real beam, θ_A, which also was obtained in Example 6.4.

The computation of the deflection and the slope to the elastic curve under the load simply requires the computation of the shear and moment

Example 6.5

Use the conjugate beam method to compute the deflection and slope to the elastic curve under the load in the beam of Example 6.4.

Real beam with loading · Conjugate beam with curvature load

Curvature diagram · Loading diagram

STEP 1   Take the curvature diagram from Example 6.4 and use it as the elastic loading on the conjugate beam. Divide the loading diagram for the conjugate beam into the two triangular loads, compute the statically equivalent concentrated loads, and apply them at the centroids.

STEP 2   Compute the reaction, R_{cA}, in the conjugate beam.

$$\circlearrowright \Sigma M_C \colon \frac{2PL^2}{27EI} \frac{4L}{9} + \frac{PL^2}{27EI} \left(L - \frac{2L}{9} \right) - R_{cA} L = 0 \quad \therefore R_{cA} = \frac{5PL^2}{81EI}$$

STEP 3   Use a free-body diagram of the left portion of the conjugate beam and obtain the internal moment, M_{cB}, and internal shear, V_{cB}, which correspond, respectively, to the deflection of and the slope of the tangent to the elastic curve at B.

$$\circlearrowright \Sigma M_B \colon R_{cA} \frac{L}{3} - \frac{PL^2}{27EI} \left(\frac{L}{3} - \frac{2L}{9} \right) + M_{cB} = 0 \quad \therefore M_{cB} = -\frac{4PL^3}{243EI} \rightarrow v_B$$

$$\Sigma F \uparrow \colon \frac{PL^2}{27EI} - V_{cB} - R_{cA} = 0 \quad \therefore V_{cB} = \frac{PL^2}{27EI} - \frac{5PL^2}{81EI} \rightarrow \theta_B$$

in the conjugate beam at that point. The free-body diagram in the example shows how this computation is made. Note that both M_{CB} and V_{CB} are negative, indicating that the deflection of the beam is downward and the slope of the tangent to the elastic curve is downward to the right. How could the maximum deflection of this beam be obtained? (*Hint*: The slope in the real beam is ʑero at its point of maximum deflection.) For numerical computation, see the values listed at the end of Example 6.4

Example 6.6 A second and more interesting application of the conjugate beam method is presented. The construction of the conjugate beam in this example shows that its left end is free in order to obtain an internal zero moment and shear which reflects the zero deflection and slope condition at the left end of the real beam. A support must be introduced at the hinge, point B, to reflect the sudden change in slope to the elastic curve that occurs there in the real beam.

The analysis for the curvature diagram, which becomes the elastic load on the conjugate beam, is started with the computation of the reaction in the real beam at C by using equation of condition at the hinge. Both reactions of the conjugate beam are computed in step 2, although only the reaction R_{cB} is required as part of the solution process. The moment at B in the conjugate beam is negative, indicating the downward deflection at that point. The change in slope at the hinge is positive, indicating that the slope becomes more positive as one passes from the left to the right side of the hinge. For numerical computations using U.S. units, take $w = 2$ kips/ft, $L = 10$ ft, $E = 29{,}000$ ksi, and $I = 200$ in^4; for SI units, take $w = 30$ kN/m, $L = 3$ m, $E = 200$ GPa, and $I = 80 \times 10^6$ mm^4.

6.6   Axial Deformations of Beams

Beams or single members subjected to axial loads undergo axial deformations which can be computed from the axial strains. Simple prismatic truss or frame members frequently have constant, or essentially constant, internal axial forces and hence strains. Other members can have applied axial loads distributed along the length of the member, such as the member's intrinsic weight or applied axial loads on the surface of friction piles. The computation of the axial deformations can be done in a very straightforward manner using simple concepts.

In Fig. 6.10a to c, a simply supported beam is subjected to an axial load, P, at the right end and a distributed axial load, $p(x)$, along its length. The loading diagram and the axial force diagram are obtained by the processes described in Section 4.9. The axial deformations or displacements of the

Example 6.6

For the beam shown below, compute the vertical deflection at the hinge and the change in slope in the elastic curve from the left to the right side of the hinge. Take EI as constant.

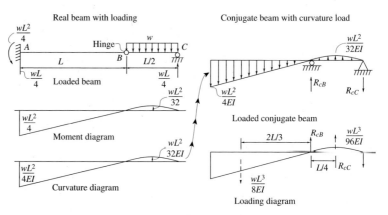

STEP 1 Compute the reactions of the real beam, then obtain the moment and curvature diagrams. In computing the reactions, the reaction at C is determined first by using a free body of the portion B–C and summing moments about B.

STEP 2 Use the curvature diagram as the elastic load on the conjugate beam. Obtain the loading diagram and compute the reactions of the conjugate beam.

$$\overset{\curvearrowright}{\Sigma M_B}: \ \frac{wL^3}{8EI}\,2\!\left(\frac{L}{3}\right) + \frac{wL^3}{96EI}\,\frac{L}{4} - R_{cC}\,\frac{L}{2} = 0 \quad \therefore R_{cC} = \frac{33wL^3}{192EI}$$

$$\Sigma F\!\uparrow: \ -\frac{wL^3}{8EI} + R_{cB} + \frac{wL^3}{96EI} - \frac{33wL^3}{192EI} = 0 \quad \therefore R_{cB} = \frac{55wL^3}{192EI}$$

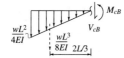

STEP 3 Take a free body of portion A–B of the conjugate beam and sum moments to obtain M_{cB}, which corresponds to the deflection v_B. The change in slope at the hinge is simply the reaction, R_{cB}, on the conjugate beam.

$$\overset{\curvearrowright}{\Sigma M_{\text{cut}}}: M_{cB} + \frac{wL^3}{8EI}\cdot 2\!\left(\frac{L}{3}\right) = 0 \rightarrow M_{cB} = -\frac{wL^4}{12EI}$$

$$\therefore v_B = -\frac{wL^4}{12EI} \qquad R_{cB} = \frac{55wL^3}{192EI} = \Delta\theta_B$$

beam are related to the axial strains by the well-known strain–displacement relation for small strains (Reference 4)

$$\frac{du}{dx} = \varepsilon \tag{6.13}$$

in which $u(x)$ is the displacement function defining axial displacements of the member at any point x along its length. The strains can be computed for linear elastic materials as σ/E, where σ is the stress and E the modulus of elasticity of the material of the member.

The variation with x of the strains in the member is shown in Fig. 6.10d and is simply obtained for linear materials as the division of the forces in the axial force diagram by AE. The displacements, $u(x)$, are obtained by integration of the function that describes the variation of strain with x and the imposition of the condition that the displacement, u, at the left end (where $x = 0$) is zero. The variation of $u(x)$ is shown in Fig. 6.10e.

The calculation of the displacement at any point, b, along the member relative to a second point, a, can be obtained from the integration of both sides of Eq. (6.13) between the points $x = a$ and $x = b$. The result is

$$u_b - u_a = \int_a^b \varepsilon \, dx = \frac{\text{area under strain diagram}}{\text{between } a \text{ and } b} \tag{6.14}$$

For linear, elastic, and prismatic members this result can be simplified still further to obtain

$$u_b - u_a = \frac{1}{AE} \int_a^b F \, dx = \frac{1}{AE} \quad \frac{\text{area under axial force}}{\text{diagram between } a \text{ and } b} \tag{6.15}$$

These concepts can be summarized as follows:

> The change in length or relative displacement between two points in an axially loaded member is equal to the area under the diagram of strain variation between those two points. If the material is linear elastic and the member prismatic, the change in length or relative displacements between the same two points is equal to the area under the diagram of axial force variation between those two points divided by AE.

The use of these observations makes the computation of the axial displacements along a member or its change of length a simple exercise of the same difficulty as finding the change in shear or moment in a beam by calculating the area under the loading or the shear diagram, respectively, as is done using Eqs. (4.4) and (4.5). The use of these concepts is illustrated in Examples 6.7 and 6.8.

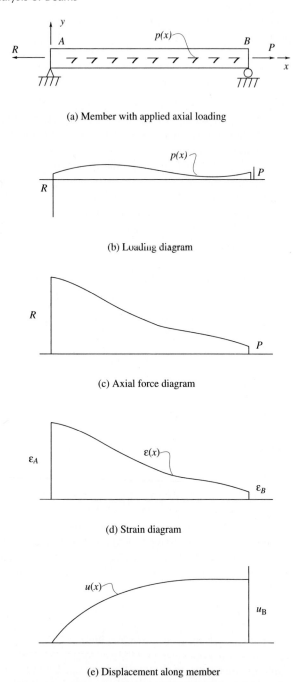

(a) Member with applied axial loading

(b) Loading diagram

(c) Axial force diagram

(d) Strain diagram

(e) Displacement along member

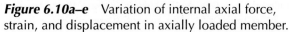

Figure 6.10a–e Variation of internal axial force, strain, and displacement in axially loaded member.

Example 6.7 A member fixed against displacement at the left end is subjected to the constant axial force per unit length, p. This might, among other possibilities, represent the situation in which a member stands vertically and is subjected to its own intrinsic weight as a loading. The change in length of the member, which is also the displacement of the right end of the member, is obtained in step 3 as the area under the axial strain diagram over the length of the member. The horizontal displacement of the center of the member is not one-half of the displacement of the right end. How can this be explained qualitatively? For numerical computation with U.S. units, take $L = 30$ ft, $p = 0.5$ kips/ft, $A = 5$ in^2, and $E = 29,000$ ksi; for SI units, take $L = 10$ m, $p = 8$ kN/m, $A = 3,000$ mm^2, and $E = 200$ GPa.

Example 6.8 This member also is fixed against displacement at the left end, but in this case the axial applied load is a force per unit length that increases linearly in magnitude as x increases to the right. This situation is a common representation of the action of a friction pile in which the friction force is assumed to increase in magnitude with depth into the ground. Again the change in length of the member is obtained in step 3 from the area under the strain diagram for the member. If the magnitudes of the applied distributed loads in this example and Example 6.7 were such that the reaction forces, R, in the two examples were equal, which member would have the greater end displacement? (Which member would have the larger area under the strain diagram?) For numerical computations with U.S. units, take $L = 30$ ft, $c = 0.04$ kip/ft^2, $A = 5$ in^2, and $E = 29,000$ ksi; for SI units, take $L = 10$ m, $c = 0.18$ kN/m^2, $A = 3,000$ mm^2, and $E = 200$ GPa.

The use of the area under strain diagrams to obtain displacements or changes in length of me3mbers is facilitated by the use of the properties of various geometric shapes that are presented in Figs. 2.7 and 6.6. Of course, direct integration as defined in Eq. (6.14) or (6.15) can always be used when a completely general situation arises.

6.7   Axial Deformations Due to Curvature

The curved, deflected shape of an initially straight horizontal member will project back on to the horizontal a length that is less than its original, as indicated in Fig. 6.11. This deformation is associated with the curvature of the deflected member and raises the question of how much axial change in length occurs because of it. A few simple geometric considerations can provide an understanding of the nature of this axial change in length.

Example 6.7

Obtain the displacement of the right end of the axially loaded member shown. Take A and E to be constant.

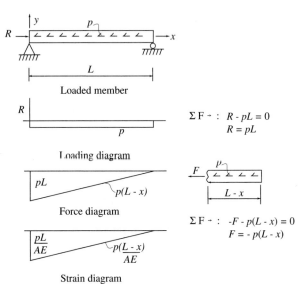

STEP 1 Compute the reaction at A, R, and draw the loading diagram. Use a free body of the right portion of the member to obtain the internal axial force and draw the axial force diagram. This step follows the procedure illustrated in Example 4.7.

STEP 2 Obtain the strain diagram by dividing the force diagram ordinates by AE.

STEP 3 Calculate the right end displacement by using Eq. (6.14) with the strain diagram or Eq. (6.15) and the force diagram.

$$u_B = \text{area under strain diagram} = \frac{1}{2}\left(-\frac{pL}{AE}\right)L = -\frac{pL^2}{2AE}$$

Example 6.8

Obtain the displacement of the right end of the member shown subjected to the linearly varying axial load. Take A and E to be constant.

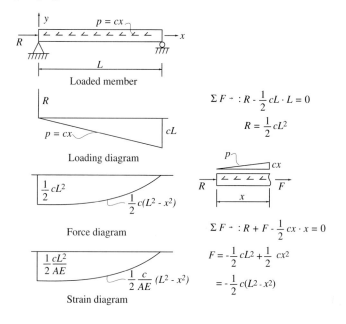

$$\Sigma F \rightarrow : R - \frac{1}{2} cL \cdot L = 0$$

$$R = \frac{1}{2} cL^2$$

Loading diagram

Force diagram

$$\Sigma F \rightarrow : R + F - \frac{1}{2} cx \cdot x = 0$$

$$F = -\frac{1}{2} cL^2 + \frac{1}{2} cx^2$$

$$= -\frac{1}{2} c(L^2 - x^2)$$

Strain diagram

STEP 1 Compute the reaction at A, R, and draw the loading diagram. Use a free body of the right portion of the member to obtain the internal axial force and draw the axial force diagram.

STEP 2 Obtain the strain diagram by dividing the force diagram ordinates by AE.

STEP 3 Calculate the right end displacement by using Eq. (6.14) with the strain diagram or Eq. (6.15) and the force diagram.

$$u_B = \text{area under strain diagram} = \frac{2}{3}\left(-\frac{cL^2}{2AE}\right)L = -\frac{cL^3}{3AE}$$

The deflected shape of a member due to the action of moments depends on the variation of the moments and hence the curvature of the member. A worst-case scenario is shown in Fig. 6.12, in which the moment and curvature, if the member is prismatic, are constant over the entire length of the member. No member can experience a greater axial change in length due to curvature than this because the curvature is maximum over its entire length. This establishes an upper limit on the axial deformation of a member due to curvature caused by transversely applied loads such as that shown in Fig. 6.11.

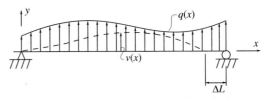

Figure 6.11   Axial shortening of loaded member due to curvature.

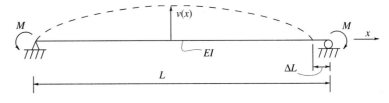

Loaded member

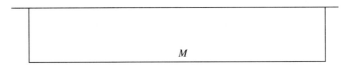

Moment diagram

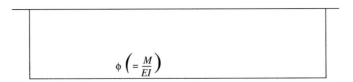

Curvature diagram

Figure 6.12   Axial shortening due to curvature of simple beam under action of a constant moment.

The change in length, ΔL, for any member, such as those shown in Figs. 6.11 and 6.12, is obtained by summing over the member length the difference between the horizontally projected lengths, dx, and the initial or undeformed lengths, ds. The geometry of the deflected member, $v(x)$, at some point along its length is shown in Fig. 6.1c. Referring to Fig. 6.1c, the relation between ds and dx and the slope of the deflected member, dv/dx, is

$$ds = \sqrt{dx^2 + dv^2} = \sqrt{1 + \frac{dv^2}{dx}}\, dx \tag{6.16}$$

The change in length, dL, for a segment of member, ds, is

$$dL = ds - dx = \sqrt{1 + \left(\frac{dv}{dx}\right)^2}\, dx - dx = dx\left[\left(1 + \frac{1}{2}\left(\frac{dv}{dx}\right)^2 - \cdots\right) - 1\right]$$
$$\approx \frac{1}{2}\left(\frac{dv}{dx}\right)^2 dx \tag{6.17}$$

where it has been recognized that the slope dv/dx is small, so that its square is the dominant term in the series expansion of the square root. If the length changes dL are summed over the entire length of the member, the result becomes

$$\Delta L = \int_0^L \frac{1}{2}\left(\frac{dv}{dx}\right)^2 dx \tag{6.18}$$

Equation (6.18) can be used to compute the change in length of any member undergoing bending deformations once the displaced shape, $v(x)$, has been defined. The relation is limited only by the assumption that the slopes dv/dx are small, which is always the case for members in trusses and frames that are stable. If the member shown in Fig. 6.12 is prismatic, the curvature of the member is $-\phi$ and constant over its entire length. Equation (6.9) may be used along with the conditions that v is zero at $x = 0$ and $x = L$ to obtain the deflection function $v(x)$. Using Eq. (6.9) and integrating twice yields

$$\frac{d^2v}{dx^2} = -\phi$$
$$\frac{dv}{dx} = -\phi x + A$$
$$v = -\frac{\phi x^2}{2} + Ax + B$$

The constants A and B can be evaluated from the conditions that $v(0) = 0$ and $v(L) = 0$. The first yields that $B = 0$ and the second that $A = \phi L/2$.

6.19 Compute the vertical deflection of point C. Neglect axial deformations ($E = 29 \times 10^3$ ksi, $I = 500$ in⁴).

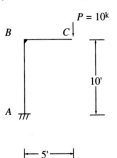

6.20 Compute the vertical deflection of C. Neglect axial deformations. ($E = 200$ GPa, $I = 200 \times 10^6$ mm⁴).

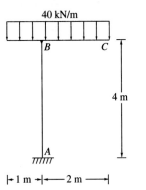

6.21 The cantilever beam loaded as shown has the moment diagram indicated. Compute the horizontal displacement of C and the rotation at B (E constant)

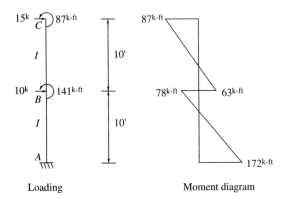

Loading   Moment diagram

6.22 Compute the deflection of the right end of the member (AE constant, $E = 29 \times 10^3$ ksi, $A = 2$ in²).

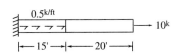

6.23 Compute the deflection of the right end of the member (AE constant, $E = 200$ GPa, $A = 100$ mm²).

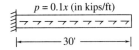

6.24 Compute the deflection of the right end of the member (AE constant).

$p = 0.1x$ (in kips/ft)

|←——— 30' ———→|

6.25   Compute the deflection of the right end of the member (*AE* constant).

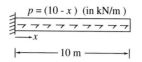

$p = (10 - x)$ (in kN/m)

x

├──────── 10 m ────────┤

6.26   Compute the displacement of the right end of the member. Show that the magnitude of *P* that makes the displacement of the right end of zero is $pL/2$ (*AE* constant).

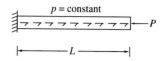

p = constant

P

├──────── L ────────┤

7

Deformation of Structures: Virtual Work

In Chapter 6, the deformations of one- or two-member beam structures under the action of applied loads are obtained directly from the differential equations that relate strains or curvatures to displacements. Once a mathematical model that is a reasonable estimate of the force–deformation relation for a material has been defined, application of the principles and techniques established in the field of mechanics of materials makes the process of calculating the deflections, elongations, and changes in slope of initially undeformed members a straightforward one. For structures with complicated geometries containing many members and multiple loading configurations, these approaches to the deformation problem are inadequate from a practical point of view and are extremely difficult to implement. Structures such as these can be analyzed using energy techniques that incorporate the principles of mechanics of materials in a mathematically elegant manner.

One of the most widely used energy techniques for the analysis of structures is the method of virtual work and complementary virtual work, which is based on the principle of conservation of energy. Its application to multiple-member structures is straightforward and practical for both hand and computer computations. The method can be used to establish the equilibrium equations for a structural member or a complete structure. It can also be used to establish the deformation behavior of a structure, and it is the latter application that will be emphasized in this chapter. The former application will be studied more extensively in a subsequent chapter, along with its use in the approximate analysis of structures based on assumed deformation behavior.

7.1   Structural Deformations _____

Structures are made from very stiff but nevertheless deformable materials. In design, the evaluation of deformations of a structure is important since they can affect its serviceability or even lead to instability of the structure under certain conditions. Structural deformations may also affect the magnitude and distribution of loads on a structure. For example, the pressure forces that act on the wing of an aircraft vary with the deflection of the wing.

The most common cause of the deformation of structures is applied loading, but deformations of structures are also caused by other actions. A temperature change causes a change in dimensions of a structure. During construction a member may be placed in the structure even though it is too long, too short, or slightly bent or curved. Supports of structures may shift or settle due to loading or other actions. Concrete members may undergo deformation due to posttensioning or shrinkage during curing. All of these actions can have a significant impact on the response of a structure, especially an indeterminate structure, where the calculation of deformations is a necessary part of its analysis. A procedure to evaluate these effects, either individually or collectively, is supplied by the complementary virtual work principle, which is a computationally powerful technique for evaluating structural behavior.

The principles of virtual work and complementary virtual work will be developed in a general form, but their application to analysis in this chapter will emphasize the evaluation of the deflection and deformation behavior of structures. The source of the deformation of a structure will be approached in a very general manner. The presentation in this chapter will be for materials having linear, elastic stress–strain relations, but any stress–strain relation can be used, as will be shown in a subsequent chapter. In applying the virtual work principles, structures fabricated from many members will be analyzed in the same manner as those comprised of only one or two members.

7.2   Principle of Virtual Work for Rigid Bodies _____

The idea behind the principle of virtual work for rigid bodies is conservation of energy. While the principle can be shown to follow from mathematical treatments of the variation of total potential energy and total complementary energy, a more physical interpretation of the principle is useful in a beginning analysis of deformations of structures. The principle for rigid bodies is stated as follows:

> If a rigid body in equilibrium under the action of any force system is subjected to a small rigid-body displacement and rotation, the work done by the force system is zero.

There are two important provisions of the principle that need to be pointed out. First, the body is *in equilibrium* under the action of the force system. The force system is, of course, comprised of concentrated forces, distributed forces, and moments or couples. The second important provision is that the body is subjected to a *small* rigid-body displacement and rotation. This implies that the geometry of the structure remains essentially unchanged after the action of the displacement system.

The principle can be proven to be correct directly from its statement and its two provisions. In an arbitrary plane body subjected to a set of forces and moments, there are three equilibrium equations. The equations set forth the relationships between the applied forces and moments in the force system. When the structure is subjected to a small rigid-body displacement and rotation, the work done by the force system as it rides through the imposed displacement will simply be the sum of the product of each force in the system times the displacement of that force along its line of action and the product of each moment times the rotation it goes through. This work expression will vanish due to the three equilibrium requirements on the force system. The proof is simple to do and is left as an exercise.

7.3 Reactions of Structures Using Virtual Work _____

The principle of virtual work for rigid bodies can be used to compute the reactions of statically determinate structures. The concept of the principle being based on equilibrium considerations is reinforced by this application. An extension of the principle can also be used as an aid in the determination of influence lines for structures, a topic addressed in Chapter 8.

Figure 7.1 The simply supported beam of Fig. 7.1a is loaded with a concentrated load. The vertical reactions can be obtained in the usual manner by summing moments about each end of the beam. Summing moment about the A end of the beam yields the upward reaction at the B end, R_B to be $P(a/L)$. Similarly, summing moments about B yields the upward reaction R_A to be $P(b/L)$.

The concept of the computation of these reactions using the virtual work principle for rigid bodies is shown in Fig. 7.1b and c. In Fig. 7.1b the beam is shown with the concentrated and the two vertical reactions acting. The force system P, R_A, and R_B represents a force system that is assumed to be in equilibrium on the structure, although the magnitudes of forces R_A and R_B are not known. To compute R_B, a small imagined or virtual displacement is assumed to occur at B and in the same direction as R_B, which causes the structure to rotate as a rigid body about A. The displacement at B is labeled δv_B, with the "δ" signifying that the displacement is both small and imagined or virtual.

The geometry of the rigid-body rotation of member A–B about A yields the vertical upward displacement $a/L(\delta v_B)$ at the point of application of the load, P. According to the principle of virtual work for rigid bodies, the work, W, done by the force system must be zero. In Fig. 7.1b the work of the reaction force, R_B, is $R_B\delta v_B$ and the concentrated load, P, is displaced upward or opposite to its direction of action, so that the work of this force is negative. The total work done is proportional to the nonzero virtual displacement, δv_B, which can be divided out of the expression to yield the equilibrium condition on R_B. A similar procedure is followed in Fig. 7.1c to obtain the magnitude of the reaction R_A. These reactions agree in magnitude and direction with those computed by using the moment equilibrium equations.

Figure 7.2 In Fig. 7.2 the use of the virtual work principle for rigid bodies is illustrated for distributed loads. Integration is required to compute the work of the general distributed load, $q(x)$. The reaction at B is calculated using the virtual work principle in Fig. 7.2b.

Figure 7.2c and d show the computation of the reaction at B for the common case of a uniformly distributed load, w. For this situation, the work done by the load can be computed as the product of the magnitude

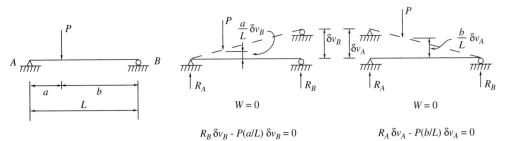

(a) Structure with concentrated load

$$R_B\,\delta v_B - P(a/L)\,\delta v_B = 0$$
$$R_B = P(a/L)$$

(b) Computation of R_B

$$R_A\,\delta v_A - P(b/L)\,\delta v_A = 0$$
$$R_A = P(b/L)$$

(c) Computation of R_A

Figure 7.1a–c   Reaction computation by virtual work principle for rigid bodies.

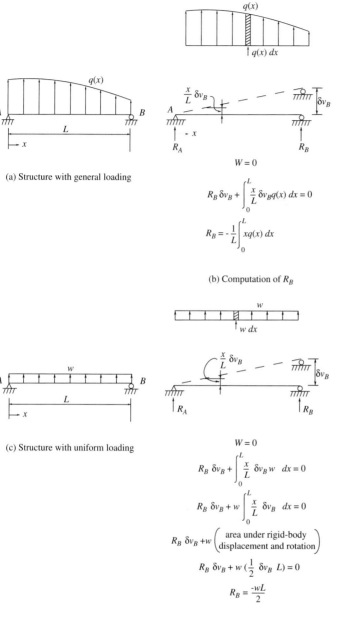

(a) Structure with general loading

$$R_B \, \delta v_B + \int_0^L \frac{x}{L} \, \delta v_B q(x) \, dx = 0$$

$$R_B = -\frac{1}{L} \int_0^L xq(x) \, dx$$

(b) Computation of R_B

(c) Structure with uniform loading

$$W = 0$$

$$R_B \, \delta v_B + \int_0^L \frac{x}{L} \, \delta v_B w \ dx = 0$$

$$R_B \, \delta v_B + w \int_0^L \frac{x}{L} \, \delta v_B \ dx = 0$$

$$R_B \, \delta v_B + w \left(\begin{array}{c} \text{area under rigid-body} \\ \text{displacement and rotation} \end{array} \right)$$

$$R_B \, \delta v_B + w \left(\frac{1}{2} \, \delta v_B \ L \right) = 0$$

$$R_B = \frac{-wL}{2}$$

(d) Computation of R_B

Figure 7.2a–d   Reaction computation by virtual work principle for rigid bodies with distributed load.

of the load, w, and the area between the displaced (in this case the rotated) and the undisplaced or original position of the body. The sign of the work is positive in this case because the direction of the displacements of the member is the same as that of w. Using the area under the displaced body simplifies the work computations for uniform loads.

The application of the ideas in Figs. 7.1 and 7.2 is presented in Example 7.1.

Example 7.1 The computation of the reactions of the beam in the example follows the same pattern in steps 2 to 4. A small imagined or virtual displacement or rotation is taken to occur in the direction of the reaction component being considered. For the calculation of R_C in step 2 a vertical displacement, δv_C, is introduced at C which causes a rigid-body rotation of B–C about B since A–B remains fixed at A. For the calculation of R_A in step 3 a vertical deflection, δv_A, is introduced at A which causes member A–B to be displaced vertically since no rotation of A–B can occur at A. Finally, for the calculation of M_A in step 4 a rotation, $\delta\theta_A$, is introduced at A which causes member A–B to rotate but not displace at A. An examination of each displaced geometry provides all the parameters needed to compute the total work of the force system. Because the distributed load is uniform, the area under the displaced member B–C is used to compute the work done by it. The horizontal reaction at A is not considered in the calculations, but it can be seen to be zero.

7.4   Principle of Virtual Work for Deformable Bodies _____

Since all bodies are deformable, applying the principle of virtual work to them is very important in engineering computations. For deformable bodies the principle is stated as follows:

> If a deformable body in equilibrium under the action of any force system is subjected to a set of small compatible deformations, the external work done by the force system is equal to the internal work of deformation done by the internal stresses in equilibrium with the external force system.

There are three important provisions in the statement of this principle. First, the force system is *in equilibrium* both *externally* and *internally*. Second, the set of deformations are *small*, so that no significant change of geometry occurs. Finally, the deformations of the structure are *compatible*,

Example 7.1

Obtain the reactions of the beam system show using the virtual work principle for rigid bodies.

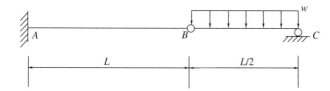

STEP 1 The structure is statically determinate and stable by inspection.

STEP 2 Compute the vertical reaction, R_C, by introducing a small imagined vertical displacement, δv_C, at C. The total external work on the structure must be zero. Segment B–C rotates about the hinge as shown. The external work of the distributed load is computed using the area under the rotated segment B–C, as illustrated in Fig. 7.2d.

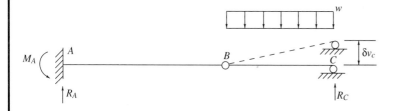

$$R_C \delta v_C - w\left(\frac{1}{2}\delta v_C \frac{L}{2}\right) = 0 \quad \therefore R_C = \frac{wL}{4}$$

STEP 3 Compute the vertical reaction, R_A, by introducing a small imagined vertical displacement, δv_A, at A.

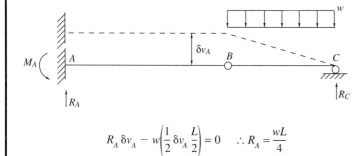

$$R_A \delta v_A - w\left(\frac{1}{2}\delta v_A \frac{L}{2}\right) = 0 \quad \therefore R_A = \frac{wL}{4}$$

Example 7.1 (continued)

STEP 4   Compute the moment reaction, M_A, by introducing a small imagined rotation, $\delta\theta_A$, at A. This produces a vertical displacement, $\delta\theta_A L$, at B because the imagined rotation at A is very small.

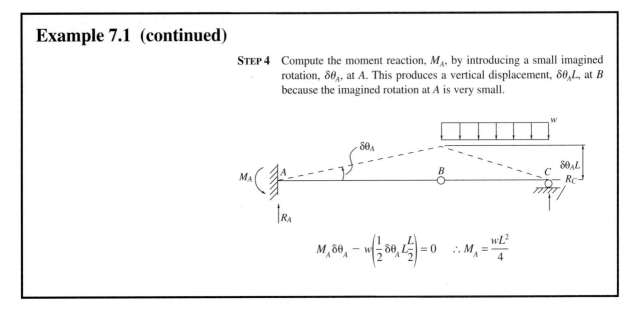

$$M_A\,\delta\theta_A - w\!\left(\frac{1}{2}\,\delta\theta_A\,L\frac{L}{2}\right) = 0 \qquad \therefore M_A = \frac{wL^2}{4}$$

which means that the structure does not break apart or displace away from points of support. These three provisions must always be satisfied in any application of the principle.

This principle can be developed from a mathematical consideration of potential and complementary energy, which is beyond the scope of this introductory treatment (see, e.g., Reference 14). However, the concept of conservation of work and energy provides a means of understanding the principle from a physical point of view. The following extended discussion is not a mathematically rigorous proof of the principle, but rather is intended to be a thoughtful evaluation of the principle.

Conservation of Energy

Consider first a geometrically stable structure that is supported against any free displacement or rotation. Any set of forces and moments will be resisted by the reactions of the structure so that rigid-body motion cannot occur. The applied loading and the reactions developed together constitute a force system in equilibrium on the structure. Internally, stresses will develop that are in equilibrium with the applied force system.

Now consider a set of small compatible deformations applied to the structure. The forces and moments acting at the supports of the structure will not undergo any displacement or rotation in the direction of their action since the deformations are compatible and do not cause the structure to break free from its supports. However, the other applied forces and moments will do work, this work being the external work computed as the sum of the products of forces with displacements and moments with rotations.

When the compatible deformations are imposed, the internal forces in the structure, being the stresses over the area on which they act, undergo displacements due to the strains in the structure and hence do internal work of deformation. The deformations are compatible, which means that the structure stays together during the deformation process. The internal work that is done represents stored internal energy in the structure that is recoverable. For example, if the imposed deformation system were removed, the structure would return to its original configuration before the imposition of the deformations.

Taking the entire process together, the concept of conservation of work and energy can be applied. The external work of the applied forces and the internal work of deformation or stored energy must sum to zero since no energy is gained or lost in the process. This leads to the conclusion that the external work must equal the internal work of deformation or stored energy.

Now consider a deformable structure that is not constrained or supported against free displacements and rotations. In this case the external forces and moments in the force system acting on the structure are in equilibrium by themselves, just as it was for the case of the rigid body considered in Section 7.2. When a set of small compatible deformations is applied to this structure, the external forces and moments undergo displacements and rotations that can be divided into two components. One component constitutes a set of rigid-body displacements and rotations for the structure and force system acting on it, and, because of the principle of virtual work for rigid bodies discussed in the preceding section, the external work done by this component of the deformation system is zero.

The remaining component of the deformation system is a set of displacements and rotations that arise directly from the actual deformations of the body. There is net external work done by the forces and moments of the force system as they ride through this component of the displacements and rotations of the deformation system. As before, the internal work is due only to the internal deformation of the structure. Again it follows from the concept of conservation of work and energy that the work done on the system for this component of the deformation system also must be zero, and hence the external work of the force system must equal the internal work of deformation done by the stresses arising from the force system. Putting the two components of the deformation system together leads to the same conclusion—that the external work of a force system in equilibrium on a deformable body is equal to the internal work of deformation.

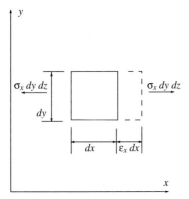

$$dW_i = \sigma_x \varepsilon_x \, dx \, dy \, dz$$
$$= \sigma_x \varepsilon_x \, dV$$

(a) Work of normal stresses

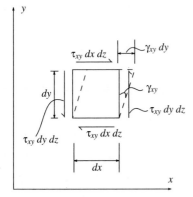

$$dW_i = \tau_{xy} \gamma_{xy} \, dx \, dy \, dz$$
$$= \tau_{xy} \gamma_{xy} \, dV$$

(b) Work of shear stresses

Figure 7.3a–b Internal work done by an internal normal and shear stress.

Equations of the Virtual Work Principles

It is useful to develop a mathematical statement of the principle of virtual work for deformable bodies, which henceforth will be called virtual work. The external work is simply the sum of the products of force times displacement in the direction of the force and of moment times rotation in the direction of the moment. The internal work of deformation is also the sum of the product of force times displacement in the direction of the force but needs to be examined in greater detail.

Figure 7.3 shows a differential element of a plane body that is subjected to a normal stress, σ_x, in Fig. 7.3a and a shear stress, τ_{xy}, in Fig. 7.3b. The stresses arise from some force system that is in equilibrium on the body, and the deformations (normal and shear strains) arise from some imposed set of small compatible deformations. The internal normal force $\sigma_x \, dy \, dz$ goes through the displacement $\varepsilon_x \, dx$ when the deformations are applied. Similarly, the shear force $\tau_{xy} \, dx \, dz$ goes through the displacement $\gamma_{xy} \, dy$. The internal work done in each case is computed as dW_i in the figure.

In a three-dimensional body there are three normal stresses and three shear stresses, so that the total internal work of deformation for a differential element is the sum of six independent quantities. The four additional force terms are $\sigma_y \, dx \, dz$, $\sigma_z \, dx \, dy$, $\tau_{yz} \, dx \, dz$, and $\tau_{xz} \, dy \, dz$, and the four additional displacement terms are $\varepsilon_y \, dy$, $\varepsilon_z \, dz$, $\gamma_{yz} \, dy$, and $\gamma_{xz} \, dx$. These contribute internal work of $(\sigma_y \varepsilon_y + \sigma_z \varepsilon_z + \tau_{yz} \gamma_{yz} + \tau_{xz} \gamma_{yz}) \, dx \, dy \, dz$ to dW_i. For simplicity in presentation, the internal work is presented simply as the product of a normal force arising from the normal stress, σ, times a corresponding displacement arising from the normal strain, ε, and a shear force arising from the shear stress, τ, times a corresponding displacement arising from the shear strain, γ. The differential internal work becomes

$$dW_i = (\sigma\varepsilon + \tau\gamma)\,dx\,dy\,dz = (\sigma\varepsilon + \tau\gamma)\,dV \tag{7.1}$$

It is understood that the normal and shear stresses and strains each can account for up to three contributions to dW_i. The total internal work done is the sum for all differential elements in the body. For infinitessimal elements this summation means integration over the complete volume, V, of the body. Integrating Eq. (7.1) yields

$$W_i = \int_V (\sigma\varepsilon + \tau\gamma)\,dV \tag{7.2}$$

The mathematical statement of the virtual work principle can now be presented. The internal work given by Eq. (7.2) arises from the stresses σ and τ associated with any force system, no matter what its origin, and the strains ε and γ associated with any deformation system, regardless of its origin. The only requirements are that the force system acting on the structure is in both external and internal equilibrium, and the deformation system consists of a set of small, compatible deformations.

There are two systems, a force system and a deformation system, that act on the structure. Depending on the nature and origin of these systems, the principle divides into two parts. For the first part, consider a structure in equilibrium under the action of a set of actual or "real" applied forces and moments which is then subjected to a set of imagined or "virtual" deformations that are small and compatible. To identify the displacements and strains associated with the virtual deformation system, a "δ" is placed in front of them. For these actions, the virtual work principle is stated mathematically as

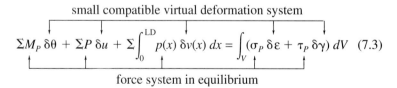

$$\Sigma M_P\,\delta\theta + \Sigma P\,\delta u + \Sigma\int_0^{LD} p(x)\,\delta v(x)\,dx = \int_V (\sigma_P\,\delta\varepsilon + \tau_P\,\delta\gamma)\,dV \tag{7.3}$$

force system in equilibrium

The right-hand side of Eq. (7.3) is the internal virtual work done by the real stresses which are subjected to the virtual strains $\delta\varepsilon$ and $\delta\gamma$, which follows directly from the definition of internal work given in Eq. (7.2). The left-hand side of Eq. (7.3) is the external work done by summation of the real concentrated forces, P, and moments, M_P, as they ride through the virtual displacements, δu in the direction of P, and virtual rotations, $\delta\theta$ in the direction of M_P at the points of application of the forces and moments. The summation of the integral term on the left side of Eq. (7.3) is the external work done by any real distributed loadings, $p(x)$, that ride through virtual displacements, $\delta v(x)$, in the direction of $p(x)$, integrated over the length of the distributed loadings, LD. Equation (7.3) can also be obtained from mathematical treatments of total potential energy, and this form of the principle is called simply *virtual work*. The principle of virtual work in this form is used to derive equilibrium conditions for structures and for approximate analysis based on assumed displacement behavior.

The second part of the principle is used in deflection analysis. For this part an imagined or "virtual" force system in equilibrium is acting on the structure when the structure is subjected to a set of actual or "real" deformations. In this part the virtual force system and corresponding virtual stresses in equilibrium are identified by a "δ" in front. This gives the mathematical statement

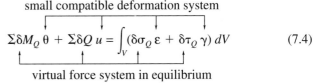

$$\sum \delta M_Q\, \theta + \sum \delta Q\, u = \int_V (\delta \sigma_Q\, \varepsilon + \delta \tau_Q\, \gamma)\, dV \qquad (7.4)$$

The right-hand side of the equation represents the internal virtual work of deformation taken directly from Eq. (7.2), and the left-hand side represents the external work of the virtual force system. The summation on the concentrated virtual forces, δ_Q, and virtual moments, δM_Q, is taken over all points where a force or moment is applied including reaction points. The displacements u and rotations θ are at the point of application and in the direction of the corresponding δQ and δM_Q. Equation (7.4) can also be obtained from mathematical treatments of total complementary energy and is referred to as *complementary virtual work*.

This form of the virtual work principle is used in deflection calculations and is referred to in that application by some authors simply as "virtual work" rather than its more mathematically accurate term, complementary virtual work. Because of the extensive use of Eq. (7.4) and forms that are derived from it and the fact that it is one of the two parts of the virtual work principle, the term *complementary virtual work* will be used for all deflection computations in this book. This eliminates any potential for confusion over which part of the virtual work principle is to be used in a particular situation.

Conceptually, Eq. (7.4) simply says that the external work of the forces and moments in the virtual force system equals the internal work of deformation done by the virtual stresses. As will be seen in Sections 7.6 and 7.7, judicious selection of a virtual force system will enable the calculation of any deflection or rotation at any point in a structure. Although simple in concept, Eq. (7.4) and subsequent equations derived from it provide a powerful tool for the analysis of the deformations of structures.

7.5   Internal Virtual Work for Prismatic Members

A prismatic member is defined as a member in which one dimension, its length, is very large in proportion to the other two dimensions and the cross section remains constant along its length. The stresses developed in a plane prismatic member that is subjected to the actions of axial force and in-plane bending are the normal stress σ_x and the shear stress τ_{xy}. In this section only the internal virtual work associated with the normal stresses will be developed. Internal work associated with the shear stress due to bending is usually quite small in proportion to that of the normal stress and usually can be neglected. Internal work due to the shear stresses will be considered in a subsequent section. The internal work in Eq. (7.4) reduces to

$$W_i = \int_V \delta\sigma_x \, \varepsilon_x \, dV \qquad (7.5)$$

Equations (7.4) and (7.5) give the internal virtual work, independent of the stress–strain relation of the material. To obtain a more useful expression for the internal work, a relation between internal normal stresses and internal moment and axial force actions must be established. The relations between normal strains in a member and the bending moments and axial forces that cause them cannot be established without knowing the form of the stress–strain relation for the material. The relations can be obtained for any known stress–strain relationship, but in this treatment only a linear-elastic stress–strain relation will be used.

Consider the prismatic member shown in Fig. 7.4a, which is in equilibrium under the action of some virtual force system. The force system gives rise to an axial tension force δF_Q in the member and a moment along the member $\delta M_Q(x)$. Because the member is prismatic, made from a linear elastic material and assumed to be loaded in the x–y plane, which is a plane of symmetry for the cross section, the virtual normal stresses can be obtained from the mechanics of materials expressions in Table 1.1 as

$$\delta\sigma_x = \frac{\delta F_Q}{A} - \frac{\delta M_Q(x)y}{I} \qquad (7.6)$$

where A is the cross-sectional area, I is the moment of inertia with respect to the centroidal axis, and y is measured upward from the neutral axis. The negative sign in the second term is due to the fact that in the presence of positive $\delta M_Q(x)$, the strains above the centroidal axis where y is positive are compressive and hence negative.

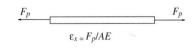

$$\varepsilon_x = F_p/AE$$

Axial strains
due to P loads

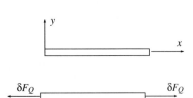

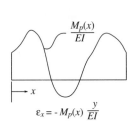

$$\varepsilon_x = -M_p(x)\,\frac{y}{EI}$$

Curvature due to P loads
and corresponding strain

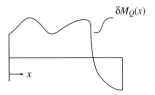

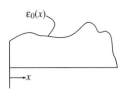

Moment distribution

Strain due to other causes

(a) Force system

(b) Deformation system strains

Figure 7.4a–b   Virtual force system and real deformation systems
for a prismatic member.

The deformations of the member come from two possible actions. A load
system, P, can cause axial and bending strains, ε_p, in the member which are
known as mechanical strains. The second source of strains is designated ε_0,
to indicate that they arise from some other action. For one-dimensional behavior in a prismatic member made with linear elastic material, the relation
between stress and strain is simply

$$\sigma = E\varepsilon \quad \text{or} \quad \varepsilon = \frac{\sigma}{E} \tag{7.7}$$

Equation (7.7) states that the strains in the member due to mechanical deformation can be obtained by dividing the stresses corresponding to the P

loading system by E. As shown in Fig. 7.4b, the strains due to the internal axial force, F_P, are F_P/AE and the strains due to the internal moment, M_P, are the curvature, M_P/EI multiplied by $-y$. The mechanical strains become

$$\varepsilon_P = \frac{F_P}{AE} - \frac{M_P(x)y}{EI} \tag{7.8}$$

The strains due to other causes, ε_0, are as yet unspecified but could vary both along the axis of the member in some arbitrary manner as shown in Fig. 7.4b, as well as in the y direction through the depth of the member. Examples of strains from other actions or causes are temperature and shrinkage strains and strains due to fabrication errors in the members of the structure. Taking all possible normal strains to act simultaneously, the set of small compatible deformations is the single strain ε_x, which comes from Eq. (7.8), and the ε_0 strain:

$$\varepsilon_x = \varepsilon_P + \varepsilon_0 = \frac{F_P}{AE} - \frac{M_P(x)y}{EI} + \varepsilon_0 \tag{7.9}$$

Substituting Eqs. (7.6) and (7.9) into Eq. (7.5) yields the result

$$W_i = \int_V \left[\frac{\delta F_Q}{A} - \frac{\delta M_Q(x)y}{I} \right] \left[\frac{F_P}{AE} - \frac{M_P(x)y}{EI} + \varepsilon_0 \right] dV \tag{7.10}$$

Integration over the volume of the member can be broken down into integration along the length of the member and integration over the area of the cross section of the member. This converts Eq. (7.10) to the form

$$
\begin{aligned}
W_i = \int_0^L \Bigg\{ &-\int_A \frac{\delta F_Q}{A} \frac{M_P(x)y}{EI} \, dA - \int_A \frac{\delta M_Q(x)y}{I} \frac{F_P}{AE} \, dA \\
&+ \int_A \left[\frac{\delta F_Q F_P}{A^2 E} + \frac{\delta M_Q(x) M_P(x)y^2}{EI^2} \right] dA \\
&+ \int_A \left[\frac{\delta F_Q}{A} - \frac{\delta M_Q(x)y}{I} \right] \varepsilon_0 \, dA \Bigg\} \, dx
\end{aligned} \tag{7.11}
$$

Each of the four integrals over the area of the cross section, A, inside the braces has parameters independent of the area integration. In particular, $M_P(x)$, $\delta M_Q(x)$, δF_Q, F_P, A, E, and I are not dependent on the area and can be brought across the area integral sign. This reduces the first two area integrals inside the braces to the integral over the area of the cross section, $\int_A y \, dA$, which is zero because the x axis is the centroidal axis. The two terms inside the third integral reduce to integrals over the area of $\int_A dA$ and $\int_A y^2 \, dA$, respectively, which become the area, A, and the moment of inertia, I, of the cross section. The fourth and last integral inside the braces cannot

255

be evaluated since the variation of ε_0 is not specified. Equation (7.11) reduces to the form

$$W_i = \int_0^L \left[\frac{\delta F_Q \, F_P}{AE} + \frac{\delta M_Q(x) \, M_P(x)}{EI} \right] dx + W_{i0} \tag{7.12}$$

where

$$W_{i0} = \int_0^L \int_A \left[\frac{\delta F_Q}{A} - \frac{\delta M_Q(x) y}{I} \right] \varepsilon_0 \, dA \, dx \tag{7.13a}$$

If, in Eq. (7.13a), the strain ε_0 is defined as

$$\varepsilon_0 = b_0 + a_0 y \tag{7.13b}$$

where b_0 and a_0 are constants or a function of x only, Eq. (7.13a) becomes, upon substitution and expansion,

$$W_{i0} = \int_0^L \left[\int_A b_0 \frac{\delta F_Q}{A} \, dA - \int_A b_0 \frac{\delta M_Q(x) y}{I} \, dA + \int_A a_0 \frac{\delta F_Q}{A} y \, dA \right.$$
$$\left. - \int_A a_0 \frac{\delta M_Q(x) y^2}{I} \, dA \right] dx \tag{7.13c}$$

As in Eq. (7.11), the middle two integrals over the area vanish while the first and last integrals over the area become the area, A, and moment of inertia I, respectively. This reduces Eq. (7.13c) to the form

$$W_{i0} = \int_0^L \left[b_0 \, \delta F_Q - a_0 \, \delta M_Q(x) \right] dx \tag{7.13d}$$

The first term inside the integrals in Eqs. (7.12) and (7.13d) can be evaluated when δF_Q, b_0, and F_P are constant along the length of the member, a situation that is by far the most common. For that case, the final form for the internal virtual work then becomes

$$W_i = \delta F_Q \frac{F_P}{AE} L + \int_0^L \delta M_Q(x) \frac{M_P(x)}{EI} \, dx + W_{i0} \tag{7.14a}$$

where

$$W_{i0} = b_0 \, \delta F_Q \, L - \int_0^L a_0 M_Q(x) \, dx \tag{7.14b}$$

and where it is understood that δM_Q, M_P, and possibly a_0 are functions of x. Equations (7.13) and (7.14) constitute the internal virtual work expression for a prismatic member, ignoring the contribution of shear stresses and shear deformation.

7.6   Application of the Complementary Virtual Work Principle to Deflection Analysis of Trusses

The deformation analysis of trusses involves only axial forces and axial strains. The internal work expression, Eq. (7.14), reduces to just the first term, and in Eq. (7.13) the second term in the brackets vanishes since there are no moments or curvatures present in truss structures. The application of these reduced forms of the internal virtual work for the calculation of the deflections of trusses is discussed in several examples. Example 7.2 illustrates the method by which the vertical displacement of joint L_1 of the simple truss can be computed for the loading shown. It is extremely important in this first example to follow the steps and identify the force system that acts on the structure when the deformations occur. It is also important to identify the source of the deformations of the structure and to obtain the correct deformation system. Example 7.3 extends this computation to the deflection analysis of a truss analyzed in Examples 4.1 and 4.2. The organization of the calculations in a spreadsheet reduces errors in the procedure and also enables panel lengths and height and the cross-sectional area of the members of the truss to vary.

> **Example 7.2**   The computation of a single displacement of the truss requires two systems to act. The first system is the virtual force system, which must be acting prior to the occurrence of the deformations of the truss. The second is the deformation system due to the applied 80-kip load, which causes the members of the structure to deform. The force system must be created for the analysis. In this example, the vertical displacement of L_1 is to be calculated; consequently, a virtual force δQ is imagined to act at L_1 in a vertical direction. This force then creates the reactions and internal member forces for the structure that are shown in the sketch labeled "Force system." The applied load δQ at L_1 and the reactions at L_0 and L_2 represent the externally applied virtual force system in equilibrium on the structure. The internal member forces are obtained in a joint-by-joint analysis such as that described in Section 3.4. These member forces are in equilibrium with the applied loads.
>
> The deformations of the structure are caused by the action of the 80-kip load at U_1. These deformations take the form of a horizontal displacement of joint L_2, horizontal and vertical displacements of L_1 and U_1, and axial strains in the members of the truss. The member strains are obtained by calculating the forces in all the members of the truss and

257

then dividing by the AE value for each member. The diagram labeled "Deformation system" summarizes these observations.

The next step in the analysis is to compute the external and internal virtual work done by the virtual force system when the real deformations associated with the 80-kip load occur. Equations (7.13) and (7.14) give the internal work of deformation with the F_P/AE in those equations coming from the 80-kip load. The W_{i0} of Eqs. (7.13) and (7.14) is zero since strains in the members are due only to the 80-kip force. The external work is the sum of the product of δQ with the unknown displacement v and the two reaction forces with their corresponding displacements. In this case the only external work is the $\delta Q v$ at L_1.

Setting the external and internal work equal gives an expression involving the vertical displacement of L_1, v, and the virtual force system, δQ. Equation (7.14) gives the internal work for a single member; therefore, the internal work for a structure composed of several members is obtained by adding the internal work of each member given by Eq. (7.14) into a single sum. Note that both the right and left sides of the complementary virtual work expression are multiplied by δQ, which, since it is nonzero, can be divided out, leaving the final expression for the vertical deflection of L_1 as v.

Example 7.3 The truss loaded as shown in Example 3.2 is analyzed for the deflection of the bottom joints L_1 and L_2 using the spreadsheet approach presented in Example 3.3. The spreadsheet in this example is displayed with the computations for the deflection, v, of L_2. However, the spreadsheet is constructed so that the deflection of L_1 can be computed simply by entering a 0 in cell H3 and a −1 in cell G3. This gives a downward displacement of 0.130 ft for L_1. With this spreadsheet, the truss can be analyzed for all member forces and bottom-chord joint vertical displacements for any loads that act along the bottom chord. Panel lengths, truss height, and area of members can be changed with appropriate entries in the spreadsheet. Note that the formulas entered in B6 to B18 in Example 3.3 are the same formulas that are entered cells F6 to F18, so that they may be copied from column B to column F.

Example 7.4 Some deformations caused by construction conditions and temperature changes in the structure of Example 7.2 are considered in Example 7.4. Again the desired deformation is the vertical displacement of L_1, so the imagined or virtual force system that is used for this calculation is the same as in Example 7.2. Now the deformation system changes for each separate action on the structure.

In part (a) the support displacement of L_2 ¼ in. downward is the only deformation occurring and there are no strains in the members of the truss. In this case the internal work is zero and the external work by itself

Example 7.3

Expand the spreadsheet presented in Example 3.3 so that the deflection of either joint L_1 or L_2 of the truss loaded as in Example 3.2 can be computed by the complementary virtual work approach. The area of each truss member is indicated by (·) in square inches. Take $E = 29{,}000$ ksi.

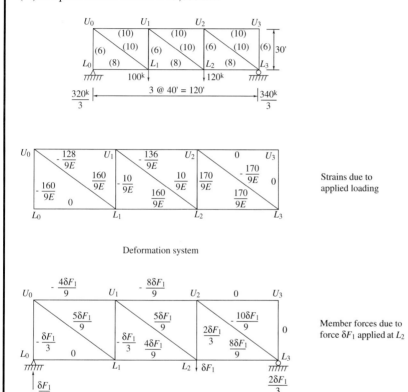

Strains due to applied loading

Deformation system

Member forces due to force δF_1 applied at L_2

Virtual force system δF_1 at L_2

Example 7.3 (continued)

STEP 1   Obtain the member strains (F_p/AE) due to the applied loading by dividing the member forces calculated in Example 3.2 by the area, A, of each member and E.

STEP 2   Create the imagined or virtual force system. Take the virtual force δF_1 to act downward at L_2, compute the reactions, and use the method of joints to compute all the member forces. These are summarized in the sketch.

STEP 3   Calculate the vertical deflection of L_2, v, using Eq. (7.14). As in Example 7.2, only the first term on the right-hand side of Eq. (7.14) is nonzero.

$$W_e = W_i \quad \therefore \delta F_1 v = \Sigma F_1 \frac{F_p}{AE} L$$

Spreadsheet Construction

The spreadsheet of Example 3.3 is enlarged to accommodate the additional data needed for the deformation calculation and the analysis for reactions and member forces due to the force system, δF_1. Member areas are added in cells C6 to C18 and the member lengths in cells D6 to D18 are taken from cell B1, D1, or F1 (for example, the length in cell D6 = +D1). The strains in cells E6 to E18 are computed using the forces in the B column, the areas in the C column, and the modulus of elasticity from cell H1 (for example, the strain in cell E8 = B8/C8/H1).

The virtual force system calculations appear in columns F through I and rows 2 to 4 and column F, rows 6 to 18. The computation of the reactions in cells F4 and I4 is identical to the calculations in the corresponding cells B4 and E4 [for example, the reaction in cell F4 = –(3 · F3 + 2 · G3 + H3)/3]. Similarly, computation of the forces in cells F6 to F18 is identical to the calculations in the corresponding cells B6 to B18 [for example, the force in the member for cell F11 = F7 − (B1/F1) · F8]. The formulas in the cells in the B column can be copied into the F column.

Example 7.3 (continued)

Computations for the virtual force system permit the calculation of deflection at L_2 (with the -1 in cell H3 as shown) or L_1 if the applied virtual force is at L_1. Note that the magnitude of δF_1 is taken to be 1 in the spreadsheet. The internal work calculation for each member is completed in cells G6 to G18 (for example, cell G8 = F8 · E8 · D8). Summing the values in cells G6 to G18 yields the internal work on the right side of Eq. (7.14) in cell G19 and also the displacement of L_2 since δF_1 has unit magnitude.

	A	B	C	D	E	F	G	H	I
1	PANEL:	40	HEIGHT:	30	DIAGONAL:	50	E:	29000	
		APPLIED LOADS ON TRUSS			VIRTUAL FORCE SYSTEM			δF1	
2	PANEL POINT	L0	L1	L2	L3	L0	L1	L2	L3
3	LOAD	0	-100	-120	0	0	0	-1	0
4	REACTIONS	106.67	—	—	113.33	0.33	—	—	0.67
5	MEMBER	FORCE (k)	AREA (sq. in.)	LENGTH (feet)	STRAIN	VIRTUAL FORCE δFI	W1		
6	L0-U0	-106.67	6.00	30.00	-6.13E-04	-0.33	6.13E-03		
7	L0-L1	0.00	8.00	40.00	0.000E+00	0.00	0.00E+00		
8	U0-L1	177.78	10.00	50.00	6.130E-04	0.56	1.70E-02		
9	U0-U1	-142.22	10.00	40.00	-4.90E-04	-0.44	8.72E-03		
10	L1-U1	-6.67	6.00	30.00	-3.83E-05	-0.33	3.83E-04		
11	L1-L2	142.22	8.00	40.00	6.130E-04	0.44	1.09E-02		
12	U1-L2	11.11	10.00	50.00	3.831E-05	0.56	1.06E-03		
13	U1-U2	-151.11	10.00	40.00	-5.21E-04	-0.89	1.85E-02		
14	L2-U2	113.33	6.00	30.00	6.513E-04	0.67	1.30E-02		
15	L2-L3	151.11	8.00	40.00	6.513E-04	0.89	2.32E-02		
16	U2-L3	-188.89	10.00	50.00	-6.51E-04	-1.11	3.62E-02		
17	U2-L3	0.00	10.00	40.00	5.881E-20	-0.00	-5.2E-35		
18	L3-U3	0.00	6.00	30.00	1.144E-19	0.00	2.29E-34		

$$\Sigma\,[\delta F1 \times (F_p/AE) \times L] \;=\; 0.135121 \quad = v \text{ (in feet)}$$

Example 7.4

For the structure of Example 7.2, compute separately the vertical deflection of L_1, v, due to each action below:

(a) Support settlement of L_2 of ¼ in. down.
(b) Temperature change of $-50°F$ in members L_0–L_1 and L_1–L_2.
(c) Member L_0–U_1 is fabricated ⅛ in. too short.

Take $E = 29,000$ ksi and $\alpha = \dfrac{1}{150,000}\dfrac{\text{strain}}{°F}$.

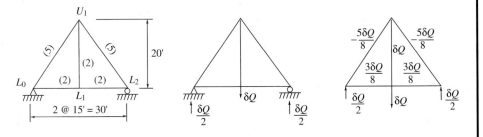

Force system

STEP 1 Because the vertical deflection of L_1, v, is wanted in all three parts, step 1 is the same as step 1 of Example 7.2. The force system from Example 7.2 is repeated above.

Deformation System (a)

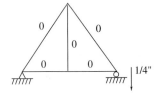

Member strains

(a) **STEP 2** No loading or other causes of strain in the members.

STEP 3 Use Eq. (7.14). $W_{i0} = 0$. Also $\delta M_Q = 0$ and all F_p/AE are zero, thus making the right-hand side of Eq. (7.14) zero.

$$\underbrace{\frac{\delta Q}{2}(0)}_{L_0} + \underbrace{\delta Q v}_{L_1} + \underbrace{\frac{\delta Q}{2}\left(\frac{-1}{4}\cdot\frac{1}{12}\right)}_{L_2} = 0$$

$$\therefore v = \frac{1}{96} = 0.01042 \text{ ft}$$

Note: The ¼-inch settlement at L_2 is negative because it is opposite in direction to the force, $\delta Q/2$.

264

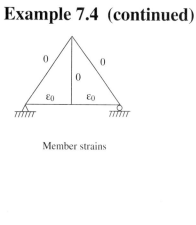

Example 7.4 (continued)

Deformation System (b)

(b) STEP 2 No loading. The deformation is due only to temperature strain.

$$\varepsilon_0 = \alpha\,\Delta T = \left(\frac{1}{150.000}\right)(-50) = \frac{-1}{3000}$$

Member strains

STEP 3 Use Eq. (7.14). $\delta M_Q = 0$ and $F_p/AE = 0$.

$$W_{i0} = \delta Q \varepsilon_0 L \text{ for } \varepsilon_0 \text{ constant over length } L:$$

$$\delta Qv = \underbrace{\left(\frac{3\delta Q}{8}\right)}_{L_0}\underbrace{\left(\frac{-1}{3000}\right)\cdot 15}_{L_0-L_1} + \underbrace{\left(\frac{3\delta Q}{2}\right)\left(\frac{-1}{3000}\right)\cdot 15}_{L_1-L_2}$$

$$\therefore v = \frac{-3}{800} = -0.00375 \text{ ft}$$

Deformation System (c)

(c) STEP 2 No loading. The deformation is due only to fabrication error strain.

$$\varepsilon_0 = \left(\frac{-1}{8}\right)\left(\frac{1}{25}\right)\left(\frac{1}{12}\right) = \frac{-1}{2400}$$

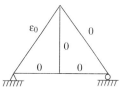

Member strains

STEP 3 Use Eq. (7.14). $\delta M_Q = F_p/AE = 0$; $W_{i0} = \delta Q \varepsilon_0 L$.

$$\delta Qv = \underbrace{-5\frac{\delta Q}{8}}_{L_1} \cdot \underbrace{\left(\frac{-1}{2400}\right)\cdot 25}_{L_0-U_1} = 5\frac{\delta Q}{768}$$

$$\therefore v = 0.000651 \text{ ft}$$

establishes the vertical displacement of L_1. Note again that the virtual force system δQ vanishes at the end of the calculation.

The temperature change in the two lower members L_0–L_1, and L_1–L_2 in part (b) gives rise to internal work, W_{i0}, as defined in Eq. (7.13). For this case ε_0 is constant through the depth of the cross section [a_0 in Eq. (7.13b) is zero] and over the length of the member, so that b_0 is equal to the coefficient of linear expansion, α, times the change in temperature, ΔT. Since the temperature change is negative (the temperature of the member has dropped from some ambient value), the strain $\varepsilon_0 = b_0$ is also negative. The only internal work for this case is the W_{i0} defined in Eq. (7.14b), and the deflection of L_1, v, is computed directly, with the virtual force system again disappearing. The negative sign for v in this calculation indicates that the direction of v is opposite to the direction of δQ acting at L_1. Thus the temperature change causes an upward displacement of joint L_1.

Finally, in part (c), a fabrication error in the member L_0–U_1 during construction gives a vertical displacement at L_1. The strain ε_0 for the fabrication error is computed by taking the change in length of the member and dividing by the correct length of the member (25 ft). The strain is constant through the depth of the cross section and along the length and is negative because the member is fabricated shorter than the required length. Thus, from Eq. (7.13b), $\varepsilon_0 = b_0 = -1/2400$. The only internal work in Eq. (7.14) is due to the constant ε_0, and again the virtual force system disappears from the calculation.

7.7   Application of the Complementary Virtual Work Principle to Deflection Analysis of Beams and Frames

The analysis of the deflection behavior of beams and frames by application of the complementary virtual work principle does not involve any new concepts but does require evaluating the interaction of moment and curvature diagrams. Examples 7.5 and 7.6 show how the technique is applied to beams and frames, respectively. In beams, the deformations are limited to curvatures due to bending, while axial deformations are also present in frames.

Example 7.5 To compute the rotation of the left end of the simply supported beam, an imagined or virtual force system consisting of a virtual moment, δA, is applied to the left end of the beam. The reactions due to δA are computed and a moment diagram is drawn with the variation of the moment being defined by the variable x. This constitutes the force system for the problem and is labeled "Force system" in the example.

Example 7.5

Compute the rotation of the left end θ_A. Take EI as constant.

STEP 1 Select the force system. Use an imagined or virtual moment, δA, at A to reflect rotation at the joint.

STEP 2 Analyze the structure and draw the moment diagram due to δA. This constitutes the force system shown.

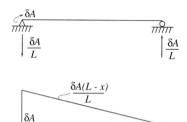

$$\delta A(L - x) \over L$$

δA

x

Moment diagram

Force system

w

A

B

$\dfrac{wL}{2}$ $\dfrac{wL}{2}$

L

$M = \dfrac{wLx}{2} - \dfrac{wx^2}{2}$

x

Moment diagram

$\dfrac{M}{EI} = \dfrac{Wx}{2EI}(L - x)$

x

Curvature diagram

Deformation system

STEP 3 Calculate the member deformations in the structure due to real loading. First compute the reactions and draw the moment diagram due to w and then create a curvature diagram by dividing the moment by EI. The deformation system is as shown.

STEP 4 Use Eq. (7.14) to evaluate external and internal work. [*Note*: $W_{i0} = 0$ and $F_p/AE = 0$ (no axial forces or strains).

$$W_e = W_i$$

$$\delta A\theta_A + \underbrace{\frac{\delta A}{L}(0)}_{A} + \underbrace{\frac{\delta A}{L}(0)}_{B} = \int_0^L \frac{wx(L - x)}{2EI}\left[\frac{\delta A(L - x)}{L}\right]dx$$

$$\delta A\theta_A = \frac{w\delta A}{2EI}\int_0^L (L^2x - 2Lx^2 + x^3)\,dx$$

$$= \frac{w\delta A}{2EIL}\left(\frac{L^2x^2}{2} - \frac{2Lx^3}{3} + \frac{x^4}{4}\right)\Big|_0^L = \delta A\frac{wL^3}{2EI}\left(\frac{1}{12}\right)$$

$$\theta_A = \frac{wL^3}{24EI}$$

267

The deformations of the structure are caused by the uniform load w, which creates the moment diagram for a uniform load as shown. The curvature diagram, which depicts the internal deformations of the member, is obtained by dividing the moment diagrams by EI and is defined by the variable x. The deformation of the structure is caused by the loading, w, and the resulting curvature and is labeled "Deformation system" in the example. The internal work is computed from Eq. (7.14), which has only the bending deformation term (the M_P/EI term), as there are no axial forces or strains and there is no internal work, W_{i0}. As always, the virtual force system δA vanishes from the calculation, giving the desired rotation of the left end. What is the rotation of the other (right) end of the beam? For numerical computations using U.S. units, take $L = 12$ ft, $w = 0.1$ kip/ft, $E = 2000$ ksi, and $I = 170$ in^4; for SI units, take $L = 4$ m, $w = 1.5$ kN/m, $E = 12$GPa, and $I = 70 \times 10^6$ mm^4.

Example 7.6 The imagined force system, δB, is established in step 1 and reflects the desired displacement quantity, the horizontal deflection, u, of the right support. The variations of the moments in the members of the frame due to δB are defined by the coordinates r for B–D, s for A–C, and t for C–D as shown. The reactions, axial forces, and moment diagrams are labeled "Force system," with the axial forces shown in parentheses on the moment diagram. In step 3 the deformation system due to the applied load, P, with its moment variations and axial force distributions is established. The curvature diagrams with axial strains in parentheses are defined with the same coordinates, r, s, and t, for each member as in the moment diagrams in the force system. All of these deformations are presented in the sketches labeled "Deformation system." In the application of Eq. (7.14), the W_{i0} term is again zero. While there are axial force and axial deformations in this structure, it turns out that no internal work arises from their presence. Note also that the only integrals that need to be evaluated are for members A–C and C–D. With the vanishing of the virtual force system in the final calculation, the horizontal displacement of the base is obtained. For numerical computations using U.S. units, take $L = 15$ ft, $P = 2$ kips, $E = 29{,}000$ ksi, $I = 50$ in^4, and $A = 4$ in^2; for SI units, take $L = 5$ m, $P = 10$ kN, $E = 200$ GPa, $I = 20 \times 10^6$ mm^4, and $A = 2500$ mm^2.

Use of the complementary work principle to obtain selected deflections or rotations of beams and frames due to support settlement, temperature change, or fabrication error follows the same steps as those used for the truss in Example 7.4. The key to the use of this principle for frames is to obtain the curvature diagrams and member axial strains for the deformation system and the moment diagram and member axial forces for the force system.

Example 7.6

Obtain the horizontal displacement, u, of joint B of the frame shown.

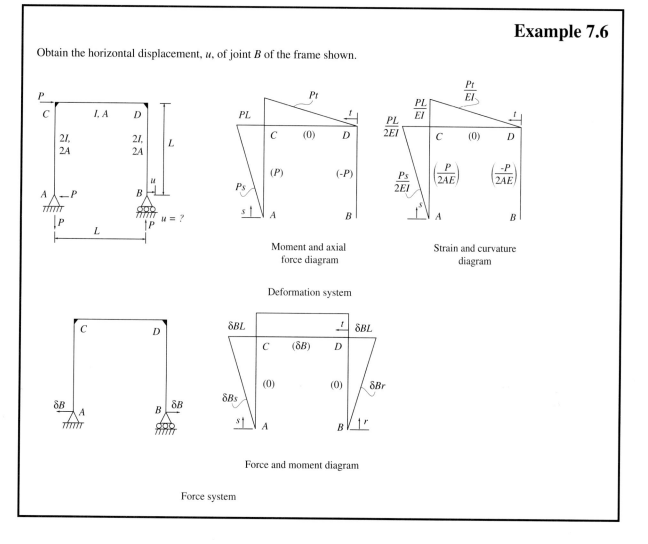

Deformation system

Moment and axial
force diagram

Strain and curvature
diagram

Force and moment diagram

Force system

Example 7.6 (continued)

STEP 1   Establish the imagined or virtual force system, δB, which reflects the desired horizontal displacement, u, at B.

STEP 2   Analyze the structure for the reactions and draw the force and moment diagrams for δB.

STEP 3   Compute the real deformations of members. From loading obtain the reactions and the force and moment diagrams. Then establish the curvature diagrams by dividing moments by EI and axial strain distribution by dividing axial force by AE.

STEP 4   Use Eq. (7.14) to obtain internal and external work.

(Note: $W_{i0} = 0$ since the strains due to other causes are zero.)

$$W_e = W_1$$

$$\overset{A}{\frown} \; \overset{B}{\frown} \; \overset{A-C}{\frown} \; \overset{C-D}{\frown} \; \overset{B-D}{\frown}$$
$$\delta B(0) + \delta Bu = (0)\frac{P}{2AE}L + \delta B(0) + (0)\frac{-P}{2AE}L$$

$$\overset{A-C}{\frown} \qquad \overset{C-D}{\frown} \qquad \overset{B-D}{\frown}$$
$$+ \int_0^L (\delta Bs)\frac{Ps}{2EI}\,ds + \int_0^L (\delta BL)\frac{Pt}{EI}\,dt + \int_0^L \delta Br(0)\,dr$$

$$\delta Bu = \frac{\delta BP}{2EI}\int_0^L s^2\,ds + \frac{\delta BLP}{EI}\int_0^L t\,dt = \frac{\delta BPL^3}{6EI} + \frac{\delta BPL^3}{2EI}$$

$$u = \frac{2}{3}\frac{PL^3}{EI}$$

7.8   Use of Integration Charts to Compute Internal Work _____

In Examples 7.5 and 7.6 the internal work calculation required the integration of the product of the variation of the moment from the force system and the curvature from the deformation system. The calculation is straightforward but tedious, with many opportunities for committing errors. It is possible with a computational aid to eliminate the integration step in the internal work calculation.

The moment and the curvature diagrams must be constructed prior to the work calculation. These diagrams are always drawn for each individual member of a structure and usually take a very simple form. The convention established in Chapter 5 of always drawing the moment diagrams, and hence the curvature diagrams, on the compression side of each member is strictly followed. In Example 7.5 the moment diagram for the force system is triangular, while the curvature diagram for the deformation system is a simple parabolic shape. In Example 7.6 the moment and curvature diagrams associated with the force and deformation systems are all either triangular or rectangular in shape for each member.

An examination of the diagrams in these examples shows that they are uniquely defined by a single maximum ordinate and by their geometric shape. If a diagram is uniquely defined by its shape and a maximum ordinate, the integration of the product of two such defined diagrams over the same length must also be uniquely defined in terms of the maximum ordinates in the diagrams and the length over which the integration occurs.

Table 7.1 Table 7.1 provides a computational aid for the integration of the product of two functions which are presented graphically as diagrams. The functions $g_1(x)$ and $g_2(x)$ are uniquely defined in an interval from 0 to 1 by the shape of the diagram that appears in the left column or top row of the chart with one to four lettered ordinates. Each intersection of a row and column of the chart gives the value of the integration of the product of the functions $g_1(x)$ in the row and $g_2(x)$ in the column over a *unit* length. For example, the function $g_2(x)$ in the column numbered 1 is simply $g_2(x) = A(1-x)$, while the function $g_1(x)$ in the row numbered 6 is $g_1(x) = Cx^2$. The integration of the product $g_1(x) g_2(x)$ over the interval $x = 0$ to $x = 1$ is given by the intersection of the sixth row and first column of the table, which has the value $AC/12$. This result can easily be verified by direct integration from 0 to 1 of the product $g_1(x)g_2(x) = Cx^2A(1 - x)$.

A variety of functions and the integration of their product over a unit interval are given in the table. The coordinates r and t used in some of the functions always have magnitude less than 1 and are used to define

Table 7.1 Integration Chart for Internal Work Computations

$$\int_0^1 g_1(x)g_2(x)\,dx$$

$g_1(x)$ \ $g_2(x)$	1	2	3	4	5	6
1	$AC/3$	$C(2A+B)/6$	$AC/6$	$AC(1+t)/6$	$CM/3$	$AC/2$
2	$AC/6$	$C(A+2B)/6$	$-AC/6$	$AC(2-t)/6$	$CM/3$	$AC/2$
3	$AC(2-r)/6$	$C[A(2-r)+B(1+r)]/6$	$AC(1-2r)/6$	$AC(1-r^2-t^2)/[6(1-r)(1-t)]$	$CM(1+r-r^2)/3$	$AC/2$
4	$-AC[3(1-r)^2-1]/6$	$C\{A[1-3(1-r)^2]+B(3r^2-1)\}/6$	$-AC(12r^2-6r+1)/6$	$AC(3r^2+t^2-1)/[6(1-t)]$ $r<(1-t)$	$-CM(4r^3-6r^2+1)/3$	$AC(2r-1)/2$
5	$AC/20$	$C(A+4B)/20$	$-3AC/20$	$AC(2-t)[1+(1-t)^2]/20$	$2CM/15$	$AC/4$
6	$AC/12$	$C(A+3B)/12$	$-AC/6$	$AC[1+(1-t)(2-t)]/12$	$CM/5$	$AC/3$
7	$AC/5$	$C(4A+B)/20$	$3AC/20$	$AC(1+t)(1+t^2)/20$	$2CM/15$	$AC/4$
8	$AC/4$	$C(3A+B)/12$	$AC/6$	$AC[1+t(1+t)]/12$	$CM/5$	$AC/3$
9	$ACr(3-r)/6$	$Cr[A(3-r)+Br]/6$	$ACr(3-2r)/6$	$AC[(1-r)^2+t(t+3r-2)]/6rt$ $r<(1-t)$	$CMr^2(2-r)/3$	$ACr/2$
10	$AC(1-r)^2/6$	$C(1-r)[A(1-r)+B(2+r)]/6$	$AC(1+r-2r^2)/6$	$AC[(1-r)(2-t)-3r(1-t)+r^2]/[6(1-r)(1-t)]$ $r<(1-t)$	$CM(1-r)^2(1+r)/3$	$AC(1-r)/2$
11	$A(2C+D)/6$	$[A(2C+D)+B(C+2D)]/6$	$A(C-D)/6$	$A[C(1+t)+D(2-t)]/6$	$M(C+D)/3$	$A(C+D)/2$
12	$AC/6$	$C(A-B)/6$	$AC/3$	$AC(2t-1)/6$	0	0
13	$AC(1-r)/6$	$C(1-r)(A-B)/6$	$AC(1-r)/3$	$AC(1-2t)[r^3-t(1-t)]/[6t(1-t)(1-2r)]$ $r\le t\le(1-r)$	0	0
14	$5AC/12$	$C(5A+3B)/12$	$AC/6$	$AC[5-(1-t)(2-t)]/12$	$7CM/15$	$2AC/3$
15	$AC/4$	$C(3A+5B)/12$	$-AC/6$	$AC[5-t(1+t)]/12$	$7CM/15$	$2AC/3$
16	$A(C+2M_1)/6$	$[A(C+2M_1)+B(D+2M_1)]/6$	$A(C-D)/6$	$A[Ct^2+D(1-t^2)+2M_1(1+t-t^2)]/6$	$M(C+4M_1+D)/15$	$A(C+4M_1+D)/6$
17	$A(13C+36E+9F+2D)/120$	$[A(13C+36E+9F+2D)+B(13D+36F+9E+2C)]/120$	$A[11(C-D)+27(E-F)]/120$	$A[(1+t)(2C+9E+36F+13D)-27(1-t)t^2(C-3E+3F-D)+9t^2(C+2E-7F+4D)-60Dt]/120$	$M[C+D+9(E+F)]/30$	$A[C+D+3(E+F)]/8$

points where sudden changes in slope or magnitude occur in several of the $g_1(x)$ or $g_2(x)$. For example, t defines the point of change of slope for the $g_2(x)$ function in column 4, and r defines the point where a sudden change occurs in the magnitude of the $g_1(x)$ function in row 4. Point m is always taken as the midpoint of the interval whenever it is used. Note that in rows 5, 6, 7, 8, 14, and 15, the slope to the function $g_1(x)$ is zero at one end of the *unit* interval. A few of the shapes defined in Table 7.1 need more than a single ordinate to define them uniquely. The trapezoids in column 2 and row 11 require two ordinates to define the shape, one at each end. In row 16 a general second-degree polynomial requires three ordinates to define it uniquely, the two end ordinates and the center ordinate. Similarly, the general third-degree polynomial in row 17 is completely defined by the two end ordinates, C and D, and the ordinates E and F at the one-third points.

Use of Table 7.1 very frequently eliminates the need to integrate the product of the moment and curvature functions in the internal work computations. For example, to obtain the result of a required integration, one of the diagrams is located in the left column or top row, while the other diagram is located in the top row or left column, respectively. Intersection of the row and column gives the row–column combination position in the table, which has the value of the integral over a *unit* interval. By multiplying the tabular value by the length of the interval of integration, the final desired result is obtained. Use of this chart is illustrated in Examples 7.7 to 7.9.

Example 7.7 Rotation of the left end of the same uniformly loaded simply supported beam considered in Example 7.5 is repeated. Computation of the moment diagram for the force system and the curvature diagram for the deformation system are the same as before. In step 4, the integral from Eq. (7.14) is shown, but then the mathematical definition of the variation with the coordinate, x, of the moment in the force system and curvature in the deformation system is replaced under the integration sign by the two diagrams that graphically represent this variation over the interval of interest. Referring to Table 7.1, it can be seen that the triangular diagram with maximum ordinate C which corresponds to δA at the left side appears in row 1, and the parabolic diagram with maximum ordinate M which corresponds to $(wL^2/8EI)$ appears in column 5. The intersection of this row and column, position 1–5 in the table, yields the value of the integration of these two functions over a *unit* interval as $CM/3$. For an interval of length, L, this tabular value must be multiplied by L to yield the result $CM/3L$. The positive value for the rotation θ_A indicates that it is in the same direction as the assumed virtual moment δA. See Example 7.5 for numerical values for computation.

Example 7.7

Rework Example 7.5 using the integration chart, Table 7.1.

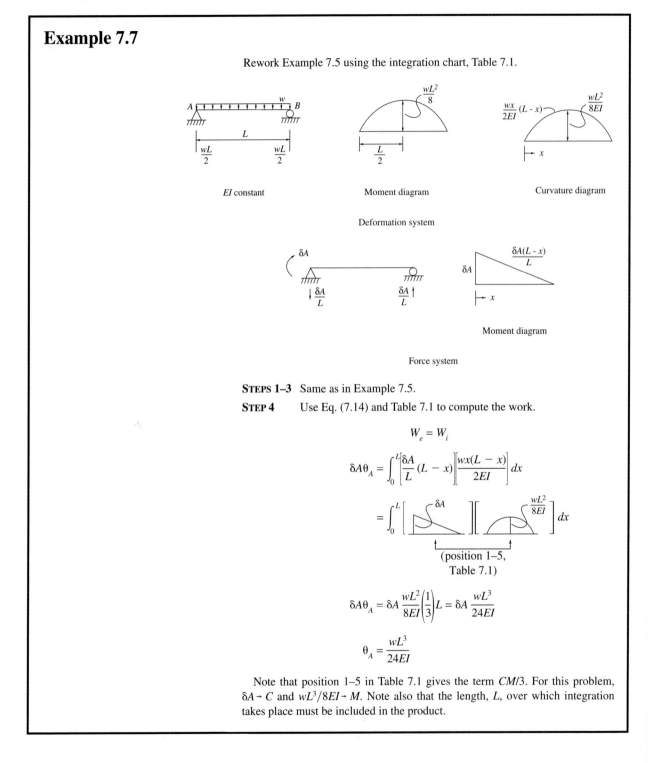

STEPS 1–3   Same as in Example 7.5.

STEP 4   Use Eq. (7.14) and Table 7.1 to compute the work.

$$W_e = W_i$$

$$\delta A \theta_A = \int_0^L \left[\frac{\delta A}{L}(L-x) \right] \left[\frac{wx(L-x)}{2EI} \right] dx$$

$$= \int_0^L \left[\qquad \right] \left[\qquad \right] dx$$

(position 1–5, Table 7.1)

$$\delta A \theta_A = \delta A \frac{wL^2}{8EI}\left(\frac{1}{3}\right)L = \delta A \frac{wL^3}{24EI}$$

$$\theta_A = \frac{wL^3}{24EI}$$

Note that position 1–5 in Table 7.1 gives the term $CM/3$. For this problem, $\delta A \rightarrow C$ and $wL^3/8EI \rightarrow M$. Note also that the length, L, over which integration takes place must be included in the product.

In Example 7.7 the moment due to the force system moment δA causes compression on the top of the beam. Similarly, the curvature due to the moment caused by the uniform load causes the top of the member to be in compression. This made the sign of the maximum ordinates in these two diagrams the same (they both are positive). The internal work computation using Table 7.1 assumes that the maximum ordinate(s) of diagrams (A, B, C, D, E, F, and M) are positive as shown, and signs in the table are correct for these conditions. However, the tabular expressions are valid for negative as well as positive values of the ordinates and will give correct values for the integrals if the ordinates in the expressions are entered with the proper sign.

The determination of the signs of the ordinates in internal virtual work computations is most easily accomplished by noting on which side of the reference line moment and curvature diagrams are plotted. When they are on the same side, as is the case in Example 7.7, both ordinates can be thought of as being positive. If the two diagrams are on opposite sides, the ordinate of one of them can be considered to be negative and the tabular expression will give the correct value when that ordinate is entered with its negative sign in the tabular expression.

Examples 7.8, 7.9 Example 7.8 repeats Example 7.6 but uses Table 7.1 to evaluate the internal work. Example 7.9 shows how Table 7.1 is used when there is a break in one of the diagrams. The deflection of the center of a simply supported, uniformly loaded beam is calculated. In using the expression from position 3–5 of Table 7.1, the value of the coordinate r that defines the location the break in slope in the diagram is required. The value of r is always less than 1 and is computed as the ratio of the distance to the break in slope from the left side (beginning) of the diagram to the total length of the diagram. In this case, r is the ratio of $L/2$ to L, or simply $1/2$. See Examples 7.5 and 7.6 for numerical values for computations.

7.9   Summary and Limitations

The virtual work principle, which is based on the concept of conservation of energy, is derived and its application in structural analysis illustrated. The principle of virtual work for rigid bodies yields the conditions of equilibrium for rigid bodies and can be used to obtain the reactions of statically determinate structures. For deformable bodies the principle requires the computation of internal work due to deformations of the body as well as the work done by the externally applied loads.

The virtual work principle has two parts. The principle of virtual work is based on a system of forces in equilibrium that is subjected to a set of small virtual compatible deformations. It is used in equilibrium calculations and can also be used for displacement-based approximate methods.

Example 7.8

Rework Example 7.6 using the integration chart, Table 7.1.

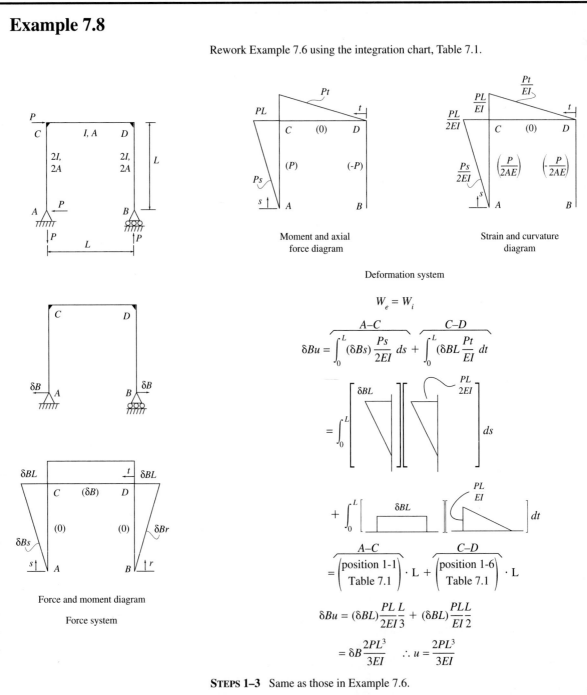

Moment and axial
force diagram

Strain and curvature
diagram

Deformation system

$$W_e = W_i$$

$$\delta Bu = \overbrace{\int_0^L (\delta Bs)\frac{Ps}{2EI}\,ds}^{A-C} + \overbrace{\int_0^L (\delta BL)\frac{Pt}{EI}\,dt}^{C-D}$$

Force and moment diagram

Force system

$$\delta Bu = (\delta BL)\frac{PL}{2EI}\frac{L}{3} + (\delta BL)\frac{PLL}{EI}\frac{1}{2}$$

$$= \delta B\frac{2PL^3}{3EI} \qquad \therefore u = \frac{2PL^3}{3EI}$$

STEPS 1–3 Same as those in Example 7.6.

STEP 4 Use Eq. (7.14) and Table 7.1 to compute the internal work.

Example 7.9

For the structure of Example 7.5, compute the vertical deflection of the center of the beam.

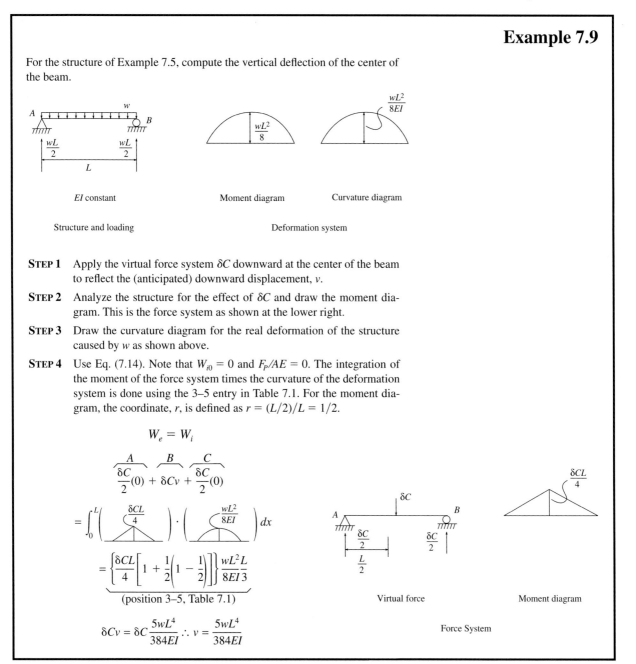

EI constant

Structure and loading

Moment diagram

Curvature diagram

Deformation system

STEP 1   Apply the virtual force system δC downward at the center of the beam to reflect the (anticipated) downward displacement, v.

STEP 2   Analyze the structure for the effect of δC and draw the moment diagram. This is the force system as shown at the lower right.

STEP 3   Draw the curvature diagram for the real deformation of the structure caused by w as shown above.

STEP 4   Use Eq. (7.14). Note that $W_{i0} = 0$ and $F_P/AE = 0$. The integration of the moment of the force system times the curvature of the deformation system is done using the 3–5 entry in Table 7.1. For the moment diagram, the coordinate, r, is defined as $r = (L/2)/L = 1/2$.

$$W_e = W_i$$

$$\overbrace{\frac{\delta C}{2}(0)}^{A} + \overbrace{\delta C v}^{B} + \overbrace{\frac{\delta C}{2}(0)}^{C}$$

$$= \int_0^L \left(\frac{\delta C L}{4} \right) \cdot \left(\frac{wL^2}{8EI} \right) dx$$

$$= \underbrace{\left\{ \frac{\delta C L}{4} \left[1 + \frac{1}{2}\left(1 - \frac{1}{2}\right) \right] \right\} \frac{wL^2 L}{8EI \, 3}}_{\text{(position 3–5, Table 7.1)}}$$

$$\delta C v = \delta C \frac{5wL^4}{384EI} \quad \therefore \; v = \frac{5wL^4}{384EI}$$

Virtual force

Moment diagram

Force System

The principle of complementary virtual work is based on a virtual force system in equilibrium that is subjected to a set of deformations derived from physically defined actions on a structure. It is used extensively to compute the deflections and rotations at specific points in structures. The material of structures being analyzed by the principles can have either linear or nonlinear stress–strain relations.

The internal work associated with the complementary virtual work principle for linear elastic materials is defined in Eqs. (7.12) to (7.14). The principle is then used to compute the deflections and rotations of structures subjected to the action of loads, temperature and fabrication error strains, and support movements. The analysis of both trusses and frames is illustrated using this principle. For frames and beams the computation of internal work requires integration. Table 7.1 presents a set of algebraic expressions that can be used to avoid the integration in the internal work computation for nearly all practical situations.

The virtual work principles provide powerful mathematical methods of analysis, but they do have limitations. These limitations are:

- The deformation analysis of trusses is subject to all of the same limitations and assumptions that are listed at the end of Chapter 3.
- Because of the use of simple bending theory in the principles, the deformation analysis of beams and frames is subject to all of the same limitations and assumptions that are listed at the end of Chapter 5 and in Section 6.2.
- The internal work expressions presented in Eqs. (7.12) to (7.14) are for linear elastic materials only.
- All material deformations must be continuous, without cracks or gaps.
- All structural deformations must be continuous, without any violation of the conditions of constraint at points of support.
- The distribution and orientation of all loads acting on a structure are not changed by deformations of the structure.
- The orientation of all reaction components at supports is unchanged by the action of applied loads or any movement of the support.
- The movements that may occur at supports are small and do not affect the initial geometry of the structure.

These limitations seem to be extensive, but are met for nearly all practical situations. Increased sophistication in the mathematical models can eliminate some or even all of these limitations, but the economics of the analysis and increased effort that is required are not warranted except in special circumstances. Some questions that should be raised in reviewing deformations that are produced in any analysis are:

- What is the effect of neglecting shear deformations, or under what circumstances should they be included?
- How much are deformations affected by the errors in the mathematical models in the near vicinity of concentrated loads and rigid joints?
- What is the effect on deformations of joints in a structure not being completely rigid?
- What differences exist between the deformations computed by other energy techniques, such as Castigliano's theorem and those computed by complementary virtual work?

PROBLEMS

7.1   Compute the vertical deflection of L_1 of the truss due to the applied load of 100 kN [E = 200 GPa; ($\cdot$) = area in cm^2].

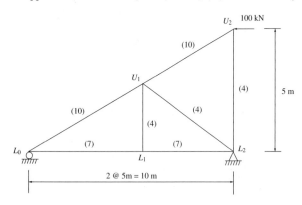

7.2   For the truss shown:
  (a)   Compute the vertical deflection of L_2.
  (b)   If U_0–U_1 is 0.5 in. too short, what are the vertical and horizontal deflections of L_2 due to the 0.5-in. fabrication error? E = 25,000 ksi for the cable member U_0–U_1; E = 29,000 ksi for all others; ($\cdot$) = area in in^2.

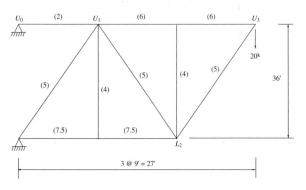

7.3   For the truss shown, compute the magnitude and direction of the deflection of U_3 due to the support settlements only (E = 10 × 10^3 ksi). Support movements are L_0: 0.01 ft to left, L_2: 0.02 ft down.

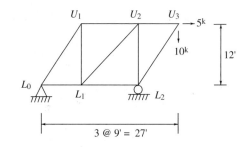

7.4   For the truss of Problem 7.1, compute the magnitude and direction of the deflection of U_2 due to the support movements only:

L_0: 60mm down
L_2: 30 mm left

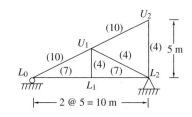

7.5 For the truss shown (E constant), the areas are:

Verticals: A
Chords: $1.2A$
Diagonals: $1.5A$

(a) Indicate what steps are necessary to compute the horizontal deflection at C due to the applied load P. Do *not* carry out the computations.
(b) Someone says that the answer to part (a) is zero. Comment as to whether or not you agree with that answer.
(c) The following support movements occur:

A: 0.2 in. right and 0.3 in. down
B: 0.2 in. down and 0.1 in. left
Compute the horizontal deflection of C.

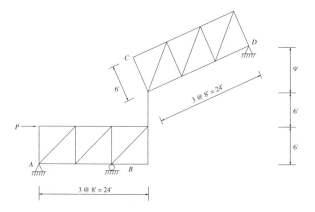

7.6 Compute the vertical deflection at L_2 of the truss due to the applied loading ($E = 29{,}000$ ksi).

Chords: $A = 5$ in^2
Verticals: $A = 3$ in^2
Diagonals: $A = 4$ in^2

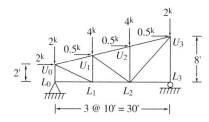

7.7 Compute the vertical deflection at L_2 of the truss due to the applied loading ($E = 200$ GPa).

Chords: $A = 3500$ mm^2
Verticals: $A = 1900$ mm^2
Diagonals: $A = 2600$ mm^2

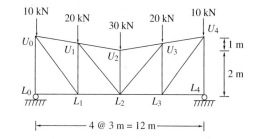

7.8 For the structure of Problem 7.6, compute the vertical deflection at L_2 due to a temperature drop in all top chord members of 40°F ($\alpha = 6.5 \times 10^{-6}$ strain/°F).

7.9 For the structure of Problem 7.7, compute the vertical deflection at L_2 due to a temperature drop in all top chord members of 25°C ($\alpha = 11.7 \times 10^{-6}$ strain/°C).

7.10 Compute the horizontal movement at L_8 of the truss due to a vertical settlement downward of 60 mm at joint L_4.

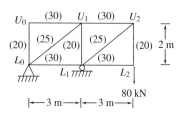

7.11 Compute the horizontal displacement at U_0 of the truss due to the applied load [$G = 200$ GPa; $(\cdot) =$ area in cm^2].

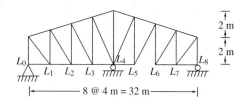

7.12   Compute the vertical deflection at L_{11} of the truss due to a fabrication error in member L_5–L_6 which makes the member ⅛ in. too short.

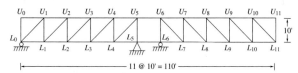

7.13   Compute the vertical deflection at L_1 of the truss due to a fabrication error in member L_3–L_4 which makes the member 40 mm too long.

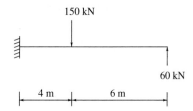

7.14   Find the location and compute the magnitude of the maximum downward displacement of the beam shown (EI constant).

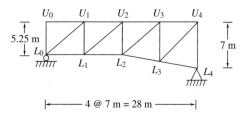

7.15   Find the vertical deflection and rotation change of slope of the left end, a (EI constant).

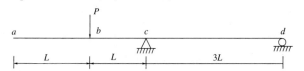

7.16   Compute the vertical deflection at points a and c and the rotation at point d. Note that the moment of inertia of the beam is $2I$ from a to d, and I from d to e (E constant).

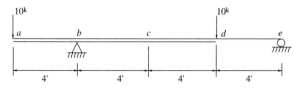

7.17   For the frame shown, compute the vertical deflection of a (E constant). Include both bending and axial forces in the virtual work computation. Indicate the percentage of the final vertical deflection at a that is due to axial force effects alone. (*Hint*: Convert all dimensions to inches before carrying out calculations.)

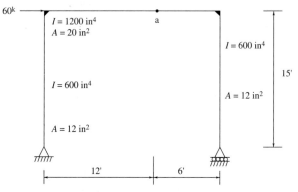

7.18   For the frame shown, compute the vertical deflection of point c. Neglect axial deformations.

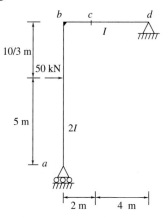

7.19   For the frame shown, compute the horizontal displacement of C. Neglect axial deformations (EI constant).

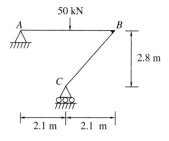

282

7.20 For the frame shown, compute the horizontal displacement of point d. Neglect axial deformations ($I = 300$ in^4, $E = 29,000$ ksi).

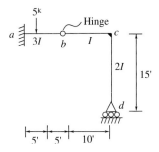

7.21 Compute the vertical deflection and end rotation of point d. Note that members a–e and e–c are pin-connected. (Member $ABCD$ only: $E = 2 \times 10^3$ ksi, $I = 3456$ in^4; members AE and EC: $E = 29 \times 10^3$ ksi.)

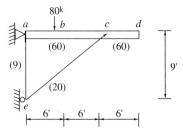

7.22 For the frame shown, compute the vertical deflection of point c. Neglect axial deformations ($E = 200$ GPa, $I = 250$ mm^4).

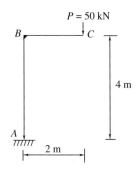

7.23 The plane frame shown experiences a temperature increase of $+40°$ F in member A–B. Compute the horizontal displacement of joint C. (All members: $E = 29 \times 10^3$ ksi, $I = 100$ in^4, $A = 8$ in^2, $\alpha = 6.5 \times 10^{-6}$ strain/$°$F.)

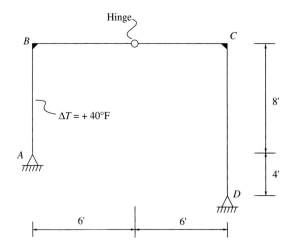

7.24 For the frame shown, obtain the horizontal deflection of C and rotation at D due to the applied loads. Neglect axial deformations (E constant).

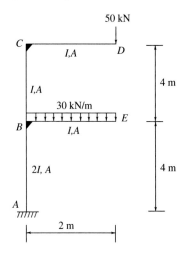

COMPUTER PROBLEMS

For the problems that follow, obtain the graphical results requested, using one of the following methods:

(a) Create a spreadsheet and use the results to obtain the graphs.

(b) Write a small computer program that obtains the results requested and create a file that can be plotted to obtain the graphs.

(c) Use available software programs that will do the analysis requested. Run the program as many times as required to obtain the results requested and create a file that can be plotted to obtain the graphs.

C7.1 Obtain a plot of $u_2/(wb^4/EI)$ versus β, where u_2 is the horizontal displacement of joint 2 of the plane frame shown. Note the hinge at 3 and vary β over the range $0 \le \beta \le 1$ for values of h/b of 0.5, 1, 1.5, and 2. Neglect axial deformations.

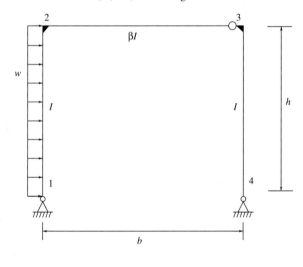

C7.2 Obtain a plot of $u_2/(wb^4/EI)$ versus a/b, where u_2 is the horizontal displacement of joint 2 of the plane frame shown. Note the hinge in member 2–3 and vary a/b over the range $0 \le a/b \le 1$ for values of h/b of 0.5, 1, 1.5, and 2. Neglect axial deformations.

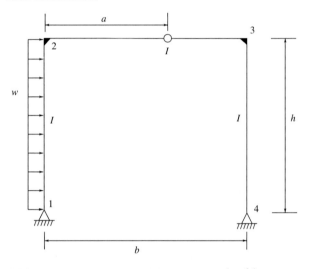

C7.3 Repeat Problem C7.1 for a plot of $u_2/(Pb^3/EI)$ versus β for the frame with the loading w replaced by the concentrated force P acting to the right at 2 as shown below.

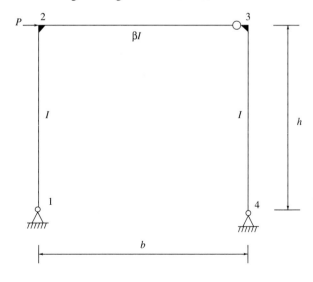

C7.4 Repeat Problem C7.2 for a plot of $u_2/(Pb^3/EI)$ versus a/b for the frame with the loading w replaced by the concentrated force P acting to the right at 2 as shown.

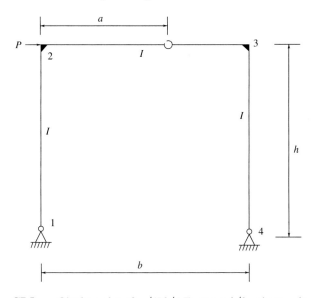

C7.5 Obtain a plot of $v_2/(Ph/AE)$ versus b/h, where v_2 is the vertical displacement of joint L_2 of the plane truss shown (E constant). Vary b/h over the range $0.5 \le b/h \le 2$. The area of the members is:

Chords: A
Verticals: $0.5A$
Diagonals: $1.2A$

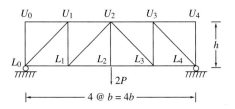

C7.6 Obtain a plot of $v_2/(Ph/AE)$ versus α, where v_2 is the vertical displacement of joint L_2 of the plane truss shown (E constant). Vary α over the range $0 \le \alpha \le 1.5$ for values of h/b of 0.25, 0.50, 0.75, and 1. The areas of the members are:

Chords: A
Verticals: $0.5A$
Diagonals: $1.2A$

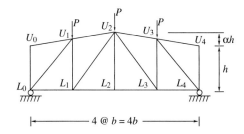

C7.7 Repeat Problem C7.5 with the area of the diagonals being $0.5A$, A, $1.2A$, and $1.5A$.

C7.8 Repeat Problem C7.6 with the area of the chords being $1.5A$, and the area of the other members unchanged.

285

8

Influence Lines for Statically Determinate Structures

In establishing mathematical models to evaluate the integrity of a proposed design, the uncertainties of the material properties and the general construction procedures must be assessed and a judgment made of their significance in the analysis of structural response. To an extent the magnitude of these uncertainties can be controlled through testing programs for the structural materials and careful monitoring of the construction process.

An equally important part of the analysis of the behavior of a structure is the loading to which the structure is subjected. Not only is the magnitude of the loads needed, but also the distribution and location of those loads on the structure. A great deal of uncertainty is associated with both the magnitude and the location of live loads on a structure. The designer must envision the worst possible combination of load magnitude and location to which the structure will be subjected during its lifetime. Only then can a realistic mathematical model be established to review the safe performance of the structure.

In the design of bridges and buildings the placement of live loading to produce the maximum stress in a member, or at a point along a member, is part of the analysis process. Configurations of live loads that produce maximum deflections are also important to that process. Experience and intuition are valuable tools of the structural engineer when undertaking these analyses, but the understanding and use of influence lines is also a great aid. In this chapter influence lines are defined and their use in assessing maximum structural response in statically determinate structures is explored. Influence lines are also used in the design of statically indeterminate structures, but the concept and use of influence lines are presented more easily with determinate structures. These ideas are extended to indeterminate systems in Chapter 13.

8.1   Definition of an Influence Line

In Chapter 1 the different types of loads to which structures are subjected were introduced and discussed. In Chapters 2 to 5 techniques were presented to enable analysis of statically determinate beams, trusses, and frames under the action of a fixed combination and configuration of concentrated and distributed loads. These analyses produced the magnitudes of the reactions of structure and the distribution of internal actions required by static equilibrium. The important and interesting question of how the magnitude of a reaction or an internal action changes because of a change in position of a single concentrated load acting on a structure is answered through the establishment and use of influence lines.

A simple definition of an influence line can be stated as follows:

> An influence line is a graphical presentation of the variation of the magnitude of a force, moment, or deflection at a single fixed point in a structure as a function of position of an applied unit load on a structure.

The two key concepts of the definition are that the force, moment, or deflection is to be measured *at a fixed point* in that structure and that the *unit load varies in position* over a specified path in the structure. Implied in the definition and use of the influence line concept is the assumption that the displacements of the structure are so small that its geometry is unchanged due to any loading. This definition is also valid for statically indeterminate structures, so for all structures:

> Influence lines for a structure are based on the assumption that the geometry of the structure is essentially unchanged by the actions of all loads that are applied to the structure.

Influence lines for deflections are not used very frequently and the remainder of this chapter is devoted to influence lines for force or moment actions.

The creation and use of influence lines is best introduced and illustrated for statically determinate beams. After developing the technique for drawing influence lines for reactions, moment at a point, and shear at a point in a beam, the use of influence lines to obtain reactions, moments, or shears at a point due to a general loading is explained. Finally, the drawing of influence lines for trusses, beams, and girder structures and the positioning of groups of loads for maximum effects is illustrated.

8.2   Influence Lines for Beams

Example 8.1 shows how the influence lines for the reactions and the moment and shear at a specified internal point for a simply supported beam are obtained. While the influence line for any force or moment action always can be written in mathematical form, a presentation of the influence line in graphical form is preferable. The influence line diagram can be constructed simply by noting the manner of variation of the force or moment action and then drawing it accordingly. The concept of obtaining influence lines follows directly from equilibrium considerations in the free body, which isolates the desired action.

Example 8.1   The influence lines for the reactions at A and B of the beam are obtained simply by summing moments about the B and A reaction points, respectively, of the whole beam. The location of the unit load is defined by the coordinate, x, which can have any value from 0 to $3L/2$. The resulting expression for the reactions are in fact functions of x, but only the nature of the variation of the reaction with position of the unit load is noted. Since in this example the variation of the reactions with position of the unit load is linear, the influence line is drawn simply by taking two convenient positions of the unit load (say, $x = 0$ and $x = L$), and computing the magnitude of the reaction of those points, plotting them in the influence line diagram, and drawing a straight line through them over the length of the load path on the structure, which is from A to D. Expressions (1) and (2) in steps 1 and 2 are plotted as the influence lines for the reactions and are labeled with numbers (1) and (2).

In drawing an influence line diagram, it is important to note the limits of applicability of each expression derived from the free-body diagram. This concept is introduced when the influence lines for the moment and shear at C are obtained. The free body for the moment at C, for example, is not of the entire structure. The free-body diagram for M_C is for the A–C part of the beam and also shows the unit load as present. Since the unit load can act anywhere on the structure, it will actually only appear in this free-body diagram when it is positioned somewhere between A and C. However, when the unit load's position is to the right of C it will

Example 8.1

Obtain the influence line for the reaction at A and B and the shear and moment at C.

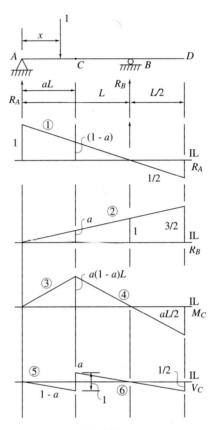

STEP 1   *Influence line for R_A:* Use the entire structure as a free body and sum moments about B.

$$\Sigma M_B\colon -R_A L + 1(L - x) = 0$$

$$R_A = 1 \cdot \frac{L - x}{L} \quad \therefore R_A \text{ linear} \qquad (1)$$

Example 8.1 (continued)

STEP 2   *Influence line for R_B:* Use the entire structure as a free body and sum moments about A.

$$\Sigma M_A:\ 1 \cdot x - R_B L = 0$$

$$R_B = 1 \cdot \frac{x}{L} \quad \therefore R_B \text{ linear} \qquad (2)$$

STEP 3   *Influence line for M_C:* Cut the beam at C and isolate the free body to the left. Note that the unit load may not be in the free body. Assume first that the unit load is *in* the free body. Show M_C and V_C in a positive sense and use equilibrium equations.

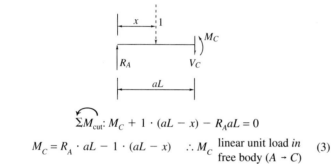

$$\Sigma M_{\text{cut}}:\ M_C + 1 \cdot (aL - x) - R_A aL = 0$$

$$M_C = R_A \cdot aL - 1 \cdot (aL - x) \quad \therefore M_C \ \begin{array}{l}\text{linear unit load } in\\ \text{free body } (A \rightarrow C)\end{array} \qquad (3)$$

When the unit load is *outside* the free body, it simply vanishes from the equilibrium expression.

$$M_C = R_A aL \quad \therefore M_C \text{ linear unit load } outside \text{ free body } (C \rightarrow D) \ (4)$$

STEP 4   *Influence line for V_C:* Use the same free body as for M_C above. Obtain an expression for V_C when the unit load is *in* the free body and *outside* the free body.

$$\Sigma F\uparrow:\ R_A - 1 - V_C = 0 \qquad V_C = R_A - 1 \quad \therefore V_C \ \begin{array}{l}\text{linear unit load } in\\ \text{free body } (A \rightarrow C)\end{array} \ (5)$$

$$V_C = R_A \quad \therefore V_C \text{ linear unit load } outside \text{ free body } (C \rightarrow D) \qquad (6)$$

not appear in the free-body diagram and can be said to be outside the free-body diagram. Thus the equilibrium expression for M_C has two forms, one when the unit load is *in* the free body [expression (3)] and a second when it is *outside* the free body [expression (4)]. The limits and form for each expression are noted and the appropriate expression used when drawing the influence line. A similar procedure is used with the same free-body diagram to obtain the influence line for V_C, which yields expressions (5) and (6). The changes in the influence lines for M_C and V_C occur because the unit load passes from the left of C to the right of C as it moves from A to B along the structure. The consequence of this is that two different equilibrium expressions are obtained for M_C and V_C, which depend on the location of the unit load with respect to point C.

The four influence lines presented in Example 8.1 give a complete picture of the variation of the reactions at A and B and the internal shear and moment at C for any position of the unit load on the structure. The units associated with the influence lines are the force unit of the unit load for the reaction and the shear influence lines and the force unit of the unit load multiplied by the length unit used in the analysis for the moment influence line. Consider, for example, the reaction R_A. As a unit load moves across the structure from A to C to B to D, the magnitude of R_A, which has the same force unit as the unit load, varies from 1 to $(1 - a)$ to 0 to $-\frac{1}{2}$. The $-\frac{1}{2}$ indicates that the magnitude of R_A is $\frac{1}{2}$ but that the direction of the reaction is down rather than up as shown in the sketch of the complete structure.

A single look at the influence line diagrams indicates how one would place a concentrated load on the structure to obtain the maximum magnitude of some particular action. To obtain the largest reaction at A due to a concentrated load, the load would be placed at A. The largest reaction at B occurs when the load is placed at D. The largest moment at C would occur when the load is placed at C or D, depending on the value of a. The largest shear occurs when, depending on the value of a, the load is placed just to the left of C, just to the right of C, or at D. Truly it can be said that an influence line provides the structural engineer with valuable insight and information about the response of a structure to loading.

A more complicated structure is analyzed for influence lines in Example 8.2. Here the presence of the hinge introduces additional changes in the shape of the influence line. A general step-by-step analysis procedure is presented in the example that is helpful in the calculations necessary for obtaining and drawing influence lines.

Example 8.2 The structure is first divided into regions, each region being defined by the ends of the member, presence of a reaction component, a continuity release (such as a hinge), or a point in a member where a stress quantity is to be computed. This division of a structure into regions is different from the division used for shear and moment diagrams in Chapter 4. There the regions that define the variation of shear and mo-

Example 8.2

Obtain the influence line for the vertical reactions at *A* and *B* and the moment and shear at *D*.

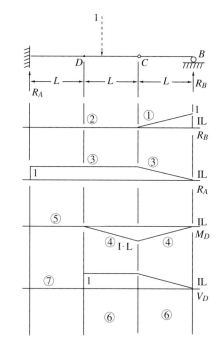

STEP 1 *Influence line for R_B:* Take the free body *C* to *B*.

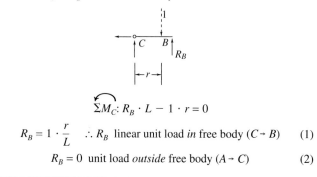

$$\Sigma M_C: R_B \cdot L - 1 \cdot r = 0$$

$$R_B = 1 \cdot \frac{r}{L} \quad \therefore R_B \text{ linear unit load } in \text{ free body } (C \rightarrow B) \quad (1)$$

$$R_B = 0 \text{ unit load } outside \text{ free body } (A \rightarrow C) \quad (2)$$

293

Example 8.2 (continued)

STEP 2 *Influence line for R_A:* Use the entire structure as the free body.

$$\Sigma F\uparrow:\ R_A - 1 + R_B = 0$$

$$R_A = 1 - R_B \quad \therefore R_A \text{ linear } (A \rightarrow B) \tag{3}$$

STEP 3 *Influence line for M_D:* Use the free body to the right of the cut through D.

$$\overset{\curvearrowright}{\Sigma M}_{\text{cut}}:\ M_D + 1t - R_B \cdot 2L = 0$$

$$M_D = 2LR_B - 1t \quad \therefore M_D \ \begin{array}{l}\text{linear unit load } in \\ \text{free body } (D \rightarrow B)\end{array} \tag{4}$$

$$M_D = 2LR_B \quad \therefore M_D \ \begin{array}{l}\text{linear unit load outside} \\ \text{free body } (A \rightarrow D)\end{array} \tag{5}$$

STEP 4 *Influence line for V_D:* Take the same free body as that used for M_D and sum forces vertically.

$$\Sigma F\uparrow:\ V_D - 1 + R_B = 0$$

$$V_D = 1 - R_B \quad \therefore V_D \text{ linear unit load } in \text{ free body } (D \rightarrow B) \tag{6}$$

$$V_D = -R_B \quad \therefore V_D \text{ linear unit load } outside \text{ free body } (A \rightarrow D) \tag{7}$$

ment are determined directly by the nature of the loading diagram, which is fixed by the specified configuration of the applied loading. In the present case the loading is not fixed, but variable, due to the unspecified position of the unit load. In this example there are three regions in which influence line diagrams can take different forms, shown by the vertical lines running through the diagrams. They are defined by the ends of the member, the location of the hinge, and point D, where the internal moment and shear are required. In Example 8.1 there are also three regions, defined by the ends of the member, the reaction at B, and point C, where the internal moment and shear are required. Again they are delineated by the vertical lines running through the diagrams.

The next step is to obtain the influence line for the reactions. This is a set of influence lines that usually requires very little effort to draw. In the present example the equation of condition for the hinge at C is used immediately to obtain the influence line for the reaction at B. In step 1 the influence line for R_B is determined to be piecewise linear, defined by the two expressions labeled (1) and (2) that come from the free-body diagram of C–B.

In step 2 the influence line for the reaction at A is obtained from a free body of the entire structure and is labeled (3). Note that the expression (3) for R_A is called linear, but is actually piecewise linear due to the piecewise linear nature of the influence line for R_B. The influence line for R_A is constructed in two stages. First it is drawn for the region A–D–C, where R_B is zero in the equation for R_A, which makes R_A constant and equal to 1. The second stage is for region C–B, where R_B, which varies linearly from 0 at C to 1 at D, is subtracted from 1 in expression (3). As can be seen, the influence lines for the reactions have different linear variations in regions A–D–C and C–B which affect the form of the influence lines for the shear and moment at D. The influence lines for the reactions will have the same force unit as that of the unit load.

Finally, the free body D–C–B is taken to isolate the desired force or moment actions and equilibrium is used to obtain their variation as a function of position of the unit load. The presence or absence of the unit load in the free body is noted for the appropriate range of its positions. The expression labeled (4) for M_D is called linear but is actually piecewise linear because of the piecewise linear character of the influence line for R_B in regions D–C and C–B. The final portion of the influence line for M_D in the region A–D is obtained from expression (5) for the condition of the unit load outside the free body. The units of the influence line are the force unit of the unit load and the units of length of L. For the same reasons stated for the influence line for the moment, M_D, the influence line for V_D is also piecewise linear in the same regions. Both influence lines for shear and moment at D have been obtained in terms of reaction (R_B in this example), which shows that the influence lines for reactions are required as part of any influence line calculation. For numerical computation in U.S. units, take $L = 6$ ft; for SI units, take $L = 2$ m.

In Examples 8.1 and 8.2, the influence lines are all linear or piecewise linear. This is a consequence of the fact that these structures are statically determinate. As shown later, statically determinate structures, be they beams, frames, or trusses, will all have piecewise linear influence lines for all force or moment actions. Indeterminate structures will have influence lines that are generally curved in nature, as shown in a later chapter.

8.3 Use of the Influence Line to Obtain the Magnitude of a Force or Moment Action for a General Loading _____

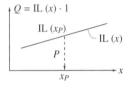

$Q = P\,\text{IL}\,(x_P)$

(a) Single concentrated load

$Q = \sum\limits_{1=i}^{m} P_i\,\text{IL}\,(x_i)$

(b) Several concentrated loads

Figure 8.1a–b
Calculation of a force or moment action due to concentrated loads.

Having an influence line for a force or moment action at a specific point significantly reduces the effort required to compute the magnitude of that force or moment action for a general loading. As seen in Fig. 8.1a, due to a single concentrated load of magnitude P, the value of a force or moment action, Q, is simply the product of P and the ordinate of the influence line for Q, IL(x), at the point of application of the load, x_P.

$$Q = P\text{IL}(x_P) \quad \text{single load} \tag{8.1}$$

This follows from the definition of the influence line that is based on a concentrated load of unit magnitude. It is assumed in Eq. (8.1) that the unit loads used to obtain IL(x) and P have the same force unit and direction of action. In Fig. 8.1b a number of loads, m, of magnitude P_i act at the m points x_i. The magnitude of Q due to all loads acting simultaneously is simply the sum of the magnitude of each load acting alone, as given in Eq. (8.1). This yields the mathematical statement

$$Q = \sum_{i \approx i}^{m} P_i IL(x_i) \quad m \text{ loads} \tag{8.2}$$

The effect of an arbitrary disturbed load $q(x)$ over a range of x from a to b can be obtained by direct integration. In Fig. 8.2 a differential slice of the distributed load, $q(x)\,dx$, is equivalent to a concentrated load, and hence the concept of summation of effects expressed by Eq. (8.2) can be used to obtain Q if the summation is replaced by integration over the range of x from a to b. This becomes

$$Q = \int_a^b q(x)IL(x)\,dx \quad \text{arbitrary distributed load} \tag{8.3}$$

The force unit of Q in Eq. (8.3) is the same as the force unit of $q(x)$. Although Eq. (8.3) defines the most general case, it is instructive to look at the common case where the distributed load $q(x)$ is of constant magnitude, w. As can be seen in Eq. (8.3), if $q(x)$ is replaced with w, it can be taken across the integration sign since it is constant. The resulting integral is simply the area under the influence line diagram. Thus

$$Q = \int_a^b wIL\,(x)\,dx = w \int_a^b IL\,(x)\,dx$$
$$= w \,[\text{area under IL}(x) \text{ between } a \text{ and } b] \qquad (8.4)$$

as shown in Fig. 8.3.

Now that the method of calculation of a force or moment action, Q, has been established using influence lines, it is a simple matter to extend the ideas to include the calculation of the absolute maximum possible value of Q for a given loading. For a single concentrated load, the maximum value of Q is given when the load acts on the structure at a point where the ordinate to the influence line for Q is an absolute maximum.

Distributed loads can act over any length of the structure. In design where the distributed load is treated as a live load (i.e., it can act over one or more different portions of the structure), the assumption is made that at some time in the life of the structure the load will be distributed on the structure in such a manner that it will cause the maximum magnitude of a force or moment action. For a uniform load, w, the maximum value of Q is obtained when the distributed load is placed on those portions of a structure such that the area, either positive or negative, under the influence line for Q is a maximum. The maximum value of Q due to the combination of a concentrated load and a uniform load of variable length is simply the sum of the maxima of the same sign due to each load type computed separately.

The discussion above shows how the structural engineer can tell exactly where to place loads and combinations of loads to achieve a maximum effect once an influence line for that effect has been established. The placement for maximum effect of only a single concentrated load has been discussed above, but the placement of multiple concentrated loads is more complicated and will be considered in a later section. The great advantage of influence lines is that once they are established, the magnitude and sense of a stress quantity can be obtained for any arbitrary loading with relative ease.

A simple example of the application of influence lines to compute maximum values of reactions and the internal moment in a structure due to both concentrated and distributed loads is shown in Example 8.3.

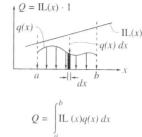

$$Q = \int_a^b IL\,(x)q(x)\,dx$$

Figure 8.2   Calculation of a force or moment action due to an arbitrary distributed load.

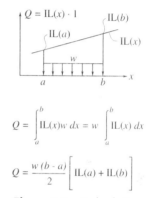

$$Q = \int_a^b IL(x)w\,dx = w \int_a^b IL(x)\,dx$$

$$Q = \frac{w\,(b-a)}{2}\left[IL(a) + IL(b) \right]$$

Figure 8.3   Calculation of a force of moment action due to a uniformly distributed load.

Example 8.3   The influence lines for the reaction, R_A, and the moment, M_C, of the structure of Example 8.1 are shown. A concentrated load of magnitude P, and a uniform distributed load of magnitude w and variable length are placed on the structure to create maximum positive and negative values of these two force actions.

In part (1) the concentrated load, P, is placed at the maximum positive and negative ordinates of the influence lines to obtain the maximum positive and negative values of R_A and M_C. In part (2), the uniform load, w,

Example 8.3

Use the influence lines for R_A and M_C from Example 8.1 for the following calculations:

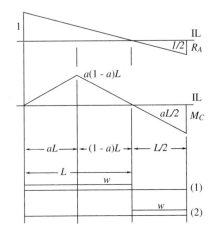

1. For a concentrated load, P, obtain the maximum positive and negative R_A and M_C. Use Eq. (8.1).

$$(R_A)_{\text{max. pos.}} = P \cdot 1 = P$$

$$(R_A)_{\text{max. neg.}} = P\left(-\frac{1}{2}\right) = -\frac{P}{2}$$

$$(M_C)_{\text{max. pos.}} = P\Big(a(1 - a)L\Big) = Pa(1 - a)L$$

$$(M_C)_{\text{max. neg.}} = P\left(\frac{-aL}{2}\right) = -P\frac{aL}{2}$$

Note: For concentrated loads the maximum positive and negative ordinate of each influence line are used in Eq. (8.1).

Example 8.3 (continued)

2. For a uniform load w of variable extent, obtain the maximum positive and negative values of R_A and M_C. Use Eq. (8.4).

$(R_A)_{\text{max. pos.}}$: Use load distribution configuration (1) above

$$(R_A)_{\text{max. pos.}} = \frac{1}{2} \cdot w \cdot 1 \cdot L = \frac{wL}{2}$$

$(R_A)_{\text{max. neg.}}$: Use load distribution configuration (2) above

$$(R_A)_{\text{max. neg.}} = \frac{1}{2} \cdot w \cdot \frac{L}{2} \cdot \left(\frac{-1}{2}\right) = \frac{-wL}{8}$$

$(M_C)_{\text{max. pos.}}$: Use load distribution configuration (1) above

$$(M_C)_{\text{max. pos.}} = \frac{1}{2}wLa\,(1 - a)L = wa(1 - a)\frac{L^2}{2}$$

$(M_C)_{\text{max. pos.}}$: Use load distribution configuration (2) above

$$(M_C)_{\text{max. neg.}} = \frac{1}{2}w\frac{L}{2} \cdot \left(\frac{-aL}{2}\right) = \frac{-waL^2}{8}$$

is assumed to act on the structure in one of the two distributions, (1) or (2) shown. Load distribution (1) corresponds to the regions in the influence line diagrams that have positive ordinates. Load distribution (2) corresponds to the regions in the influence line diagrams that have negative ordinates. To obtain the maximum positive value of R_A due to w, w must be distributed over that portion of the influence line for R_A that is positive, which corresponds to distribution (1). Using Eq. (8.4), the maximum positive value of R_A is $wL/2$, as indicated. For the maximum negative value of R_A due to w, distribution (2) is used. Again, Eq. (8.4) yields the value $-wL/8$ for R_A, which corresponds to distribution (2). Similar considerations yield the maximum positive and negative values of the moment M_C.

8.4   Influence Lines for Trusses

In Chapter 3 it was established that loads are applied to the panel points of trusses by means of a system of stringers and floor beams. Because loads are applied only at panel points of trusses, the drawing of any influence line for a truss can be undertaken simply by applying the unit load successfully at each panel point of the truss and then computing the force action. This technique is tedious and unnecessary if the load-carrying actions of a truss are understood.

Figure 3.3 shows the floor system for a highway truss bridge, which is typical of the construction of trusses to carry loads over long spans, although the floor system may connect to the top rather than bottom panel points of the truss. The floor system of the truss is constructed from floor beams that are perpendicular to the two main trusses and connect the panel point of one truss to the corresponding panel point of the other truss. The connection is moment free, so that the floor beam can be considered to be simply supported at the two panel points. Running parallel to the trusses between the floor beams are stringers which are connected in a moment-free manner to the floor beams. The stringers are considered to be simply supported beams resting on the floor beams. The deck of the structure then rests on the stringers.

The action of the floor system can be described in the following manner. Any load applied to a floor beam is carried to each of the two trusses in proportion to its location on the floor beam. The load appears at the panel points of each truss. Any load applied to a stringer is carried by proportion to each of the floor beams at the ends of the stringer and then to the four panel points of the trusses to which the two floor beams are connected. Any load applied to the deck between stringers and floor beams is assumed to go by the action of the deck proportionately to the stringers, then proportionately to the floor beams, and finally proportionately to the panel points of the truss. For every point on the deck of the truss, there is a load path for a concentrated load to follow to panel points of the two main trusses.

The process just described shows how a three-dimensional pattern of loading is resolved into a vertical loading on the two main trusses of a bridge. The design of these trusses for the vertical loading is then considered to be a two-dimensional problem. In drawing influence lines for a truss, it is convenient to take the unit loads to act in the plane of the truss.

The action of the floor system provides a linearly varying change in the magnitude of the load that appears at a panel point as the load moves parallel to the axis of the truss between that panel point and adjacent panel points. For example, if a unit load acts on the floor beam that connects to panel point L_1 in Fig. 3.3, it appears in the truss as a load at L_1 only. As the load moves from the floor beam at L_1 to the floor beam at L_2, the effect is for the floor beam at L_1 to carry a linearly decreasing proportion of the unit load while the floor beam at L_2 carries simultaneously a linearly increasing proportion of the load. When the unit load arrives at the floor beam connected to L_2, all of the load enters at panel L_2. These observations are very important to the process of drawing influence lines for trusses.

In design the influence lines for the forces in different members of a truss are wanted. The procedure that is followed to obtain these influence lines is nearly the same as that for obtaining the influence lines for shear and moment at some point in a beam. An important difference in developing the influence lines for trusses is the treatment of the unit load in the free body, which is used to isolate the force in a particular truss member. Conceptually, the unit load can either be in the free body or outside the free body as with beams, but since the free body of the truss does not include the floor system, an important refinement must be introduced. For a position of the unit load on the floor system between panel points, the effect of the unit load will enter the truss at two panel points, as discussed above. Since a free body of the truss is created by a section between panel points, the unit load may be positioned on the floor system so that the panel points where it enters the truss are both in the free body, both outside the free body, or with one in the free body and one outside it. In these situations, the unit load is said to be entirely in the free body, entirely outside the free body, or partially in the free body, respectively. These ideas are shown in Example 8.4.

Example 8.4 In the first step the influence lines for the reactions of the truss are obtained. The structure can be considered to act as a single rigid member in obtaining those influence lines. Next, the panel points are identified as dividing points for each of the six regions of the structure. Finally, using the method of sections technique presented in Chapter 3, a section is passed through the structure, isolating a free body with the member force of interest.

The influence line for F_a is obtained from the free body containing the panel points L_0, L_1, and L_2. This free body will have the unit load entirely in it as long as the unit load is acting on the floor system between L_0 and L_2 because the floor system will always transmit the effect of the unit

Example 8.4

For the truss shown, obtain influence lines for R_{L0}, R_{L6}, F_a, and F_b.

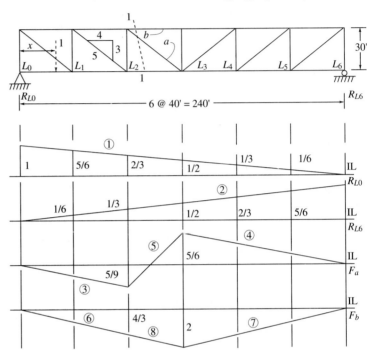

STEP 1 Obtain the influence line for R_{L0} and R_{L6} by using the entire structure as a free body.

$$\sum M_{L6}: 1 \cdot (240 - x) + R_{L0} \cdot 240 = 0 \rightarrow R_{L0} = 1 \cdot \frac{240 - x}{240}$$

$$\therefore R_{L0} \text{ linear } (L_0 \rightarrow L_6) \tag{1}$$

$$\sum M_{L0}: R_{L0} \cdot 240 - 1 \cdot x = 0 \rightarrow R_{L6} = 1 \cdot \frac{x}{240}$$

$$\therefore R_{L6} \text{ linear } (L_0 \rightarrow L_6) \tag{2}$$

Example 8.4 (continued)

STEP 2 Obtain the influence line for F_a. Take section 1–1 and isolate the free body to the left. Resolve F_a into horizontal and vertical components. Consider both the case of the unit load *entirely in* the free body and *entirely outside* the free body.

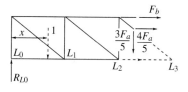

Section 1 – 1

Unit load *entirely in* free body:

$$F\uparrow: R_{L0} - 1 - 3 \cdot \frac{F_a}{5} = 0 \qquad F_a = 5\frac{R_{L0} - 1}{3}$$

$$\therefore F_a \text{ linear } (L_0 \to L_2) \tag{3}$$

Unit load *entirely outside* free body:

$$F_a = 5\frac{R_{L0}}{3} \quad \therefore F_a \text{ linear } (L_3 \to L_6) \tag{4}$$

$$\text{Connect } L_2 \text{ and } L_3 \text{ ordinates} \tag{5}$$

STEP 3 Obtain the influence line for F_b. Use the same free body as for F_a calculation and sum moments about L_3.

Unit load *entirely in* free body:

$$\overset{\curvearrowleft}{\Sigma M_{L3}}: -R_{L0} \cdot 120 + 1 \cdot (120 - x) - F_b \cdot 30 = 0$$

$$F_b = 1 \cdot \frac{120 - x}{30} - 4R_{L0} \quad \therefore F_b \text{ linear } (L_0 \to L_2) \tag{6}$$

Unit load *entirely outside* free body:

$$F_b = -4 \cdot R_{L0} \quad \therefore F_0 \text{ linear } (L_3 \to L_6) \tag{7}$$

Connect L_2 and L_3 ordinates. \hfill (8)

load to one or two of the panel points L_0 to L_2. As the unit load moves to the right of L_2, its effect is not entirely shown in the free body, because the floor system transmits a portion of the unit load to panel point L_3, which is not part of the free body. When the unit load reaches L_3, the effect of the unit load at L_2 goes to zero, because the floor beam at L_3 carries the entire load. When the unit load is between L_3 and L_6, it does not enter the truss at any of the panel points from L_0 to L_2, so it is entirely outside the free body.

The influence line for F_a can be drawn from L_0 to L_2 for the unit load entirely in the free body and from L_3 to L_6 for the unit load entirely outside the free body. When the unit load is in the panel L_2–L_3, it is only partially in the free body. However, as discussed above, the floor system transmits the effect of the unit load to L_2 in a manner that is linearly proportional to its distance from L_2, and hence the variation of the force in F_a also varies linearly. The influence line for F_a is completed simply by drawing a straight line between the ordinates of the influence at L_2 and L_3. The influence line for F_b is obtained in exactly the same manner. The influence lines for the reactions and the member axial forces all carry the force unit of the unit load.

8.5   Influence Lines for Beam-and-Girder Structures

Beam-and-girder structures are identical in behavior to a truss structures, in that the girders have loads brought to them by a floor system at panel points. In fact, comparing Figs. 4.5 and 3.3, the label "girder" in Fig. 4.5 and the label "truss" in Fig. 3.3 are interchangeable since the floor system shown in those figures are the same. With identical floor systems the analysis of beam-and-girder systems for influence lines parallels that for trusses. Examples 8.5 and 8.6 illustrate the procedure.

Example 8.5 A simple beam-and-girder system is shown in which the influence lines for the shear in panel BC and the moment at panel point D of the girder are wanted. The floor system is shown as a series of simply supported beams so that the loads can enter the girder only at the panel points labeled A to F. The influence lines for the reactions of the girder at A and F are drawn first, and then with the appropriate free bodies the internal moment and shear at the specified points in the girder are obtained. Just as with trusses, care must be taken to note when the unit load is entirely in a free body, partially in a free body, and entirely outside a free body. The panel for which the load is only partially in the free body can be treated as in trusses (i.e., the ordinates at each end of the panel are connected by a straight line). This is illustrated by the portions labeled (5) and (8) in the influence lines for V_{BC} and M_D, respectively. The force unit is the same as the unit load. For the moment, M_D, the

Example 8.5

Obtain the influence lines for the reactions at A and F, shear in panel B–C and moment at panel point D.

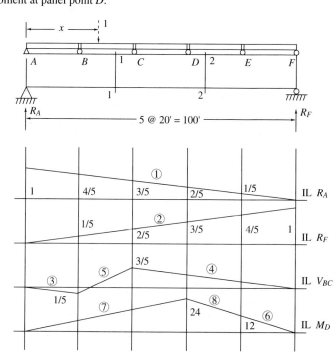

STEP 1 Obtain the influence lines for R_A and R_B. Use the entire structure as the free body and sum moments about F for R_A and about A for R_B.

$$\Sigma M_F\colon R_A \cdot 100 - 1 \cdot (100 - x) = 0$$

$$R_A = \frac{1 \cdot (100 - x)}{100} \quad \therefore R_A \text{ linear } (A \to F) \tag{1}$$

$$\Sigma M_A\colon 1 \cdot x - R_F \cdot 100 = 0$$

$$R_F = \frac{1 \cdot x}{100} \quad \therefore R_F \text{ linear } (A \to F) \tag{2}$$

Note the correspondence of the equation numbers with the influence line diagrams above.

305

Example 8.5 (continued)

STEP 2 Obtain the influence line for V_{BC}. Take section 1–1 to obtain the shear in panel B–C. Sum forces vertically and consider both the case of the unit load *entirely in* and *entirely outside* the free body. Note that F_A and F_B are replaced by the unit load when it is entirely in the free body.

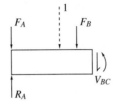

Section 1 – 1

Unit load *entirely in* free body:

$$V_{BC} = R_A - 1 \quad \therefore V_{BC} \text{ linear } (A \rightarrow B) \tag{3}$$

Unit load *entirely outside* free body:

$$(F_A = F_B = 0): V_{BC} = R_A \quad \therefore V_{BC} \text{ linear } (C \rightarrow F) \tag{4}$$

Connect the ordinates at B and C to complete the influence line. (5)

STEP 3 Obtain the influence line for M_D. Take section 2–2 to obtain the moment at D by summing the moments about the cut. Consider both the case of the unit load *entirely in* and *entirely outside* the free body. Note that F_E and F_F are replaced by the unit load when it is entirely in the free body.

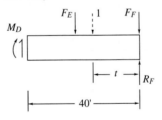

Section 2–2

Unit load *entirely in* free body:

$$\Sigma M_D: M_D + 1 \cdot (40 - t) - R_F \cdot 40 = 0$$
$$M_D = 40R_F - 1 \cdot (40 - t) \quad \therefore M_D \text{ linear } (E \rightarrow F) \tag{6}$$

Unit load *entirely outside* free body ($F_E = F_F = 0$):

$$M_D = 40R_F \quad \therefore M_D \text{ linear } (A \rightarrow D) \tag{7}$$

Connect the ordinates at D and E to complete the influence line. (8)

force unit is the same as the unit load, but the length unit is feet, since the dimensions of the structure are given in feet.

Example 8.6 A more complicated beam-and-girder system problem is shown in which a discontinuous floor system can give rise to a sudden change or jump in the ordinates of an influence line. Both the influence line for the shear, V_{BC}, and the moment, M_C, experience a sudden jump in ordinates at point G in the floor system. The cause of the jump is the sudden change in load path followed by the unit load from its point of application on the structure, through the floor system and onto the floor beams connected to the panel points. The movement of the unit load from the left side to the right side of the break or gap in the floor system at G causes a sudden change of the load path to the girder panel points from points A and B when the load is to the left of G to points C and D when it is to the right of G. The force unit for all influence lines is the force unit of the unit load. The length unit for the influence lines for M_C is feet.

8.6 Influence Lines Using Virtual Work and the Müller–Breslau Principle

In Section 7.2 the virtual work principle for rigid bodies was used to obtain the reactions of statically determinate structures. With the selection of an appropriate imagined rigid body displacement or rotation, the reactions of a statically determinate structure with a unit load acting on it can be obtained and related to the influence lines for those reactions. The concept is developed by using the beam shown in Figure 8.4a with a unit load acting a distance, x, from the left support. This is the same structure that is analyzed in Example 8.1 to illustrate the computation of influence lines.

Figure 8.4 The beam is in equilibrium under the action of the applied unit load and the reactions R_A and R_B. In Fig. 8.4b an upward displacement at B of magnitude 1 is imposed, or imagined to occur, which causes the member to rotate about A as a rigid body and develop an upward displacement at the unit load of $v_{1B}(x)$. Applying the principle of virtual work for rigid bodies, the work done by the force system must be zero, which leads to the value of the reaction to be $R_B = 1 \cdot v_{1B}(x)$. The displacement $v_{1B}(x)$ is a function of x because the displacement is defined to act at the point of application of the unit load, which has its position defined by the coordinate x. In Fig. 8.4b, the displacement $v_{1B}(x)$ defines the displaced shape of the beam AC due to the unit displacement at B. Since the magnitude of R_B is defined for every position of the unit load between A and C, the expression $R_B = 1 \cdot v_{1B}(x)$ is by definition the influence line for R_B and is simply the deflected shape of the structure multiplied by a unit force.

Example 8.6

Consider the beam-and-girder system of Example 8.5 but with the floor system configured with a gap at point G.

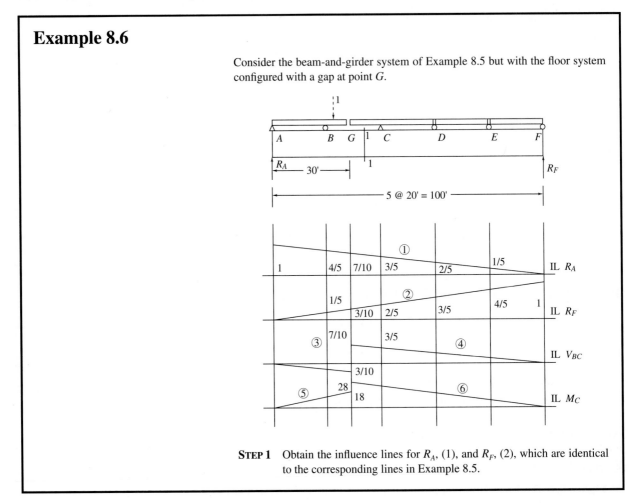

STEP 1 Obtain the influence lines for R_A, (1), and R_F, (2), which are identical to the corresponding lines in Example 8.5.

Example 8.6 (continued)

STEP 2   Obtain the influence lines for V_{BC} and M_C. Cut section 1–1 just to the left of C and isolate the free body with V_{BC} and M_C acting. Sum forces vertically for V_{BC} and moments about cut for M_C. Consider both cases of unit load *entirely in* and *entirely outside* the free body.

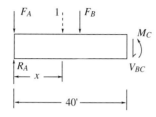

Section 1–1

Unit load *entirely in* free body: F_A and F_B replaced by unit load.

$$\Sigma F \uparrow : R_A - 1 - V_{BC} = 0$$
$$V_{BC} = R_A - 1$$
$$\therefore V_{BC} \text{ linear } (A \rightarrow G \text{ left}) \tag{3}$$

Unit load *entirely outside* free body: $F_A = F_B = 0$

$$V_{BC} = R_A$$
$$\therefore V_{BC} \text{ linear } (G \text{ right } \rightarrow F) \tag{4}$$

$$\Sigma M : M_C + 1 \cdot (40 - x) - 40 R_A = 0$$
$$M_C = 40 R_A - 1 \cdot (40 - x)$$
$$\therefore M_C \text{ linear } (A \rightarrow G \text{ left}) \tag{5}$$

$$M_C = 40 R_A$$
$$\therefore M_C \text{ linear } (G \text{ right} \rightarrow F) \tag{6}$$

The sudden jump in the influence line ordinates is due to the abrupt change in load path from supports A and B when the unit load is to the left of G to supports C and D when the unit load is to the right of G.

The influence line for reaction R_A is obtained by the same process in Fig. 8.4c. A comparison of these influence lines with those obtained in Example 8.1 for the same reactions shows that they are identical. The use of this technique, which is based on a particular application of the virtual work principle, provides a means of obtaining the influence lines for the reactions of a statically determinate structure. The introduction of a unit displacement at a force reaction or a unit rotation at a moment re-

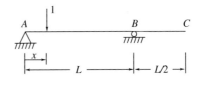

(a) Structure with unit load

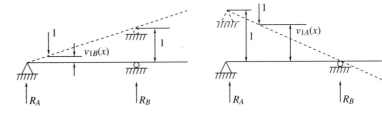

Unit displacement at B	Unit displacement at A
$W = 0$	$W = 0$
$R_B \cdot 1 - 1 \cdot v_{1B}(x) = 0$	$R_A \cdot 1 - 1 \cdot v_{1A}(x) = 0$
$R_B = 1 \cdot v_{1B}(x)$	$R_A = 1 \cdot v_{1A}(x)$
Virtual work computation	Virtual work computation

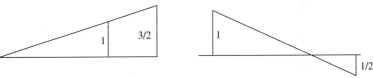

Influence line R_B Influence line R_A

(b) Use of virtual work principle to obtain influence line for R_B

(c) Use of virtual work principle to obtain influence line for R_A

Figure 8.4a–c Influence lines for reactions obtained by using the virtual work principle for rigid bodies.

action will cause the structure to deflect into the shape of the influence line for that particular reaction quantity. The influence lines are all linear or piecewise linear because the structure is statically determinate and a displacement or rotation at a reaction point produces only rigid-body displacements and rotations of the structure.

The concept of using the displaced form of a structure due to the introduction of some unit deformation as a means of obtaining influence lines for reactions can be extended to influence lines for internal actions such as moments and shears in beams and axial forces in trusses. The technique is based on the Müller–Breslau principle, which for statically determinate structures is obtained from the principle of virtual work for rigid bodies. The *Müller–Breslau principle* can be stated as follows:

> In a statically determinate structure the influence line for any force or moment action is the deflected shape of the structure that is obtained by removing the restraint corresponding to the particular action and introducing a unit deformation in the positive sense of that action.

In applying the principle to statically determinate structures, the release of a continuity restraint creates a mechanism in the structure which makes it geometrically unstable. However, the introduction of a specified unit deformation at the point of release of the restraint defines the exact form of the mechanism to be some combination of rigid-body translations and rotations. As a result, the influence lines for determinate structures are always simply straight lines.

Figure 8.5   The influence lines for a reaction and the moment and shear at an interior point in a simple beam are drawn using the Müller–Breslau principle. The structures with the releases of continuity of deformation are shown in Fig. 8.5b. Also shown are the reaction force, R_B, the internal shear force, V_C, and the internal moment, M_C, required in each case at the point of release to maintain equilibrium of the structure due to the presence of the unit load. The unit deformation of Fig. 8.5c completely defines for each case the deflected shape of the structure, which is one or a combination of two rigid-body rotations of the released structure. It is easy to show that if the principle of virtual work for rigid bodies is applied to the deformations of Fig. 8.5c in the same way it was in Fig. 8.4b and c, the values of R_B, V_C, and M_C, respectively, are given as the unit load times the corresponding displacement it experiences in each case. The displacement of the unit load due to the imposed relative unit displacement or rotation defines the displaced shape of the structure in Fig. 8.5c and the corresponding influence line in Fig. 8.5d.

Application of the Müller–Breslau principle provides a means of quickly determining the shape of the influence line for any force or moment action in the structure of Fig. 8.5. Note, for example, that the slopes of the right and left segments of the influence line for shear at C must be equal since slope continuity at C was not released when the unit deformation corresponding to the internal shear at C was introduced. The deformations shown in Fig. 8.5c have been exaggerated for clarity, but the Müller–Breslau principle for statically determinate structures and the virtual work principle for rigid bodies on which it is based require that the deformations introduced are small, so that the geometry of the structure is not altered significantly.

The Müller–Breslau principle for statically determinate structures can also be applied to obtain influence lines for trusses.

(a) Structure for which influence lines are to be drawn

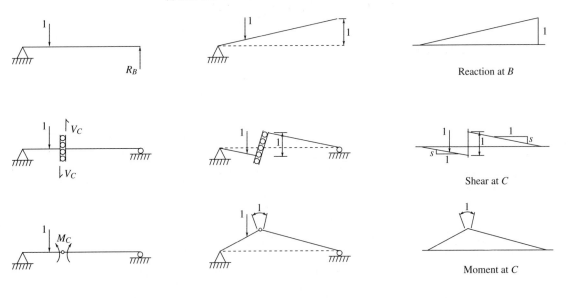

(b) Structure with restraint released

(c) Deflected shape of structure after imposing of unit deformation

(d) Influence line for force or moment action

Figure 8.5a–d Application of the Müller–Breslau principle for influence lines in a simple beam.

Figure 8.6 Several influence lines are sketched in Fig. 8.6 for a simple truss. Perhaps the easiest way to visualize the form of the influence lines is to consider the member of the truss for which the influence line is wanted as being removed from the structure and replaced with the member force required to maintain equilibrium with the applied unit load, as shown in Fig. 8.6b. The truss is then made up of two rigid sections which are joined in some manner at the panel where the member is "removed." When a deformation corresponding to axial tension in the

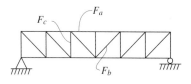

(a) Structure and members for which influence lines are to be drawn
for unit load applied along bottom chord

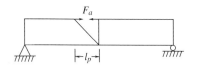

Restraint for member *a* released

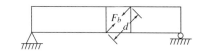

Restraint for member *b* released

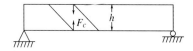

Restraint for member *c* released

(b)

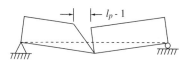

Unit deformation at *a*

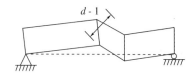

Unit deformation at *b*

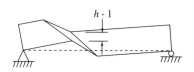

Unit deformation at *c*

(c)

Influence line F_a

Influence line F_b

Influence line F_c

(d)

Figure 8.6a–d   Influence lines for a statically determinate truss.

313

deleted member is introduced, the distance between the points of attachment of the member is reduced by a unit displacement, causing the panel to deform and the attached rigid sections simply to rotate. The deflected position of the truss lower chord, which contains the panel point by which the unit load enters the structure, defines the influence line. If the loads enter through the top panel points, the deflected position of the top chord defines the influence line.

The application of the Müller–Breslau principle to the analysis of beam-and-girder systems to obtain the influence lines for the girders is similar to that for trusses. Owing to the identical character of the floor beam and stringer system for the two structures, the influence lines for shear and moment in a girder have the same form as the influence lines for the forces F_b and F_a, respectively, in Fig. 8.6. Compare the influence lines for forces F_a and F_b in Example 8.4 with the influence line for V_{BC} and M_D in Example 8.5. When using the Müller–Breslau principle for beam-and-girder systems, the deflected shape of the structure used for influence lines is that of the floor system, because the unit load travels along the stringers. It is important to recognize the path followed by the unit load along a structure when using the Müller–Breslau principle, or sudden jumps in the influence line ordinates such as those in Example 8.6 may be overlooked.

The actual ordinates to the influence for the force or moment actions for statically determinate structures as shown in Figs. 8.5 and 8.6 can be obtained from geometric considerations and the unit magnitude of the imposed deformation. This works very well for the influence lines for reactions, but for other force or moment actions it is usually easier to sketch the shape of the influence line and then place the unit load at one or two points on the structure where the magnitude of a significant influence line ordinate is needed. The analysis of the structure for these one or two positions of the unit load is easily carried out, and the key ordinates can then be used to complete the influence line.

8.7 Absolute Maximum Moment and Shear

The discussions in previous sections have dealt with single influence lines for various force or moment actions. For trusses, the influence line for each member provides all the information needed to compute the member force for any loading acting anywhere on the structure. This fixes the force in the member due to dead load, and the design problem is to compute the maximum member force that can occur for the expected live loading configurations, a problem that has been studied in previous sections.

The design problem for beams and girders is more complicated because the design of the member is dependent on both the maximum magnitude of shear or moment and the point in the member where that maximum occurs. For each point in a beam there is a particular distribution of live loading combined with the dead loading that causes a maximum shear or moment at that point. In the common case of a member that does not vary in size along its length, the design of the member is determined by the maximum value of shear or moment that occurs at any point in the member. The problem is to find the points where the maximum value of shear and moment is the greatest. A further complication is that the loading on a member that creates a maximum moment or shear at one point may be due to some type of distribution of a continuous live loading, while at another point it may be due to a series of concentrated live loads.

A continuous member has an infinite number of influence lines for internal shear or moment because each point in the member has its own influence line. For a simply supported beam, an envelope of the maxima of all possible influence lines for shear or moment can be found as shown in Fig. 8.7. These envelopes are constructed by obtaining the maximum ordinate of an influence line for a particular position in the member, such as point C a distance aL for A in the beam of Example 8.1, and plotting the maxima as a function of position along the member.

Envelopes for influence lines are helpful in locating potential points where the maximum shear or moment may occur, but the envelopes become complicated for beams that have geometries other than the simply supported case. In addition, the maximum ordinate of the envelope only provides the location in the structure where the absolute maximum magnitude of shear or moment will occur for a single concentrated load. The more common case of loadings that are a series of concentrated loads or a combination of uniform and concentrated loads will not give the absolute maximum shear or moment when used with the influence line that has as its maximum ordinate the maximum ordinate of the envelope. The envelopes for the case of a fixed dead loading and a predetermined form of a live loading can be constructed and used for design. Reference 44 has a more extended discussion of envelopes.

In the design of statically determinate highway or railroad bridges, the absolute maximum values of shear are always found adjacent to supports. The same observation is nearly always valid for other types of statically determinate structures unless there are unusual configurations of the structure or its live loading. In statically determinate structures, the absolute maximum values of moment will occur near the center of the span or possibly at the supports in structures that have cantilevered segments. There is no general rule that can be applied to find the specific point where the maximum moment will occur.

For a simply supported beam, some more definitive observations can be made. The absolute shear will occur adjacent to a support. For a uniform distributed loading or a single concentrated load, the absolute maximum

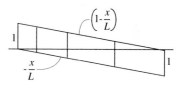

(a) Simply supported beam

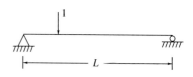

(b) Envelope of maximum ordinates of all influence lines for shear

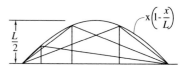

(c) Envelope of maximum ordinates of all influence lines for moment

Figure 8.7a–c Envelopes of maximum ordinates for all influence lines for shear and moment in a simply supported beam.

315

moment will occur at the center of the beam and with the concentrated load at the center. These observations can be derived directly from a careful examination of the envelopes of the influence lines for shear and moment for the simply supported beam shown in Fig. 8.7.

When the loading that is acting on a simply supported beam is a series of concentrated loads, the moment diagram for any position of the loading is a series of straight lines. This means that the maximum moment for any position of the loading must be at the point of application of one of the concentrated loads. The position of the loading on the span and some guidance toward the determination of which load the maximum absolute moment will occur under can be obtained from the following analysis.

In Fig. 8.8 a series of concentrated loads, P_i, $i = 1, \ldots, n$, act on the structure such that the position of the resultant, R, of these loads is located a distance r from the center of the span. The distance from R to an adjacent load, P_k, which is assumed for convenience to act on the opposite side of the center of the beam from R, is designated s. The reaction R_1 is determined by summing moments about the right reaction point with the result that $R_1 = (L/2 - r)(R/L)$. Taking a section through the point where the load P_k is acting, the moment about that point yields as the equation for the moment, M_k:

$$M_k + \sum_{i=1}^{k-1} t_i P_i - R_1\left(\frac{L}{2} - s + r\right) = 0$$

Substituting for R_1 and solving yields the result

$$M_k = \frac{R}{L}\left[\left(\frac{L^2}{4} - r^2\right) - s\left(\frac{L}{2} - r\right)\right] - \sum_{i=1}^{k-1} t_i \cdot P_i$$

The position of the loading system is defined by r. To find the position of the loading which will produce the maximum moment, M_k, under the load P_k, the equation for M_k is differentiated with respect to r and set equal to zero. The result is

$$\frac{dM_k}{dr} = \frac{R}{L}(-2r + s) - 0 = 0 \quad \therefore r = \frac{s}{2}$$

The value of r indicates that the maximum value of the moment M_k under the load P_k is obtained when the loading is positioned such that the center of the beam is located halfway between the resultant and the load P_k. Since P_k can be any load in the series of concentrated loads, the result defines the position of the loading which will give the maximum value of the moment under any of the loads P_i. It is assumed in the calculation that all of the concentrated loads are acting on the beam for any position of the loading.

With the position of the loading defined to give the maximum moment in the beam under any load, P_i, computation of that moment can be completed

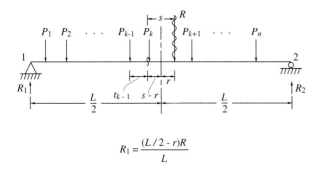

$$R_1 = \frac{(L/2 - r)R}{L}$$

(a) Beam and reaction R_1 for loading positioned so that the center
of the beam is between the resultant R and the load P_k

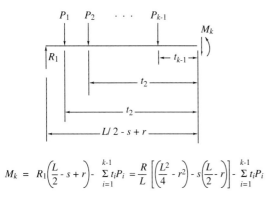

$$M_k = R_1\left(\frac{L}{2} - s + r\right) - \sum_{i=1}^{k-1} t_i P_i = \frac{R}{L}\left[\left(\frac{L^2}{4} - r^2\right) - s\left(\frac{L}{2} - r\right)\right] - \sum_{i=1}^{k-1} t_i P_i$$

(b) Free body giving value of moment M_k under load P_k

Figure 8.8a–b   Investigation of the positioning of a series of concentrated loads to produce the absolute maximum moment in a simply supported beam.

in one of two ways. The reaction R_1 can be obtained and the moment under the load computed from equilibrium. The other possibility is to draw the influence line for the moment at the position of the load P_i and use the influence line with Eq. (8.2) to obtain the moment.

The procedure just described gives the maximum moment that can occur in the beam under any load in the series of concentrated loads but does not give the absolute maximum moment that will occur in the beam due to the loading. The investigation of more than one position of the loading may be needed to obtain the absolute maximum. In nearly all practical situations, the absolute maximum moment will occur under the heavier of the two loads that are on either side of the position of the resultant, R, or the loading. In the special case where the position of the resultant coincides with

317

one of the loads, P_j, the absolute maximum moment may occur under P_j and when P_j is at the center of the beam.

8.8   Summary and Limitations

Influence lines are graphical presentations of the variation of force or moment actions, such as reaction components or forces or moments at some point in a structure as a function of position of a unit load that acts along some predetermined path across the structure. For statically determinate structures, the influence lines are all linear or piecewise-linear functions. The influence lines can be used to position concentrated loads or distributed loads so as to create a maximum value of a force or moment action.

Influence lines are constructed through the use of equilibrium equations which express the magnitude of a force or moment action as a function of position of the unit load on the structure. Influence lines for reactions are nearly always constructed first, with influence lines for forces in truss members, shears, and moments in beams obtained as a second step. Examples of the construction of influence lines for trusses, beams, and beam-and-girder systems are presented.

An extension of the principle of virtual work for rigid bodies provides a means of constructing influence lines. The Müller–Breslau principle for statically determinate structures is developed, which enables the sketching of the shape of influence lines. Using this principle, the magnitude of key ordinates for the influence lines can be established from geometric considerations alone, although the usual equilibrium calculations for positions of the unit load at key ordinate locations can also be used.

The calculation of the magnitude of force or moment actions using influence lines is established in Eq. (8.2) for concentrated loads, Eq. (8.3) for a general distributed load, and Eq. (8.4) for uniformly distributed loads. To obtain the maximum positive or negative magnitude of a force or moment action due to a distributed live load, the load is placed on the structure in all regions where the influence line ordinates are positive for the positive magnitude, or, for the negative maximum, in all regions of the structure where the influence line ordinates are negative. To obtain the maximum magnitude for a concentrated load, the load is placed at the maximum influence line ordinate.

The absolute maximum value of shear or moment in a beam is difficult to establish. In Section 8.7 it is stated that shear usually is maximum adjacent to a support. For uniform loads and single concentrated loads, the absolute maximum moment is in the center of spans or adjacent to a support when the structure has cantilevered segments. A loading that is made up of a series of concentrated loads presents additional difficulties in determining the absolute maximum moment in a beam. For the special case of a simply supported beam, the absolute maximum moment in the beam is shown to occur under a concentrated load in the series, with the loads positioned so that the center of the beam is midway between the resultant of the loading and one of the large-magnitude concentrated loads near and usually adjacent to the resultant.

Influence lines provide the engineer with a valuable tool for design considerations and for understanding structural behavior. Material properties are not required for establishing and using influence lines for statically determinate structures, so these influence lines are valid for structures made from linear or nonlinear materials. Other assumptions and limitations on the establishment and use of influence lines are the following:

- All of the limitations listed at the end of Chapters 2 to 5 apply to all influence line calculations.
- Use of the Müller–Breslau principle is based on the assumptoin that the unit displacement that is imposed on the structure is small, so that the changes in geometry caused by it can be computed using the dimensions of the original undeformed structure and the assumption of very small rotations.
- The positioning, configuration, and distribution of all live loads on the structure in order to create the maximum value of some force or moment action represents a probable situation of loading at least once during the lifetime of the structure.

Some questions that should be in the mind of engineers when they use influence lines in the design process are:

- What is the effect on the magnitude of a force or moment action of a concentrated load not acting at a dimensionless point on a structure but, in fact, being distributed over a short but finite length of the structure?
- In trusses and beam-and-girder systems, how well does the idealization of the deck–stringer–floor beam system represent the true structural action of the system in bringing the action of the loads on the structure to the main structural elements?

PROBLEMS

8.1   Compute and draw the influence lines for:

(a)   Reactions at A and B (upward)

(b)   Moment at B

(c)   Shear just to the left of B

(d)   Moment in center of span A–B

8.2   Using the influence lines from problem 8.1 and (i) an 80-kN concentrated force and (ii) a 16-kN uniform load of variable length, compute:

(a)   The maximum reaction at B

(b)   The maximum moment at B

(c)   The maximum shear to the left of B

(d)   The maximum positive and negative moment at the center of span A–B

8.3   Proceed in the following order: Draw (a) the influence line for the vertical reaction at A, (b) the vertical reaction at E, (c) the shear at D, and (d) the moment at D.

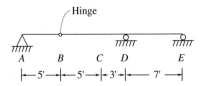

8.4   Obtain the following influence lines for beam A–B–C–D:

(a)   Vertical reaction at C

(b)   Moment at A

(c)   Shear just to the left of C

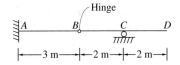

8.5   For the bridge structure shown, drawn the influence line for:

(a)   Vertical reaction at A

(b)   Vertical reaction at B

(c)   Moment in the beam at B

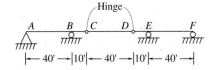

8.6   The Influence line for some action, Q, is shown. Compute the maximum positive and negative values of Q for the combined effects of w and P.

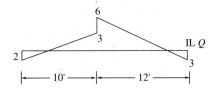

8.7 The influence line for some action is shown. Using the 50–kN concentrated load and the 25–kN/m uniform load, obtain the maximum positive and negative magnitudes of the action.

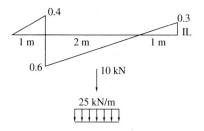

8.8 The influence line for action Q is shown. For a uniform live load of 3 kips/ft and a concentrated force of 50 kips, compute:

(a) Maximum positive value of Q

(b) Maximum negative value of Q

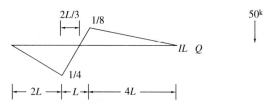

8.9 Draw the influence lines for the reaction at D, the shear in panel B–C, and the moment at panel point C.

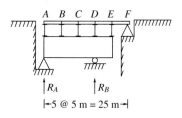

8.10 Draw the influence line for M_D, V_{DE}, M_C, and V_{BC} in the girder A–F.

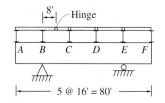

8.11 Find the influence lines for the shear in panel A–B and the moment at C.

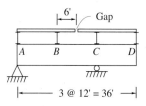

8.12 Find the influence lines for the shear in panel C–D and the moment at C.

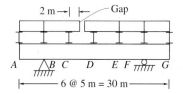

8.13 The trussed bridge shown has its deck attached to the top chord. Obtain the following influence lines:

(a) Reactions at L_1, L_5, and L_{11}

(b) Force in members a, b, and c

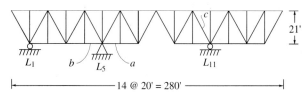

8.14   The truss shown can be subjected to a concentrated load of 50 kN and a uniformly distributed load of variable length of 7.5 kN/m. Obtain the maximum possible compression force in member U_3–U_4 and the maximum tension force in member U_2–L_3.

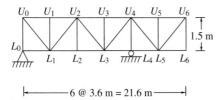

6 @ 3.6 m = 21.6 m

8.15   The highway truss shown has a floor system along the bottom chord. Note the expansion joint at the center of panel 2–3. Obtain the influence lines for the truss reactions and for tension in the indicated members.

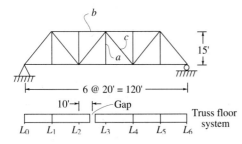

8.16   The influence line for axial force in a truss member is shown. For the concentrated load of 250 kN and a distributed load of 90 kN/m, compute the maximum tension and compression that will occur in the member.

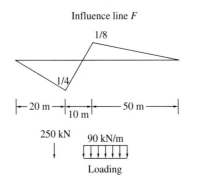

Influence line F

1/8

1/4

20 m   10 m   50 m

250 kN   90 kN/m

Loading

8.17   The truss structure shown is loaded along its top chord. Draw the influence line for the force in member a.

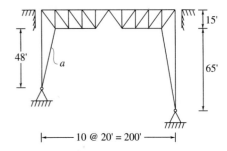

48' a 65'   15'

10 @ 20' = 200'

8.18   In the K–truss shown, obtain the influence lines for the force in the three members indicated. Assume that the load is applied to the bottom chord.

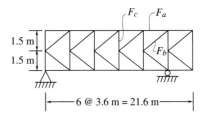

F_c F_a F_b

1.5 m
1.5 m

6 @ 3.6 m = 21.6 m

8.19   For the truss shown, draw influence lines for the members indicated. Assume that the load is applied to the bottom chord.

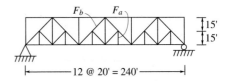

F_b F_a

15'
15'

12 @ 20' = 240'

8.20   The plane truss shown is loaded along the bottom chord. Draw the influence line for:

(a)   The reaction at C
(b)   The force in member a
(c)   The force in member b

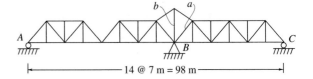

b a

A B C

14 @ 7 m = 98 m

8.21   Draw influence lines for forces in members a and b for the truss shown. Loading occurs along the top chord of the truss and is vertical.

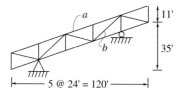

8.22   For the structure shown, draw the influence line for each of the reaction, and for the moments at B and E.

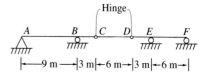

9

Deformation Analysis of Nonprismatic Beams and Beams of Nonlinear Material

In Chapter 6 bending deformations of prismatic beams made from linear elastic materials were computed utilizing simple bending theory. The use of nonprismatic members in structures is relatively common. Figure 9.1a shows a three-span continuous steel bridge with haunched girders such as those sketched in Fig. 9.1e, and Fig. 9.1b shows nonprismatic 100-ft-high piers for the interstate bridge over the Kishwaukee River. Extensions of simple bending theory to the deformation analysis of nonprismatic beams and beams with nonlinear materials is possible, but limitations on the accuracy of the computations need to be understood. For nonprismatic beams made with linear materials, this involves a review of the applicability of the theory to the variation in the geometry of the cross section along the length of the member. For the bending of beams made from nonlinear materials, this requires an evaluation of the assumption in simple bending theory that plane sections remain plane and normal strains vary linearly through the depth of the member.

Axial deformations of nonprismatic beams made with linear materials require an evaluation of the assumption in Chapter 6 of constant normal strains through the depth of the cross section. This same assumption

Example 9.1

Use the curvature-area theorems to obtain the rotation, θ, and the displacement, v, of the right end of the cantilever beam.

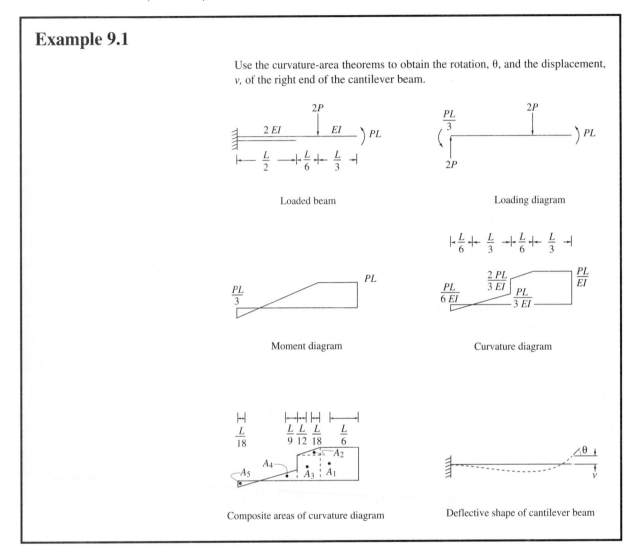

Loaded beam

Loading diagram

Moment diagram

Curvature diagram

Composite areas of curvature diagram

Deflective shape of cantilever beam

Example 9.1 (continued)

$$A_1 = \frac{PL}{EI}\frac{L}{3} = \frac{PL^2}{3EI} \qquad A_2 = \frac{1}{2}\left(\frac{PL}{3EI}\right)\frac{L}{6} = \frac{PL^2}{36EI} \qquad A_3 = \frac{2PL}{3EI}\frac{L}{6} = \frac{PL^2}{9EI}$$

$$A_4 = \frac{1}{2}\left(\frac{PL}{3EI}\right)\frac{L}{3} = \frac{PL^2}{18EI} \qquad A_5 = \frac{1}{2}\left(-\frac{PL}{6EI}\right)\frac{L}{6} = -\frac{PL^2}{72EI}$$

STEP 1 Obtain the curvature diagram from the loading and moment diagrams.

STEP 2 Divide the curvature diagram into a composite of simple geometric figures, compute areas, and locate centroids.

STEP 3 Obtain θ by using Eq. (6.10) over the complete length of the beam.

$$\theta = \Sigma A_i = \frac{PL^2}{EI}\left(\frac{1}{3} + \frac{1}{36} + \frac{1}{9} + \frac{1}{18} - \frac{1}{72}\right) = \frac{37PL^2}{72EI}$$

STEP 4 Obtain v by using Eq. (6.12) over the complete length of the beam.

$$v = A_1\frac{L}{6} + A_2\left(\frac{L}{3} + \frac{L}{18}\right) + A_3\left(\frac{L}{3} + \frac{L}{12}\right) + A_4\left(\frac{L}{2} + \frac{L}{9}\right) + A_5\left(L - \frac{L}{18}\right)$$

$$v = \frac{PL^2}{EI}\left[\frac{1}{3}\cdot\frac{L}{6} + \frac{1}{36}\cdot\frac{7L}{18} + \frac{1}{9}\cdot\frac{5L}{12} + \frac{1}{18}\cdot\frac{11L}{18} + \left(-\frac{1}{72}\right)\cdot\frac{17L}{18}\right] = \frac{173\,PL^3}{1296EI}$$

9.2   Use of Numerical Integration in Deformation Calculations

In situations where the variations of section properties is gradual, as shown in Fig. 9.1e and f, the curvature and strain functions can become very complicated and direct integration must be used. In computing the deformations of beams having nonlinear material, curvature functions may become exceedingly complicated. In these situations the integrals may be difficult or tedious to evaluate, or in some cases a closed-form expression may not exist for the integral.

An alternative approach is to evaluate the integrals using numerical integration. For many applications, the error of numerical integration is sufficiently small so as to be of no practical significance. Computer programs for structural analysis that incorporate nonprismatic members also use some form of numerical integration for members with gradually varying section properties.

Many numerical integration formulas exist, but several that are useful are listed below. The simplest but least accurate numerical integration formula is the *trapezoidal rule*. For an interval a to b, the value of the integral of a function $f(x)$ is given approximately as

$$\int_a^b f(x)\, dx \approx \frac{b-a}{2} [f(a) + f(b)] \tag{9.1}$$

This formula is based on the idea that the function $f(x)$ can be approximated as a linear function in the interval from a to b and that the value of the integral is the area under that linear function. This obviously would give very poor results in most situations since $f(x)$ is not likely to be linear or nearly linear. A means of improving the results is to divide the interval from a to b into a series of n segments of equal lengths, h, where $h = (b - a)/n$. Then the integration can be obtained from the expression known as the *extended trapezoidal rule*.

$$\int_a^b f(x)\, dx \approx h\left[\frac{f(a)}{2} + f(a + h) + f(a + 2h) + \cdots \right.$$
$$\left. + f(b - h) + \frac{f(b)}{2}\right] \tag{9.2}$$

The concept here is that $f(x)$ is approximated by n piecewise-linear functions over the n segments in the interval from a to b and that the sum of the areas under the linear functions in these segments gives the desired integral. If n is sufficiently large, any degree of accuracy is attainable. A shortcoming of this approach is that it requires many values or evaluations of the function $f(x)$.

Another and generally more accurate approach to the numerical integration problem is to use *Simpson's rule*. This gives an approximate value for the integral of $f(x)$ of

$$\int_a^b f(x)\, dx \approx \frac{b-a}{6}\left[f(a) + 4f\left(\frac{a+b}{2}\right) + f(b)\right] \quad (9.3)$$

The expression is based on the concept that $f(x)$ can be approximated as a parabola of second degree which is defined by the values of the function $f(x)$ at a and b and in the middle of the interval from a to b. The area under this parabola is given by the right-hand side of Eq. (9.3), which is then an approximation of the desired integral. This usually gives better results than what is obtained in Eq. (9.2) for $n = 2$, which requires the same number of values of the function $f(x)$. Still better results can be obtained by using *Simpson's extended rule*,

$$\int_a^b f(x)\, dx \approx \frac{h}{3}[f(a) + 4f(a+h) + 2f(a+2h) + 4f(a+3h) + \cdots$$
$$+ 2f(b-2h) + 4f(b-h) + f(b)] \quad (9.4)$$

where $h = (b-a)/n$ and n is an *even* number of intervals. Analogous to Eq. (9.2), the concept behind Eq. (9.4) is that $f(x)$ is approximated by $n/2$ piecewise-continuous parabolic segments of length $2h$.

While there are many additional numerical integration formulas, two more worthy of mention are of the *Gaussian type* and take the form for two and three points, respectively:

$$\int_a^b f(x)\, dx \approx \frac{b-a}{2}\left[f\left(\frac{a+b}{2} - c\right) + f\left(\frac{a+b}{2} + c\right)\right] \quad (9.5)$$

where

$$c = \frac{b-a}{6}\sqrt{3}$$

and

$$\int_a^b f(x)\, dx \approx \frac{b-a}{18}\left[5f\left(\frac{a+b}{2} - c\right)\right.$$
$$\left. + 8f\left(\frac{a+b}{2}\right) + 5f\left(\frac{a+b}{2} + c\right)\right] \quad (9.6)$$

where

$$c = \frac{b-a}{10}\sqrt{15}$$

333

The two-point Gauss integration formula of Eq. (9.5) is based on computing the area under the linear function over the interval from a to b that is obtained by passing a straight line through the two indicated points. In the three-point Gauss integration formula (9.6), the area over the interval from a to b under a parabola that passes through the three indicated points provides the approximation of the integral. The location of the points that define the first- and second-degree curves used in Eqs. (9.5) and (9.6), respectively, is based on a process that attempts to minimize the error of the integration. Additional information and other numerical integration expressions appear in Reference 19.

The accuracy of these numerical integration formulas is somewhat related to the fact that exact results are produced by:

- Eq. (9.1) if $f(x)$ is constant or a linear function
- Eq. (9.2) if $f(x)$ is a set of piecewise-constant or piecewise-linear functions
- Eqs. (9.3) and (9.5) if $f(x)$ is a polynomial of degree three or less
- Eq. (9.4) if $f(x)$ is a set of piecewise polynomials of degree three or less
- Eq. (9.6) if $f(x)$ is a polynomial of degree five or less

The choice of numerical integration formula is left open, as each of these formulas has advantages and disadvantages. For continuous functions such as the curvature due to a distributed load acting over the whole length of a prismatic or gradually varying nonprismatic member, Eq. (9.6) generally gives good results. The same is true for piecewise-continuous functions or functions that have step discontinuities if the integration is applied separately to the regions defined by the points of discontinuity of the function or its slope. Equation (9.2) or (9.4) also works well when the number of intervals is such that the points of discontinuity in the function or its slope all coincide with the points of evaluation of the function $f(x)$ in those equations. The application of these numerical integration formulas is illustrated in the following example.

Example 9.2   The computation of the deflection of the center of a doubly tapered simply supported beam with a uniform load is shown. The curvature function is piecewise continuous because of the change in the variation of the moment of inertia at the center of the beam. Symmetry of the structure, curvature diagram, and moment diagram of the force system used in the complementary virtual work analysis simplifies the computations. The integration is reduced to a single integral, J, of the function $g(x)$ over one-half of the structure. In this particular case, the integral can be evaluated exactly for comparison with the numerical

334

Example 9.2

Use the complementary virtual work method to obtain the deflection at the center of the doubly tapered, simply supported beam. The moment of inertia is given by $I(x) = I_0(1 + 4x/L)$ and E is constant.

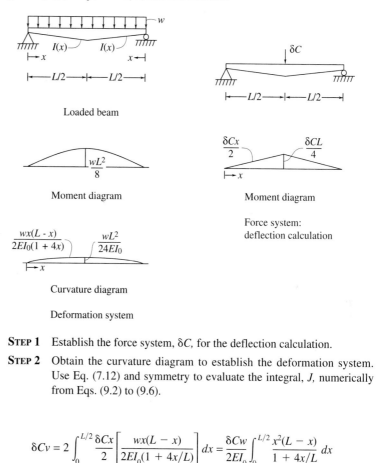

Loaded beam

Moment diagram

Moment diagram

Curvature diagram

Force system: deflection calculation

Deformation system

STEP 1 Establish the force system, δC, for the deflection calculation.

STEP 2 Obtain the curvature diagram to establish the deformation system. Use Eq. (7.12) and symmetry to evaluate the integral, J, numerically from Eqs. (9.2) to (9.6).

$$\delta Cv = 2 \int_0^{L/2} \frac{\delta Cx}{2} \left[\frac{wx(L-x)}{2EI_0(1 + 4x/L)} \right] dx = \frac{\delta Cw}{2EI_0} \int_0^{L/2} \frac{x^2(L-x)}{1 + 4x/L} dx$$

$$v = \frac{w}{2EI_0} J \quad \text{where} \quad J = \int_0^{L/2} \frac{x^2(L-x)}{1 + 4x/L} dx = \int_0^{L/2} g(x) \, dx$$

Example 9.2 (continued)

x	$(L - x)$	x^2	$1+4x/L$	$g(x)$
0	L	0	1	0
$L/4 - L/20\sqrt{15}$	$0.94365L$	$0.003175L^2$	1.2254	$0.002445L^3$
$L/4 - L/12\sqrt{3}$	$0.89434L$	$0.011165L^2$	1.4226	$0.007019L^3$
$L/4$	$0.75L$	$0.0625L^2$	2.0000	$0.023438L^3$
$L/4 + L/12\sqrt{3}$	$0.60566L$	$0.155502L^2$	2.5774	$0.036542L^3$
$L/4 + L/20\sqrt{15}$	$0.55635L$	$0.196825L^2$	2.7746	$0.039466L^3$
$L/2$	$0.50L$	$0.25L^2$	3.0000	$0.041667L^3$

J	Exact $= 0.01104L^4$ $\therefore$ Exact $v = wJ/(2EI_0)$	$= 0.00552wL^4/EI_0$
J By Eq. (9.2)	$L/4(1/2 \cdot 0 + 0.023438 + 1/2 \cdot 0.041667)L^3$	$= 0.01107L^4$
J By Eq. (9.3)	$L/12(0 + 4 \cdot 0.023438 + 0.041667)L^3$	$= 0.01128L^4$
J By Eq. (9.5)	$L/4(0.007019 + 0.036542)L^3$	$= 0.01089L^4$
J By Eq. (9.6)	$L/36(5 \cdot 0.002445 + 8 \cdot 0.023438 + 5 \cdot 0.039466)L^3$	$= 0.01103L^4$

Maximum error: 2.2% → Eq. (9.3)

integration results. The numerical integration expressions in Eqs. (9.2), (9.3), (9.5), and (9.6) are used, with the function $g(x)$ being evaluated at three points in all but Eq. (9.5), where only two points are used. The numerical integration results compare favorably with the exact result, with the various errors illustrating the type of accuracy that can be expected. Similar results would be obtained in computations for other deformation quantities.

Example 9.3 The use of Eqs. (9.2) and (9.4) provide a spacing of points that matches the discontinuity of slope in the curvature function and the moment function in the complementary virtual work calculation. Again the exact value of the integral, J, can be obtained for comparison with the numerical results. Note that the function $g(x)$ is defined differently for the left and right halves of the member. The error associated with Eq. (9.4) is only 1%, but Eq. (9.2) requires more than four intervals to provide reasonable accuracy.

The use of numerical integration in Examples 9.2 and 9.3 illustrates the capability of this approach in the evaluation of integrals. In Example 9.2 all of the numerical integration expressions gave excellent results with little computational effort. Example 9.3 illustrates the need to recognize discontinuities in slope or steps in the functions when using numerical integration. If Eqs. (9.5) and (9.6) had been used over the entire length of the member, the results would not have been meaningful since no recognition would have been made of the discontinuous slope in the function $g(x)$ in the center of the member. However, if the interval were divided into two segments of length $L/2$, Eqs. (9.5) and (9.6) would give very good results with a slight increase in computational effort. Also, if the number of intervals in the numerical integration computation is increased to six from the four shown in Example 9.3, the error in J using Eq. (9.2) drops from 11% to 5%.

9.3 Axial Deformations of Nonprismatic Members

The procedures used to calculate the axial deformations of nonprismatic members are exactly the same as those for prismatic members. The assumption of constancy of strain over any cross section is satisfied in prismatic segments of nonprismatic members, but deviations from this assumption will be present over a length of member approximately equal to its width or depth if a sudden change of cross section occurs. Similar to the situation for bending deformations of such members, the error associated with assuming that the strain suddenly changes at points of discontinuities in cross-sectional area is small and can be neglected for practical design considerations. For solid members with gradually varying width or

Example 9.3

Use the complementary virtual work method to obtain the deflection at the center of the singly tapered, simply supported beam. The moment of inertia is given by $I(x) = I_0(1 + 4x/L)$ and E is constant.

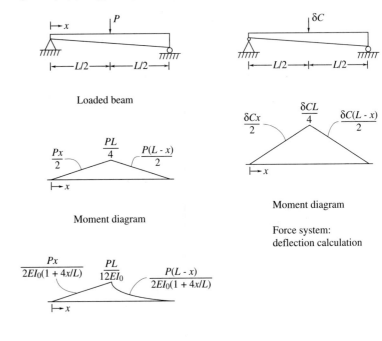

Loaded beam

Moment diagram

Force system: deflection calculation

Moment diagram

Curvature diagram

Deformation system

STEP 1   Establish the force system, δC, for the calculation of v.

STEP 2   Obtain the curvature diagram to establish the deformation system and use Eq. (7.12). Evaluate the integral, J, numerically using Eqs. (9.2) and (9.4) and four intervals $(h = L/4)$.

$$\delta C v = \int_0^{L/2} \frac{\delta C x}{2} \left[\frac{Px}{2EI_0\,(1 + 4x/L)} \right] dx$$

$$+ \int_{L/2}^{L} \frac{\delta C(L - x)}{2} \left[\frac{P(L - x)}{2EI_0\,(1 + 4x/L)} \right] dx$$

$$v = \frac{P}{4EI} J$$

Example 9.3 (continued)

where

$$J = \int_0^L g(x)\,dx \rightarrow g(x) = \frac{x^2}{(1 + 4x/L)} \qquad 0 \le x \le \frac{L}{2}$$

$$= \frac{(L - x)^2}{1 + 4x/L} \qquad \frac{L}{2} \le x \le L$$

x	$L - x$	x^2	$(L - x)^2$	$1 + 4x/L$	$g(x)$
0	L	0	L^2	1	0
$L/4$	$0.75L$	$0.0625L^2$	$0.5625L^2$	2.0000	$0.031250L^2$
$L/2$	$0.50L$	$0.25L^2$	$0.25L^2$	3.0000	$0.083333L^2$
$3L/4$	$0.25L$	$0.5625L^2$	$0.0625L^2$	4.0000	$0.015625L^2$
L	0	L^2	0	5.0000	0

$J:$ Exact $= 0.02921L^3$ $\therefore$ Exact $v = P\,\dfrac{J}{4EI_0} = 0.007302P\,\dfrac{L^3}{EI_0}$

J by Eq. (9.2): $\dfrac{L}{4}[0 + 0.03125 + 0.08333 + 0.015625 + 0]L^2 = 0.0326L^3$

J by Eq. (9.4): $\dfrac{L}{12}[0 + 4 \cdot 0.03125 + 2 \cdot 0.08333$

$$+ 4 \cdot 0.015625 + 0] \cdot L^2 = 0.0295L^3$$

Minimum error in J: 1.0% → Eq. (9.4)

depth, the strain and stress distribution is not constant on the cross section, but the error is less that 1% if the rate of change of area along the member axis is less than about 25% (see Reference 40).

The following example illustrates the calculations of the change of length of an individual member due to a distributed load acting along the length of the member.

Example 9.4 The tubular member varies in diameter along its length while the wall thickness is constant. Numerical integration of Eq. (6.14) is used in this problem and the results are compared with the exact values of the integral for different t/D ratios. The details of the numerical computations are not provided, but the procedure is the same as that used in Examples 9.2 and 9.3. The maximum error always occurs when using Eq. (9.2) with $n = 2$, but the error is not of any practical significance.

9.4   Axial Deformations of Members of Nonlinear Materials

The definition of a stress–strain relation for materials has previously been limited to the simple linear relation $\sigma = E\varepsilon$. A variety of nonlinear stress–strain relations are observed in materials, and typical idealizations of them are shown in Fig. 9.2. Any one of these can be used as the basis for illustrating the computation of axial deformations. This computation requires that the stress at any point in a member be converted to the corresponding axial strain. Use of a nonlinear stress–strain relation, which makes this conversion an easy one, speeds the computation process.

A stress–strain relation which is a modified form of one known as the Ramberg–Osgood relation (see References 18 and 27) provides a convenient and reasonably realistic representation of the stress–strain behavior

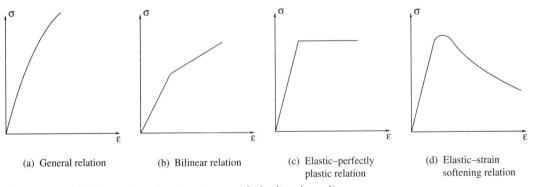

(a) General relation   (b) Bilinear relation   (c) Elastic–perfectly plastic relation   (d) Elastic–strain softening relation

Figure 9.2a–d   Examples of various types of idealized nonlinear stress–strain relationships.

Example 9.4

The tapered tube of constant wall thickness is subjected to a uniform applied force per unit length, p. Compute the displacement of the right end. Take E as constant, and compute A.

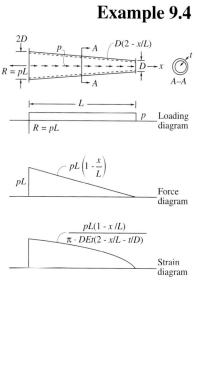

A–A

$$A = \frac{\pi[D^2(x) - (D(x) - 2t)^2]}{4} = \frac{\pi[D^2(x) - D^2(x) + 4D(x)t - 4t^2]}{4}$$

$$= \pi t[D(x) - t]$$

$$D(x) = D\left(2 - \frac{x}{L}\right) \qquad \therefore A = \pi Dt\left(2 - \frac{x}{L} - \frac{t}{D}\right)$$

STEP 1 Compute the reaction at the left end and construct the loading, force, and strain diagrams.

STEP 2 Use Eq. (6.14) to compute the displacement of the right end of the member. The area under the strain diagram can be computed by numerical integration using Eqs. (9.2) ($n = 2$), (9.3), (9.5), and (9.6) to compute J as listed in the table below.

$$u = \int_0^L \varepsilon\, dx = \int_0^L \frac{pL\,(1 - x/L)\,dx}{\pi DEt\,[2 - t/D - x/L]}$$

$$= \frac{pL^2}{\pi DEt}\int_0^L \frac{(1 - x/L)\;d(x/L)}{2 - t/D - x/L} = \frac{pL^2}{\pi DEt}J$$

t/D	Exact	Eq. (9.2)	Eq. (9.3)	Eq. (9.5)	Eq. (9.6)	Maximum Error (%)
0.05	0.31683	0.3006	0.3154	0.3178	0.3169	5.1
0.10	0.32751	0.3102	0.3258	0.3286	0.3275	5.3
0.15	0.33895	0.3203	0.3370	0.3402	0.3390	5.5
0.20	0.35126	0.3312	0.3490	0.3527	0.3514	5.7

341

of a material such as steel, which sometimes is idealized by the relation shown in Fig. 9.2c. The relation is presented in the nondimensional form as

$$\frac{\sigma}{\sigma_0} = \frac{\varepsilon/\varepsilon_0}{[1 + (\varepsilon/\varepsilon_0)^n]^{1/n}} \tag{9.7}$$

where the three parameters σ_0, ε_0, and n define the shape of the relation. The relation also has the property that it can easily be inverted to give strain in terms of stress and has the form

$$\frac{\varepsilon}{\varepsilon_0} = \frac{\sigma/\sigma_0}{[1 - (\sigma/\sigma_0)^n]^{1/n}} \tag{9.8}$$

In these equations σ_0 represents the maximum stress the material is capable of sustaining under an applied loading and corresponds, for example, to the yield stress in a mild steel; ε_0 is a reference strain corresponding closely to the strain at which yielding of a mild steel occurs and is used to establish the initial slope, E, of the stress–strain relation from the definition $\varepsilon_0\sigma_0/E$; and n defines the shape of the relation and in particular how fast the stress reaches σ_0 with the increase in strain ε.

Figure 9.3 Equation (9.7) is presented for values of n of 1, 2, and 4. The idealized stress–strain of Fig. 9.2c is also shown for reference. The stress–strain relation (9.7) is symmetric for even values of n, although symmetry of the relation can be preserved for odd values of n if the term in the denominator of Eq. (9.7) is taken as an absolute value. As n increases, the form of the stress–strain relation rapidly approaches the idealized one of Fig. 9.2c. The actual stress–strain relation of a mild steel is not much different than Eq. (9.7) with $n = 4$.

There is no restriction on the value of ε or $\varepsilon/\varepsilon_0$ in Eq. (9.7), but clearly the value of σ/σ_0 in Eq. (9.8) is limited to values of less than 1. For $n = 2$ the value of σ is approximately $0.71\sigma_0$ for $\varepsilon/\varepsilon_0 = 1$ and $0.89\sigma_0$ for $\varepsilon/\varepsilon_0 = 2$. For $n = 4$ the value of σ is approximately $0.84\sigma_0$ for $\varepsilon/\varepsilon_0 = 1$ and $0.98\sigma_0$ for $\varepsilon/\varepsilon_0 = 2$.

The slope of the stress–strain relation is obtained by differentiation of Eq. (9.7), and when the value of the initial slope, E, is taken as σ_0/ε_0 yields the result

$$\frac{d\sigma}{d\varepsilon_0} = \frac{\sigma_0/\varepsilon_0}{[1 + (\varepsilon/\varepsilon_0)^n]^{(n+1)/n}} = \frac{E}{[1 + (\varepsilon/\varepsilon_0)^n]^{(n+1)/n}} \tag{9.9}$$

which, for $n = 2$, gives approximately $0.35E$ for $\varepsilon/\varepsilon_0 = 1$ and $0.089E$ for $\varepsilon/\varepsilon_0 = 2$, where E is the slope to the stress–strain relation for $\varepsilon/\varepsilon_0 = 0$. For $n = 4$, the slope from Eq. (9.9) is $0.42E$ at $\varepsilon/\varepsilon_0 = 1$ and $0.029E$ at $\varepsilon/\varepsilon_0 = 2$.

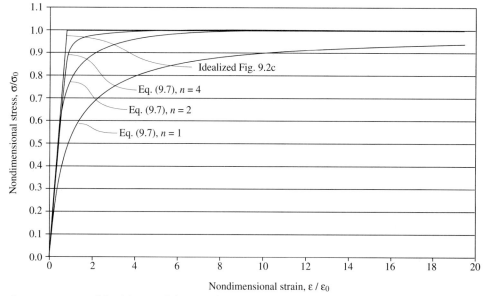

Figure 9.3   Modified form of the Ramberg–Osgood stress–strain relation given by Eq. (9.7).

The use of Eq. (9.7) and its inverted form, Eq. (9.8), to compute the elongation of a single member can be illustrated with the use of Eq. (6.14). The following example is a repeat of Example 6.7 but with the stress–strain relation defined in Eqs. (9.7) and (9.8) with $n = 2$.

Example 9.5 The member is loaded with the constant applied load, $p = 60$ kN/m, which causes the stress at the reaction to reach 214 MPa $= 0.864\sigma_0$. The loading and force diagrams appear as in Example 6.7, but the strain diagram is obtained by calculating the stress from the internal member force and using Eq. (9.8). The end displacement is found as the area under the strain diagram, which can be obtained by direct or numerical integration. By direct integration the displacement is 4.28 mm. Using numerical integration and evaluating the strains at only three points with Eqs. (9.2), (9.3), (9.5), and (9.6), the end displacement becomes 4.92, 4.48, 4.16, and 4.25 mm, respectively, giving a minimum error of 0.5% with Eq. (9.6). For comparison with linear elastic material, the results of Example 6.7 yields a displacement of 3.21 mm. For computations with U.S. units, take $L = 20$ ft, p $= 4$ kips/ft, $E = 29,000$ ksi, $\sigma_0 = 36$ ksi, $\varepsilon_0 = 0.001241$, and $A = 2.6$ in^2.

The deformation analysis of trusses was presented in Chapter 7 using the complementary virtual work approach. This approach can be used with

Example 9.5

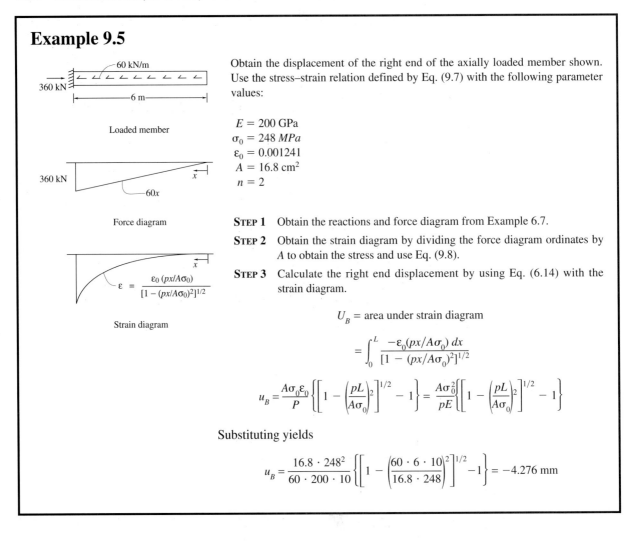

360 kN

Loaded member

360 kN

$-60x$

Force diagram

$\varepsilon = \dfrac{\varepsilon_0\,(px/A\sigma_0)}{[1-(px/A\sigma_0)^2]^{1/2}}$

x

Strain diagram

Obtain the displacement of the right end of the axially loaded member shown. Use the stress–strain relation defined by Eq. (9.7) with the following parameter values:

$E = 200$ GPa
$\sigma_0 = 248$ *MPa*
$\varepsilon_0 = 0.001241$
$A = 16.8$ cm^2
$n = 2$

STEP 1   Obtain the reactions and force diagram from Example 6.7.

STEP 2   Obtain the strain diagram by dividing the force diagram ordinates by A to obtain the stress and use Eq. (9.8).

STEP 3   Calculate the right end displacement by using Eq. (6.14) with the strain diagram.

$$U_B = \text{area under strain diagram}$$

$$= \int_0^L \frac{-\varepsilon_0(px/A\sigma_0)\,dx}{[1-(px/A\sigma_0)^2]^{1/2}}$$

$$u_B = \frac{A\sigma_0\varepsilon_0}{P}\left\{\left[1-\left(\frac{pL}{A\sigma_0}\right)^2\right]^{1/2}-1\right\} = \frac{A\sigma_0^2}{pE}\left\{\left[1-\left(\frac{pL}{A\sigma_0}\right)^2\right]^{1/2}-1\right\}$$

Substituting yields

$$u_B = \frac{16.8 \cdot 248^2}{60 \cdot 200 \cdot 10}\left\{\left[1-\left(\frac{60 \cdot 6 \cdot 10}{16.8 \cdot 248}\right)^2\right]^{1/2}-1\right\} = -4.276 \text{ mm}$$

nonlinear materials if the internal work expression, Eq. (7.5), is modified to incorporate the nonlinear stress–strain relation. Using, for an axially loaded member, the virtual stress $\delta F_Q/A$ of Eq. (7.6) for $\delta\sigma_x$, Eq. (7.5) becomes

$$W_i = \int_V \delta\sigma_x \varepsilon \, dV = \int_0^L \int_A \delta\frac{F_Q}{A} \varepsilon \, dA \, dx = \int_0^L \delta F_Q \varepsilon \, dx \qquad (9.10)$$

where the integration over the area of the cross section can be executed since both the stress and the strain are assumed constant everywhere in the cross section. The strain, ε, in the member from the deformation system is ε_P, which is derived from the stress, σ_P, the stress in the member due to the axial force F_P. Using the strain–stress relation, Eq. (9.8), assuming that the area, A, the axial force δF_Q, and the stress σ_P are constant along the length of the member and substituting into Eq. (9.10) yields the result

$$W_i = \int_0^L \delta F_Q \varepsilon_P \, dx = \int_0^L \delta F_Q \frac{\varepsilon(\sigma_P/\sigma_0)}{[1 - (\sigma_P/\sigma_0)^n]^{1/n}} \, dx$$

$$= \frac{\delta F_Q \varepsilon_0(\sigma_P/\sigma_0)L}{[1 - (\sigma_P/\sigma_0)^n]^{1/n}} \qquad (9.11)$$

Recognizing that $\sigma_P = F_P/A$, where F_P is the axial force in the member due to the applied loads on the structure, the final form of Eq. (9.11) becomes

$$W_i = \frac{\delta F_Q \varepsilon_0(F_P/A\sigma_0)L}{[1 - (F_P/A\sigma_0)^n]^{1/n}} \qquad (9.12)$$

which is the form that is directly applicable to deflection calculations. Note that the ratio ε_0/σ_0 in the numerator of Eq. (9.12) is simply $1/E$, where E is the initial slope to the stress–strain relation. To use Eq. (9.12) properly, n must be even or, if it is odd, the absolute value of F_P must be used in the denominator of the work expression.

Computation of the deflections of a truss with a nonlinear material such as that described in Eq. (9.7) is illustrated in Example 9.6. This truss is the same one analyzed in Examples 3.2 and 7.3, where internal member forces are determined from equilibrium and deformations are computed from a linear elastic stress–strain relation. The truss is statically determinate, so that internal member forces are determined from equilibrium as in the previous examples, but deformations or internal member strains are determined from the nonlinear relation, Eq. (9.8). The use of the complementary virtual work expression, Eq. (9.12), enables the computation of the deflections of the truss for a nonlinear stress–strain relation.

Example 9.6 The truss shown is loaded with the same loads as in Examples 3.2 and 7.3, but with the additional option that the magnitude of the loads can be varied by the parameter, β, which multiplies the applied loads. The condition that $\beta = 1$ corresponds to the "design" load, or the magnitude of load used in the preceding examples. The spreadsheet used in the previous examples is modified to include the parameters n, σ_0, and ε_0, describing the nonlinear stress–strain relation. The strains in the members in column E are computed using Eq. (9.8). The spreadsheet is shown with $n = 2$ and $\beta = 1$ and indicates that the displacement at L_2 is 0.1555 ft downward. This compares with the 0.1351-ft displacement obtained in Example 7.3 using the linear stress–strain relation with material modulus $E = 29,000$ ksi.

As illustrated in Example 9.6, computation of deflections of a truss with a nonlinear material does not require the introduction of any new concepts, merely the establishment of an appropriate form for the internal work.

Figure 9.4 The deflection behavior of the truss in Example 9.6 is presented as a function of the ratio of the applied load to the design load, which is characterized by the parameter β in the spreadsheet of the example. The increase in deflection of the structure with the increase of the applied loads is shown for the material of the structure having a linear

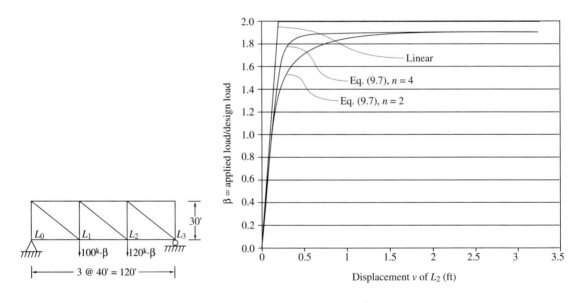

(a) Loaded truss   (b) Force and displacement relation

Figure 9.4a–b Variation of the vertical displacement of joint L_2 of the truss of Example 9.6 with the applied load for three different stress–strain relations.

Example 9.6

Expand the spreadsheet presented in Example 7.3 to accommodate a nonlinear material with the stress–strain relation defined in Eq. (9.7). Compute the vertical deflection of joint L_2 of the truss loaded as in Example 7.3 using the complementary virtual work approach and the internal work expression (9.12). The parameters for the stress–strain relation are

$E = 29{,}000$ ksi
$\sigma_0 = 36$ ksi
$\varepsilon_0 = 0.001241$
$n = 2$

Stresses, σ_P, due to applied loading; strains from Eq. (9.8)

$$\varepsilon = \frac{\varepsilon_0(\sigma_P/\sigma_0)}{[1-(\sigma_P/\sigma_0)^2]^{1/2}}$$

Deformation system

Member forces due to force δF_1 applied at L_2

Vital force system δF_1 at L_2

STEP 1 Obtain the member strains due to the applied loading by dividing the member forces calculated in Example 7.3 by the area, A, and using Eq. (9.8).

STEP 2 Take the virtual force system, δF_1, from Example 7.3.

Example 9.6 (continued)

STEP 3 Calculate the vertical deflection of L_2, v, using Eq. (9.12) for the internal work calculation. Modify the spreadsheet of Example 7.3 by adding a row of parameter values, σ_0, ε_0, and n in order to compute the nonlinear strains. A parameter, β, is also added which multiplies the applied loads so that a load–displacement relation can be obtained from the analysis.

Spreadsheet Construction

The spreadsheet of Example 7.3 is enlarged by adding row 2 with the values of σ_0, ε_0, and n that are needed to calculate the strains from the nonlinear stress–strain relation. The parameter β is also added to multiply the applied loads, and the contents of cells B5, C5, D5, and E5 are multiplied by the values of β in cell I2. The strains in column E are obtained from Eq. (9.8) using the stresses computed as the ratio of the forces in column B to the areas in column C. The internal work for each member in column G is computed from Eq. (9.12) as the product of the member lengths in column D, strains in column E, and the virtual forces in Column F. Because the magnitude of δF_1 in column F is taken as 1, the sum of the internal work for each member in cell G21 is the deflection of the truss in feet.

	A	B	C	D	E	F	G	H	I
1	PANEL:	40	HEIGHT:	30	DIAGONAL:	50	E: (in ksi)	29000	
2	n:	2	σ_0:	(in psi)	36000	ε_0:	0.001241	β:	1
3			APPLIED LOAD ON TRUSS				VIRTUAL FORCE SYSTEM δF1		
4	PANEL POINT	L0	L1	L2	L3	L0	L1	L2	L3
5	LOAD	0	-100	-120	0	0	0	-1	0
6	REACTIONS	106.67	—	—	113.33	0.33	—	—	0.67
7	MEMBER	FORCE	AREA	LENGTH	STRAIN	VIRTUAL FORCE δFI	W_1		
8	L0-U0	-106.67	6.00	30.00	-7.05E-04	-0.33	7.05E-03		
9	L0-L1	0.00	8.00	40.00	0.00E+00	0.00	0.00E+00		
10	U0-L1	177.78	10.00	50.00	7.05E-04	0.56	1.96E-02		
11	U0-U1	-142.22	10.00	40.00	-5.34E-04	-0.44	9.49E-03		
12	L1-U1	-6.67	6.00	30.00	-3.83E-05	-0.33	3.83E-04		
13	L1-L2	142.22	8.00	40.00	7.05E-04	0.44	1.25E-02		
14	U1-L2	11.11	10.00	50.00	3.83E-05	0.56	1.06E-03		
15	U1-U2	-151.11	10.00	40.00	-5.74E-04	-0.89	2.04E-02		
16	L2-U2	113.33	6.00	30.00	7.65E-04	0.67	1.53E-02		
17	L2-L3	151.11	8.00	40.00	7.65E-04	0.89	2.72E-02		
18	U2-L3	-188.89	10.00	50.00	-7.65E-04	-1.11	4.25E-02		
19	U2-U3	0.00	10.00	40.00	5.88E-20	-0.00	-5.2E-35		
20	L3-U3	0.00	6.00	30.00	1.14E-19	0.00	2.29E-34		

$$\Sigma\ [\delta F1 \times G(F/AE)] = 0.15553 = v\ (in\ feet)$$

stress–strain relation or the nonlinear relation of Eq. (9.7) for two different values of n. As indicated in the discussion of Eq. (9.7), higher values of n make the shape of the stress–strain relation more like that of the elastic–perfectly plastic material in Fig. 9.3c. This is reflected in Fig. 9.4, as a load–deflection curve for $n = 4$ follows the linear stress–strain relation for a larger portion of the loading sequence.

There are two interesting observations to be made and commented on in Fig. 9.4. First, the deflections become extremely large as the load ratio approaches 1.90588. This is due to the fact that the stress in members L_2–U_2, L_2–L_3, and U_2–L_3 becomes 36 ksi as the loads on the structure approach 1.90588 times the design load. The steel members of the structure will yield at this stress and the deflections of the structure will increase without any additional increase in load, causing the structure to collapse. The second observation is that the deflections of the structure are only about 0.5 ft at the point where the structure is about to collapse. This deflection is still sufficiently small so that the assumption of no significant change in geometry of the structure due to loading can still be used without introducing a large error in the analysis.

9.5   Bending Deformations of Members of Nonlinear Materials

The calculation of deflections and rotations of beams that are made from nonlinear materials requires a reexamination of the relations between stress and strain and moment and curvature. The treatment of this nonlinear problem will be limited to stress–strain relations that are symmetric in tension and compression, cross sections that are doubly symmetric, and for the case of bending action without axial forces. In fact, all of the assumptions and limitations discussed in Chapter 6 about bending action are applicable herein except for the use of a nonlinear rather than a linear material.

The first step in the computation of the deformations of beams is to establish a moment–curvature relation for the material. When the material is linear elastic, Eq. (6.9) gives this relation as

$$M = EI\phi \qquad (9.13)$$

If the nonlinear stress–strain relation, Eq. (9.7), is to be used for bending, the location of the neutral axis in the cross section and a relation between the internal normal stresses and the applied moment must be found in order to obtain the moment–curvature relation. For ease of development, a rectangular cross section will be used along with Eqs. (9.7) and (9.8) with $n = 2$. The assumption in simple bending theory that plane sections before deformation remain plane after deformation is observed to be applicable to bending deformations of beams with nonlinear materials and will be used.

The latter assumption determines that the variation of the normal strains, ε_x, is linear through the depth and can be taken from Eq. (6.6) to be

$$\varepsilon_x = \frac{-y}{R} = -y\phi \tag{9.14}$$

In Fig. 9.5a a rectangular cross section of height, h, and width, b, is shown with the neutral axis, or axis of zero strain located a distance $\bar{y}$ from the bottom of the cross section. Taking the origin of the coordinate system at this point, the strain variation is defined by Eq. (9.14) and the stress variation is defined by Eq. (9.7) as shown in Fig. 9.5b and c. For a bending moment acting by itself, there can be no net axial force, F, on the cross section, so equilibrium requires that

$$F = 0 = \int_A \sigma \, dA = \int_{-\bar{y}}^{h-\bar{y}} \sigma b \, dy = \int_{-\bar{y}}^{h-\bar{y}} \frac{b\sigma_0(\varepsilon/\varepsilon_0)}{\sqrt{1 + (\varepsilon/\varepsilon_0)^2}} \, dy$$

$$= b\sigma_0 \int_{-\bar{y}}^{h-\bar{y}} \frac{(-y\phi/\varepsilon_0)}{\sqrt{1 + (-y\phi/\varepsilon_0)^2}} \, dy$$

Integrating and evaluating the resulting expression at the two limits yields

$$\frac{-b\sigma_0\varepsilon_0^2}{\phi^2}\left[\sqrt{1 + \frac{\phi^2(h-\bar{y})^2}{\varepsilon_0^2}} - \sqrt{1 + \frac{\bar{y}^2\phi^2}{\varepsilon_0^2}}\right] = 0$$

which is zero when y is equal to $h/2$. This is an expected result since the cross section is symmetric and the stress–strain relation is also symmetric. The same results would be obtained for any doubly symmetric cross section.

With the location of the neutral axis established, the internal moment developed from stresses on the cross section can be developed from the equilibrium requirement

$$M = \int_A \sigma y \, dA = \int_{-h/2}^{h/2} y\sigma b \, dy = \int_{-h/2}^{h/2} \frac{b\sigma_0 y(\varepsilon/\varepsilon_0)}{\sqrt{1 + (\varepsilon/\varepsilon_0)^2}} \, dy$$

$$= b\sigma_0 \int_{-h/2}^{h/2} \frac{y(-y\phi/\varepsilon_0)}{\sqrt{1 + (-y\phi/\varepsilon_0)^2}} \, dy$$

where M is the applied moment on the cross section. Performing the integration and evaluating the resulting expression at the limits yields

$$M = -b\sigma_0\left(\frac{\varepsilon_0}{\phi}\right)^2\left[\frac{h\phi}{2\varepsilon_0}\sqrt{1 + \left(\frac{h\phi}{2\varepsilon_0}\right)^2} - \sinh^{-1}\frac{h\phi}{2\varepsilon_0}\right]$$

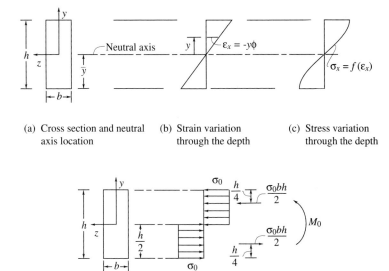

(a) Cross section and neutral
    axis location

(b) Strain variation
    through the depth

(c) Stress variation
    through the depth

(d) Stress distribution on cross section when
    the ultimate moment, M_0, acts

Figure 9.5a–d   Location of neutral axis and stress distribution
at ultimate moment in a beam of rectangular cross section
made from a material with a nonlinear stress–strain relation
and subjected to pure bending.

in which the identity

$$\ln(z + \sqrt{z^2 + 1}) = \sinh^{-1} z$$

has been used and $\ln(\,\cdot\,)$ is the natural logarithm function.

The stress–strain relation, Eq. (9.7), has a limiting value of stress, σ_0, which develops as the strains become large. In Eq. (9.14) the curvatures will become large as the strains become large, so as the curvature increases the stresses in the cross section become σ_0 in tension and compression on the bottom and top, respectively, of the beam. Further increases in curvature result in the σ_0 stresses developing closer and closer to the center of the section until the limiting state of stress shown in Fig. 9.5d is reached. At this point the maximum moment capacity of the cross section is obtained and is defined to be M_0. From the stress distribution in Fig. 9.5d it is easy to show that M_0 is given by

$$M_0 = \frac{\sigma_0 bh^2}{4} \tag{9.15}$$

A reference curvature, ϕ_0, is defined as the curvature expressed in the linear elastic moment–curvature relation, Eq. (9.13), which corresponds to the maximum moment M_0. For a rectangular cross section, $I = bh^3/12$, and for the stress–strain relation, Eq. (9.7), $\varepsilon_0 = \sigma_0/E$. With these relations and the definition of M_0 in Eq. (9.15), the reference curvature ϕ_0 becomes for the rectangular cross section, from Eq. (9.13),

$$\phi_0 = \frac{M_0}{EI} = -\frac{\sigma_0 \cdot b \cdot h^2/4}{Ebh^3/12} = -\frac{3\sigma_0}{Eh} \quad \text{or} \quad \sigma_0 = -3\frac{\varepsilon_0}{h} \qquad (9.16)$$

where the negative sign is required because a positive moment and positive curvature induce a negative strain on the portion of the cross section where y is positive [see Eq. (9.14)].

Using the definitions in Eqs. (9.15) and (9.16), the final expression for the relation between moment and curvature for a rectangular cross section with the nonlinear material defined by Eq. (9.7) with $n = 2$ becomes

$$M = M_0 \left(\frac{2}{3}\frac{\phi_0}{\phi}\right)^2 \left[\frac{3}{2}\frac{\phi}{\phi_0}\sqrt{1 + \left(\frac{3}{2}\frac{\phi}{\phi_0}\right)^2} - \sinh^{-1}\left(\frac{3}{2}\frac{\phi}{\phi_0}\right)\right] \qquad (9.17)$$

The form of the moment–curvature relation in Eq. (9.17) is shown in Fig. 9.6. Although Eq. (9.17) is developed from a stress–strain relation that is consistent with behavior of a material such as mild steel, the equation is complicated to use and suffers from the disadvantage that it cannot be inverted to give the curvature in terms of the moment. A nonlinear moment–curvature relation that is analogous in form to the stress–strain relation, Eq. (9.7), is

$$M = \frac{M_0(\phi/\phi_0)}{[1 + (\phi/\phi_0)^n]^{1/n}} \qquad (9.18)$$

Figure 9.6 Equations (9.17) and (9.18) for $n = 2$ and 4 and a form of moment–curvature relation similar to the idealized stress–strain relation in Fig. 9.3c are illustrated. The shape of the moment–curvature relation, Eq. (9.18), for $n = 2$ is very similar to that of Eq. (9.17), which is for a rectangular cross section. For a W- or I-shaped cross section, Eq. (9.18) with $n = 2$ provides a very close approximation of those moment–curvature relations, which can be improved by using $n = 4$. The figure shows that the use of Eq. (9.18) provides a reasonable moment–curvature relation for idealizing bending of beams beyond the elastic limit.

For the remainder of the calculations in this section, Eq. (9.18) will be used as the nonlinear moment–curvature relation. The inverse of this relation is

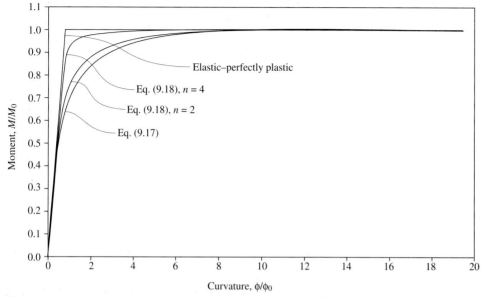

Figure 9.6 Moment–curvature relationships for three different assumptions of nonlinear material behavior.

$$\phi = \frac{\phi_0(M/M_0)}{[1 - (M/M_0)^n]^{1/n}} \qquad (9.19)$$

In Eqs. (9.18) and (9.19) the parameters M_0 and ϕ_0 are defined so that the initial slope to the moment–curvature relation is EI, where E is the initial ($\varepsilon = 0$) modulus of elasticity of the material and I is the moment of inertia of the cross section assuming that the material is linear elastic. Thus

$$M_0 = EI\phi_0 \qquad (9.20)$$

The moment, M_0, is the maximum moment the cross section is able to sustain when the cross section is fully stressed to a stress of $\pm\sigma_0$ and ϕ_0 is the elastic curvature that corresponds to M_0, as defined in Eq. (9.20). In Eq. (9.19) the moment M cannot exceed the moment M_0, which is the maximum moment the cross section can carry. Also, in both Eqs. (9.18) and (9.19), n must be even for the relation to be symmetric in positive and negative bending, or if n is odd, the curvature and moment terms in the denominators of these equations must be taken as absolute values.

Using the moment–curvature relations (9.18) and (9.19), the rotation and vertical displacement at the end of a cantilever beam are computed in Example 9.7.

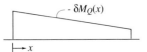

(a) Moment in force system

(b) Curvature in deformation system

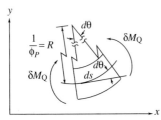

(c) Deformation of beam segment

$$dW_i = \delta M_Q \, d\theta$$

(d) Internal work dW_i

Figure 9.7a–d Internal work of moment δM_Q due to curvature in deformation system.

Example 9.7 The vertical load on the end of the cantilever beam creates a linear moment diagram that is transformed into a curvature diagram through the use of Eq. (9.19). The end slope or rotation and displacement are obtained using the curvature-area theorems, although the conjugate beam method can also be used. The expressions for θ and v will reduce to those for a linear material at very small values of the maximum moment PL. If the square root function in the θ expression and the square root and inverse sine functions in the v expression are expanded in a Taylor series, for very small values of PL/M_0 and with the substitution of the relation between M_0 and ϕ_0 in Eq. (9.20), the θ rotation becomes $PL^2/2EI$ and the v displacement becomes $-PL^3/3EI$. If the material of the member had been taken as totally elastic, the slope and displacement of the member would be 0.01398 rad and 1.118 in. Using numerical integration with no more than three evaluations of the curvature to obtain the integrals yields slopes of 0.0353, 0.0288, 0.0201, and 0.0214 rad and deflections of 3.755, 2.823, 1.688, and 1.865 in. from Eqs. (9.2), (9.3), (9.5), and (9.6), respectively. The minimum error in these results is with Eq. (9.6) and is 4% for the slope and 5.6% for the deflection.

It is interesting to note again in this example that the vertical deflection and slope of the left end of the member are still very small, even though the maximum moment at the support is nearly 97% of that which would cause the beam to collapse. The assumption of small slopes and deflections is still valid in the mathematical model of bending deformations. It must be pointed out that the expressions obtained for the slope and deflection are valid for all values of the maximum moment, PL, less than M_0, but cannot be used in the limit when PL equals M_0 because the denominators in the integral expressions for θ and v become zero.

To compute the deflections and rotations of more complicated structures, it is useful to establish an internal work expression for use with nonlinear moment–curvature relations. In Fig. 9.7a, a moment, $\delta M_Q(x)$, associated with some virtual or imagined force system is acting on a member of length L. A deformation system curvature, $\phi_P(x)$, shown in Fig. 9.7b, causes the member to deform and internal work is done by the moment $\delta M_Q(x)$. An infinitessimal length, ds, of the deformed member is shown in Fig. 9.7c with the moment δM_Q acting. Similar to the action shown in Fig. 6.2c, the curvature of the deformation system acting on the member causes a change in slope, $d\theta$, and hence a relative rotation of the same amount of the right side of the element relative to the left. Internal work is done by the moment δM_Q as it rotates through the angle $d\theta$ which, in turn, is related to the curvature, ϕ_P, by Eq. (6.1). The total internal work for the

Example 9.7

For the cantilever beam shown, obtain the slope and deflection of the left end due to the applied load, P. Assume that the material follows the moment–curvature curve shown in Eqs. (9.18) and (9.19) with $n = 2$. Evaluate the slope and deflection for a steel W 12×16 beam. For W 12×16:

$$M_0 = 720 \text{ kips-in}$$

$$EI = 2.987 \times 10^6 \text{ kips-in.}^2$$

$$\phi_0 = \frac{M_0}{EI} = 0.000241/\text{in}$$

$$P = 5.8 \text{ kips}$$

$$L = 10 \text{ ft} = 120 \text{ in}$$

$$\frac{PL}{M_0} = 5.8 \cdot \frac{120}{720} = 0.967$$

Moment diagram

$$\phi = \frac{\phi_0 \, (Px/M_0)}{[1 - (Px/M_0)^2]^{1/2}}$$

Curvature diagram

STEP 1 Obtain the moment and curvature diagrams. Use Eq. (9.19) to establish the curvature variation.

STEP 2 Use the first curvature-area theorem, Eq. (6.10), to compute the slope, θ, at the left end.

$$\theta = -\int_0^L \phi \, dx = -\int_0^L \frac{-\phi_0(Px/M_0) \, dx}{[1 - (Px/M_0)^2]^{1/2}} = \frac{-\phi_0 M_0}{P}\left\{ \left[1 - \left(\frac{Px}{M_0}\right)^2\right]^{1/2} \right\}_0^L$$

$$= \frac{-\phi_0 M_0}{2}\left\{ \left[1 - \left(\frac{PL}{M_0}\right)^2\right]^{1/2} - 1 \right\}$$

Example 9.7 (continued)

STEP 3 Use the second curvature-area theorem, Eq. (6.12), to compute the deflection, v, at the left end.

$$v = \int_0^L x\phi \, dx = \int_0^L \frac{-x\phi_0(Px/M_0) \, dx}{[1 - (Px/M_0)^2]^{1/2}}$$

$$= -\frac{\phi_0 M_0^2}{P^2} \left\{ -\frac{Px}{2M_C} \left[1 - \left(\frac{Px}{M_0}\right)^2 \right]^{1/2} + \frac{1}{2} \sin^{-1} \frac{Px}{M_0} \right\}_0^L$$

$$v = \frac{\phi_0 M_0^2}{2P^2} \left\{ \frac{PL}{M_0} \left[1 - \left(\frac{PL}{M_0}\right)^2 \right]^{1/2} - \sin^{-1} \frac{PL}{M_0} \right\}$$

Substituting the numerical values, the slope and deflection become

$$\theta = -0.000241 \cdot \frac{720}{5.8} \{[1 - (0.967)^2]^{1/2} - 1\} = 0.02230 \text{ rad}$$

$$v = 0.000241 \cdot \frac{720^2}{2 \cdot 5.8^2} \{0.967[1 - (0.967)^2]^{1/2} - \sin^{-1}(0.967)\}$$

$$= -1.981 \text{ in.}$$

member is the summation over the length of the member of the work in elements of length, ds. This leads to the final expression for the internal work:

$$W_i = \int_0^L \delta M_Q \, d\theta = \int_0^L \delta M_Q \frac{d\theta}{ds} \, ds = \int_0^L \delta M_Q \, \phi_P \, ds = \int_0^L \delta M_Q \, \phi_P \frac{ds}{dx} \, dx$$

$$= \int_0^L \delta M_Q \, \phi_P \sqrt{1 + \left(\frac{dv}{dx}\right)^2} \, dx \approx \int_0^L \delta M_Q \, \phi_P \, dx \qquad (9.21)$$

in which the geometric relations established in Fig. 6.1c have been used. Additionally, the assumption of small displacements and slopes is used so that the square of the slope, dv/dx, can be neglected in relation to 1.

Equation (9.21) can be used to compute the internal work for any moment–curvature relation. For linear elastic materials, the curvature ϕ is defined as M_P/EI in Eq. (6.9), which can then be used to replace the ϕ_P in Eq. (9.21). This converts Eq. (9.21) to the same expression for internal work due to curvature as that given in Eq. (7.14). The application of Eq. (9.21) to the calculation of the displacement of the center of a simply supported beam is demonstrated in Example 9.8.

Example 9.8 The moment variation in the member is parabolic and symmetric with respect to the center. As a consequence, the curvature diagram is also symmetric, as is the moment diagram for the virtual force system. The integration over the length of the member can be reduced to twice the integral over the left half of the member. The evaluation of the integrals is done numerically and is based on the use of two or three points in the interval from 0 to $L/2$. The exact integration cannot be performed for the problem, so the "exact" value presented is obtained by using Eq. (9.4) with a large number of intervals. The value of the vertical displacement is for a member which has been loaded so that the maximum moment has reached 97% of its ultimate capacity. If the member were the W 12 × 16 of Example 9.7 with a span of 10 ft or 120 in., which is about the maximum for such a small section as this, the vertical displacement is $v = 0.2593 \cdot 0.000241 \cdot (120)^2 = 0.90$ in. Again this is a small displacement of the beam and does not significantly alter its initial geometry.

The examples for bending deformations with nonlinear materials have all been single-member structures without axial loads. The complementary virtual work calculations for structures in which either axial forces or bending moments exist in the members of the structure can be carried out by summing the internal work of axial forces as computed by Eq. (9.12)

Example 9.8

Obtain the deflection of the center of the uniformly loaded, simply supported beam. Assume that the material of the beam follows the moment–curvature relation defined by Eqs. (9.18) and (9.19) with $n = 2$, and the maximum moment in the center of the beam is $0.97M_0$.

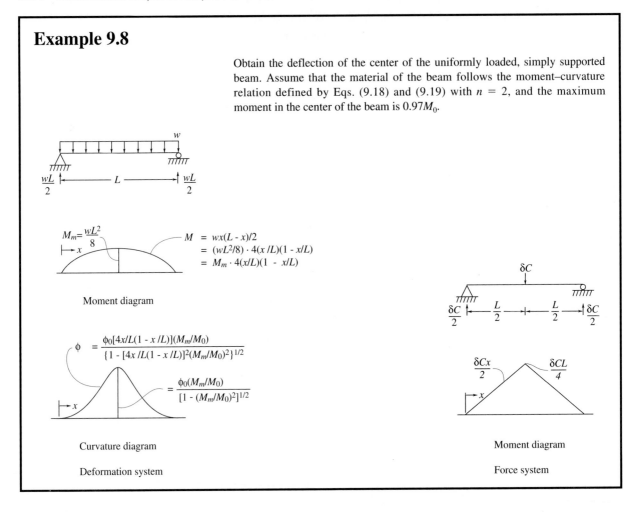

$$M = wx(L - x)/2$$
$$= (wL^2/8) \cdot 4(x/L)(1 - x/L)$$
$$= M_m \cdot 4(x/L)(1 - x/L)$$

$$\phi = \frac{\phi_0[4x/L(1 - x/L)](M_m/M_0)}{\{1 - [4x/L(1 - x/L)]^2(M_m/M_0)^2\}^{1/2}}$$

$$= \frac{\phi_0(M_m/M_0)}{[1 - (M_m/M_0)^2]^{1/2}}$$

Moment diagram

Curvature diagram

Deformation system

Moment diagram

Force system

358

Example 9.8 (continued)

STEP 1 Establish the force system and the moment diagram due to δC. Note that the moment diagram is symmetric.

STEP 2 Obtain the moment diagram due to the uniform load, w, and express it in terms of the maximum moment $M_m = wL^2/8$. The curvature diagram is calculated from Eq. (9.19) and is symmetric.

STEP 3 Apply the complementary virtual work principle using Eq. (9.21) to compute the internal work.

$$W_e = W_i$$

$$\delta Cv = 2\int_0^{L/2} \delta C \frac{x}{2}\phi\, dx = 2\int_0^{L/2} \delta C \frac{x}{2}\frac{\phi_0[4x/L(1-x/L)]\,(M_m/M_0)\,dx}{\{1-[4x/L(1-x/L)]^2(M_m/M_0)^2\}^{1/2}}$$

$$v = \phi_0 L^2 \int_0^{L/2} \frac{(x/L)[4x/L(1-x/L)](M_m/M_0)\,d(x/L)}{\{1-[4x/L(1-x/L)]^2(M_m/M_0)^2\}^{1/2}}$$

Evaluate the integral numerically using Eqs. (9.2), (9.3), (9.5), and (9.6) using two or three points in the interval.

Eq. (9.2): $0.3156\phi_0 L^2$ Eq. (9.3): $0.2546\phi_0 L^2$

Eq. (9.5): $0.2535\phi_0 L^2$ Eq. (9.6): $0.2656\phi_0 L^2$

"Exact": $0.2593\phi_0 L^2$ Minimum error: 1.8% [Eq. (9.3)]

and the internal work of bending moments as computed by Eq. (9.21). If a member has both axial force and bending actions occurring simultaneously, a new expression has to be derived for internal work calculations because there is no longer a complete independence of axial force and bending. Attempting to sum the internal work from Eqs. (9.12) and (9.21) for such a member creates an error in the calculation.

9.6   Summary and Limitations

The deformation analysis of nonprismatic members presented in this chapter involves the use of the principles already established in Chapters 6 and 7. Both axial and bending deformations are calculated based on the assumption that axial strains can be computed as the stress divided by E and the curvatures can be computed as the moment divided by EI. For nonprismatic members with gradually varying cross-sectional properties, these assumptions do not create any significant errors. In fact, for nearly all practical situations the errors are negligible or less than the errors arising from the uncertainties about material properties and location and magnitude of loads.

For the common case of nonprismatic members where the cross-sectional properties change abruptly at some point in the member, the deformations are not modeled accurately by the techniques presented in this chapter. In a distance equal to the depth of the member on either side of the point of abrupt change, the strains and curvatures cannot be calculated from the simple principles of mechanics of materials. The errors, however, in replacing these deformations with the ones computed from simple principles are not large in practical problems, being on the order of 3 to 6%.

Several numerical integration formulas are presented in Section 9.2. These formulas enable the evaluation of integrals that are difficult, or perhaps impossible, to obtain exactly. The integration of the strain and curvature expressions for nonprismatic materials and members with nonlinear materials is illustrated in succeeding sections. The numerical integration expressions are not difficult to use and are representative of the many such formulas that are available. Although errors in the values of integrals obtained by numerical integration occur, the magnitude of these errors can be assessed in situations that require it by using Eq. (9.2) or (9.3) with a large value of the number of intervals, n, over the range of integration.

Axial deformations of members with nonlinear materials are defined and illustrated for a stress–strain relation, Eq. (9.8), that approximates an elastic–perfectly plastic material such as a mild steel. This same relation is used to develop the internal work expression, Eq. (9.12), so that the complementary virtual work principle can be used to compute the deflections of plane trusses. Computations of deformations of structures using Eqs. (9.8), (9.9), and (9.12) are illustrated, but the same computations can be carried out for a different stress–strain relation with the establishment of equations that correspond to the ones used herein. It is shown in Example 9.6 that even when the loads on a truss are sufficiently large to bring the structure to the threshold of collapse, the deflections are still small enough to satisfy the assumption of no significant change in geometry of the structure under loading.

The use of a nonlinear stress–strain relation for the calculation of bending deformations is developed from basic principles of mechanics of materials using the assumption that plane sections remain plane. The resultant moment–curvature relation, Eq. (9.17), is very complicated and cannot be inverted to obtain the curvature in terms of the moment. As a consequence, bending deformations of members with a nonlinear material are presented in terms of the moment–curvature relation, Eq. (9.18), which has a simpler mathematical form than that of Eq. (9.17), but which still shows essentially the same variation of moment with curvature. The moment–curvature relation is representative of one for a cross section such as the W section for a mild steel. Equation (9.18) is also used to develop an internal work expression, Eq. (9.21), which can be used with the complementary virtual work principle to compute deflections or rotations of any structure where the members are subject to bending in the absence of axial forces. Calculations of deflections and rotations of members that are loaded to the point of incipient collapse in Examples 9.7 and 9.8 illustrate that deflections remain small and the assumption of small slopes and deflections in the simple bending theory is not violated.

The concepts presented in this chapter are an extension of those of previous chapters, in which prismatic members with linear materials were analyzed with an appropriate mathematical model. There are still many limitations and assumptions that are required to use the mathematical models developed in this chapter. Specifically:

- The deformation analysis of trusses is subject to all of the same limitations and assumptions that are listed at the end of Chapter 3 except that the material can be nonlinear.

- Because of the use of simple bending theory principles, the deformation analysis of beams and frames is subject to all of the same limitations and assumptions that are listed at the end of Chapter 5 and in Section 6.2 except that members can be nonprismatic and have materials with nonlinear stress–strain relations.

- The internal work expressions presented in Eqs. (9.12) and (9.21) are limited to the action of axial forces alone in Eq. (9.12) and bending alone in Eq. (9.21).

- All material deformations must be continuous, without any cracks or gaps.

- All structural deformations must be continuous, without any violation of the conditions of constraint at points of support.
- The distribution and orientation of all loads acting on a structure are not changed by the deformations of the structure.
- The orientation of all reaction components at supports is unchanged by the action of applied loads or any movement of the support.
- If movements occur at supports, they are small and do not affect the initial geometry of the structure.
- The principle of superposition is no longer valid for structures that have nonlinear materials.
- Sudden changes in cross-sectional properties in nonprismatic members are not so large that the assumed simple deformation behavior from mechanics of materials is greatly in error.

Some questions that structural engineers should be considering when using nonprismatic members or members made from nonlinear materials are:

- How much error is developed in the mathematical model of a nonprismatic member if an abrupt change in cross-sectional property is very large?
- What kinds of assumptions are made in computer programs that perform deformation analysis of members that are nonprismatic or are made with nonlinear materials?
- How can a mathematical model be developed for members made from nonlinear material that experience the simultaneous action of axial force and bending?
- What is the effect of nonlinear materials on the stability of members under the action of bending or combined axial force and bending?
- What is the effect on the moment–curvature relation if a material has a nonlinear stress–strain relation that is not symmetric in tension and compression?

PROBLEMS

9.1(a) Compute the vertical displacement of the right end *(E* constant).

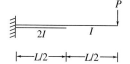

(b) Assume that the inertia, *I,* varies as shown below and compute the vertical displacement of the right end *exactly.*

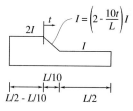

9.2 Repeat Problem 9.1, but evaluate the integral over the segment of length $L/10$, where *I* varies linearly using Simpson's rule, Eq. (9.3), the trapezoidal rule, Eq. (9.2), with $n = 2$, and Gaussian quadrature, Eqs. (9.5) and (9.6).

9.3 Compute the center deflection of the beam in Example 9.3 by using Gaussian quadrature, Eqs. (9.5) and (9.6). Apply numerical integration expression over the left and right halves of the beam and sum for the results. Compare with the exact values.

9.4 Compute the deflection using the trapezoidal rule and Simpson's rule with four intervals. Use symmetry of the problem and take *E* to be constant.

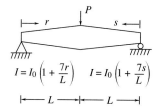

9.5 Obtain the end displacement rotation by using numerical integration and compare with the exact results. Use the

trapezoidal rule and Simpson's rule with four intervals. [*E* constant, $I = I_0(1 + 7x/l)$].

Exact Values:

$$v_1 = -0.057083 \frac{Pl^3}{EI_0}$$

$$\theta_1 = +0.10042 \frac{Pl^2}{EI_0}$$

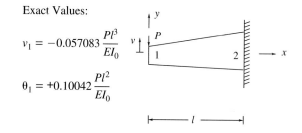

9.6 Repeat Problem 9.5 with the two- and three-point Gaussian quadrature, Eqs. (9.5) and (9.6).

9.7 Obtain the end displacement and rotation by using numerical integration and compare with the exact results. Use the trapezoidal rule and Simpson's rule with four intervals. [*E* constant $I = I_0(1 + 7x/l)$].

Exact values:

$$v_1 = -0.019732 \frac{wl^4}{EI_0}$$

$$\theta_1 = 0.028854 \frac{wl^4}{EI_0}$$

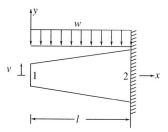

9.8 Repeat Problem 9.7 with the two- and three-point Gaussian quadrature, Eqs. (9.5) and (9.6).

363

9.9 The cantilever beam is made with a W section of linearly varying depth and constant modulus E.

(a) Determine the moment of inertia of the beam as a function of x.

(b) For the loading shown, obtain an approximate value of the deflection of the left end of the beam. Use Simpson's rule to evaluate any integrals in the computation process.

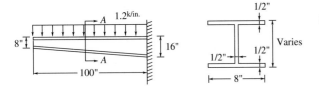

9.10 The cantilever beam has a rectangular cross section of linearly varying depth and constant modulus E.

(a) Determine the moment of inertia of the beam as a function of x.

(b) For the loading shown, obtain an approximate value of the deflection of the left end of the beam. Use Simpson's rule to evaluate the integrals in the computation process.

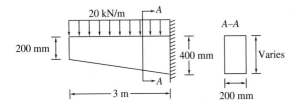

9.11 Compute the vertical displacement of the left end of the beam.

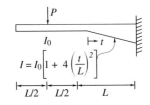

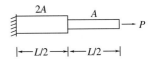

9.12 (a) Compute the displacement of the right end (E constant).

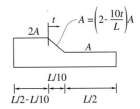

(b) Assume that the area, A, varies as shown below and compute the displacement of the right end *exactly*.

$$A = \left(2 - \frac{10t}{L}\right)A$$

9.13 Repeat Problem 9.12, but evaluate the integral over the segment of length $L/10$, where A varies linearly using Simpson's rule, Eq. (9.3), the trapezoidal rule, Eq. (9.2), with $n = 2$, and Gaussian quadrature, Eqs. (9.5) and (9.6).

9.14 For the member with the area varying as shown, obtain the end displacement. Compute the result by evaluating all integrals exactly *(E constant)*.

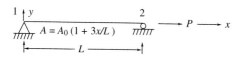

9.15 Compute the end displacement at 2 due to *P*. Use numerical integration *(E constant)*.

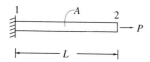

9.16 Repeat Problem 9.15, but replace the concentrated load *P* with a uniformly distributed load *p* over the length of the member.

9.17 The stress–strain relation for the member below is defined by Eqs. (9.7) and (9.8) with $n = 4$. Compute the end displacement of the member in terms of $\varepsilon_0 L$ if $P = 0.9\sigma_0 A$.

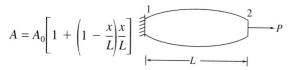

9.18 The stress–strain relation for the member below is defined by Eqs. (9.7) and (9.8) with $n = 4$. Compute the end displacement in terms of $\varepsilon_0 L$ if $p = 0.9\sigma_0 A/L$.

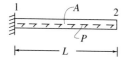

9.19 Compute the horizontal displacement of U_1 due to *P* in terms of $\varepsilon_0 L$. Use the stress–strain relation of Eqs. (9.7) and (9.8) with $n = 2$ and take $P = 0.57\sigma_0 A$ *(A constant for all members)*.

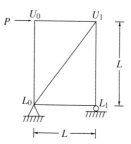

9.20 The stress–strain relation for the material of the member is given as

$$\sigma = \sigma_0 \frac{\varepsilon}{\varepsilon_0}\left[1 - \frac{1}{3}\left(\frac{\varepsilon}{\varepsilon_0}\right)^2\right]$$

For $P = 0.95\sigma_0 A$, compute the displacement of the right end of the member in terms of $\varepsilon_0 L$. (Hint: Assume values of $\varepsilon = U_2/L$ and solve by trial and error.)

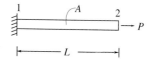

9.21 Compute the rotation and deflection of the left end of the cantilever beam in terms of $\phi_0 L$ using the moment–curvature relation defined by Eqs. (9.18) and (9.19) with $n = 2$. Take $w = 1.94 M_0/L^2$. Use numerical integration *(M_0 constant)*.

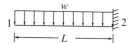

9.22   Repeat Problem 9.21 using $n = 4$ in Eqs. (9.18) and (9.19).

9.23   Compute the rotation and deflection of the left end of the cantilever beam in terms of $\phi_0 L$ using the moment–curvature relation defined by Eqs. (9.18) and (9.19) with $n = 2$. Take $M_0 = M_0$ for the left half, $M_0 = 2M_0$ for the right half, and ϕ_0 constant over the length of the member.

$$P = 1.94 \frac{M_0}{L}$$

M_0 $2M_0$

$\longleftarrow L/2 \longrightarrow\!\longleftarrow L/2 \longrightarrow$

9.24   Compute the rotation and deflection of the left end of the member in terms of $\phi_0 L$ using the moment–curvature relation defined by Eqs. (9.18) and (9.19) with $n = 2$. Take the variation of M_0 in those expressions as shown and assume that ϕ_0 is constant over the length of the member.

$$M_0 = M_0\left(1 + \frac{x}{L}\right)$$

$$P = 1.94 \frac{M_0}{L}$$

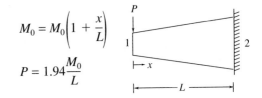

$\longleftarrow L \longrightarrow$

9.25   Repeat Problem 9.23, but replace the concentrated load P with the uniform load w, where $w = 3.88 M_0/L^2$.

9.26   Repeat Problem 9.24, but replace the concentrated load P with the uniform load w, where $w = 3.88 M_0/L^2$.

COMPUTER PROBLEMS

For the problems that follow, obtain the graphical results requested using one of the following methods:

(a) Create a spreadsheet and use the results to obtain the graphs.

(b) Write a small computer program that obtains the results requested and create a file that can be plotted to obtain the graphs.

(c) Use available software programs that will do the analysis requested. Run the program as many times as required to obtain the results requested and create a file that can be plotted to obtain the graphs.

C9.1   Obtain a plot of the vertical deflection at 2 versus a/L for $0 \leq a/L \leq 1$ (E constant).

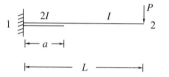

C9.2   Repeat Problem C9.1 with a uniform load on the beam.

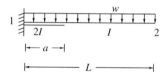

C9.3   Obtain a plot of the vertical deflection at 2 versus α for $1 \leq \alpha \leq 3$. The depth varies linearly along the beam (E and b constant).

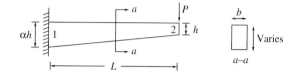

C9.4   Repeat Problem C9.3 with a uniform load on the beam.

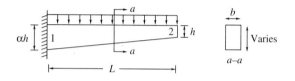

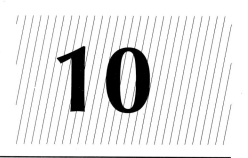

10

Analysis of Statically Indeterminate Structures

Determination of the geometric stability and static determinacy of beams, trusses, and frames was studied in Chapters 1 to 5. For geometrically stable structures, the internal distributions of axial and shear forces and moments as well as the reaction components were obtained from use of the equations of static equilibrium. In design it is important to know the magnitudes of these external and internal actions in order to verify safe performance of the material of these structures when they are subjected to every possible antic-ipated loading during their lifetimes.

In Chapters 6, 7, and 9, methods of deformation analysis of structures were established and illustrated. Deformations of structures are important in design because they indicate how well a structure resists deformation under the action of service loadings. The deformations are important for both aesthetic and functional reasons. If, under service loadings, structures deflect to the point where it is noticeable to a casual observer or where cracks form in nonstructural walls or ceilings, their performance would be judged unacceptable, even though the structures are quite safe from mater-ial failure or collapse. Deflections can also affect sensitive equipment so that it operates either improperly or not at all.

The calculation of structural deformations requires knowledge of the dis-tribution of the internal strains in members that arise from axial forces and the curvatures that arise from moments. A complete analysis of a structure includes its deformations but must start with calculation from equilibrium of the magnitudes of all internal actions and the values of all reaction components. This is a simple task for statically determinate structures, but cannot be done in statically indeterminate structures. The analysis of these structures requires a mathematical model that incorporates both structural deformation and equilibrium considerations.

Statically indeterminate structures are very common and have many desirable behavioral properties. In this chapter procedures are presented whereby statically indeterminate structures can be analyzed completely for all internal member actions and deformations. The procedures are based on the techniques already established for the analysis of statically determinate structures. Indeterminate structures can be analyzed with the same degree of accuracy as determinate ones by combining the equations of static equilibrium with a deformation analysis.

10.1 Statically Indeterminate Structures

The analysis of statically indeterminate, or more simply, indeterminate structures requires the execution of a three-step process. First, the structure must be converted into a statically determinate form by releasing one or more restraints on deformation behavior, such as the displacement restraint at a support or the continuity of slope in a member. At each restraint that is released, an unknown action will be introduced, such as a force at a released support displacement restraint or an internal moment at a point of release continuity of slope. All actions on this statically determinate structure can now be obtained from equilibrium.

The evaluation of deformation behavior is used in the remaining two steps. In the second step a force–deformation relation is introduced so that internal strains due to axial forces and curvatures due to bending moments can be calculated. The third and final step is to evaluate the deformations of the structure and enforce compatible deformation behavior. The analysis in these three steps is carried out in the statically determinate form of the structure by methods of analysis studied in the previous chapters. Any method of evaluating deformation behavior can be used in the third step, but the complementary virtual work method described in Chapter 7 is ideally suited for this task and will be used exclusively in this chapter.

Several methods of analysis can be used in indeterminate analysis, but only two will be explored in this treatment. The first, the method of superposition, is an intuitively straightforward procedure that emphasizes the physical behavior of the structure. The second, the method of consistent deformation, is somewhat more general and easier to implement but is better understood after the superposition method has been mastered.

Indeterminate trusses, beams, and frames are treated separately in this chapter, but the concepts of indeterminate analysis are not different for each type of structure. A structure is indeterminate because the combination of reaction components, internal member forces, or internal member moments exceeds the number of equations of equilibrium. The degree of indeterminacy, p, of geometrically stable trusses is given by Eq. (3.5) and of geometrically stable beams and frames by Eq. (5.2). A degree of indeterminacy of p indicates that there are an excess of p restraints beyond what is needed to maintain a load system on a structure in stable static equilibrium.

When any single reaction component that restrains a displacement or rotation is added to a geometrically stable and statically determinate structure as shown, for example, in Fig. 3.9c, that reaction component creates a restraint that is not needed for the structure to carry applied loading. The restraint is unnecessary, and it, along with the corresponding reaction component, is called redundant. If a member is added between two existing joints of a geometrically stable and statically determinate truss, as shown in Fig. 3.9b, the force developed in that member and the restraint on deformations of the truss are unnecessary for proper performance of the truss and the member is therefore called redundant. The addition of a member between two existing joints of a frame creates three redundants, because the axial force, shear force, and bending action each constitute a restraint on the deformation behavior of the frame. In summary, the value of p, the degree of indeterminacy of a structure, is increased by one for each restraint that is added to a geometrically stable structure.

The addition of restraints to a structure also affects the manner in which the structure behaves. When $p = 0$, there is only one way a load system acting on a structure can be resisted. That is, the equations of static equilibrium determine uniquely the distribution of external force and moment reactions and the internal distribution and magnitude of the axial forces, shears, and moments in the members of the structure. The addition of one restraint to the structure introduces an alternative way and, from the standpoint of the equilibrium equations, an unnecessary or redundant way for the applied loads to be resisted. Because the load is resisted by both of these ways simultaneously, the equations of static equilibrium will no longer uniquely define the external and internal distributions of forces and moments and the structure is statically indeterminate. This concept is illustrated in Fig. 10.1.

Figure 10.1 A beam for which p is 1 from Eq. (5.2) is shown. One unknown force, moment, or reaction component represents a redundant constraint of the structure. For example, the vertical deflection restraint at B or the rotation restraint at A can be considered as redundant. Releasing either one of those restraints creates two different geometrically stable and statically determinate structures capable of carrying the applied

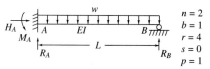

$$n = 2$$
$$b = 1$$
$$r = 4$$
$$s = 0$$
$$p = 1$$

Figure 10.1   Simple indeterminate structure.

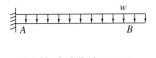

(a) Vertical displacement
restraint at *B* removed

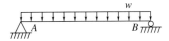

(b) Rotation restraint at
A removed

Figure 10.2a–b   Statically
determinate forms of the
structure in Fig. 10.1.

uniform load, as shown in Fig. 10.2. Other possible restraints that could be released are shown in Fig. 10.3. However, care must be exercised in the selection of restraints since the release of a restraint that is not redundant, as shown in Fig. 10.4, creates a geometrically unstable structure.

Although there is only one redundant restraint for the structure shown in Fig. 10.1, there is no unique redundant for the structure. A careful review of Figs. 10.2 and 10.3 shows that some of the different statically determinate forms created by the released redundant restraint may be easier than others to analyze.

The analysis of indeterminate structures is based on creating a geometrically stable, statically determinate form of the structure by selecting and releasing one (or more) redundant restraints(s). The determinate structure is then analyzed for specific deformations associated with the released restraint(s). With the experience of solving indeterminate structures will come a sense of selection of redundant restraints that simplifies the subsequent analysis. No general guidelines will be set forth for redundant selections until a later section, after some experience with the analysis of indeterminate structures has been gained.

10.2   Analysis by Superposition

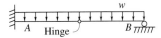

(a) Vertical displacement
restraint at *A* removed

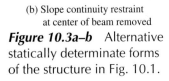

(b) Slope continuity restraint
at center of beam removed

Figure 10.3a–b   Alternative
statically determinate forms
of the structure in Fig. 10.1.

The structure shown in Fig. 10.1 will be analyzed by releasing the vertical deflection restraint imposed by the roller support reaction at *B* as shown in Fig. 10.2a.

Figure 10.5 The reactions and internal moments due to the uniform load, *w*, are obtained from equilibrium as shown in Fig. 10.5a. From equilibrium the upward force, *R*, applied at *B* yields the reactions and internal moment shown in Fig. 10.5b. Because the structure will deform under these two loadings, a downward deflection, v_W, due to *w* and an upward deflection, v_R, due to *R* will occur at *B*. If the loadings *w* and *R* are applied simultaneously, the net downward vertical deflection of *B* will be $(v_w - v_R)$.

There is no vertical deflection of *B* in the indeterminate structure of Fig. 10.1 due to the displacement restraint produced by the reaction force, R_B. In Fig. 10.5, the magnitude of v_W is set by *w*, but the magnitude of v_R depends on the magnitude of *R*. If the magnitude of *R* creates a vertical deflection v_R such that the sum $(v_R - v_w)$ is zero, the effect of *w* and *R* acting together on the structure in Fig. 10.5 will be the same as *w* and R_B acting together on the structure of Fig. 10.1. The requirement that the vertical displacement at *B* be zero in Fig. 10.5 makes the deformations of that structure identical to, or compatible with, the deformations of the indeterminate structure in Fig. 10.1. Thus the condition stated mathematically as

$$v_R - v_w - 0 \qquad (10.1)$$

is known as a compatibility equation. It simply states that the superposition of the effects of the uniform load w and the vertical force R acting on the determinate structure of Fig. 10.5 must produce a zero vertical displacement of its right end. Equation (10.1) provides an expression for calculation of the magnitude of R.

The introduction of the moment–curvature relation enables the computation of v_w and v_R. The complementary virtual work method will be used to obtain them.

Figures 10.6, 10.7 Using complementary virtual work, the deflections $v_w = wL^4/8EI$ in Fig. 10.6 and $v_R = RL^3/3EI$ in Fig. 10.7 are obtained. All calculations are based on the actions occurring on the structure of Fig. 10.5. The integral of the product of the moment diagram with the curvature diagram based on Eq. (7.12) is carried out using Table 7.1 in both figures. The computation in Fig. 10.6 uses the expression in position 8–1 of Table 7.1 and position 1–1 of Table 7.1 in Fig. 10.7.

Substituting these values into Eq. (10.1) and solving gives the final magnitude of R, which is equal to the value of the reaction R_B for the structure of Fig. 10.1.

$$R = R_B = \frac{3wL}{8} \qquad (10.2)$$

The final moment diagram for the structure of Fig. 10.1 is obtained by direct superposition of the moment diagrams due to w and R in Fig. 10.5,

Figure 10.4   Improper release of the structure in Fig. 10.1, giving unstable structure.

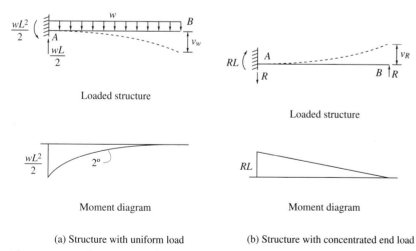

(a) Structure with uniform load   (b) Structure with concentrated end load

Figure 10.5a–b   Structural deformations and moment diagrams for the two load systems acting on a statically determinate form of the structure of Fig. 10.1.

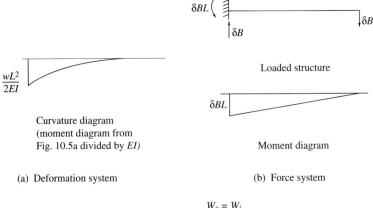

(a) Deformation system

Curvature diagram
(moment diagram from
Fig. 10.5a divided by *EI*)

(b) Force system

Moment diagram

Loaded structure

$$W_e = W_i$$

$$\delta B \, v_w = \int_0^L \left(\boxed{} \atop \delta BL \right) \left(\boxed{} \atop \dfrac{wL^2}{2\,EI} \right) dx = \dfrac{wL^2}{2EI}\,(\delta BL)\,\dfrac{L}{4}$$

(position 8–1, Table 7.1)

$$\therefore \; v_w = \dfrac{wL^4}{8EI}$$

(c) Complementary virtual work calculation of v_w

Figure 10.6a–c Calculation of v_W in Fig. 10.5a by complementary virtual work.

using the magnitude of R given by Eq. (10.2). This calculation is shown in Fig. 10.8.

In summary, the analysis of the indeterminate structure of Fig. 10.1 is a three-step process. Step 1 is the creation of a statically determinate form of the structure by releasing the vertical displacement restraint at B and using equilibrium to calculate reactions and internal moments as illustrated in Fig. 10.5. Step 2 is the introduction of the moment–curvature relations, which enabled the vertical deflections v_w and v_R to be calculated in Figs. 10.6 and 10.7. Step 3 is the use of the compatibility condition of Eq. (10.1) to calculate the reaction force R_B in Eq. (10.2).

The process just described is known as indeterminate analysis by the method of superposition. Indeterminate problems are solved by the direct addition, or superposition, of forces, moments, and deflections caused by two or more loading systems acting on a statically determinate structure. Two important assumptions are implicit to this method. First, all analyses of the determinate structure are based on its original, undeformed geometry. Second, the calculation of deflection of the structure is carried out indepen-

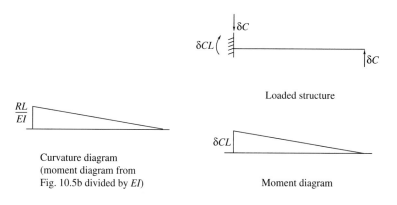

Loaded structure

(a) Deformation system

Curvature diagram
(moment diagram from
Fig. 10.5b divided by *EI*)

(b) Force system

Moment diagram

$$W_e = W_i$$

$$\delta C v_R = \int_0^L \left(\underbrace{\quad}_{\delta CL} \right) \left(\underbrace{\quad}_{\dfrac{RL}{EI}} \right) \, dx = (\delta C\, L)\left(\dfrac{RL}{EI}\right)\dfrac{L}{3}$$

(position 1–1, Table 7.1)

$$\therefore \; v_R = \dfrac{RL^3}{3EI}$$

(c) Complementary virtual work calculation of v_R

Figure 10.7a–c   Calculation of v_R in Fig. 10.5b by complementary virtual work.

dently for each load system using linear, elastic material. These considerations lead to the following statement:

> The method of superposition may be used in any structure in which the material is linear elastic and the deformations are sufficiently small so that its geometry after deformation is essentially unchanged.

Application of the superposition technique to an indeterminate truss is shown in Example 10.1.

Example 10.1 Selection of the release of axial continuity in member U_1–U_2 rather than the vertical displacement restraint at one of the reaction joints simplifies analysis of the truss. The release of axial action in

Example 10.1

Obtain the reactions and member forces in the indeterminate truss due to the 200-kip load. Take $E = 29,000$ ksi and the area in square inches of members as follows: chords 4, verticals 3, and diagonals 5.

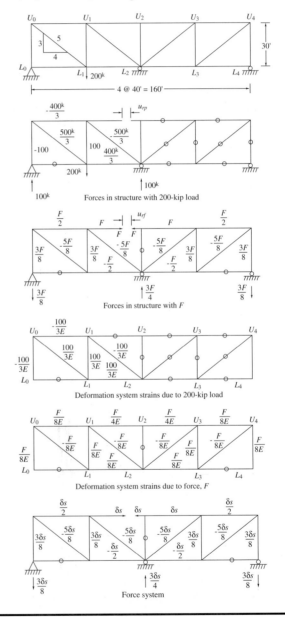

Forces in structure with 200-kip load

Forces in structure with F

Deformation system strains due to 200-kip load

Deformation system strains due to force, F

Force system

Example 10.1 (continued)

STEP 1   Release continuity of axial deformation of member U_1–U_2, and reduce the structure to two simply supported trusses. Under the action of various loads, the release introduces a gap in the member, U_r. For the 200-kip load, the gap is U_{rp}, while for the force F applied at the gap it is U_{rf}. No gap exists in the indeterminate structure, so the superposition of the two gaps must be zero, or

$$U_{rf} + U_{rp} = 0$$

This is the compatibility equation for solution of the unknown force, F.

STEP 2   Compute the two deformations at the gap. First analyze the truss for the member forces and strains due to the 200-kip load and repeat the analysis for the force, F. Next, establish a force system δS that reflects the gap displacements. These deformations and the force system member forces are summarized in the table below.

1	2	3	4	5	6	7	8	9
MEMBER	LENGTH	AREA	FORCE SYSTEM FORCE	DEFORMATION SYSTEM STRAINS (Fp / AE)		INTERNAL WORK Wi		FINAL FORCES (K)
				FOR	FOR	DUE TO 200K	DUE TO F	
	L	A	δFs	200K	F			
L0L1	40	4	0	0	0	0	0	0
L1L2	40	4	-0.5 δS	33.333 /E	-0.13 F/E	-666.67 $\delta S/E$	2.5 $\delta S*F/E$	120.33
L2L3	40	4	-0.5 δS	0	-0.13 F/E	0	2.5 $\delta S*F/E$	-13.01
L3L4	40	4	0	0	0	0	0 $\delta S*F/E$	0
U0U1	40	4	0.5 δS	-33.33 /E	0.125 F/E	-666.67 $\delta S/E$	2.5 $\delta S*F/E$	-120.3
U1U2	40	4	1 δS	0	0.25 F/E	0	10 $\delta S*F/E$	26.016
U2U3	40	4	1 δS	0	0.25 F/E	0	10 $\delta S*F/E$	26.016
U3U4	40	4	0.5 δS	0	0.125 F/E	0	2.5 $\delta S*F/E$	13.008
L0U0	30	3	0.375 δS	-33.33 /E	0.125 F/E	-375 $\delta S/E$	1.406 $\delta S*F/E$	-90.24
L1U1	30	3	0.375 δS	33.333 /E	0.125 F/E	375 $\delta S/E$	1.406 $\delta S*F/E$	109.76
L2U2	30	3	0	0	0	0	0	0
L3U3	30	3	0.375 δS	0	0.125 F/E	0	1.406 $\delta S*F/E$	9.7561
L4U4	30	3	0.375 δS	0	0.125 F/E	0	1.406 $\delta S*F/E$	9.7561
U0L1	50	5	-0.625 δS	33.333 /E	-0.13 F/E	-1041.7 $\delta S/E$	3.906 $\delta S*F/E$	150.41
U1L2	50	5	-0.625 δS	-33.33 /E	-0.13 F/E	1041.67 $\delta S/E$	3.906 $\delta S*F/E$	-182.9
U3L2	50	5	-0.625 δS	0	-0.13 F/E	0	3.906 $\delta S*F/E$	-16.26
U4L3	50	5	-0.625 δS	0	-0.13 F/E	0	3.906 $\delta S*F/E$	-16.26
			TOTAL Wi =	Σ =		-1333.3 $\delta S/E$	51.25 $\delta S*F/E$	

Example 10.1 (continued)

Step 3   Compute U_{rp} and U_{rf} by applying the complementary virtual work technique. For U_{rp} due to the 200-kip load and using the sum of the internal work terms in column 7 of the table, the complementary virtual work becomes

$$W_e = W_i$$

or

$$\delta S\, U_{rp} = \Sigma\, (\delta F_s)\left(\frac{F_p L}{AE}\right) = -1333.3 \cdot \frac{\delta S}{E} \quad \therefore\ U_{rp} = \frac{-1333.3}{E}$$

For U_{rf} due to force F and using the sum of the internal work terms in column 8 of the table, the complementary virtual work becomes

$$W_e = W_i$$

or

$$\delta S\, U_{rf} = \Sigma\, (\delta F_s)\left(\frac{F_p L}{AE}\right) = 51.25 \cdot \delta S \frac{F}{E} \quad \therefore\ U_{rf} = 51.25 \frac{F}{E}$$

Use the superposition equation established in step 1 to compute the final value of F.

$$U_{rp} + U_{rf} = 0$$

$$\therefore\ -\frac{1333.3}{E} + 51.25\frac{F}{E} = 0 \ \rightarrow F = 26.02 \text{ kips}$$

Step 4   Compute the final reactions and member forces by superposition of member forces and reactions due to the 200-kip load and F calculated in step 1. The member force results are summarized in column 9 of the table and both member forces and reactions in the sketch below.

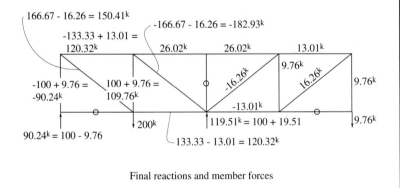

Final reactions and member forces

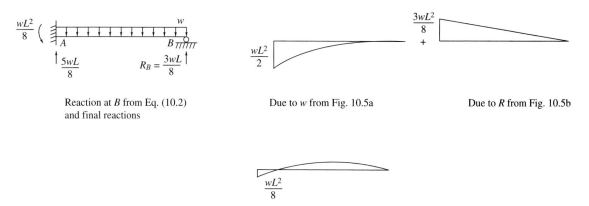

| Reaction at B from Eq. (10.2) and final reactions | Due to w from Fig. 10.5a | Due to R from Fig. 10.5b |

Final

Figure 10.8   Final reactions and use of superposition to obtain final moment diagram for indeterminate structure of Fig. 10.1.

the member effectively divides the truss into two structures. This isolates the deformations due to the 200-kip load to the left side of the structure and makes the deformations of the truss caused by the axial force, F, in member U_1–U_2 symmetric because of the symmetry of the structure. The effort required to compute the member axial strains is significantly reduced. The relative displacements of the ends of the member at the cut are computed by the complementary virtual work method using a force system with equal forces on each side of the cut in member U_1–U_2. The calculations are organized in tabular form as shown using a commercial spreadsheet application software program. The use of the compatibility equation established in step 1 gives the magnitude of F. Superposition of the member forces due to F and the 200-kip load is done in the spreadsheet and is summarized in the sketch at the end of the example.

A two-span beam that is indeterminate to the first degree is analyzed in Example 10.2.

Example 10.2 The redundant restraint selected in this example is the continuity of rotation at B. This divides the structure into two parts and isolates deformation effects. Again, the addition or superposition of the deformation effect for each action gives the condition for compatible behavior of the structure. The compatibility equation is established in step 1, ensuring the continuity of slope at the hinge introduced at B. Two relative rotations at B are computed by the complementary virtual work method. The magnitude of the moment at B is obtained from solution of the compatibility equation and the final moment diagram from the superposition of the moments due to w and M_B. For numerical computations with U.S. units take $L = 20$ ft and $w = 4$ kips; for SI units, take $L = 7$ m and $w = 60$ kN/m.

379

Example 10.2

Obtain the final moment diagram for the two-span beam shown below.

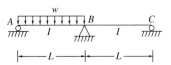

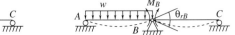

Indeterminate structure (E constant)   Statically determinate form of structure

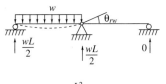

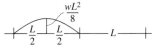

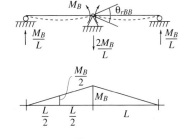

Moment   Moment

Curvature   Curvature

Moment and curvature diagrams   Moment and curvature diagrams
for loading w for moment M_B

STEP 1 Release the slope continuity restraint at B to create a structure that is two statically determinate simply supported beams. The release introduces a relative rotation of the right and left sides of hinge at B, θ_{rBB}. Superposition of the relative rotations caused by the load system, w, and load system M_B gives the compatibility equation for relative rotation at B:

$$\theta_{rB} = \theta_{rw} + \theta_{rBB} = 0$$

Satisfaction of this equations enables the moment M_B to be computed.

STEP 2 Create the δM force system to compute the relative rotations θ_{rw} and θ_{rBB} at B. Obtain the moment diagram for this force system and use the curvatures for the w loading as the deformation system for θ_{rw} and the curvatures for the moment M_B as the deformation system for θ_{rBB}.

Example 10.2 (continued)

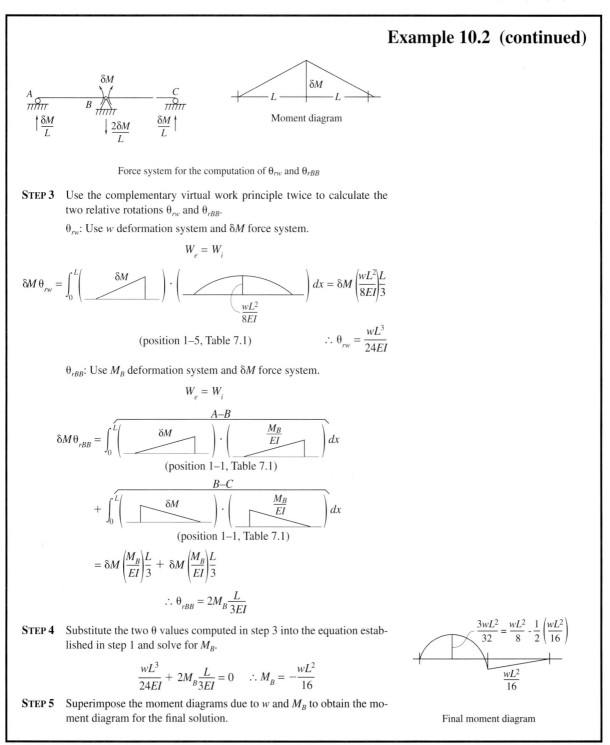

Force system for the computation of θ_{rw} and θ_{rBB}

STEP 3 Use the complementary virtual work principle twice to calculate the two relative rotations θ_{rw} and θ_{rBB}.

θ_{rw}: Use w deformation system and δM force system.

$$W_e = W_i$$

$$\delta M\,\theta_{rw} = \int_0^L \left(\underset{\delta M}{\diagup} \right) \cdot \left(\underset{\dfrac{wL^2}{8EI}}{\frown} \right) dx = \delta M \left(\dfrac{wL^2}{8EI} \right)\!\dfrac{L}{3}$$

(position 1–5, Table 7.1) $\qquad \therefore \theta_{rw} = \dfrac{wL^3}{24EI}$

θ_{rBB}: Use M_B deformation system and δM force system.

$$W_e = W_i$$

$$\delta M\,\theta_{rBB} = \int_0^L \overbrace{\left(\underset{\delta M}{\diagup} \right) \cdot \left(\underset{\dfrac{M_B}{EI}}{\diagup} \right)}^{A-B} dx$$

(position 1–1, Table 7.1)

$$+ \int_0^L \overbrace{\left(\underset{\delta M}{\diagdown} \right) \cdot \left(\underset{\dfrac{M_B}{EI}}{\diagdown} \right)}^{B-C} dx$$

(position 1–1, Table 7.1)

$$= \delta M \left(\dfrac{M_B}{EI} \right)\!\dfrac{L}{3} + \delta M \left(\dfrac{M_B}{EI} \right)\!\dfrac{L}{3}$$

$$\therefore \theta_{rBB} = 2M_B \dfrac{L}{3EI}$$

STEP 4 Substitute the two θ values computed in step 3 into the equation established in step 1 and solve for M_B.

$$\dfrac{wL^3}{24EI} + 2M_B \dfrac{L}{3EI} = 0 \quad \therefore M_B = -\dfrac{wL^2}{16}$$

STEP 5 Superimpose the moment diagrams due to w and M_B to obtain the moment diagram for the final solution.

$$\dfrac{3wL^2}{32} = \dfrac{wL^2}{8} - \dfrac{1}{2}\left(\dfrac{wL^2}{16} \right)$$

$$\dfrac{wL^2}{16}$$

Final moment diagram

Examples 10.1 and 10.2 illustrate the necessary steps for the solution. First, the redundant restraint is identified and released to create a statically determinate and geometrically stable structure. Associated with a displacement release is an unknown redundant force and with a rotation release an unknown redundant moment. The redundants provide a means of restoring the continuity of deformation at points of release of the restrains. Equilibrium is then used to calculate all reactions and internal member forces and moments. The second step is to introduce the force–deformation or moment–curvature relations and compute the deflections or rotations due to each load system acting on the structure. The third step is to use the equation of compatibility, which superimposes the deflections or rotations caused by the different load systems acting on the determinate structure. The compatibility equation can then be solved for the magnitude of the redundant force or moment. The final forces and moment diagrams of the indeterminate structure are obtained by superimposing the effect of each load system acting on the structure.

10.3   Selection of Redundant Restraints

The redundant restraints for a structure are not unique, as was illustrated for the structure in Fig. 10.1. Four different releases are shown in Figs. 10.2 and 10.3. The question of which redundant restraint should be selected for a structure has no simple answer, but certain general guidelines can be established.

1. The restraints released must not make the structure unstable.
2. The redundants should isolate, insofar as possible, load effects in the statically determinate structure.
3. If the structure has any symmetry, the releases should be selected to take advantage of that symmetry.
4. If portions of the determinate structure can be made symmetric by redundant selection, that selection should be used.

These guidelines were followed in Examples 10.1 and 10.2, which reduced the amount of computation to a modest level. Every indeterminate structure is different, so there cannot be rules about how to select redundants, but these guidelines are helpful, and the experience gained as more and more structures have been analyzed will provide the best insight into redundant selection. Note how these guidelines are followed in the examples in succeeding sections.

10.4   Method of Consistent Deformations

In the method of superposition a compatibility equation is established for each restraint of continuity of deformation released in an indeterminate structure. A very similar approach is used in the method of consistent deformation, but a significant difference in the method makes it more general in application. As in the superposition method of analysis, a determinate form of an indeterminate structure is created by the release of one or more redundant restraints and the introduction of an unknown redundant force or moment at each point of release. If this statically determinate structure is subjected to the same system of external applied loads as the original indeterminate structure, deformations of the determinate structure will be the same as the deformations of the indeterminate structure when the unknown redundant force(s) or moment(s) acting at the point(s) of restraint release attain the value(s) that restore the deformation continuity.

To introduce this method, examine the indeterminate structure of Fig. 10.1 in the context of the statically determinate form shown in Fig. 10.2a and reproduced with applied loading in Fig. 10.9.

Figure 10.9 The deformations of the structure in Fig. 10.9a are caused by the uniform load w and the applied force R, just as the deformations of the indeterminate structure of Figs. 10.1 and 10.9b are caused by the same loading actions. The only difference between the structure in Fig. 10.1 or Fig. 10.9b and 10.9a is that the latter structure is free to displace vertically at B. If the vertical displacement of B, v_B, in Fig. 10.9a is required to be zero, the deformation behavior of this statically determinate structure will be identical to or consistent with the deformation behavior of the original indeterminate structure in Fig. 10.9b. The computation of the vertical displacement, v_B, in Fig. 10.9a and requiring it to be zero are the basis of the method of consistent deformations.

Implementation of this method of analysis is illustrated with a complete three-step analysis of the structure of Fig. 10.1. First the redundant restraint is selected to be the vertical displacement at B with the corresponding

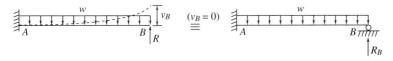

(a) Determinate structure   (b) Indeterminate structure

Figure 10.9a–b   Equivalent deformation behavior between determinate and indeterminate structures.

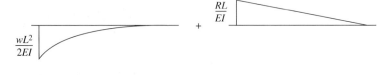

Curvature due to w Curvature due to R

Figure 10.10   Deformation system for statically determinate structure of Fig. 10.9a.

redundant force R as shown in Fig. 10.9a. This creates a statically determinate structure that can be analyzed for reactions and internal moments.

The second step in the process is to analyze the determinate structure for the vertical deflection, v_B, at B by introducing the moment–curvature relation and using the complementary virtual work method. The vertical deflection at B is caused by the load w and force R acting *simultaneously* on the structure. The effect of these loads is to produce a curvature diagram which, for convenience of computation, is shown in Fig. 10.10 as the sum of two curvature diagrams, one associated with w and the other with R. This constitutes the deformation system acting on the determinate structure. To compute v_B, the imagined or virtual force system shown in Fig. 10.11 is used. Applying the complementary virtual work principle yields the equation for v_B as summarized in Fig. 10.12. This expression is valid for any values of w and R.

Finally, the determinate structure is required to have a zero vertical displacement at $B(v_B = 0)$, which makes deformations of the structure consistent or compatible with the deformation behavior of the original indeterminate structure. The resulting equation is exactly the same as Eq. (10.2) and yields the magnitude of the unknown redundant R.

The significant difference between the method of consistent deformations and the method of superposition is that in the former the action of the applied load system *(w* in Fig. 10.9a) and the redundant load system *(R* in Fig. 10.9a) is treated as occurring *simultaneously* rather than separately. If the load w on the indeterminate structure in Fig. 10.1 is applied slowly from a magnitude of zero to its final value w, the reaction force R_B at B will also

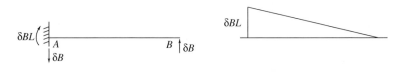

Loaded structure   Moment diagram

Figure 10.11   Force system to obtain the vertical deflection of B of the statically determinate structure of Fig. 10.9a.

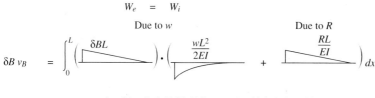

$$W_e \;=\; W_i$$

$$\delta B\, v_B \;=\; \int_0^L \left(\underbrace{\frac{\delta BL}{}}_{}\right)\cdot\left(\underbrace{\frac{wL^2}{2EI}}_{}\right) \;+\; \left(\underbrace{\frac{RL}{EI}}_{}\right) dx$$

(position 8–1, Table 7.1)   (position 1–1, Table 7.1)

$$= \; -(\delta BL)\frac{wL^2}{2EI}\,\frac{L}{4} \;+\; (\delta BL)\!\left(\frac{RL}{EI}\right)\!\frac{L}{3}$$

$$\therefore \quad v_B \;=\; -\,\frac{wL^4}{8EI} \;+\; \frac{RL^3}{EI}$$

Figure 10.12 Complementary virtual work calculation for vertical deflection of B of the statically determinate structure of Fig. 10.9a.

increase proportionally with w from zero to its final value R_B. Similarly, if the same loading procedure for w is followed in the structure of Fig. 10.9a, while always maintaining that v_B be zero, the value of R will also increase proportionally with w in the exact same way that it does in the indeterminate structure. The concept of the method of consistent deformations can be summarized as follows:

> The method of consistent deformations is based on the concept that the statically determinate form of an indeterminate structure created by the release of one or more continuities of deformation can be made to deform in exactly the same manner as the indeterminate structure. The essential requirements of the method are that all actions that both the determinate and indeterminate structures are subjected to must occur simultaneously and that the deformations at the point(s) of release in the determinate structure must match the deformations of the indeterminate structure at the same point(s).

The completely parallel response of the determinate and indeterminate forms of the structure permits the extension of the method of consistent deformations to the analysis of structures having nonlinear elastic materials. This extension is explored in a later section.

Application of this method to the truss of Example 10.1 is shown in Example 10.3. The effect of a temperature change in two members and a fabrication error in a third is considered in addition to the 200-kip load.

Example 10.3 Because of the fabrication error in member U_1–L_1 and a temperature change in members L_2–L_3 and L_3–L_4 being added to the

385

Example 10.3

For the truss of Example 10.1, obtain reactions and final member forces due to the 200-kip load, temperature change, and fabrication error indicated.

$E = 29,000$ ksi

area in square inches:

Chords: 4

Verticals: 3

Diagonals: 5

$$\alpha = \frac{1}{150,000} \frac{\text{Strain}}{°F}$$

$\Delta T = -60°F$ in members L_2–L_3 and L_3–L_4

Member U_1–L_1 fabricated ¼ in. too long.

$$\varepsilon_{0t} = -\frac{60}{150,000}$$

$$= -\frac{1}{2500}$$

$$\varepsilon_{0t} = \frac{1/4}{30 \cdot 12}$$

$$= \frac{1}{1440}$$

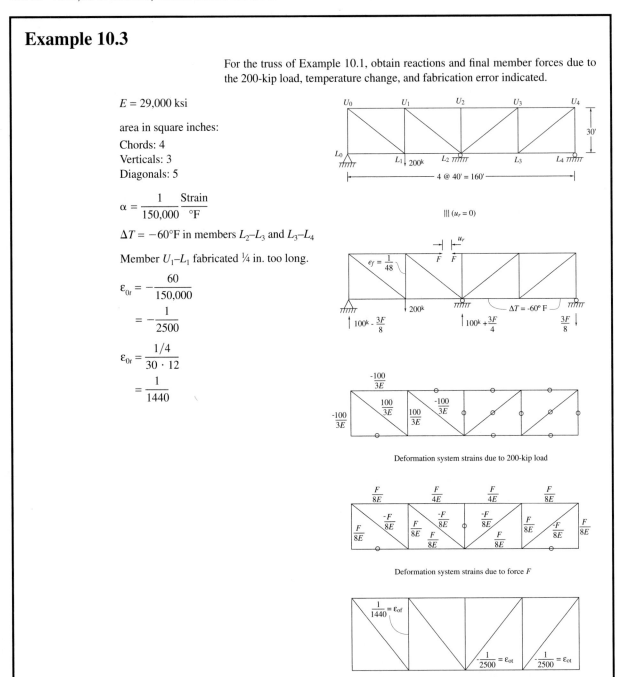

Deformation system strains due to 200-kip load

Deformation system strains due to force F

Deformation system strains due to other causes
(temperature change and fabrication error)

Example 10.3 (continued)

Step 1   As in Example 10.1, release the continuity of axial deformation of member U_1–U_2, creating the statically determinate form of the structure with a relative displacement at the gap of U_r.

Step 2   Establish the force system, the same as that used in Example 10.1, to compute the relative displacement U_r.

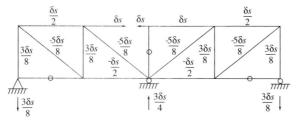

Forces system

Step 3   The deformation system consists of the deformations caused by the 200-kip system of loads, the F system of loads, and the member deformations caused by the fabrication error and the temperature change. These deformations are summarized in the three labeled sketches shown above. In addition, all deformations are entered in the table below. The deformations due to the 200-kip load are in column 5, due to F in column 6, due to the fabrication error in column 7, and due to the temperature change in column 8. Note that for constant strain ε_0, the deformation for a member is simply $\varepsilon_0 L$ in the expression for W_{i0} in Eq. (7.13).

1	2	3	4	5	6	7	8	9	10	11	12	13
MEMBER	LENGTH	AREA	FORCE SYSTEM FORCE	DEFORMATION SYSTEM STRAINS				INTERNAL WORK Wi DUE TO				FINAL FORCES (K)
				(Fp / AE)		ε_{of} FAB. ERROR	ε_{ot} TEMP. CHG.	200K	F	FAB. ERROR	TEMP. CHG.	
				DUE TO 200K	DUE TO F							
	L	A	δFs									
L0L1	40	4	0	0	0	0	0	0	0	0	0	0
L1L2	40	4	-0.5 δS	33.333 /E	-0.13 F/E	0	0	-667 δS/E	2.5 δS*F/E	0	0	124.8
L2L3	40	4	-0.5 δS	0	-0.13 F/E	0	-0.0004	0	2.5 δS*F/E	0	0.008 δS	-8.53
L3L4	40	4	0	0	0	0	-0.0004	0	0	0	0	0
U0U1	40	4	0.5 δS	-33.33 /E	0.125 F/E	0	0	-667 δS/E	2.5 δS*F/E	0	0	-125
U1U2	40	4	1 δS	0	0.25 F/E	0	0	0	10 δS*F/E	0	0	17.07
U2U3	40	4	1 δS	0	0.25 F/E	0	0	0	10 δS*F/E	0	0	17.07
U3U4	40	4	0.5 δS	0	0.125 F/E	0	0	0	2.5 δS*F/E	0	0	8.534
L0U0	30	3	0.375 δS	-33.33 /E	0.125 F/E	0	0	-375 δS/E	1.406 δS*F/E	0	0	-93.6
L1U1	30	3	0.375 δS	33.333 /E	0.125 F/E	0.0007	0	375 δS/E	1.406 δS*F/E	0.008 δS	0	106.4
L2U2	30	3	0	0	0	0	0	0	0	0	0	0
L3U3	30	3	0.375 δS	0	0.125 F/E	0	0	0	1.406 δS*F/E	0	0	6.401
L4U4	30	3	0.375 δS	0	0.125 F/E	0	0	0	1.406 δS*F/E	0	0	6.401
U0L1	50	5	-0.63 δS	33.333 /E	-0.13 F/E	0	0	-1042 δS/E	3.906 δS*F/E	0	0	156
U1L2	50	5	-0.63 δS	-33.33 /E	-0.13 F/E	0	0	1042 δS/E	3.906 δS*F/E	0	0	-177
U3L2	50	5	-0.63 δS	0	-0.13 F/E	0	0	0	3.906 δS*F/E	0	0	-10.7
U4L3	50	5	-0.63 δS	0	-0.13 F/E	0	0	0	3.906 δS*F/E	0	0	-10.7
Σ =								-1333 δS/E	51.25 δS*F/E	0.008 δS	0.008 δS	

Example 10.3 (continued)

STEP 4 Compute U_r by applying the complementary virtual work method and take the results of the summation from the last row of the table. All deformations are considered to occur simultaneously. Note that the two 0.008 δS values in the last row of columns 11 and 12 in the table represent rounded values of the terms $\delta S/128$ and $\delta S/125$, respectively.

$$W_e = W_i + W_{i0}$$

$$\delta S\, U_r = \overbrace{\Sigma\, (\delta F_s)\left(F_p \frac{L}{AE}\right)}^{200 \text{ kips}} + \overbrace{\Sigma\, (\delta F_s)\left(F_p \frac{L}{AE}\right)}^{F} + \overbrace{W_{i0}}^{\text{fab. +} \atop \text{temp.}}$$

$$= -1333 \frac{\delta S}{E} + 51.25\, \delta S \frac{F}{E} + \frac{\delta S}{128} + \frac{\delta S}{125}$$

$$U_r = -\frac{1333}{E} + 51.25 \frac{F}{E} + \left(\frac{1}{128} + \frac{1}{125}\right)$$

The requirement that $U_r = 0$ gives

$$F = 26.02 - 4.42 - 4.53 = 17.07 \text{ kips}$$

STEP 5 Compute final member forces by adding the member forces for the 200-kip load and F. The results are displayed in column 13 of the table and in the diagram below of a truss with reactions.

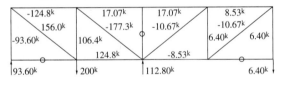

analysis for the 200-kip applied load, the compatibility equation for the computation of the redundant force F involves the simultaneous application of deformations from the 200-kip force system, the F force system, the temperature strains in members L_2–L_3 and L_3–L_4, and the fabrication error in member U_1–L_1. The internal work associated with the complementary virtual work calculation from Eq. (7.14) includes a nonzero contribution from the W_{i0} term. The final value for the redundant force F is the sum of 26.02 kips due to the 200-kip load, -4.42 kips due to the fabrication error, and -4.53 kips due to the temperature change. The contribution to the final value of F due to the fabrication error and the temperature change are each about 17% of that due to the 200-kip load. As can be seen, the magnitude of the effects of temperature changes and fabrication errors in indeterminate structures are not insignificant and must be considered in design.

The structure of Example 10.2 is modified to have a fixed support at A and an increase of the moment of inertia of AB to $2I$ from I. The structure is now 2° indeterminate $(p = 2)$.

Example 10.4   Although this structure is 2° indeterminate, the computational effort in only slightly increased over that required for Example 10.2, which illustrates the economy of computation in using the method of consistent deformations. As in Example 10.2, two applications of the complementary virtual work method are all that are needed to obtain the final compatibility equations. The key to the solution is *simultaneous* application of the deformations associated with the applied load, w, and the two redundant moments, M_A and M_B. For numerical calculations with U.S. units take $L = 20$ ft and $w = 4$ kips/ft; for SI units take $L = 7$ m and $w = 60$ kN/m.

Use of the method of consistent deformations for analysis of a laterally loaded simple frame is shown in Example 10.5, in which the effects of axial deformations are included in the analysis.

Example 10.5   Both axial and bending deformations are included in the calculation of the value of the horizontal reactive force H. Computation of the bending deformation in Eq. (7.14) is greatly aided through the use of Table 7.1. The position in that table of the expressions for the integrals is indicated. It is interesting to note that the contribution to the internal work of the axial deformations is only 0.24% of the total, which also means that it is only 0.24% of the value of H. In small frames such as these, the effects on computations of axial deformations are often neglected because they are indeed very small.

The structure shown in Fig. 10.13 is typical of the construction of a small industrial building. The three columns of the end frame are pin supported at their base and the connection of the center column to the horizontal beam

Example 10.4

Analyze the structure by the method of consistent deformations.

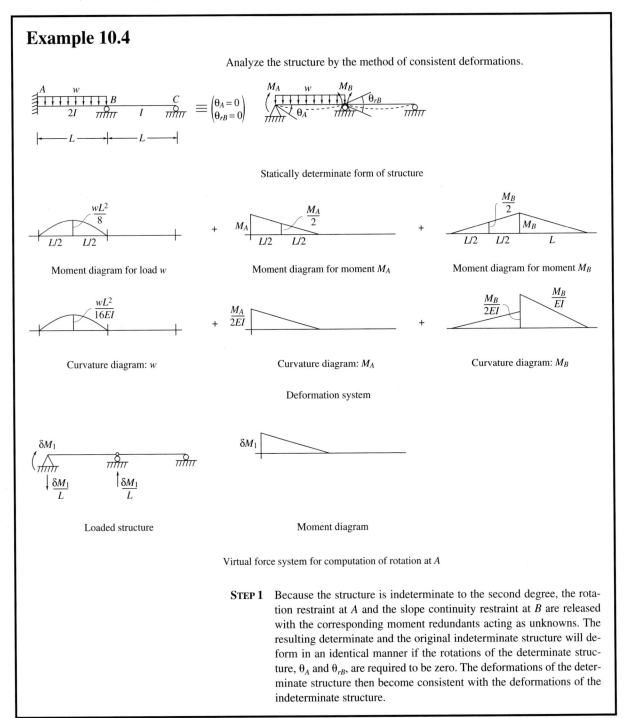

Statically determinate form of structure

Moment diagram for load w Moment diagram for moment M_A Moment diagram for moment M_B

Curvature diagram: w Curvature diagram: M_A Curvature diagram: M_B

Deformation system

Loaded structure   Moment diagram

Virtual force system for computation of rotation at A

STEP 1 Because the structure is indeterminate to the second degree, the rotation restraint at A and the slope continuity restraint at B are released with the corresponding moment redundants acting as unknowns. The resulting determinate and the original indeterminate structure will deform in an identical manner if the rotations of the determinate structure, θ_A and θ_{rB}, are required to be zero. The deformations of the determinate structure then become consistent with the deformations of the indeterminate structure.

Example 10.4 (continued)

STEP 2 Compute the rotation θ_A. The virtual force system δM_1 is established on the determinate structure for this computation. The deformation system is the sum of the deformations due to w, M_A, and M_B as shown in the curvature diagrams. All three deformations occur simultaneously, so the complementary virtual work computation is made on that basis with the aid of the expressions in Table 7.1. From Eq. (7.14), noting that only the member AB contributes to the internal work:

$$W_e = W_i$$

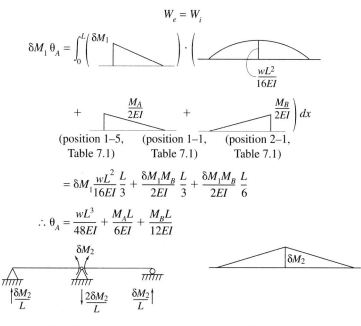

(position 1–5, (position 1–1, (position 2–1,
Table 7.1) Table 7.1) Table 7.1)

$$= \delta M_1 \frac{wL^2}{16EI} \frac{L}{3} + \frac{\delta M_1 M_B}{2EI} \frac{L}{3} + \frac{\delta M_1 M_B}{2EI} \frac{L}{6}$$

$$\therefore \theta_A = \frac{wL^3}{48EI} + \frac{M_A L}{6EI} + \frac{M_B L}{12EI}$$

Loaded structure   Moment diagram

Virtual force system for the computation of relative rotation at B

STEP 3 Compute the relative rotation θ_{rB}. The virtual force system δM_2 is established for this computation. Again the deformation system is the sum of the curvatures due to w, M_A, and M_B. All deformations occur simultaneously and the complementary virtual work computation using Eq. (7.14) becomes, with the aid of Table 7.1:

Example 10.4 (continued)

$$W_e = W_i$$

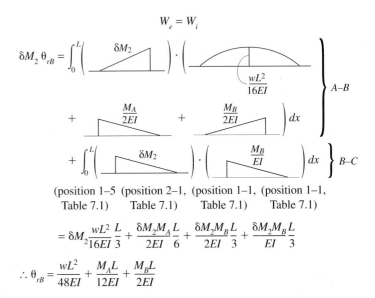

(position 1–5 (position 2–1, (position 1–1, (position 1–1,
  Table 7.1)   Table 7.1)   Table 7.1)   Table 7.1)

$$= \delta M_2 \frac{wL^2}{16EI} \frac{L}{3} + \frac{\delta M_2 M_A}{2EI} \frac{L}{6} + \frac{\delta M_2 M_B}{2EI} \frac{L}{3} + \frac{\delta M_2 M_B}{EI} \frac{L}{3}$$

$$\therefore \theta_{rB} = \frac{wL^2}{48EI} + \frac{M_A L}{12EI} + \frac{M_B L}{2EI}$$

STEP 4 Impose compatibility. The deformations of the determinate structure will be consistent with the indeterminate structure only if, *simultaneously,* θ_A and θ_{rB} are zero. This yields the two equations

$$\frac{M_A L}{6EI} + \frac{M_B L}{12EI} = -\frac{wL^3}{48EI}$$

$$\frac{M_A L}{12EI} + \frac{M_B L}{2EI} = -\frac{wL^3}{48EI}$$

solving

$$M_A = -\frac{5wL^2}{44}$$

$$M_B = -\frac{wL^2}{44}$$

STEP 5 Draw the final moment diagrams. The moment diagrams associated with each separate action can be added together. From the moment diagrams in step 1, the moment at the center of AB is $wL^2/8 + M_A/2 + M_B/2$.

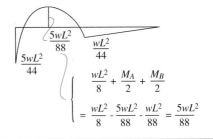

$$\frac{wL^2}{8} + \frac{M_A}{2} + \frac{M_B}{2}$$

$$= \frac{wL^2}{8} - \frac{5wL^2}{88} - \frac{wL^2}{88} = \frac{5wL^2}{88}$$

Example 10.5

For the frame shown below, obtain the reactions and final moment diagram by the method of consistent deformations. Take $E = 200$ GPa, $I = 208 \times 10^6$ mm, and $A = 64.5$ cm^3.

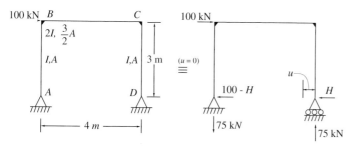

Original indeterminate structure and determinate form

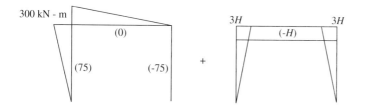

Moment diagrams for loaded structure (axial forces in parentheses)

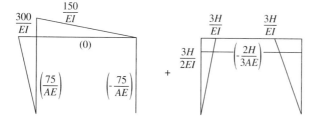

Deformation system curvature and strains (in parentheses)

STEP 1   The structure is made statically determinate by releasing the redundant horizontal restraint at D. The determinate and indeterminate structures are identical when $u = 0$.

STEP 2   Create the virtual force system to compute the horizontal displacement u at D. Compute axial forces and draw moment diagrams for the force

393

Example 10.5 (continued)

system. Curvatures and strains for the deformation system are computed. The deformation system is due to simultaneous application of the 100-kN load and the redundant force H. Applying the complementary virtual work expression, Eq. (7.14), gives

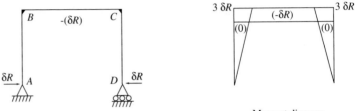

Moment diagram
(axial forces in parentheses)

Force system

$$W_e = W_i$$

$$\delta Ru = \overbrace{(0)\left(\frac{75}{AE}\right) \cdot 3}^{A-B \text{ (axial)}} + \overbrace{(-\delta R)\left(\frac{-2H}{3AE}\right) \cdot 4}^{B-C \text{ (axial)}} + \overbrace{(0)\left(\frac{-75}{AE}\right) \cdot 3}^{C-D \text{ (axial)}}$$

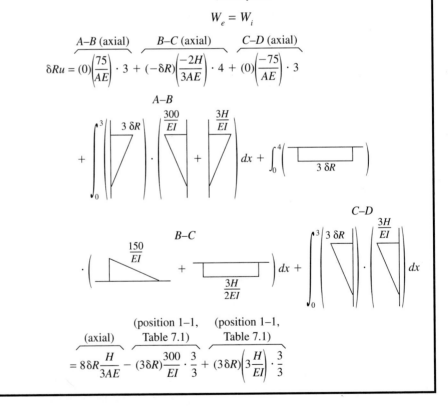

$$= 8\delta R\frac{H}{3AE} - \overbrace{(3\delta R)\frac{300}{EI} \cdot \frac{3}{3}}^{\text{(position 1–1, Table 7.1)}} + \overbrace{(3\delta R)\left(3\frac{H}{EI}\right) \cdot \frac{3}{3}}^{\text{(position 1–1, Table 7.1)}}$$

394

Example 10.5 (continued)

$$+ (3\delta R)\overbrace{\left(\frac{3H}{2EI}\right) \cdot 4}^{\substack{\text{(position 1–6,}\\ \text{Table 7.1)}}} - (3\delta R)\overbrace{\frac{150}{EI} \cdot \frac{4}{2}}^{\substack{\text{(position 1–6,}\\ \text{Table 7.1)}}} + (3\delta R)\overbrace{\left(3\frac{H}{EI}\right) \cdot \frac{3}{3}}^{\substack{\text{(position 1–1,}\\ \text{Table 7.1)}}}$$

$$\therefore u = 8\frac{H}{3AE} - \frac{1800}{EI} + 36\frac{H}{EI}$$

$$= H\frac{36 + (8/3)(I/A)}{EI} - \frac{1800}{EI}$$

$$= H\frac{36 + (8/3)(208/64.5)(1/100)}{EI} - \frac{1800}{EI}$$

$$u = \frac{36.086H - 1800}{EI}$$

(Note that axial force contribution to the deflection, u, is 0.086/36 or only 0.024%.)

STEP 3   Require deformation of the structure to be consistent with the indeterminate structure (i.e., $u = 0$).

$$0 = \frac{36.086H - 1800}{EI} \rightarrow H = 49.88 \text{ kN}$$

STEP 4   Compute the moments due to H, add them to the moments due to the 100-kN load, and draw the final moment diagram.

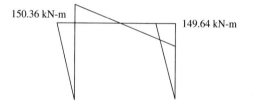

150.36 kN-m   149.64 kN-m

Final moment diagram

395

Figure 10.13 Two-bay industrial building. (Courtesy Bruce W. Abbott.)

can be idealized as a pin. There is also a pin connection in the horizontal beam in the right bay that appears as shown in Fig. 2.2a. This structure is analyzed in the next example for a horizontal load at the top of the left column.

Example 10.6 The structure is stable, indeterminate to the first degree, and is analyzed by the method of consistent deformations. The effect of axial deformations is neglected in the computations. No particularly unusual problems arise in the analysis, and use of the integration chart, Table 7.1, simplifies the calculations.

As all these examples illustrate, the key to the analysis of indeterminate structures is the identification of the redundant restraints that can be released to create a geometrically stable, statically determinate structure. Once the restraint releases have been identified, the analysis can proceed by the superposition method or the method of consistent deformations. The

Example 10.6

For the industrial building of Fig. 10.13, obtain the final moment diagram for a horizontal load P. Neglect axial deformations.

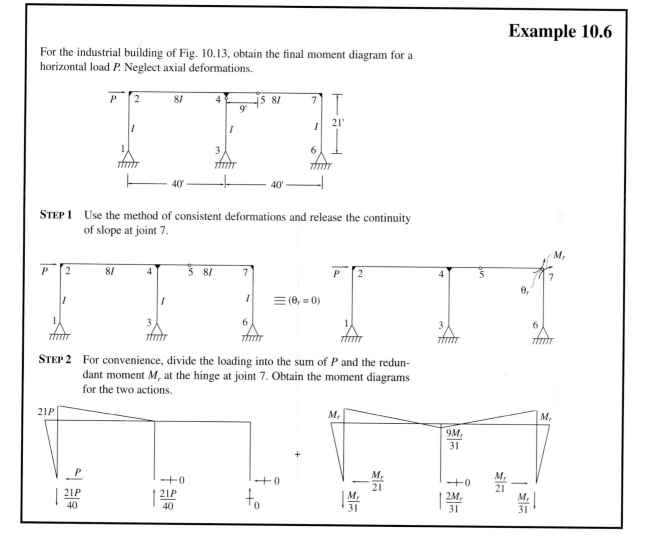

STEP 1 Use the method of consistent deformations and release the continuity of slope at joint 7.

STEP 2 For convenience, divide the loading into the sum of P and the redundant moment M_r at the hinge at joint 7. Obtain the moment diagrams for the two actions.

397

Example 10.6 (continued)

STEP 3 Establish the virtual force system, δR, and create the force and deformation system moments and curvatures.

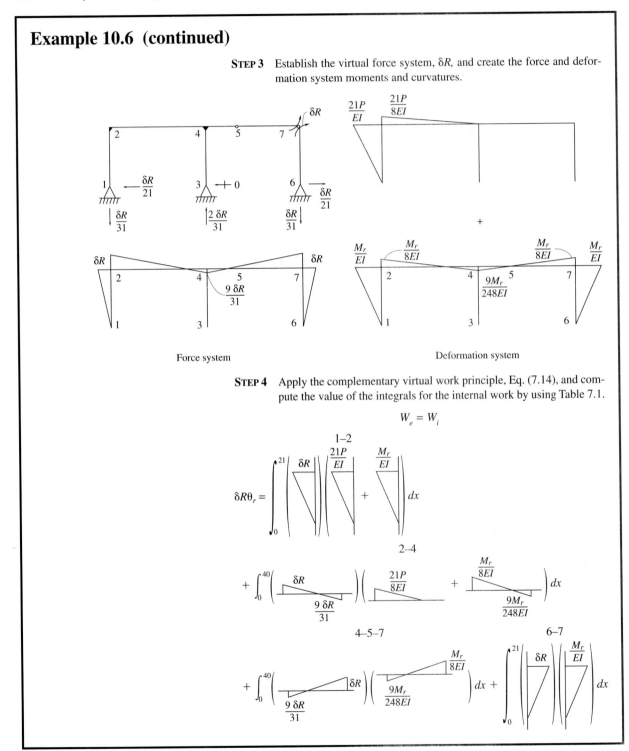

Force system

Deformation system

STEP 4 Apply the complementary virtual work principle, Eq. (7.14), and compute the value of the integrals for the internal work by using Table 7.1.

$$W_e = W_i$$

Example 10.6 (continued)

$$\delta R\theta_r = \underbrace{\frac{1}{3} \cdot \delta R \cdot \frac{21P}{EI} \cdot 21}_{\substack{\text{(position 1–1,} \\ \text{Table 7.1)}}} + \underbrace{\frac{1}{3} \cdot \delta R \cdot \frac{M_r}{EI} \cdot 21}_{\substack{\text{(position 1–1,} \\ \text{Table 7.1)}}} + \underbrace{\frac{1}{6} \cdot \frac{21P}{8EI}\left(2\delta R - \frac{9\delta R}{31}\right) \cdot 40}_{\substack{\text{(position 1–2,} \\ \text{Table 7.1)}}}$$

$$+ \underbrace{\frac{1}{6}\left[\frac{M_r}{8EI}\left(2R - \frac{9\delta R}{31}\right) - \frac{9M_r}{248EI}\left(\delta R - 2 \cdot \frac{9\delta R}{31}\right)\right] \cdot 40}_{\text{(position 11–2, Table 7.1)}}$$

$$+ \underbrace{\frac{1}{6}\left[\frac{-9M_r}{248EI}\left(-2 \cdot \frac{9\delta R}{31} + \delta R\right) + \frac{M_r}{8EI} - \left(\frac{-9\delta R}{31} + 2\delta R\right) \cdot 40\right]}_{\text{(position 11–2, Table 7.1)}}$$

$$+ \underbrace{\frac{1}{3}\delta R \cdot \frac{M_r}{EI} \cdot 21}_{\substack{\text{(position 1–1,} \\ \text{Table 7.1)}}}$$

$$\therefore \theta_r = \frac{P}{EI}\left(\frac{10{,}969}{62}\right) + \frac{M_r}{EI}\left(\frac{47{,}992}{2883}\right)$$

STEP 5 Require deformations of the statically determinate structure to be consistent with the indeterminate structure by setting θ_r equal to zero and solving for M_r. Draw the final moment diagram by adding the moments due to M_r to the moments due to P.

$$\theta_r = 0 \quad \therefore M_r = -\frac{10{,}969P}{62} \cdot \frac{2883}{47{,}992} = -10.63P$$

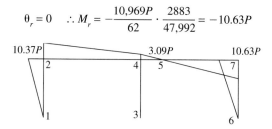

Final moment diagram

399

latter method is more general and is preferred for indeterminate analysis. More extensive treatments of indeterminate analysis can be found in Reference 22, 24, 25, 32, and 43.

10.5   Analysis of Symmetric Structures

The concept of dividing a structure that is symmetric into two parts and then analyzing one of those two parts was presented in Section 1.10. For indeterminate structures, this reduction of a structure into symmetric halves can reduce the analysis effort substantially.

Figure 10.14 Consider the two-span continuous beam shown with uniform loading over both spans. The same beam is shown in Fig. 1.10a, where because of symmetry of the structure and loading, it divides into two beams, each with a fixed support corresponding to the symmetric rotation condition at the center of the original structure. The analysis of the

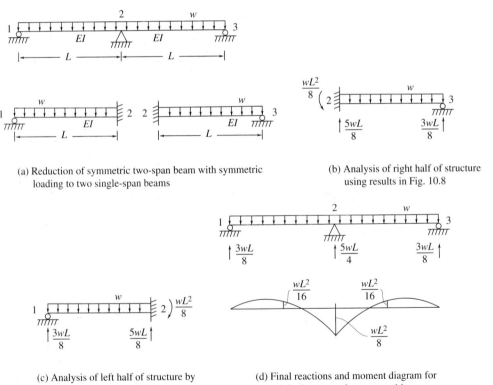

(a) Reduction of symmetric two-span beam with symmetric loading to two single-span beams

(b) Analysis of right half of structure using results in Fig. 10.8

(c) Analysis of left half of structure by symmetry with right half

(d) Final reactions and moment diagram for original structure by superposition

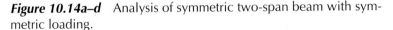

Figure 10.14a–d   Analysis of symmetric two-span beam with symmetric loading.

structure in Fig. 10.14a can thus be reduced to the analysis of a single-member structure as opposed to a structure with two members.

The beam shown in Fig. 10.14b is the same structure analyzed earlier in Fig. 10.1. The results of that analysis are presented in Fig. 10.8 and repeated as the results shown in Fig. 10.14b. The beam in Fig. 10.14c is the symmetric counterpart to the one in Fig. 10.14b, and the use of symmetry provides the results shown. Superimposing in Fig. 10.14d the results from Fig. 10.14b and c yields the final result for the original structure. The effort in this analysis of the original symmetric structure is reduced by half. As indicated in Section 1.10, superposition of the results for the symmetrically loaded symmetric structure is valid for nonlinear materials and structures undergoing large displacements.

When a symmetric structure is loaded with an antisymmetric load as shown in Fig. 10.15, the analysis can be halved by dividing the structure into its symmetric halves.

Figure 10.15 For this particular structure, which is indeterminate to the first degree, the antisymmetry conditions on the restraints at the axis of symmetry are such that the resulting "half" structure is statically determinate. An indeterminate analysis is avoided completely in this case. As indicated in Section 1.10, the use of antisymmetry of loading on a symmetric structure is limited to structures undergoing very small displacements and having nonlinear materials that have symmetric stress–strain relations in tension and compression.

The analyses in Figs. 10.14 and 10.15 illustrate that the effort required to analyze an indeterminate structure can be reduced by taking advantage of the symmetry of the structure and the symmetry or antisymmetry of applied loading. For a general loading applied to symmetric structures made from linear materials and undergoing small deformations, a reduction in the effort required for analysis can also be realized by dividing the structure in half and doing two simple, rather than one complicated, analysis. This is illustrated in Fig. 10.16.

Figure 10.16 The four-span continuous beam is symmetric about its center, but the loading is neither symmetric nor antisymmetric. However, Fig. 10.16a demonstrates how the loading can be divided into symmetric and antisymmetric parts, which sum to the original loading. The symmetric structure subjected to a symmetric and antisymmetric loading can be divided for analysis into two structures that are half the size of the original.

The structure for the symmetric loading in Fig. 10.16b requires the analysis of a two-span beam that is indeterminate to the second degree. The moment diagram for this analysis is shown, and the symmetric moment diagram for the right half of the structure is obtained by inspection

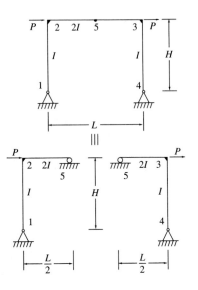

(a) Reduction of a symmetric frame with antisymmetric
loading to two simple frames

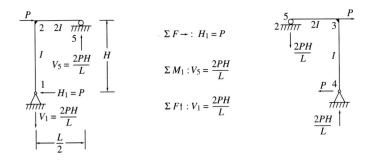

(b) Analysis of left half
by static equilibrium

(c) Analysis of right half by antisymmetry

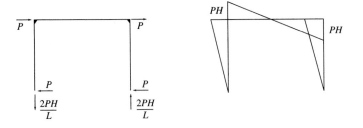

(d) Final reactions and moment diagram for
original structure by superposition

Figure 10.15a–d   Analysis of symmetric frame with antisymmetric
loading.

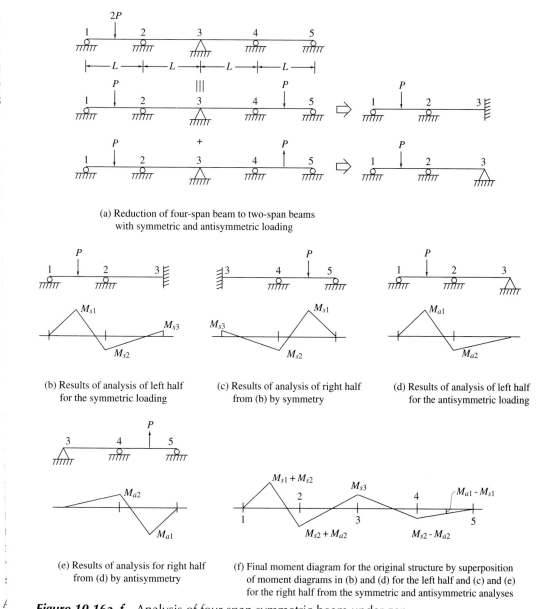

(a) Reduction of four-span beam to two-span beams
with symmetric and antisymmetric loading

(b) Results of analysis of left half
for the symmetric loading

(c) Results of analysis of right half
from (b) by symmetry

(d) Results of analysis of left half
for the antisymmetric loading

(e) Results of analysis for right half
from (d) by antisymmetry

(f) Final moment diagram for the original structure by superposition
of moment diagrams in (b) and (d) for the left half and (c) and (e)
for the right half from the symmetric and antisymmetric analyses

Figure 10.16a–f   Analysis of four-span symmetric beam under general loading by combining analyses for symmetric and antisymmetric loadings.

similar manner. Consider, for example, the three-story frame shown in Fig. 10.17a with the vertical and lateral loads. The final moment diagram and member axial forces for this loading are shown in Fig. 10.17b, the diagram and axial forces being obtained from an analysis of the structure using a computer program. Table 10.1 lists properties of the members of the frame and Table 10.2 the joint displacement and rotation values from the computer analysis. Suppose that the vertical deflection at the center of member 1–2 is desired. This problem can be treated as a single-member problem, using the method of complementary virtual work.

Consider the beam 1′–2′ in Fig. 10.18. The end moments and uniform load acting on member 1′–2′ give rise to a moment and hence curvature diagram that is identical to the one for member 1–2 in Fig. 10.17. It can thus be concluded that the bending deformations of 1–2 and 1′–2′ must be identical. If, in addition, the supports at 1′ and 2′ for member 1′–2′ ex-

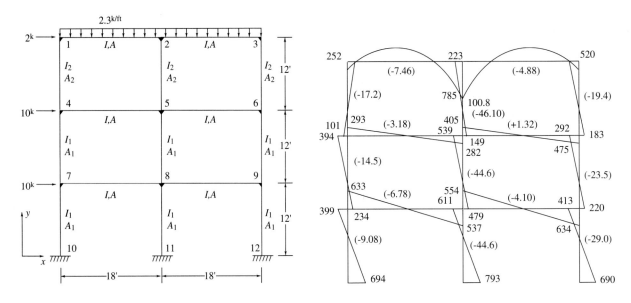

(a) Structure and applied loading

(b) Moments (kip-in.) and axial forces (kips)

Figure 10.17a–b Computer analysis of three-story two-bay frame.

Table 10.1   Member Properties for Frame of Fig. 10.16

Member	I (in^4)	A (in^2)	E (kips/in^2)
I, A	446.3	10.6	29,000.0
I_1, A_1	350.8	13.2	29,000.0
I_2, A_2	310.1	11.8	29,000.0

Table 10.2   Final Solution: Joint Displacements and Rotations

Joint	x Displacement (in.)	y Displacement (in.)	Rotations (rad)
1	0.819	−0.0161	−0.00217
2	0.814	−0.0529	−0.00028
3	0.810	−0.0279	0.00125
4	0.662	−0.0089	−0.00096
5	0.660	−0.0336	−0.00087
6	0.661	−0.0198	−0.00145
7	0.336	−0.0034	−0.00209
8	0.331	−0.168	−0.00129
9	0.328	−0.0109	−0.00196

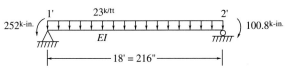
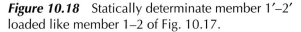

Figure 10.18   Statically determinate member 1′–2′ loaded like member 1–2 of Fig. 10.17.

perience a vertical movement equal to the vertical movement of joints 1 and 2 of the frame of Fig. 10.17, as given in Table 10.2, the complete set of deformations experienced by member 1–2 is reproduced exactly in member 1′–2′.

These observations form the basis of analysis of the midspan deflection of member 1–2. Since member 1′–2′ is a statically determinate structure, the analysis of its deflection can easily be carried out using the complementary virtual work approach.

Figure 10.19 All of the operations required to compute the vertical deflection of the center of member 1′–2′ have been carried out. The vertical deflection of the center of member 1′–2′ is, of course, also the vertical deflection of member 1–2 since the two members undergo identical deformations. The computation of the vertical deflection is shortened by use of Table 7.1 to evaluate the integrals.

As shown in Figs. 10.17 to 10.19, the computation of any deflection quantity that is not given by some analysis of an indeterminate structure can quickly be obtained by subjecting an appropriate determinate structure to the same deformations and using the complementary virtual work method.

This approach to the computation of the deflections of indeterminate structures can also be used to make an independent evaluation of the

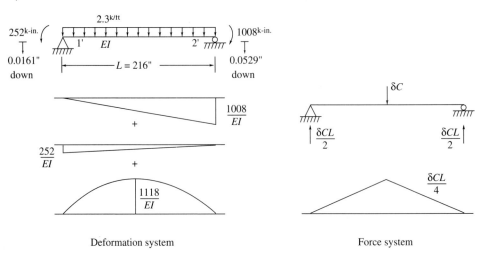

Deformation system   Force system

The complementary virtual work calculation, Eq. (7.12), including the vertical displacements of supports 1' and 2', yields (using Table 7.1 with the position of the integration expressions indicated)

$$W_e \qquad = \qquad W_i$$

$$\delta C \, v_C + \frac{\delta C}{2}(-0.0161) + \frac{\delta C}{2}(-0.0529) = \int_0^L \left(\frac{\delta CL}{4}\triangle\right)\left[\frac{1008}{EI} + \frac{252}{EI} + \frac{1118}{EI} \right] dx$$

Position 2–4, Table 7.1 ($t = 1/2$)   Position 1–4, Table 7.1 ($t = 1/2$)

$$\delta C \, v_C - 0.0345 \, \delta C = -\frac{1008}{EI}\left(\frac{\delta CL}{4}\right)\left(2 - \frac{1}{2}\right)\frac{1}{6}L \qquad -\frac{252}{EI}\left(\frac{\delta CL}{4}\right)\left(1 + \frac{1}{2}\right)\frac{1}{6}L$$

Position 3–5, Table 7.1 ($r = 1/2$)

$$+\frac{1118}{EI}\left(\frac{\delta CL}{4}\right)\left[1 + \frac{1}{2}\left(1 - \frac{1}{2}\right)\right]\frac{1}{3}L$$

$$v_C - 0.0345 = \frac{L^2}{EI}(-63 - 15.75 + 116.5) = \frac{37.71 \, L^2}{EI}$$

For $L = 216$ in., $E = 29{,}000$ ksi, and $I = 446.3$ in^4

$$v_C = 0.0345 + 0.1359 = 0.1704 \text{ in. downward}$$

Figure 10.19   Vertical deflection of center of member 1′–2′ of Fig. 10.18 and member 1–2 of Fig. 10.17.

correctness of a given analysis. Refer again to the frame analysis presented in Fig. 10.17a. Deflections and rotations of joints 1, 4, and 7 are presented in Table 10.2, but are those values correct?

Figure 10.20 A nonprismatic cantilever beam having the same member properties and lengths between labeled points as the column 1–4–7–10 in Fig. 10.17 is shown. If the cantilever beam of Fig. 10.20 is subjected to the same deformations (i.e., curvatures) as the outside column 1–4–7–10 in Fig. 10.17, it must undergo the same joint displacements and rotations. The curvature-area (moment-area) technique is used to obtain the displacements and rotations as shown. All values of displacements and rotations match those reported in Table 10.2, indicating that these deformations are consistent with the moments and curvatures presented in Fig. 10.17b.

Additional consistency checks can also be made on the computer-generated solution. In Fig. 10.21, a simple cantilever frame representing the first-story and left-side bay of the structure of Fig. 10.17 is analyzed for the rotational displacement of joint 11′ using the complementary virtual work approach.

Figure 10.21 Joint 11 of the frame in Fig. 10.17 is fully fixed against all displacement and rotation. If the computer-generated analysis is correct, the axial strains and curvatures that come from that analysis should deform frame 10′–7′–8′–11′ of Fig. 10.21 in such a manner that the rotation of 11′ is zero. The curvatures and axial strains shown are the deformation system and the force system simply the moment δM applied at 11′. The analysis by complementary virtual work of the rotation at joint 11′ is assisted by the use of the expressions in Table 7.1 for the evaluation of the integrals. The position in that table of the expression used in indicated. The rotation at 11′ and hence at 11 is computed to be "zero." The curvatures and strains used in the computation are given to only three significant figures, so that the computation cannot be expected to yield exactly zero.

The computation of the rotation in Fig. 10.21 gives a zero value at joint 11′, indicating that the solution generated by the computer analysis satisfies rotational compatibility of deformation between joints 10 and 11 along the path 10–7–8–11. Many other consistency checks can be carried out in this manner to evaluate the correctness of any proposed solution to an indeterminate problem. Use of the complementary virtual work method, the integration charts of Table 7.1, and simple statically determinate structural forms having deformations identical to the indeterminate structure make the analysis straightforward and tractable.

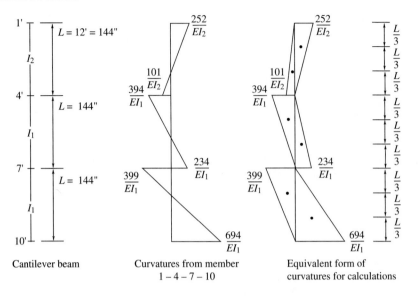

Cantilever beam   Curvatures from member   Equivalent form of
                                  $1 - 4 - 7 - 10$ curvatures for calculations

Calculate rotations using the first curvature-area theorem, Eq. (6.10). Note that the story heights, $L = 144$ in., are all the same.

$$\theta_7' = \theta_7 = -\frac{1}{2}\left(\frac{694}{EI_1}\right)L + \frac{1}{2}\left(\frac{399}{EI_1}\right)L = -\frac{295}{2EI_1}L = -0.00209 \text{ rad.}$$

$$\theta_4' = \theta_4 = -0.00209 + \frac{1}{2}\left(\frac{394}{EI_1}\right)L - \frac{1}{2}\left(\frac{234}{EI_1}\right)L = -0.00209 + \frac{80L}{EI_1} = -0.00209 + 0.00113$$

$$= -0.00096 \text{ rad}$$

$$\theta_1' = \theta_1 = -0.00096 + \frac{1}{2}\left(\frac{101}{EI_2}\right)L - \frac{1}{2}\left(\frac{252}{EI_2}\right)L = -0.00096 - \frac{151L}{2EI_2} = -0.00096 - 0.00121$$

$$= -0.00217 \text{ rad}$$

Calculate displacements using the second curvature-area theorem, Eq. (6.12). Again use story heights as $L = 144$ in.

$$u_7' = u_7 = -\frac{1}{2}\left(\frac{694}{EI_1}\right)L\frac{2}{3}L + \frac{1}{2}\left(\frac{399}{EI_1}\right)L\frac{1}{3}L = 0.336 \text{ in.}$$

$$u_4' = u_4 = \frac{1}{2}\left(\frac{234}{EI_1}\right)L\left(\frac{2}{3}\right)L - \frac{1}{2}\left(\frac{394}{EI_1}\right)L\left(\frac{1}{3}\right)L + \left(\frac{1}{2}\right)\frac{694}{EI_1}L\left(1 + \frac{2}{3}\right)L - \frac{1}{2}\left(\frac{399}{EI_1}\right)L\left(1 + \frac{1}{3}\right)L$$

$$= 0.662 \text{ in.}$$

$$u_1' = u_1 = \frac{1}{2}\left(\frac{253}{EI_2}\right)L\left(\frac{1}{3}\right)L - \frac{1}{2}\left(\frac{101}{EI_2}\right)L\left(\frac{2}{3}\right)L + \left(\frac{1}{2}\right)\frac{234}{EI_1}L\left(1 + \frac{2}{3}\right)L - \frac{1}{2}\left(\frac{394}{EI_1}\right)L\left(1 + \frac{1}{3}\right)L$$

$$+ \frac{1}{2}\frac{694}{EI_1}L\left(2 + \frac{2}{3}\right)L - \frac{1}{2}\left(\frac{399}{EI_1}\right)L\left(2 + \frac{1}{3}\right)L = 0.819 \text{ in.}$$

Figure 10.20 Deflections of left side of structure of Fig. 10.17.

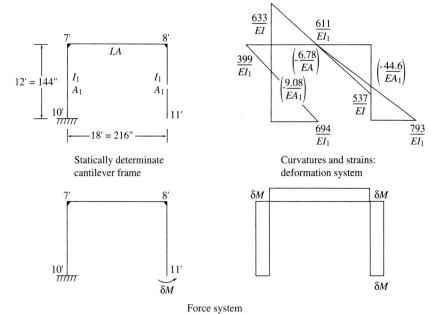

Statically determinate
cantilever frame

Curvatures and strains:
deformation system

Force system

Use the complementary virtual work principle, Eq. (7.12), and Table 7.1 to evaluate integrals to obtain the rotation of joint 11' (or 11).

$$W_e \qquad = \qquad W_i$$

Member 10–7   Member 7–8   Member 8–11

$$\delta M\,\theta'_{11} = \delta M\,\theta_{11} = \int_0^{144} \left(\begin{array}{c} \square \\ \delta M \end{array}\right)\left(\dfrac{\frac{399}{EI_1}}{\frac{694}{EI_1}}\right) dx + \int_0^{216} \left(\delta M\right)\left(\dfrac{\frac{633}{EI}}{\frac{537}{EI}}\right) dx + \int_0^{144} \left(\begin{array}{c} \square \\ \delta M \end{array}\right)\left(\dfrac{\frac{611}{EI_1}}{\frac{793}{EI_1}}\right) dx$$

$$= \qquad \frac{\text{All from position 11–6, Table 7.1}}{\dfrac{\delta M}{2}\,144\left(\dfrac{399}{EI_1} - \dfrac{694}{EI_1}\right) + \dfrac{\delta M}{2}\,216\left(\dfrac{633}{EI} - \dfrac{537}{EI}\right) + \dfrac{\delta M}{2}\,144\left(\dfrac{793}{EI_1} - \dfrac{611}{EI_1}\right)}$$

$$= \quad \delta M\left(-\dfrac{21,240}{EI_1} + \dfrac{10,368}{EI} + \dfrac{13,104}{EI_1}\right) = \delta M\left(\dfrac{10,368}{EI} - \dfrac{8136}{EI_1}\right)$$

$$= \quad \delta M(0.0000013) \approx 0 \quad \text{(within computational accuracy)}$$

Figure 10.21   Compatibility of rotation between joints 10 and 11 of the structure of Fig. 10.17.

415

10.7 Analysis of Indeterminate Structures with Nonlinear Materials

The introduction of a nonlinear stress–strain relation adds a significant complication to the analysis of indeterminate structures. The discussion in this section is limited to structures with prismatic members and having stress–strain relations for axially loaded members and moment–curvature relations for members subjected to bending that are symmetric. It is further assumed that there are no combined actions of axial forces and bending in any member of a structure. These limitations on combined axial force and bending action are very restrictive and essentially eliminate the possibility of analyzing plane frames. The concept of analysis of structures with nonlinear materials will be presented for plane trusses and beams.

With the introduction of a nonlinear material force deformation relation, the analysis of indeterminate structures can no longer be undertaken using the superposition method of Section 10.2. The method of consistent deformations of Section 10.3 is applicable to indeterminate analysis because in this method all the force actions on the statically determinate form of the structure are taken to occur simultaneously. Internal axial forces in members are the sum of the forces from loads and redundants, and the strains are computed from the stress–strain relation using the stress obtained from the combined sum of the axial forces. Similarly, in beams, curvatures are computed from the simultaneous action of the moments from loads and redundants. The complementary virtual work principle is used as the basis of evaluating all member and structural deformations.

The stress–strain relation presented in Eqs. (9.8) and (9.9) provides a reasonable nonlinear material relation for a plane truss. Example 10.9 shows how the indeterminate truss of Example 10.1 responds to the applied 200-kip load using these stress–strain relations, with the value of n in Eqs. (9.8) and (9.9) being 2.

Example 10.9 The method of consistent deformations coupled with the complementary virtual work principle provides the means of solving this problem. If the loading process of this structure is visualized as a slow increase of the applied load at L_1 until it reaches the magnitude of 200 kips, the force, F, in member U_1–U_2 as well as the other members of the structure will also slowly increase to their final values, although not in linear proportion to the 200-kip force because the structure is created from nonlinear materials. The important point to recognize is that the forces and strains in all members of the structure, at all times in the loading process, are due to simultaneous application of the 200-kip force and the member force, F.

To satisfy the consistency of deformation between the indeterminate and determinate forms of the truss, the relative displacement at the cut in

416

Example 10.9

Obtain the reactions and internal member forces in the indeterminate truss of Example 10.1. Use the nonlinear stress–strain relation of Eqs. (9.8) and (9.9) with $n = 2$. Take $\sigma_0 = 36$ ksi, $\varepsilon_0 = 36/29000 = 0.001241$, and the area of members (in square inches) as shown in parentheses.

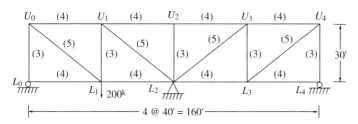

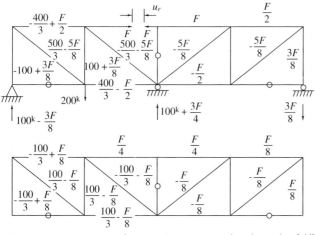

Stresses $\sigma_p = (F_p + F_f)/A$;   strains $\varepsilon = \varepsilon_0(\sigma_p/\sigma_0)/[1(\sigma_f/\sigma_0)^2]^{1/2}$
Deformation system

STEP 1 Use the method of consistent deformations, and as in Example 10.1, select the release of continuity in member U_1–U_2 with the force, F, in the member as the redundant. The force system δS is the same as before.

STEP 2 The force in the members of the statically determinate form of the truss due to the 200-kip load and F are the same as before. To obtain the strains in the members of the truss for the deformation system, the stresses due to the combined forces of the 200-kip load and F in the members are substituted into Eq. (9.9).

417

Example 10.9 (continued)

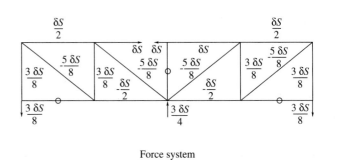

Force system

STEP 3   Use the complementary virtual work principle to compute the relative displacement, u_r, at the cut and require it to be zero. The internal work is given by Eq. (9.12). Equating the external and internal work yields

$$W_e = W_i$$

$$\delta S u_r = \Sigma\, \delta F_s \left[\frac{\varepsilon_0(\sigma_p/\sigma_0)}{[1 - (\sigma_p/\sigma_0)^2]^{1/2}} \right] L$$

where $\sigma_p = (F_p + F_f)/A$ and F_p are the forces in the members due to the 200-kip load and F_f the forces due to F as shown in the sketch.

STEP 4   Set the relative displacement, u_r, in the complementary virtual work expression to zero and solve for F. The equation is highly nonlinear, and the solution is carried out in the spreadsheet shown. The spreadsheet is a modified form of the one presented in Example 10.1. The parameter β in the spreadsheet is the multiplier of the force, F, acting at the cut in member U_1–U_2. The strains in the deformation system are computed by Eq. (9.9) and are due to the stresses caused by the sum of the forces from F and the 200-kip load. The internal work is computed as the product of the force system forces, δF_s, the deformation system strains, and the length of the member. The sum of the internal work of all members appears at the bottom of the W_i column and is essentially zero. The value of β was changed by trial and error until the internal work became "zero," which represents the solution of the problem for F. The final member forces are shown in the last column of the spreadsheet and on the sketch of the structure.

Example 10.9 (continued)

| n : | 2 | σ o: | (in ksi) 36 | ε₀: | 0.00124 | β: | 10.466 | E: 29000 |

MEMBER	LENGTH	AREA	FORCE SYSTEM FORCE	DEFORMATION SYSTEM STRAINS				INTERNAL WORK	FINAL FORCES (K)
				MEMBER 200K	FORCES FORCE		MEMBER STRAIN	Wi	
			L	A	δFs		FORCE	F	
L0L1	40	4	0	0	0		0	0	0
L1L2	40	4	-0.5 δS	133.33	-5.233		0.002418	-0.0484 δS	128.1
L2L3	40	4	-0.5 δS	0	-5.233		-4.5 E-05	0.000 δS	-5.233
L3L4	40	4	0	0	0		0	0	0
U0U1	40	4	0.5 δS	-133.33	5.233		-0.00242	-0.0484 δS	-128.1
U1U2	40	4	1 δS	0	10.466		9.05 E-05	0.00362 δS	10.466
U2U3	40	4	1 δS	0	10.466		9.05 E-05	0.00362 δS	10.466
U3U4	40	4	0.5 δS	0	5.233		4.51 E-05	0.0009 δS	5.233
L0U0	30	3	0.375 δS	-100	3.92475		-0.00242	-0.0272 δS	-96.08
L1U1	30	3	0.375 δS	100	3.92475		0.00439	0.04939 δS	103.92
L2U2	30	3	0	0	0		0	0	0
L3U3	30	3	0.375 δS	0	3.92475		4.51 E-05	0.00051 δS	3.9247
L4U4	30	3	0.375 δS	0	3.92475		4.51 E-05	0.00051 δS	3.9247
U0L1	50	5	-0.625 δS	166.67	-6.5413		0.002418	-0.0756 δS	160.13
U1L2	50	5	-0.625 δS	-166.7	-6.5413		-0.00439	0.13718 δS	-173.2
U3L2	50	5	-0.625 δS	0	-6.5413		-4.5 E-05	0.00141 δS	-6.541
U4L3	50	5	-0.625 δS	0	-6.5413		-4.5 E-05	0.00141 δS	-6.541
					TOTAL W$_i$ =	Σ =		-6.5 E-06 δS	

Final reactions and internal member forces

419

the member U_1–U_2, u_r, in the determinate truss must be zero. The force system, δS, provides the means of computing u_r by complementary virtual work, using Eq. (9.12) to evaluate the internal work with the nonlinear material.

From the concept of consistent deformation, the relative displacement, u_r, must be zero, which reduces the complementary virtual work calculation to finding the magnitude of F for which the internal work, W_i, is zero. The spreadsheet shown in the example details the computation of the strain and hence internal work for each member of the structure. The magnitude of the force F is scaled by the parameter, β, and the value of β shown is 10.466, which corresponds to the magnitude of F that makes the internal work calculated as the sum of the terms in the W_i column of the spreadsheet zero. Since the external work, W_e, is defined as $\delta S u_r$ and $W_e = W_i$ in the complementary virtual work method, the relative displacement u_r is zero.

The value of the redundant force in the member U_1–U_2 has changed from 26.02 kips in Example 10.1 to 10.47 kips in Example 10.9. The reason for the large change is due to the fact that the forces in members U_1–L_2 and U_1–L_1 of the linear elastic solution of Example 10.1 cause stresses that are in excess of the yield stress of the material, $\sigma_0 = 36$ ksi, in Example 10.9. As a result, the strains in these member become large and the forces and stresses in surrounding members are increased to maintain equilibrium. These members are stressed at 34.6 ksi in Example 10.9 and have strains that are nearly four times ε_0. The nonlinear character of the material stress–strain relation has a significant impact on the internal distribution of member forces. The 200-kip load represent a load that is close to the one that would cause this structure to collapse. The use of a nonlinear moment–curvature relation is illustrated in Example 10.10.

Example 10.10 The two-span continuous beam is symmetric and symmetrically loaded. As a result, the structure can be divided in half and is reduced to the analysis of a singe-member indeterminate beam. The moment–curvature relation defined in Eqs. (9.18) and (9.19) with $n = 2$ provides a reasonable approximation of the behavior of a W cross section fabricated from a mild steel. The use of the method of consistent deformations establishes the condition that the deformation behaviors of the determinate and indeterminate forms of the structure are identical when the vertical displacement of the left end of the determinate structure is zero. The curvatures in the determinate form of the structure are obtained through the moment–curvature relation, Eq. (9.19), from the moments due to the simultaneous application of the uniform load, w, and the vertical force, R.

Complementary virtual work is used to calculate the vertical displacement with the internal work being evaluated from expression (9.21). The

Example 10.10

Obtain the reactions and final moment diagram for the uniformly loaded two-span beam. The material is nonlinear and has a moment–curvature relation defined by Eqs. (9.18) and (9.19) with $n = 2$. Take the ratio $wL^2/M_0 = 11.2$, where M_0 is the maximum moment capacity of the cross section.

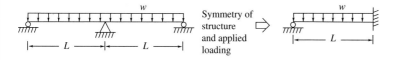

STEP 1   Take advantage of the symmetry of the structure and applied loading to reduce the structure to a propped cantilever beam. Use the method of consistent deformations and reduce the structure to the cantilever beam with an unknown upward load applied to the left end. Obtain moments and then curvatures from Eq. (9.19) to identify the deformation system.

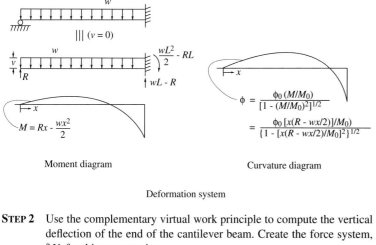

$$\phi = \frac{\phi_0 (M/M_0)}{[1 - (M/M_0)^2]^{1/2}}$$

$$= \frac{\phi_0 [x(R - wx/2)]/M_0}{\{1 - [x(R - wx/2)/M_0]^2\}^{1/2}}$$

Moment diagram   Curvature diagram

Deformation system

STEP 2   Use the complementary virtual work principle to compute the vertical deflection of the end of the cantilever beam. Create the force system, δU, for this computation.

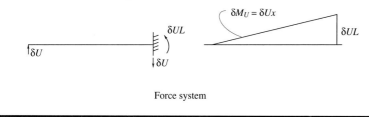

$$\delta M_U = \delta U x$$

Force system

421

Example 10.10 (continued)

STEP 3   Apply the complementary virtual work principle and evaluate the internal work using Eq. (9.21) and the moment and curvature expressions for the cantilever beam. The method of consistent deformation establishes the condition that the deflection of the end of the cantilever beam must be zero for the simultaneous application of w and R.

$$W_e = W_i$$

$$\delta Uv = \int_0^L \delta M_U \phi \, dx = \int_0^L \frac{\delta Ux \, \phi_0 \, [x(R - wx/2)/M_0] \, dx}{\{1 - [x(R - wx/2)/M_0]^2\}^{1/2}}$$

Rearranging the terms and recognizing the δU can be dropped out of the expression yields the deflection, v, as

$$v = \int_0^L \frac{\phi_0 \, Rx^2/M_0 \, (1 - wx/2R)] \, dx}{\{1 - [Rx/M_0 \, (1 - wx/2R)]^2\}^{1/2}}$$

To evaluate the integral, make the substitution $z = x/L$, which changes the integral to the form

$$v = \phi_0 \int_0^L \frac{(RL^3/M_0)z^2(1 - wL/2Rz)] \, dz}{\{1 - [RL/M_0 z (1 - wL/2Rz)]^2\}^{1/2}}$$

The condition that $v = 0$ yields an expression that can be used to solve for the unknown value of R. A further simplification of the integral is possible by recognizing that R, M_0, and L are independent of the integration and by substituting $r = RL/M_0$. The integral becomes

$$v = 0 = \int_0^L \frac{z^2 (1 - wL^2/2M_0z/r)] \, dz}{\{1 - [rz (1 - wL^2/2M_0 z/r)]^2\}^{1/2}}$$

Substituting the value of $wL^2/M_0 = 11.2$, the integral can be evaluated numerically using Eq. (9.4). This yields a nonlinear equation in r that can be solved by any convenient means for r. For $n = 4$ in Eq. (9.4), $r = 4.623$, which yields a value of $R = 0.41277wL$. For $n = 32$, the value of r is 4.6005 and $R = 0.4076wL$. The final moment diagram is shown below for $R = 0.4128wL$.

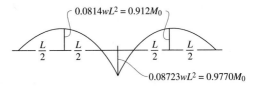

condition that the displacement, v, be zero establishes an integral that is best evaluated by numerical integration. The nonlinear character of the function being integrated encumbers the solution procedure, but a simple computer program or the use of a spreadsheet to do the numerical integration with Eq. (9.4) will accomplish the task. The solution shows that the final moments reach about 98% of M_0 at the center support and more that 91% at the center of the spans. For numerical calculations with U.S. units, take $L = 20$ ft and $M_0 = 300$ kip-ft; for SI units, take $L = 6$ m and $M_0 = 400$ kN-m.

The method of consistent deformations provides a means of analyzing the response of indeterminate structures made from nonlinear materials. The complementary virtual work principle has been used to calculate deformations, but other methods are available. In Example 10.10 the second curvature-area theorem or the conjugate beam method could have been used to compute the vertical deflection of the beam once the curvature diagram is established. Examples 10.9 and 10.10 illustrate that the resulting mathematical models of the structures are characterized by highly nonlinear equations that require extensive calculations for solution. The wide availability of computers is making the analysis of nonlinear problems more practical. The nonlinear material deformation relations used in this section are representative of what can be done to idealize material behavior but are not intended to limit choices for any particular structure.

10.8   Summary and Limitations

The analysis of statically indeterminate structures is based on their deformation behavior. The method of superposition of deflections is illustrated as a means of analyzing indeterminate structures made from linear materials. A more general approach is presented called the method of consistent deformations which is extended in a later section to the analysis of indeterminate structures with nonlinear materials. The computation of deflections or rotations of structures is carried out by using the complementary virtual work principle, although other methods can also be used.

Indeterminate analysis is based on the use of a statically determinate form of the indeterminate structure. The determinate structure is created by releasing the continuity of deformation or reaction constraint at one or more points in the indeterminate structure. The number of releases is equal to the degree of indeterminacy of the structure, and each release has a force or moment action associated with it. These actions are referred to as redundants, because their presence in the structure is not needed to provide the necessary conditions of equilibrium. The redundant force or moment actions act with unknown magnitude on the determinate form of the structure.

One or more deformation analyses of the determinate structure establishes the equations, called equations of compatibility, that enable the computation of the magnitudes of the redundant actions.

The selection of redundant constraints is shown to be most effective if they can isolate the actions of loads on the statically determinate form of the structure. The goal of redundant selection is to reduce as much as possible the work of the deformation analysis. When more than one redundant exists in a structure, the analysis requires the solution of a set of simultaneous equations. The use of symmetry of a structure can reduce the work in indeterminate analysis, although for a general loading on a symmetric structure, the structure must be made of linear materials and be undergoing small displacements in order for the analysis to take advantage of the symmetry. For generally loaded symmetric structures, the use of symmetry reduces the degree of indeterminacy in the resulting analysis, or in some cases eliminates it entirely.

The analysis of indeterminate structures for deflections or rotations can be carried out on any stable determinate form of that structure. As illustrated in the examples and figures, the deformation analysis of an indeterminate structure for which the distribution of internal and external actions has been obtained is no more complicated than the analysis of a statically determinate structure. The use of simple deformation calculations as a means of checking the analysis of an indeterminate structure is also illustrated.

In the final section of the chapter, an analysis of structures with nonlinear materials is presented. The method of consistent deformations provides the mechanism for this analysis, and the procedure is no different from that for structures with linear materials. The resulting compatibility equations are nonlinear, which requires considerably more effort to solve, but the use of computer programs or spreadsheets makes the solutions practical.

The assumptions and limitations of indeterminate analysis are the same as those associated with determinate analysis. A summary of these limitations is presented below.

- The deformation analysis of trusses is subject to all of the same limitations and assumptions that are listed at the end of Chapter 3 except that the material can be nonlinear.

- Because of the use of simple bending theory principles, the deformation analysis of beams and frames is subject to all of the same limitations and assumptions that are listed at the end of Chapter 5 and in Section 6.2 except that members can have materials with nonlinear stress–strain relations.

- The analysis of structures with nonlinear materials is limited to those where the action of axial forces occurs in the absence bending or where bending action occurs in the absence of axial forces.

- All material deformations must be continuous, without any cracks or gaps.
- All structural deformations must be continuous, without any violation of the conditions of constraint at points of support.
- The distribution and orientation of all loads acting on a structure are not changed by the deformations of the structure.
- The orientation of all reaction components at supports is unchanged by the action of applied loads or any movement of the support.
- If movements occur at supports, they are small and do not affect the initial geometry of the structure.
- The principle of superposition is no longer valid for structures that have nonlinear materials.
- The analysis of symmetric structures under general loading is possible only for those structures where the principle of superposition of deformations is valid.

Some questions that are worthy of consideration when using the principles presented in this chapter are the following:

- If a stress–strain or moment–curvature relation is defined by a set of data points, can an indeterminate analysis still be carried out for a structure made from that material?
- What modifications, if any, are required for the analysis of indeterminate structures with nonlinear materials?
- The analysis of an indeterminate structure with nonlinear materials may yield more that one solution. How can the solution be checked to verify that it is the correct one?
- Should computer programs that provide the internal distributions of forces and moments in an indeterminate structure also provide some set of displacements or rotations of the structure, and why or why not?

PROBLEMS

10.1 For the truss shown, compute the reactions and final member forces. Area of members in square inches is indicated in the parentheses on each member. Take E as constant.

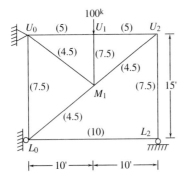

10.2 In the truss loaded as shown find the final force in member a–d. [*Note:* $(\cdot)$ = area in square centimeters of member; take E as constant.]

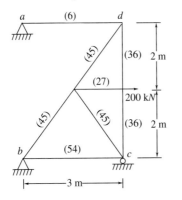

10.3 For the king post truss shown, compute the member forces and draw the moment diagram for member A–C. (*Note:* The hinges at A, B, and C are effective only for truss members A–D, B–D, and C–D. Member properties are as follows:

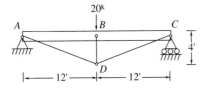

Member Properties

Member	E (ksi)	A (in²)	I (in⁴)
AC	1,450	130	1,460
BD	1,450	40	—
AD	29,000	1.5	—
CD	29,000	1.5	—

10.4 Determine the horizontal deflection of point c of the truss due to the applied load *(E constant)*.

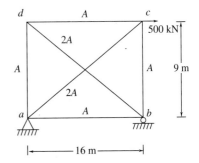

10.5 Find the bar forces in the structure shown *(E constant)*.

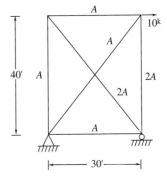

10.6 Analyze and draw the final moment diagram ($E = 29 \times 10^6$ psi).

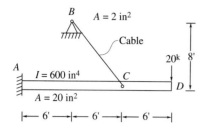

10.7 For the loading shown and the forcing of member e–g into place even though it is 3 mm too short, determine the bar force in member d–f (except for members shown, $A = 80$ cm^2, $E = 200$GPa).

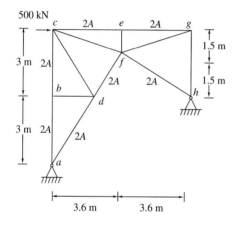

10.8 The truss shown is loaded with the 120-kip load. Attached is the solution for the final member forces (in kips). Compute the *change* in the member forces if the temperature drops 50°F in members A–C and B–C, and the support B moves 0.01 ft to the left. For the structure, $E = 29 \times 10^3$ ksi and $\alpha = 6.5 \times 10^{-6}/°F$.

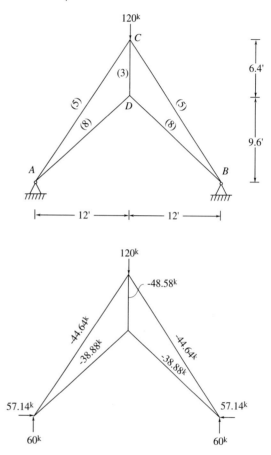

427

10.9   The plane truss is loaded with a 150-kN load at joint 3. Note that member 2–3 has a different material modulus of elasticity. Member areas in square centimeters are shown in parentheses on each member. Obtain the reaction forces.

Member 2–3: $E = 200$ GPa
All other members: $E = 70$ GPa

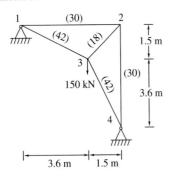

10.10   Determine the horizontal reactions for the plane truss ($E = 29 \times 10^3$ ksi, $A = 8$ in^2 except as noted).

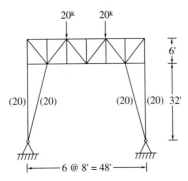

10.11   Draw the final moment diagram for the structure shown below (E constant).

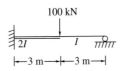

10.12   For the structure shown, calculate the deflection under the 15-kip load in terms of E and I. The moment diagram for the loading is given.

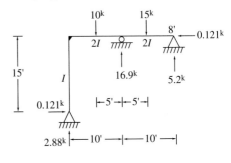

Moment diagram

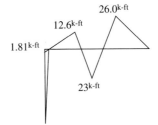

10.13   Analyze and draw the final moment diagram for the structure shown (E constant). Neglect axial deformations.

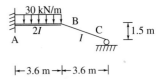

10.14   For the frame shown, compute the reactions and draw the final moment diagram. Neglect axial deformations (E constant).

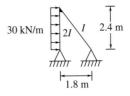

10.15 The two-span continuous beam shown is supported at its center (point 2) by a jack ($E = 29{,}000$ ksi, $I = 1350$ in^4).

(a) Obtain the final moment diagram for the uniform load of 5 kips/ft assuming that the supports are unmovable.

(b) To obtain a more advantageous distribution of the moments in the structure, it is proposed to raise or lower the jack at the center support until the magnitude of the moment in the center of the spans is equal to the magnitude of the moment at the center support (point 2). Determine the amount of movement that must take place at point 2 and its direction so that the conditions on the magnitude of the moments in the structure are realized.

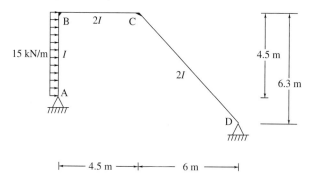

10.16 Obtain the final moment diagram for the frame loaded as shown below. Neglect axial deformations of the members (E constant).

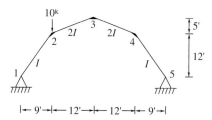

10.17 Obtain the final moment diagram for the structure shown. Neglect axial deformations.

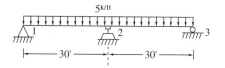

10.18 Obtain the final moment diagram for the structure shown. Neglect axial deformations.

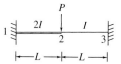

10.19 Draw the final moment diagram for the two-span beam. The material modulus is the same for all members.

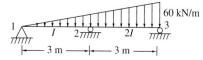

10.20 For the frame and loading shown, find the distribution of internal forces and the final moment diagram. Note that the cable is not capable of resisting moment.

Cable: $E = 25 \times 10^3$ ksi
$A = 1$ in^2
Other members:
$E = 29 \times 10^3$ ksi
$I = 100$ in^4
$A = 5$ in^2

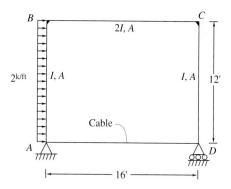

10.21   For the frame shown, determine the reactions and draw the moment diagram for all members. *(Note:* Take *EI* constant and neglect all axial deformation effects.)

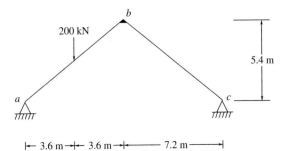

|← 3.6 m →|← 3.6 m →|←——— 7.2 m ———→|

10.22   For the beam shown:

(a) Compute the reactions and draw the final moment diagram for all members.

(b) Assume that support B settles 0.24 in. down and support A rotates 0.005 rad clockwise. With no loads acting on the structure and only the prescribed support movements, compute the reactions and draw the final moment diagram for all members *(E* = 29,000 ksi , *I* = 400 in⁴).

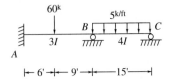

|← 6' →|← 9' →|←——15'——→|

10.23   Analyze and draw the final moment diagram for the structure shown. Neglect axial deformations.

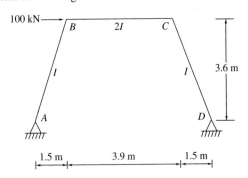

| 1.5 m | 3.9 m | 1.5 m |

10.24   Draw the final moment diagram for the structure shown. Neglect axial deformations.

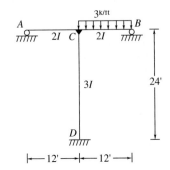

|←— 12' —→|←— 12' —→|

10.25   Draw the final moment diagram for the structure shown. Neglect axial deformations (*E* constant).

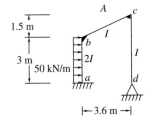

|←3.6 m→|

10.26   The crossed simply supported beams of moment of inertial *I* and 2*I* rest, one upon the other, at their centers in the unloaded state. Determine the deflection of the center of the beams when the load *P* (in the negative *z* direction) is applied (*E* constant).

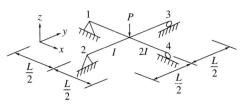

10.27 Draw the final moment diagram for the structure shown. Note hinges at *A* and *D*. Neglect axial deformations (*E* constant).

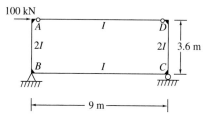

10.28 Show that the two structures are equivalent. Analyze the structure to the right. Neglect axial deformations.

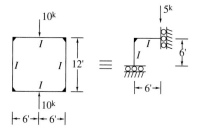

10.29 The structure shown has been analyzed by a "junior engineer" who has given you the final moment diagram indicated. Without doing the problem, check as to whether or not his solution is correct. Clearly indicate the computations you make in the check.

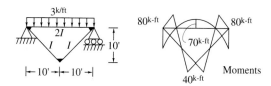

10.30 Analyze and draw the final moment diagram for the structure shown. Use the method of consistent deformations. Neglect axial deformations.

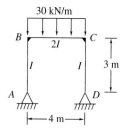

COMPUTER PROBLEMS

For the problems that follow, obtain the graphical results requested using one of the following methods:

(a) Create a spreadsheet and use the results to obtain the graphs.

(b) Write a small computer program that obtains the graphical results requested and create a file that can be plotted to obtain the graphs.

(c) Use available software programs that will do the analysis requested. Run the program as many times as required to obtain the results requested and create a file that can be plotted to obtain the graphs.

C10.1   In the truss loaded as shown, find the final force in member *a–d*. Create a computer graphical representation of the variation of the force in member *a–d* as a function of its area for β varying from 1 to 20 (*E* constant).

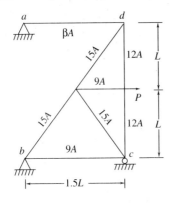

C10.2   Obtain a plot of the moment at joint 2 versus β for values of $h/b = 0.5, 1.0, 1.5, 2$, and $1 \leq \beta \leq 5$. Neglect axial deformation (*E* constant).

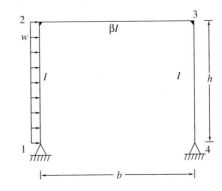

C10.3   Repeat Problem C10.2 but plot the value of the horizontal displacement at 2 versus β.

C10.4   Repeat Problem C10.2 but with the horizontal loading *w* replaced with a concentrated load *P* acting to the right at joint 2.

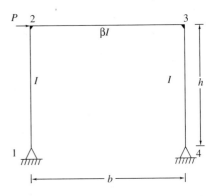

C10.5 Repeat Problem C10.3 but with the horizontal loading w replaced with a concentrated load P acting to the right at joint 2.

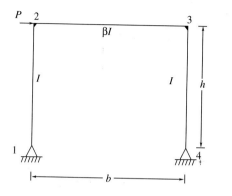

C10.6 Obtain a plot of the variation of the moment at B versus β for values of $a/L = 0.7, 1.0, 1.2, 1.5$, and $1 \le \beta \le 3$ (*E* constant).

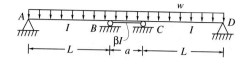

11.1   Slope-Deflection Method

The analysis of structures in the preceding chapters has been based on the equations of statics and an evaluation of the compatible deformation behavior of statically determinate structures. The approach has been to seek force distributions in the structure which satisfied equilibrium and, after the use of a suitable mathematical model of material deformation behavior, also deformed the structure in a compatible manner. These methods are known as *force methods of analysis*. They establish the equations of compatibility for indeterminate structures; which, upon solution, yield values of redundant forces or moments at specific points of release of constraints on the continuity of deformation.

The slope-deflection method is based on a technique in which the displacements and rotations of a structure are obtained from a set of equations of equilibrium at the joints of the structure. These equations, upon solution, yield the displacements and rotations of the joints of the structure. Calculation of forces and moments in the members of the structure takes place as a second step in the analysis process. This method is referred to as a *displacement method of analysis*.

When doing hand computations, the slope-deflection method is the best method for analysis of structures that are highly indeterminate and have only a few possible joint displacements and rotations. The method is easily adapted to computer methods of structural analysis. Nearly every computer program for frame and beam analysis is based on this approach, so that an understanding of the method is important to the structural engineer faced with the need to evaluate the output of a computer-generated analysis of a structure. The method is based on the relation between the transverse displacements and rotations of the ends of a member and the corresponding shear forces and end moments. A general derivation of the method is followed by its application to the analysis of beams and frames.

11.2   Slope-Deflection Equations

Member *A–B* shown in Fig. 11.1 is a typical member that is connected to joints *A* and *B* of some frame structure. The member may have some transverse loading, $q(x)$, acting on it. Due to the actions of loads acting on the structure and $q(x)$ on *A–B*, joints *A* and *B* undergo the vertical displacements and counterclockwise rotations shown in Fig. 11.2. For *A–B*, the member end displacements and rotations and the $q(x)$ loading together give rise to the member end moments and end shears shown in Fig. 11.1.

Because *A–B* represents any member in a frame, it is also subjected to axial displacements and forces. When considering bending behavior in previous chapters, it has been assumed that there is no interaction between axial forces and bending. This assumption is valid for sufficiently small axial forces, but cannot be used in general. As long as the magnitude of the axial force is less than EI/L^2 for a member that has unlimited elastic response, the error of neglecting interaction effects between axial force and bending is less than about 6%. For the treatment herein, the effect of axial forces on the transverse bending and deformation behavior of the member is neglected. In addition, the assumption is made that displacements and rotations are small, so that the geometry of the deformed structure is essentially the same as for the undeformed structure, an assumption that has been used consistently throughout previous chapters.

Since the presence of axial force is neglected, the equilibrium equation for axial forces is no longer available. Member A–B in Fig. 11.1 must be in equilibrium, but now the number of equilibrium equations is reduced to two. For the two equilibrium equations to be satisfied, only two of the four end moments and forces M_{AB}, M_{BA}, V_{AB}, and V_{BA} can be independent. Using the two equilibrium equations, V_{AB} and V_{BA} can be computed in terms of M_{AB}, M_{BA}, and $q(x)$ by summing moments about each end of the member.

The concept of treating member A–B as a simply supported beam for equilibrium considerations was developed in Section 5.5 and illustrated in Fig. 5.5. From a computational or conceptual standpoint, this is equivalent to making the reactions at the ends of the beam correspond to the shear forces V_{AB} and V_{BA} on the end of the member. This concept is developed in Fig. 11.3, in which an equivalent simply supported beam at the right of the figure is subjected to the same actions as member A–B of the structure. The computation of the relations between all moments, curvatures, and end displacements and rotations for member A–B can be done as though it were a simply supported beam.

Slope-Deflection Equations for End Moments

In what follows, two relations between the end moments M_{AB} and M_{BA} and the end rotations (slopes) and end displacements (deflections) are derived. The actions of member A–B are defined with respect to a local member coordinate system. As shown in Figs. 11.1 to 11.3, the x axis of this system is the member centroidal axis with the origin of the system at the left (A) end of the member and the y axis positive upward. The right-hand-rule sign convention is followed, so that end moments and end rotations are all positive when counterclockwise. Member displacements and any applied loading $q(x)$ are positive upward or in the positive y-coordinate direction. The usual assumptions of linear elastic material, small displacements, and rotations of simple beam theory are used.

The slope-deflection equations are derived by finding the relations between the member end rotations and displacements caused by internal

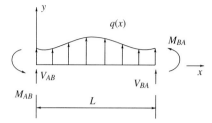

Figure 11.1 End moments and end forces acting on a single member in the positive sense.

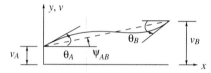

Figure 11.2 Positive end displacements and rotations.

437

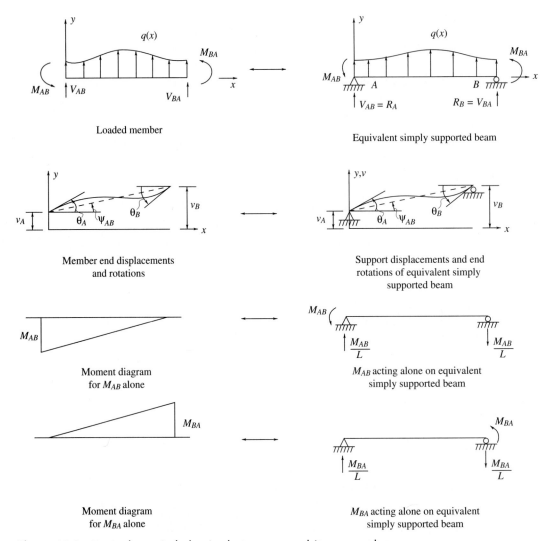

Figure 11.3 Equivalence in behavior between an arbitrary member A–B and a simple supported beam.

member deformations. These deformations are the curvatures due to M_{AB}, M_{BA}, and the applied loading $q(x)$, which are obtained by dividing the moment diagrams for each of these actions by EI and summing as shown in Fig. 11.3. The curvature diagrams constituting the deformation system for the member are shown in Fig. 11.4.

The virtual force system shown in Fig. 11.5 is applied to the equivalent simply supported beam of Fig. 11.3. When this force system is subjected to the end displacements and rotations in Fig. 11.2, it produces external work due to the product of δM_1 with θ_A and the products of v_A and v_B with the $\delta M_1/L$ reactions. With δM_1 acting, the deformation system consisting of the displacements and rotations at A and B in Fig. 11.2 and the curvatures in Fig. 11.4 is applied and the complementary virtual work expression, Eq. (7.14), yields

$$W_e = W_i$$

$$\delta M_1\,\theta_A + \frac{\delta M_1}{L}\,v_A - \frac{\delta M_1}{L}\,v_B$$

$$= \int_0^L \left[\frac{-\delta M_1}{L}(L-x)\right]\left[\frac{-M_{AB}}{EI}\frac{L-x}{L} + \frac{M_{BA}}{EI}\frac{x}{L} + \frac{M(x)}{EI}\right] dx$$

$$= \delta M_1 \frac{M_{AB}}{EI}\frac{L}{3} - \delta M_1 \frac{M_{BA}}{EI}\frac{L}{6} - \int_0^L \frac{\delta M_1}{L}(L-x)\frac{M(x)}{EI}\,dx \qquad (11.1a)$$

The two integrals associated with the curvatures due to M_{AB} and M_{BA} can be evaluated by direct integration or by use of the expressions found in position 1–1 and position 2–1, respectively, of the integration chart, Table 7.1. For the M_{AB} term, the product of the moment diagram for δM_1 of Fig. 11.5 and the curvature diagram of Fig. 11.4 is located at position 1–1 in Table 7.1, and is positive since the two diagrams are both negative. For the M_{BA} term, the product of the moment diagram for δM_1 of Fig. 11.5 and the curvature diagram of Fig. 11.4 is located at position 2–1 in Table 7.1, and the work is negative since the M_{BA} diagram is positive and the δM_1 diagram negative. Simplifying and noting that the virtual force system moment, δM_1, disappears yields

$$\theta_A - \frac{v_B - v_A}{L} = \frac{M_{AB}L}{3EI} - \frac{M_{BA}L}{6EI} - M_{00} \qquad (11.1b)$$

where

$$M_{00} = \int_0^L \frac{L-x}{L}\frac{M(x)}{EI}\,dx$$

The integral defined in M_{00} can be evaluated when the moment diagram and hence curvature diagram $M(x)/EI$ for the loading $q(x)$ on the simply supported beam is known. Note that $M(x)/EI$ is assumed to cause compression on top of the beam.

The virtual force system δM_2 shown in Fig. 11.6 is used to relate the end rotation θ_B and the end displacements v_A and v_B in Fig. 11.2 with the member deformations or curvatures. With this force system acting during application

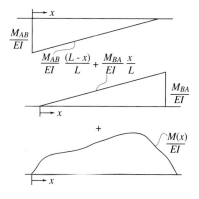

$$\frac{M_{AB}}{EI}\frac{(L-x)}{L} + \frac{M_{BA}}{EI}\frac{x}{L}$$

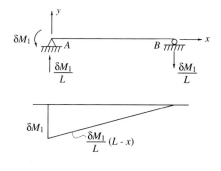

Deformation system

Figure 11.4 Curvature diagrams for member A–B of Fig. 11.3.

Force system

Figure 11.5 Virtual force system on the equivalent simple beam for calculation of the rotation at A.

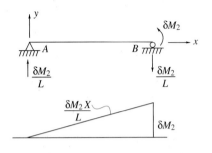

Force system

Figure 11.6 Virtual force system on equivalent simple beam for calculation of the rotation at B.

of the deformation system of Fig. 11.4, the complementary virtual work expression, Eq. (7.14), yields the result

$$\frac{\delta M_2}{L} v_A - \frac{\delta M_2}{L} v_B + \delta M_2 \theta_B$$

$$= \int_0^L \left(\delta M_2 \frac{x}{L}\right)\left[-\frac{M_{AB}}{EI}\frac{L-x}{L} + \frac{M_{BA}}{EI}\frac{x}{L} + \frac{M(x)}{EI}\right] dx$$

$$= -(\delta M_2)\frac{M_{AB}}{EI}\frac{L}{6} + \delta M_2 \frac{M_{BA}}{EI}\frac{L}{3} + \int_0^L \frac{\delta M_2}{L}x\frac{M(x)}{EI} dx \qquad (11.2a)$$

Again the two integrals associated with the curvatures due to M_{AB} and M_{BA} have been evaluated using the integration chart of Table 7.1. For the M_{AB} term, the product of the moment diagram for δM_2 of Fig. 11.6 and the curvature diagram of Fig. 11.4 is located at position 2–1 in Table 7.1, and the work is negative since the M_{AB} diagram is negative and the δM_2 diagram is positive. For the M_{BA} term, the product of the moment diagram for δM_2 of Fig. 11.6 and the curvature diagram of Fig. 11.4 is located at position 1–1 in Table 7.1, and the work is positive since both the M_{BA} diagram and the δM_2 diagram are positive. Simplifying yields

$$\theta_B - \frac{v_B - v_A}{L} = \frac{-M_{AB}L}{6EI} + \frac{M_{BA}L}{3EI} + M_0 \qquad (11.2b)$$

where

$$M_0 = \int_0^L \frac{x}{L}\frac{M(x)}{EI} dx$$

The integral defined in M_0 can be evaluated when the moment diagram and hence curvature diagram $M(x)EI$ for the loading $q(x)$ on the simply supported beam in known.

Equations (11.1) and (11.2) relate the member end displacements and rotations with the member end moments M_{AB} and M_{BA}. Equations (11.1b) and (11.2b) can be solved simultaneously for the end moments, giving the result

$$M_{AB} = \frac{2EI}{L}\left(2\theta_A + \theta_B - 3\frac{v_B - v_A}{L}\right) + FEM_{AB}$$

$$(11.3)$$

$$M_{BA} = \frac{2EI}{L}\left(\theta_A + 2\theta_B - 3\frac{v_B - v_A}{L}\right) + FEM_{BA}$$

where

$$FEM_{AB} = \frac{2EI}{L}(2M_{00} - M_0)$$

$$\tag{11.4}$$

$$FEM_{3A} = \frac{2EI}{L}(M_{00} - 2M_0)$$

Equations (11.3) are known as the slope-deflection equations for end moments. They give directly the moments on the ends of a member in terms of the slopes (rotations) and deflections of the ends of the member. The FEM_{AB} and FEM_{BA} are referred to as the fixed end moments for the member. They represent the moments that develop on the ends of a member due to the action of a transverse loading $q(x)$ if the ends of the member are fully fixed against both displacement and rotation, as shown in Fig. 11.7. If there is no intermediate loading $q(x)$, the fixed end moments are zero. They can be computed for any loading $q(x)$ from Eqs. (11.4) with the definitions of M_{00} and M_0 in Eqs. (11.1) and (11.2), respectively. The curvature function (diagram) $M(x)/EI$ appearing in these equations in obtained from the loading, $q(x)$, acting on an equivalent simply supported beam as shown in Fig. 11.3 where, as shown in Fig. 11.4, the curvature $M(x)/EI$ causes compression on the top of the member. The evaluation of the fixed end moments is illustrated a little later in Example 11.1.

In some presentations, the slope-deflection equations are expressed in terms of the chord rotation ψ_{AB}, which is shown and defined in Figs. 11.2 and 11.3. As can be seen in the figure, for small displacements and rotations the term $(v_B - v_A)/L$ in Eq. (11.3) can be replaced with ψ_{AB}, where ψ_{AB} is positive when counterclockwise. This form of the slope-deflection equations is preferred by some structural engineers, but it is not the form used in computer applications of these equations and is not the form used in this book.

Slope-Deflection Equations for End Shears

The end shears V_{AB} and V_{BA} are obtained from equilibrium. As shown in Figs. 5.5, 5.6, and 11.3, the shear, V_{AB} for example, can be obtained by direct

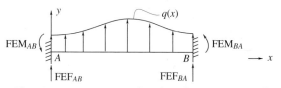

Figure 11.7 Positive fixed end moments and forces due to $q(x)$.

superposition of the reactions of the equivalent simply supported beam under the action of the three loading systems M_{AB}, M_{BA}, and $q(x)$. Note that R_{Aq} is defined as the reaction at A in the equivalent simply supported beam due to $q(x)$ acting by itself. The summation process gives for the shears

$$V_{AB} = \frac{M_{AB}}{L} + \frac{M_{BA}}{L} + R_{Aq}$$

$$V_{BA} = \frac{-M_{AB}}{L} - \frac{M_{BA}}{L} + R_{Bq}$$

(11.5)

Substituting for M_{AB} and M_{BA} from Eqs. (11.3) and (11.4) and simplifying yields the final expressions

$$V_{AB} = \frac{6EI}{L^2}\left(\theta_A + \theta_B - 2\frac{v_B - v_A}{L}\right) + \text{FEF}_{AB}$$

$$V_{BA} = -\frac{6EI}{L^2}\left(\theta_A + \theta_B - 2\frac{v_B - v_A}{L}\right) + \text{FEF}_{BA}$$

(11.6)

where

$$\text{FEF}_{AB} = \frac{6EI}{L^2}(M_{00} - M_0) + R_{Aq}$$

$$\text{FEF}_{BA} = -\frac{6EI}{L^2}(M_{00} - M_0) + R_{Bq}$$

(11.7)

These equations are defined as the slope-deflection equations for end forces. Just as in Eqs. (11.3), the end forces or end shear forces are completely defined by the rotations and displacements of the ends of the member and the effects of the transverse loading, $q(x)$. Equations (11.7) are used in problems where joint displacements occur in structures and also are part of the formulation used in computer programs that do frame analysis.

Fixed End Moments and Fixed End Forces

The FEM_{AB} and FEM_{BA} are the fixed end moments that arise from the presence of the transverse intermediate loading $q(x)$ as shown in Fig. 11.7. The integrals M_0 and M_{00} can be evaluated for any particular $q(x)$. Once known, the FEM_{AB} and FEM_{BA} moments are calculated from their definition, Eq. (11.4).

The FEF_{AB} and FEF_{BA} are the fixed end forces that arise from the presence of the transverse intermediate loading $q(x)$ as shown in Fig. 11.7. Computations of the fixed end forces follow directly from Eq. (11.7), where, as indicated earlier, the R_{Aq} and R_{Bq} are the equivalent simple beam reactions shown in Fig. 11.3 for the loading $q(x)$. Because the end forces or shears are always obtained from the two equations of equilibrium for the member A–B, FEF_{AB} and FEF_{BA} can be obtained directly in terms of the fixed end moments and transverse loading acting on member A–B in Fig. 11.7. It is left as an exercise for the reader to show that summing moments about end B in Fig. 11.7 gives

$$FEF_{AB} = \frac{FEM_{AB} + FEM_{BA}}{L} + R_{Aq}$$

and summing moments about end A gives

$$FEF_{BA} = -\frac{FEM_{AB} + FEM_{BA}}{L} + R_{Bq}$$

which are identical to the expressions in Eq. (11.7).

The calculation of fixed end forces and fixed end moments is best understood with a specific application.

Example 11.1 The fixed end forces and moments are computed for the case of a concentrated load acting a distance, a, from the left end of the member. In step 1 the load, P, is assumed to act on an equivalent simply supported beam which by equilibrium of moments gives the reactions $R_{Aq} = P(b/L)$ and $R_{Bq} = P(a/L)$. The moment and curvature diagrams, $M(x)/EI$, are easily obtained for the simply supported beam. In evaluating the integral for M_{00} in step 2, it is convenient to use Table 7.1. The form of the function $(L - x)/L$ required to use the table is shown as the triangular diagram with maximum ordinate 1 at the left end. The integral for M_{00} can be evaluated with the expression in position 1–4 of Table 7.1, where the parameter, t, is simply b/L. A similar procedure is followed in step 3 to evaluate the integral for M_0 using the diagram shown for the function x/L and the expression in position 2–4 where again, $t = b/L$. With these two integrals evaluated, the fixed end moments and fixed end forces are obtained directly from Eqs. (11.4) and (11.7) in steps 4 and 5. Since the end shear forces are dependent on the end moments and applied loading as discussed earlier, the fixed shear end forces can also be obtained from equilibrium of the original fixed-fixed beam once the fixed end moments have been computed in step 4. For example, summing moments about B and solving for FEF_{AB} gives the result

$$FEF_{AB} = \frac{FEM_{AB} + FEM_{BA} + Pb}{L} = \frac{(ab^2 - a^2b + bL^2)P}{L^3}$$

$$= \frac{(ab^2 - a^2b + ba^2 + 2b^2a + b^3)P}{L^3}$$

$$= \frac{Pb^2(L + 2a)}{L^3}$$

443

Example 11.1

Compute the fixed end moments and the fixed end forces for a concentrated load on the beam A–B.

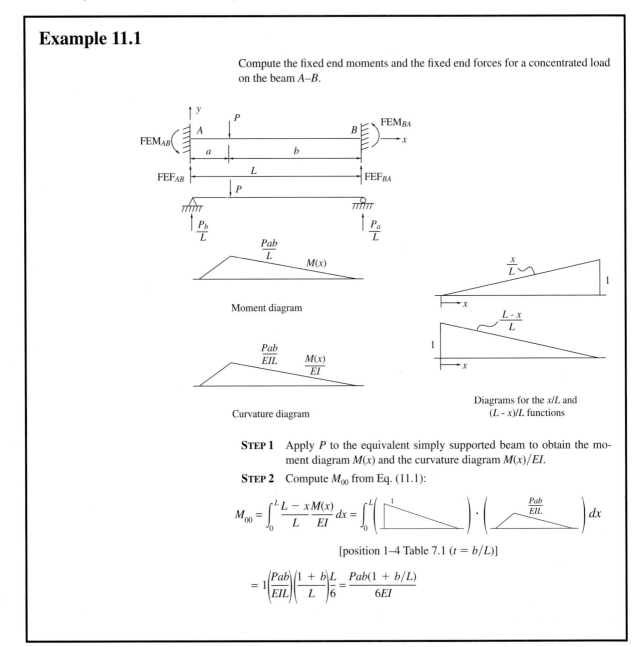

STEP 1 Apply P to the equivalent simply supported beam to obtain the moment diagram $M(x)$ and the curvature diagram $M(x)/EI$.

STEP 2 Compute M_{00} from Eq. (11.1):

$$M_{00} = \int_0^L \frac{L-x}{L} \frac{M(x)}{EI}\, dx = \int_0^L \left(\vphantom{\frac{Pab}{EIL}} \right) \cdot \left(\vphantom{\frac{Pab}{EIL}} \right) dx$$

[position 1–4 Table 7.1 $(t = b/L)$]

$$= 1\left(\frac{Pab}{EIL}\right)\left(\frac{1+b}{L}\right)\frac{L}{6} = \frac{Pab(1+b/L)}{6EI}$$

Example 11.1 (continued)

STEP 3 Compute M_0 from Eq. (11.2):

$$M_0 = \int_0^L \frac{x}{L} \frac{M(x)}{EI}\,dx = \int_0^L \left(\underset{\text{[diagram]}}{} {}^1 \right) \cdot \left(\underset{\text{[diagram]}}{\frac{Pab}{EIL}} \right) dx$$

[position 2–4 Table 7.1 ($t = b/L$)]

$$= 1\left(\frac{Pab}{EIL}\right)\left(2 - \frac{b}{L}\right)\frac{L}{6} = \frac{Pab(2 - b/L)}{6EI}$$

STEP 4 Compute the fixed end moments from Eq. (11.4):

$$\text{FEM}_{AB} = \frac{2EI(2M_{00} - M_0)}{L}$$

$$= 2\frac{EI}{L}\frac{[2(1 + b/L) - (2 - b/L)]Pab}{6EI} = \frac{Pab^2}{L^2}$$

$$\text{FEM}_{BA} = \frac{2EI(M_{00} - 2M_0)}{L}$$

$$= 2\frac{EI}{L}\frac{[(1 + b/L) - 2(2 - b/L)]Pab}{6EI} = -\frac{Pa^2b}{L^2}$$

STEP 5 Compute the fixed end forces from Eq. (11.7):

$$\text{FEF}_{AB} = \frac{6EI(M_{00} - M_0)}{L^2} + R_{Aq}$$

$$= \frac{6EI}{L^2}\frac{[(1 + b/L) - (2 - b/L)]Pab}{6EI} + \frac{Pb}{L}$$

$$= \frac{Pab}{L^2}\left(\frac{2b}{L} - 1\right) + \frac{Pb}{L} = \frac{Pb^2(L + 2a)}{L^3}$$

$$\text{FEF}_{BA} = -\frac{6EI(M_{00} - M_0)}{L^2} + R_{Bq} = -\frac{Pab}{L^2}\left(\frac{2b}{L} - 1\right) + \frac{Pa}{L}$$

$$= \frac{Pa^2(L + 2b/L)}{L^3}$$

445

upon noting that $L = a + b$. This agrees with the result in step 5. Summing moments about A also gives the same value for FEF_{BA} that is obtained in step 5.

The procedure followed in Example 11.1 can be used to find the fixed end forces and moments for any type of loading on a member. In Fig. 11.8 values are shown for several common loadings.

11.3   Application of the Slope-Deflection Equations to the Solution of Structures with No Joint Translation _____

With no joint translations taking place, the v_A and v_B terms in the slope-deflection equations are zero and rotations at points of support of beams with transverse loading are the only unknowns. All frame structures undergo some joint translation, so that a general application of the slope-deflection equations must include those effects. However, there are structures in which joint translations or displacements are due solely to axial deformations of members. For many of these structures the influence on the internal distribution of moments and forces of axial deformations is so small that the axial force deformations can be neglected (or taken to be zero). As seen in Example 10.5, the effects of axial deformations on internal moments and were about 0.2%. If the assumption is made that for the purpose of a slope-deflection analysis, the members do not deform axially, there are a number of structures in addition to beams in which joint transition can be considered to be zero. It is these structures that are treated in this section.

The first application of the slope-deflection method of solution is presented for a beam structure.

Example 11.2 The two-span beam with a uniform load has the two extreme ends fully fixed. The first step in the analysis is to establish for each member of the structure the end moments in terms of the member end displacements and rotations using Eqs. (11.3). The sense of the coordinate system of each member is shown on the diagram. The fixed end moments for the loaded span A–B are obtained from Fig. 11.8. The next step is to delete the rotations and displacements known to be zero and at the same time, for convenience, replace I/L with K in these expressions. This leaves only θ_B as unknown; that is, if the value of θ_B were known, all member end moments could be calculated from the expressions presented in steps 1 and 2.

The third step is to impose equilibrium and compatibility conditions at the joints of the structure. The deformations of the structure are all compatible, as the rotations at A and C are zero, the displacements at A, B, and C are zero, and the rotation at B is common to both members A–B and B–C. Equilibrium is satisfied at joints A and C because they are reactions. Equilibrium at joint B must also be satisfied, so it is isolated in a free-body diagram with all member end forces and moments acting in their positive sense. To aid in proper drawing of the direction of the moments M_{BA} and M_{BC} and the shear and axial forces that act on the joint, short pieces of members A–B and B–C are shown with the end moments and forces acting in the positive sense, which is tension for axial forces, vertically upward for shears, and counterclockwise for the end moments.

Force equilibrium is satisfied at the joint no matter what the magnitude of the member axial forces and shears. Axial (horizontal) force equilibrium is satisfied because the two members are attached to the reactions at A and C and can resist any axial force that may develop in the members. Similarly, equilibrium of vertical forces from member shears is satisfied by the vertical reactive force at B. However, moment equilibrium must

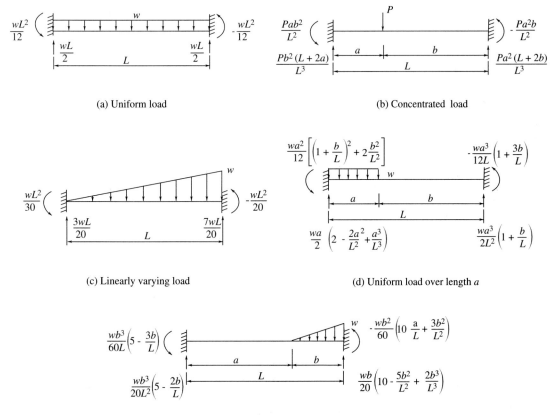

(a) Uniform load

(b) Concentrated load

(c) Linearly varying load

(d) Uniform load over length a

(e) Linearly varying load over length b

Figure 11.8a–e Fixed end forces and moments for common loadings.

Example 11.2

Use the slope-deflection method to obtain the final moment diagram. Take E constant.

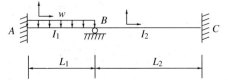

STEP 1 Write down the member end moments from Eq. (11.3). The fixed end moments can be obtained from Fig. 11.8.

Step 1	Step 2

$$M_{AB} = \frac{2EI_1}{L_1}\left[2\theta_A + \theta_B - \frac{3}{L_1}(v_B - v_A)\right] + \frac{wL_1^2}{12} \qquad = 2EK_1\theta_B + \frac{wL_1^2}{12}$$

$$M_{BA} = \frac{2EI_1}{L_1}\left[2\theta_B + \theta_A - \frac{3}{L_1}(v_B - v_A)\right] - \frac{wL_1^2}{12} \qquad = 4EK_1\theta_B - \frac{wL_1^2}{12}$$

$$M_{BC} = \frac{2EI_2}{L_2}\left[2\theta_B + \theta_C - \frac{3}{L_2}(v_C - v_B)\right] + 0 \qquad = 4EK_2\theta_B$$

$$M_{CB} = \frac{2EI_2}{L_2}\left[2\theta_C + \theta_B - \frac{3}{L_2}(v_C - v_B)\right] + 0 \qquad = 2EK_2\theta_B$$

STEP 2 The rotations θ_A and θ_C are zero due to the support conditions. Also, $v_A = v_B = v_C = 0$ from support conditions. Reduce equations and use $K_1 = I_1/L_1$ and $K_2 = I_2/L_2$.

STEP 3 Isolate joint B and write moment equilibrium. Note that force equilibrium will always be satisfied since no displacements can occur at B.

$$\overset{\curvearrowright}{\Sigma M_B}: M_{BA} + M_{BC} = 0$$

Substitute from expressions in steps 1 and 2:

$$4EK\theta_B - \frac{wL_1^2}{12} + 4EK_2\theta_B = 0$$

Example 11.2 (continued)

Solving for θ_B gives

$$\theta_B = \frac{wL_1^2/12}{4E(K_1 + K_2)}$$

STEP 4 Use the solution for θ_B and backsubstitute into moment expressions in steps 1 and 2.

$$M_{AB} = 2EK_1\theta_B + \frac{wL_1^2}{12} = \frac{2EK_1}{4E(K_1 + K_2)}\frac{wL_1^2}{12} + \frac{wL_1^2}{12} = \frac{3K_1 + 2K_2}{2(K_1 + K_2)}\frac{wL_1^2}{12}$$

$$M_{BA} = 4EK_1\theta_B - \frac{wL_1^2}{12} = \frac{4EK_1}{4E(K_1 + K_2)}\frac{wL_1^2}{12} - \frac{wL_1^2}{12} = \frac{-K_2}{K_1 + K_2}\frac{wL_1^2}{12}$$

$$M_{BC} = 4EK_2\theta_B = \frac{4EK_2}{4E(K_1 + K_2)}\frac{wL_1^2}{12} = \frac{K_2}{K_1 + K_2}\frac{wL_1^2}{12}$$

$$M_{CB} = 2EK_2\theta_B = \frac{2EK_2}{4E(K_1 + K_2)}\frac{wL_1^2}{12} = \frac{K_2}{2(K_1 + K_2)}\frac{wL_1^2}{12}$$

STEP 5 Plot the final moment diagram. For the moment diagram for A–B, use superposition of moment diagrams due to w and the two end moments acting on an equivalent simply supported beam.

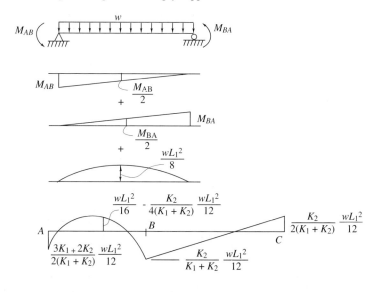

449

be enforced because the end moments M_{BA} and M_{BC} were established in step 1 independently for members $A–B$ and $B–C$. Summing moments and substituting the values of M_{BA} and M_{BC} from step 2 gives an equation involving the unknown rotation θ_B.

The fourth step is to solve for θ_B and backsubstitute into the equations of step 2 to obtain the end moments on the two members. The final step is to draw the moment diagram for each member in the structure. This process is very easy if two important ideas are kept in mind. First, the moments all act counterclockwise on the end of a member when they are positive. Second, once the moments on the ends of the member are known, the member can be treated as an equivalent simply supported beam subjected to the separate action of each end moment and, if present, an applied loading $q(x)$. The moments from all of these actions can then be summed to obtain the final moment diagram as shown in step 5. For numerical calculations with U.S. units, take $L_1 = 15$ ft, $L_2 = 20$ ft, $w = 3$ kips/ft, $I_1 = I$, and $I_2 = 1.5I$; for SI units, take $L_1 = 5$ m, $L_2 = 7$ m, $w = 50$ kN/m, $I_1 = I$, and $I_2 = 1.5I$ (constant E).

The drawing of the final moment diagram in the fifth step of Example 11.2 follows the concepts presented in Figs. 5.5 and 11.3. The moment diagrams due to M_{AB}, M_{BA}, and the uniform load acting individually on an equivalent simply supported beam have very simple forms that are easily constructed. By adding the coordinates of the ends and middle of the simple diagrams the more complicated diagram of the combined moment variation is obtained, as illustrated.

Example 11.3 treats a more complicated frame structure. Note that this structure is indeterminate to the fifth degree but an analysis by the slope-deflection method requires only the solution of two simultaneous equations. There are no joint translations in this structure since it is assumed that there are no axial deformations.

Example 11.3 In this example the same five steps are followed as in Example 11.2. In writing the slope-deflection equations, it is important to point out that the equations for the end moments are all written in the local member coordinate system. The coordinate system selected for each member is shown as the unlabeled coordinate axes. Members $A–B$ and $D–C$ have positive axes to the right and upward, while $B–C$ and $C–E$ have positive axes down and to the right. It also is important to recognize that all of these member coordinate systems have, by the right-hand rule, moments and rotations that are positive in the counterclockwise direction.

Member $B–F$ is not included in the tabulation of member end moments using the slope-deflection equations in steps 1 and 2 because the member is a cantilever beam anchored at B. As such, it is statically de-

Example 11.3

Solve the structure shown by the slope-deflection method.

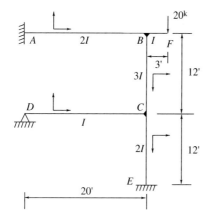

STEP 1   Write down the member end moments from Eq. (11.3). Use the conditions that the rotations θ_A and θ_E and all joint displacements are zero.

STEP 2   For member D–C immediately use the zero moment condition at D to eliminate θ_D. Also, ignore member B–F since it is statically determinate.

STEP 3   Isolate joints B and C and write the moment equilibrium equations.

STEP 4   Solve for θ_B and θ_C, and backsubstitute for member end moments.

STEP 5   Draw the final moment diagram for the structure.

STEP 1

$$M_{AB} = \frac{4EI}{20}\,\theta_B \qquad\qquad M_{BA} = \frac{8EI}{20}\,\theta_B$$

$$M_{BC} = \frac{6EI}{12}\,(2\theta_B + \theta_C) \qquad M_{CB} = \frac{6EI}{12}\,(2\theta_C + \theta_B)$$

$$M_{CE} = \frac{4EI}{12}\,(2\theta_C) \qquad\qquad M_{EC} = \frac{4EI}{12}\,\theta_C$$

Example 11.3 (continued)

STEP 2

$$M_{DC} = \frac{2EI}{20}(2\theta_D + \theta_C) = 0 \rightarrow \theta_D = -\frac{\theta_C}{2}$$

$$\therefore M_{CD} = \frac{2EI}{20}(2\theta_C + \theta_D) = \frac{3EI}{20}\theta_C$$

STEP 3, JOINT B

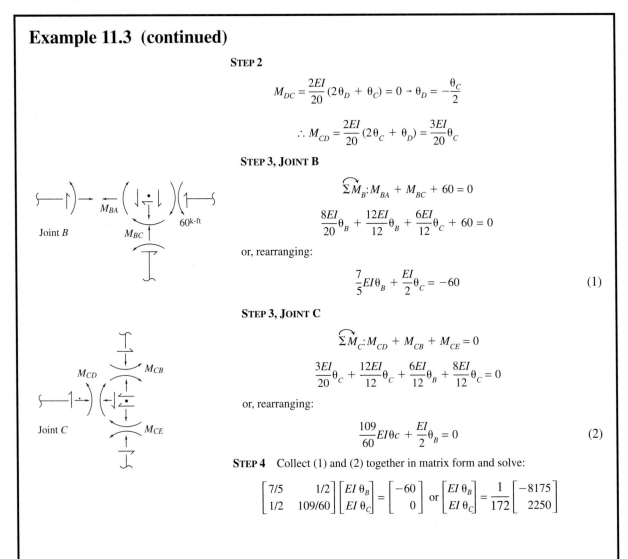

Joint B

$60^{\text{k-ft}}$

$$\overset{\curvearrowright}{\Sigma M}_B : M_{BA} + M_{BC} + 60 = 0$$

$$\frac{8EI}{20}\theta_B + \frac{12EI}{12}\theta_B + \frac{6EI}{12}\theta_C + 60 = 0$$

or, rearranging:

$$\frac{7}{5}EI\theta_B + \frac{EI}{2}\theta_C = -60 \tag{1}$$

STEP 3, JOINT C

Joint C

$$\overset{\curvearrowright}{\Sigma M}_C : M_{CD} + M_{CB} + M_{CE} = 0$$

$$\frac{3EI}{20}\theta_C + \frac{12EI}{12}\theta_C + \frac{6EI}{12}\theta_B + \frac{8EI}{12}\theta_C = 0$$

or, rearranging:

$$\frac{109}{60}EI\theta c + \frac{EI}{2}\theta_B = 0 \tag{2}$$

STEP 4   Collect (1) and (2) together in matrix form and solve:

$$\begin{bmatrix} 7/5 & 1/2 \\ 1/2 & 109/60 \end{bmatrix} \begin{bmatrix} EI\,\theta_B \\ EI\,\theta_C \end{bmatrix} = \begin{bmatrix} -60 \\ 0 \end{bmatrix} \text{ or } \begin{bmatrix} EI\,\theta_B \\ EI\,\theta_C \end{bmatrix} = \frac{1}{172}\begin{bmatrix} -8175 \\ 2250 \end{bmatrix}$$

Example 11.3 (continued)

The final member end moments become:

$$M_{AB} = \frac{4EI}{20}\theta_B \qquad = \frac{4}{20}\cdot\frac{-8175}{172} \qquad = -9.51 \text{ kip-ft}$$

$$M_{BA} = \frac{8EI}{20}\theta_B \qquad = \frac{8}{20}\cdot\frac{-8175}{172} \qquad = -19.01 \text{ kip-ft}$$

$$M_{BC} = \frac{6EI}{12}(2\theta_B + \theta_C) = \frac{6}{12}\cdot(2\cdot-8175+2250)\frac{1}{172} = -40.99 \text{ kip-ft}$$

$$M_{CB} = \frac{6EI}{12}(2\theta_C + \theta_B) = \frac{6}{12}\cdot(2\cdot2250-8175)\frac{1}{172} = -10.68 \text{ kip-ft}$$

$$M_{CE} = \frac{4EI}{12}(2\theta_C) \qquad = \frac{8}{12}\cdot\frac{2250}{172} \qquad = 8.72 \text{ kip-ft}$$

$$M_{EC} = \frac{4EI}{12}\theta_C \qquad = \frac{4}{12}\cdot\frac{2250}{172} \qquad = 4.36 \text{ kip-ft}$$

$$M_{DC} = 0 \qquad\qquad\qquad = 0$$

$$M_{CD} = \frac{2EI}{20}(2\theta_C + \theta_D) = \frac{3EI}{20}\theta_C = \frac{3}{20}\cdot\frac{2250}{172} \qquad = 1.96 \text{ kip-ft}$$

STEP 5   Draw the final moment diagram.

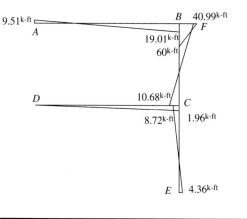

terminate and all member force actions can be obtained from equilibrium. However, the moment it brings to the joint is part of the free-body diagram for the establishment of moment equilibrium at *B*. The recognition and use of the zero moment condition at *D* in step 2 reduces the number of unknowns and speeds the computation process by reducing the number of simultaneous equilibrium equations that need to be solved.

The use of the slope-deflection method for structures with no joint displacements illustrated in Examples 11.2 and 11.3 shows the very systematic approach that makes the method applicable to computer methods of analysis. The solution procedure always follows the same sequence of steps and is not modified by the configuration of the structure or its degree of indeterminacy. First, the slope-deflection equations for end moments, Eqs. (11.3), are established for every member of the structure, taking as zero rotations at joints with fixed support restraints. Second, the moment equilibrium equations are established at every joint of the structure where rotations are unknown. Third, these equations are solved simultaneously for the joint rotations, which are then backsubstituted into the equations established in the first step to obtain the final member end moments. This was one of the very first techniques that was programmed to analyze continuous beams when digital computers were introduced in the 1950s.

11.4   Application of the Slope-Deflection Equations to the Solution of Structures with Joint Translation

The basic concepts and solution techniques in the slope-deflection method of analysis are not changed if joint displacements can occur in the structure. As before, moment equilibrium equations are required and written at joints where rotations are possible, as illustrated in previous examples. If a translation is possible at some joint, a force equilibrium equation must be written in the direction of that joint's translation. Use of the slope-deflection equations for member end forces presented in Eqs. (11.6) makes this task a simple extension of the procedure. Again the application of the procedure is best illustrated with examples.

The nonprismatic propped cantilever beam shown in Example 11.4 can also be thought of as a two-prismatic-member structure by introducing a joint at the point where the moment of inertia changes. This joint can then undergo vertical displacement as well as rotation and provides a simple structure for introduction of the slope-deflection method for structures with joint translation. Example 11.5 shows how a simple frame with joint translation is analyzed.

Example 11.4 It is important to observe that the local member coordinate systems selected for members $A–B$ and $B–C$ both have positive axes to the right and vertically upward. This ensures that the rotation and vertical displacement at B, as expressed in *each local member coordinate system*, is positive in the same direction. In step 1 of the procedure two equations for the shear forces, one on the right end of member $A–B$ and one on the left end of member $B–C$, are obtained from Eqs. (11.6) and added to the equations of the end moments since there is a free vertical joint displacement, v_B, of joint B. The fixed end forces and moments are obtained from Fig. 11.8. The displacement v_B for member $A–B$ appears as $(v_B - 0)$ in the slope-deflection equations for the end moments and shear force, V_{BA}, while it appears as $(0 - v_B)$ in the equations for end moments and shear force, V_{BC}, for member $B–C$. The reason for this is that Eqs. (11.3) and (11.6) are for a single member where the origin of the coordinate system of the member is at its left end, so v_B represents a different vertical displacement in the local coordinate system for each member. As in Example 11.3, the zero moment condition at C is used immediately to eliminate the unknown rotation at C to reduce the number of simultaneous equations.

In the third step of the procedure, when equilibrium at joint B is being established, moment and force equilibrium equations in the direction of θ_B and v_B are written. The moment and force equilibrium equations are cast in terms of the rotation θ_B and displacement v_B through substitution of the slope-deflection equations of step 1, which can then be solved for θ_B and v_B. The remaining solution operations are unchanged from previous examples. Again, as illustrated in the last step of Example 11.2, members $A–B$ and $B–C$ can be treated as simply supported beams with the known end moments acting on each end for the purpose of moment calculations and drawing of the final moment diagram. For numerical calculations with U.S. units, take $L = 15$ ft and $w = 2$ kips/ft; for SI units, take $L = 5$ m and $w = 30$ kN/m (constant E).

Example 11.5 Again the manner of selection of the local member coordinate systems shown in the sketch of the structure is to ensure the same positive sense for both members of the rotation at joint B. The horizontal displacement of joint B is taken as u_B and is a positive displacement to the left in the local member coordinate system for $A–B$. Because there is no (assumed) axial deformation of the member $B–C$, the horizontal displacement of C is also u_B. The procedure is the same as in Example 11.4, except that the axial force of member $B–C$ appears in the horizontal force equilibrium equation at joint B. Because of the displacement at C, a horizontal force equilibrium equation is written from the free body at that joint. Force equilibrium in the free body of member $B–C$ yields the axial force on each end of $B–C$ to be zero and reduces equation (2) to one

Example 11.4

Solve the structure shown by the slope-deflection method. Take E as constant.

Step 1	Step 2

$$M_{AB} = \frac{4EI}{L}\left[\theta_B - \frac{3}{L}(v_B - 0)\right] + \frac{wL^2}{12} \qquad = 4EK\theta_B - \frac{12EK}{L}v_B + \frac{wL^2}{12}$$

$$M_{BA} = \frac{4EI}{L}\left[2\theta_B - \frac{3}{L}(v_B - 0)\right] - \frac{wL^2}{12} \qquad = 8EK\theta_B - \frac{12EK}{L}v_B - \frac{wL^2}{12}$$

$$M_{BC} = \frac{2EI}{L}\left[2\theta_B + \theta_C - \frac{3}{L}(0 - v_B)\right] + \frac{wL^2}{12} \qquad = 3EK\theta_B + \frac{3EK}{L}v_B + \frac{wL^2}{8}$$

$$M_{CB} = \frac{2EI}{L^2}\left[2\theta_C + \theta_B - \frac{3}{L}(0 - v_B)\right] - \frac{wL^2}{2} \qquad = 0 \rightarrow \theta_C = \frac{\theta_B}{2} - \frac{3v_B}{2L} + \frac{wL^3}{48EI}$$

$$V_{BA} = -\frac{12EI}{L^2}\left[\theta_B - \frac{2}{L}(v_B - 0)\right] + \frac{wL^2}{2} \qquad = -12\frac{EK}{L}\theta_B + 24\frac{EK}{L^2}v_B + \frac{wL}{2}$$

$$V_{BC} = \frac{6EI}{L^2}\left[\theta_B + \theta_C - \frac{2}{L}(0 - v_B)\right] + \frac{wL}{2} \qquad = 3\cdot\frac{EK}{L}\theta_B + 3\cdot\frac{EK}{L^2}v_B + \frac{5wL}{8}$$

STEP 1   Write down the member end moments from Eq. (11.3) and the shear forces V_{BA} and V_{BC} from Eq. (11.6). Note that the rotation θ_A and the displacements v_A and v_C are zero.

STEP 2   Use the zero moment condition at C to eliminate θ_C from the equations and substitute $K = I/L$.

STEP 3   Isolate joint B and establish moment equilibrium and force equilibrium in the direction of v_B (upward).

$$\overset{\frown}{\Sigma M}_B: M_{BA} + M_{BC} = 0$$

$$8EK\theta_B - \frac{12EK}{L}v_B - \frac{wL^2}{12} + 3EK\theta_B + \frac{3EK}{L}v_B + \frac{wL^2}{8} = 0$$

Example 11.4 (continued)

or, rearranging,

$$11EK\theta_B - \frac{9EKv_B}{L} = -\frac{wL^2}{24} \tag{1}$$

$$\Sigma F\uparrow: \quad -V_{BA} - V_{BC} = 0$$

$$\frac{12EK}{L}\theta_B - \frac{24EK}{L^2}v_B - \frac{wL}{2} - \frac{3EK}{L}\theta_B - \frac{3EK}{L^2}v_B - \frac{5wL}{8} = 0$$

or, rearranging,

$$-\frac{9EK}{L}\theta_B + \frac{27EK}{L^2}v_B = -\frac{9wL}{8} \tag{2}$$

STEP 4   Collect and solve (1) and (2) for $EK\theta_B$ and EKv_B and backsubstitute values into the end moment equations formed in step 1 to obtain final values of end moments.

$$\begin{bmatrix} 11 & -9/L \\ -9/L & 27/L^2 \end{bmatrix}\begin{bmatrix} EK\theta_B \\ EKv_B \end{bmatrix} = \frac{-wL}{8}\begin{bmatrix} L/3 \\ 9 \end{bmatrix}$$

solving yields

$$\begin{bmatrix} EK\theta_B \\ EKv_B \end{bmatrix} = \begin{bmatrix} -5wL^2/96 \\ -17wL^3/288 \end{bmatrix}$$

This yields the final end moments:

$$M_{AB} = 4\left(-\frac{5wL^2}{96}\right) - \frac{12(-17wL^3/288)}{L} + \frac{wL^2}{12} = \frac{7wL^2}{12}$$

$$M_{BA} = 8\left(-\frac{5wL^2}{96}\right) - \frac{12(-17wL^3/288)}{L} - \frac{wL^2}{12} = \frac{5wL^2}{24}$$

$$M_{BC} = 3\left(-\frac{5wL^2}{96}\right) + \frac{3(-17wL^3/288)}{L} + \frac{wL^2}{8} = -\frac{5wL^2}{24}$$

$$M_{CB} = \qquad\qquad\qquad\qquad = 0$$

STEP 5   Draw the final moment diagram for the structure.

$$\frac{7wL^2}{12} \qquad\qquad \frac{5wL^2}{24}$$

Example 11.5

For the frame shown, use the slope-deflection method to obtain the final member end moments and draw the final moment diagram. Take E as constant and neglect axial deformations.

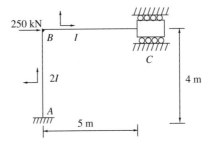

STEP 1 Write down member end moments from Eq. (11.3) and the shear force, V_{BA}, from Eq. (11.6). In writing these equations use the conditions that $\theta_A = v_A = 0$ for member A–B and that $v_B = \theta_C = v_C = 0$ for member B–C. Take the horizontal displacement of joint B to be u_B, which is also the horizontal displacement of C since there are no axial deformations.

$$M_{AB} = \frac{4EI}{4}\left[\theta_B - \frac{3}{4}(u_B - 0)\right] = EI\theta_B - 3\cdot\frac{EI}{4}u_B$$

$$M_{BA} = \frac{4EI}{4}\left[2\theta_B - \frac{3}{4}(u_B - 0)\right] = 2EI\theta_B - 3\cdot\frac{EI}{4}u_B$$

$$M_{BC} = \frac{2EI}{5}(2\theta_B) = \frac{4EI}{5}\theta_B \qquad M_{CB} = \frac{2EI}{5}\theta_B$$

$$V_{BA} = \frac{-12EI}{4^2}\left[\theta_B - \frac{2}{4}(u_B - 0)\right] = -3\cdot\frac{EI}{4}\theta_B + 3\cdot\frac{EI}{8}u_B$$

STEP 2 Isolate joint B and write the force and moment equilibrium equations. Isolate joint C and write the horizontal force equilibrium equation. Finally, isolate member B–C and use axial force equilibrium to eliminate the axial force F_{CB}.

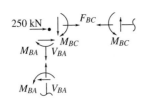

Joint B

$$\overset{\frown}{\sum M_B}:\ M_{BA} + M_{BC} = 0$$

$$2EI\theta_B - 3\cdot\frac{EI}{4}u_B + 4\cdot\frac{EI}{5}\theta_B = 0$$

Example 11.5 (continued)

or

$$14 \cdot \frac{EI}{5} \cdot \theta_B - 3 \cdot \frac{EI}{4} u_B = 0 \qquad (1)$$

$$\Sigma F \rightarrow : \; 250 + V_{BA} + F_{BC} = 0$$

$$250 - 3 \cdot \frac{EI}{4} \theta_B + 3 \cdot \frac{EI}{8} u_B + F_{BC} = 0$$

or

$$-3 \cdot \frac{EI}{4} \theta_B + 3 \cdot \frac{EI}{8} u_B = -250 - F_{BC} \qquad (2)$$

At joint C:

$$\Sigma F \rightarrow : \; -F_{CB} = 0$$

For member B–C:

$$\Sigma F \rightarrow : \; -F_{BC} + F_{CB} = 0 \quad \therefore \; F_{BC} = F_{CB} = 0$$

Joint C

Member B–C

STEP 3   Collecting the equations from the three free bodies yields two equilibrium equations. Solve these equations.

$$\begin{bmatrix} 14/5 & -3/4 \\ -3/4 & 3/8 \end{bmatrix} \begin{bmatrix} EI\,\theta_B \\ EI\,u_B \end{bmatrix} = \begin{bmatrix} 0 \\ 250 \end{bmatrix} \rightarrow \begin{bmatrix} EI\,\theta_B \\ EI\,u_B \end{bmatrix} = - \begin{bmatrix} 15{,}000/39 \\ 56{,}000/39 \end{bmatrix}$$

STEP 4   Backsubstitute into the original end moment equations in step 1.

$$M_{AB} = EI\,\theta_B - 3 \cdot \frac{EI}{4} u_B \qquad = \frac{27{,}000}{39} = 692.3 \text{ kN-m}$$

$$M_{BA} = 2EI\,\theta_B - 3 \cdot \frac{EI}{4} u_B \qquad = \frac{12{,}000}{39} = 307.7 \text{ kN-m}$$

$$M_{BC} = \frac{4EI}{5} \theta_B \qquad = \frac{-12{,}000}{39} = -307.7 \text{ kN-m}$$

Example 11.5 (continued)

$$M_{CB} = \frac{2EI}{5}\theta_B \qquad\qquad = \frac{6000}{39} = -153.8\,\text{kN-m}$$

$$V_{BA} = -3\cdot\frac{EI}{4}\theta_B + 3\cdot\frac{EI}{8}u_B \qquad\qquad = \frac{9750}{39} = 250\,\text{kN}$$

STEP 5 Draw the final moment diagram.

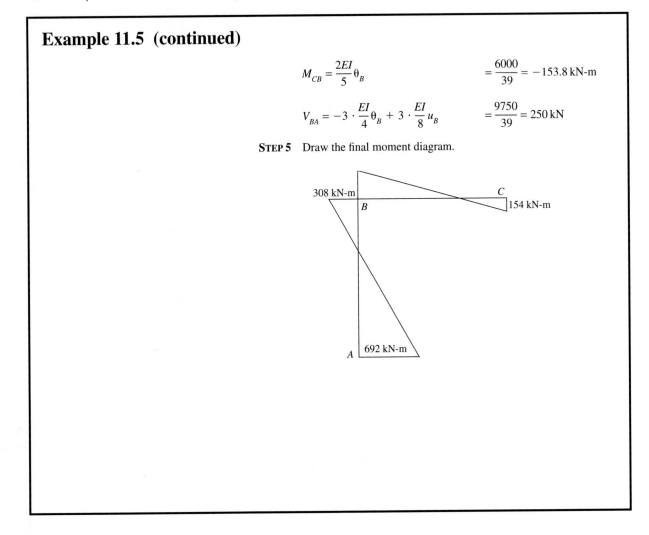

308 kN-m

B

C

154 kN-m

A

692 kN-m

involving θ_B and u_B. The solution of (1) and (2) yields negative values for θ_B and u_B. The negative value for u_B is expected since the loading causes the structure to displace to the right at B, which is opposite the defined positive sense of u_B in the local coordinate system of member AB.

Examples 11.4 and 11.5 illustrate that the sequence of steps followed in an analysis by the slope-deflection method is not modified by the configuration of the structure or its degree of indeterminacy. The programming of the method for structures that have joint translations is a little more complicated than it is for beams since the geometry of the structure is more complicated. In Chapter 17, implementation of the slope-deflection method with the inclusion of axial deformations will show that this approach can be programmed in a straightforward manner for use on a digital computer.

11.5 Analysis of Symmetric Structures

In Chapters 1 and 10 the reduction of symmetric structures to simpler forms for analysis was presented. The same type of reduction can be made for symmetric structures in the slope-deflection method of analysis. When the joints that define the ends of members fall on the axis of symmetry, there is no modification of slope-deflection equations (11.3) and (11.6) because the conditions of symmetry or antisymmetry of the loading will define the appropriate rotation and displacement of these joints. When a member crosses the axis of symmetry, the member is cut in half and only one joint of the member appears in the portion of the structure being analyzed. A suitable modification of Eqs. (11.3) and (11.6) can be made so that the rotation or displacement of the joint created on the axis of symmetry need not be considered.

Consider a member A–B as shown in Figs. 11.1 and 11.2 that crosses the axis of symmetry of a structure and assume that the structure and the member are symmetrically loaded. Because of symmetry of the displacements and rotations and of the moments and forces on the ends of the member, and following the sign convention for the slope-deflection equations, the end moment $M_{BA} = -M_{AB}$, the end force $V_{BA} = V_{AB}$, the end rotation $\theta_B = -\theta_A$, and the end displacement $v_B = v_A$. Substituting for θ_B and v_B in the first of Eqs. (11.3) and (11.6) leads to the slope-deflection equations for a member crossing the axis of symmetry of a symmetrically loaded structure, which take the form

$$M_{AB} = \frac{2EI}{L}\theta_A + \text{FEM}_{AB}$$

$$V_{AB} = \text{FEF}_{AB}$$

(11.8)

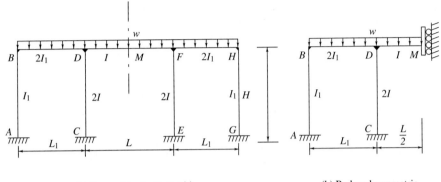

(a) Symmetric structure with
symmetric loading

(b) Reduced symmetric
structure

$$M_{DF} = \frac{2EI}{L}\left[2\theta_D + (-\theta_D) - 2\frac{v_D - (v_D)}{L}\right] + FEM_{DF} = \frac{2EI}{L}\theta_D + \frac{wL^2}{12}$$

$$V_{DF} = \frac{6EI}{L^2}\left[\theta_D + (-\theta_D) - 2\frac{v_D - (v_D)}{L}\right] + FEF_{DF} = \frac{wL}{2}$$

(c) Slope-deflection equations for member D–F

Figure 11.9a–c   Slope-deflection equations for member end moment and force for symmetrically loaded member crossing axis of symmetry of symmetrically loaded structure.

It is important to note in Eq. (11.8) that the fixed end moment and fixed end force are computed for the *whole member A–B*, and that these actions are due to a loading on the member that is *symmetric*. The concept of symmetry for a symmetrically loaded member crossing the axis of a symmetrically loaded structure is presented for member D–F in Fig. 11.9. The fixed end moment and fixed end force are those for the uniformly loaded member in Fig. 11.8a.

Now consider a member A–B as shown in Figs. 11.1 and 11.2 that crosses the axis of symmetry of a structure and assume that the structure and the member are antisymmetrically loaded. The antisymmetry of the loading causes an antisymmetric distribution of forces, moments, displacements, and rotations to develop on the ends of this member. From Figs. 11.1 and 11.2 and following the sign convention of the slope-deflection equations, the end moment $M_{BA} = M_{AB}$, the end force $V_{BA} = -V_{AB}$, the end rotation $\theta_B = \theta_A$, and the end displacement $v_B = -v_A$. Substituting for θ_B and v_B in the first of Eqs. (11.3) and (11.6) leads to the slope-deflection equations for a member crossing the axis of symmetry of an antisymmetrically loaded structure, which take the form

$$M_{AB} = \frac{6EI}{L}\left(\theta_A + 2\frac{v_A}{L}\right) + FEM_{AB}$$

$$\text{(11.9)}$$

$$V_{AB} = \frac{12EI}{L^2}\left(\theta_A + 2\frac{v_A}{L}\right) + FEF_{AB}$$

It is important to note in Eq. (11.9) that the fixed end moment and fixed end force are computed for the *whole member A–B*, and that these actions are due to a loading on the member that is *antisymmetric*. The concept of the antisymmetric behavior of a member crossing the axis of symmetry of an antisymmetrically loaded structure is presented in Fig. 11.10.

The analysis of the reduced structures of Figs. 11.9 and 11.10 by the slope-deflection method can be carried out without having to consider the

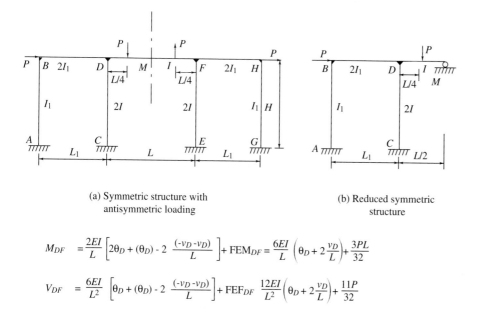

(a) Symmetric structure with antisymmetric loading

(b) Reduced symmetric structure

$$M_{DF} = \frac{2EI}{L}\left[2\theta_D + (\theta_D) - 2\frac{(-v_D - v_D)}{L}\right] + FEM_{DF} = \frac{6EI}{L}\left(\theta_D + 2\frac{v_D}{L}\right) + \frac{3PL}{32}$$

$$V_{DF} = \frac{6EI}{L^2}\left[\theta_D + (\theta_D) - 2\frac{(-v_D - v_D)}{L}\right] + FEF_{DF}\quad \frac{12EI}{L^2}\left(\theta_D + 2\frac{v_D}{L}\right) + \frac{11P}{32}$$

(c) Slope-deflection equations for member *D–F*

Figure 11.10a–c   Slope-deflection equations for member end moment and force for antisymmetrically loaded member crossing axis of symmetry of antisymmetrically loaded structure.

displacement or rotation of joint M of member D–M. The conditions of symmetry or antisymmetry of loading provide the proper slope-deflection equations for member D–M–F in terms of the rotation and displacement of joint D. Neglecting axial deformations, the analysis of the structure in Fig. 11.9 involves two equations for the rotations at B and D. Similarly, the analysis of the structure in Fig. 11.10, involves three equations for the rotations at B and D and the horizontal displacement of the top of the frame. As these figures indicate, a considerable amount of effort can be saved by using symmetry and antisymmetry of loading.

The value of symmetry of a structure and loading as an aid in analysis is illustrated in Example 11.6. The structure in the example would require the solution of eight simultaneous equations by the slope-deflection method. The use of symmetry reduces the analysis to the solution of only three simultaneous equations.

Example 11.6 Symmetry of the structure and loading reduces the structure to one in which there are only three unknown rotations at joints 2, 4, and 5. The possibility of lateral displacements of the two stories of the complete structure is eliminated because of the symmetry. The displacements of all joints of the structure are zero because of the assumption of no axial deformation, and the rotations of joints 1 and 3 are zero because they are fixed. In step 1, the local coordinate system that is used for each member in writing the slope-deflection equations is selected so that all joint rotations are positive counterclockwise. Joints M_1 and M_2 are ignored in the analysis since members 4–7 and 5–8 are assumed to behave in a symmetric manner, which is reflected by the no-rotation and no-horizontal-displacement conditions at M_1 and M_2 on the axis of symmetry. Use of Eq. (11.8) for the end moments on these members ensures proper recognition of symmetry. The computation of the fixed end moments is taken from Fig. 11.8, with the use of Fig. 11.8b twice for the two 10-kip loads on member 4–7. Steps 2 and 3 follow the usual slope-deflection method procedure and the solution for the joint rotations. The drawing of the final moment diagram for the structure in step 5 is based on the end moments for individual members computed in step 4. The diagram for member 2–4 is obtained by the superposition of the moment diagrams for the two end moments and the uniformly distributed load. Thus the center ordinate of that diagram is $2 \cdot 18^2/8 - (1/2)(30.87) - (1/2)(59.10) = 36.02$ kip-ft, as shown.

The moment diagrams for members 4–7 and 5–8 are shown for only their left half, with the diagrams for the right halves of these members, as well as the remainder of the members of the structure, being obtained by symmetry. By symmetry, the end moment for the right end of member 5–8 is -42.90 kip-ft. Proceeding in the same manner as for the

Example 11.6

For the symmetrically loaded symmetric frame, obtain the final moment diagram. Neglect axial deformations.

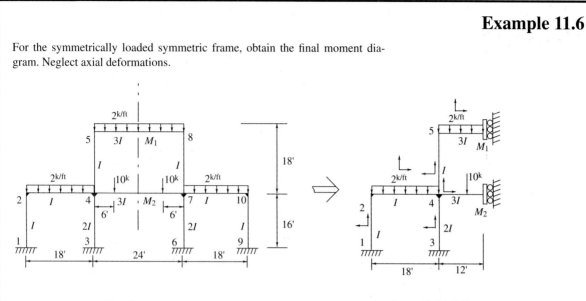

Complete structure   Symmetric left half

STEP 1 Divide the structure at the axis of symmetry and develop the analysis of the left half. Note that there are no joint translations. Write the slope-deflection equations for end moments using Eqs. (11.3) and (11.8), the latter being used for members 4–7 and 5–8, which cross the axis of symmetry. Impose the conditions of zero joint displacement and zero rotation at joints 1 and 3.

$$M_{12} = \frac{2EI}{16}\theta_2 = \frac{EI}{8}\theta_2 \qquad M_{21} = \frac{2EI}{16}2\theta_2 = \frac{EI}{4}\theta_2$$

$$M_{24} = \frac{2EI}{18}(2\theta_2 + \theta_4) + \frac{2\cdot 18^2}{12} = \frac{EI}{9}(2\theta_2 + \theta_4) + 54$$

$$M_{42} = \frac{EI}{9}(\theta_2 + 2\theta_4) - 54$$

$$M_{34} = \frac{2E\cdot 2I}{16}\theta_4 = \frac{EI}{4}\theta_4 \qquad M_{43} = \frac{EI}{2}\theta_4$$

$$M_{47} = \frac{2E\cdot 3I}{24}\theta_4 + \frac{10\cdot 6\cdot 18^2}{24\cdot 24} + \frac{10\cdot 6^2\cdot 18}{24\cdot 24} = \frac{EI}{4}\theta_4 + 45$$

$$M_{45} = \frac{2EI}{18}(2\theta_4 + \theta_5) = \frac{EI}{9}(2\theta_4 + \theta_5) \qquad M_{54} = \frac{EI}{9}(\theta_4 + 2\theta_5)$$

$$M_{58} = \frac{2E\cdot 3I}{24}\theta_5 + \frac{2\cdot 24^2}{12} = \frac{EI}{4}\theta_5 + 96$$

Example 11.6 (continued)

STEP 2 Write the moment equilibrium equations at joints 2, 4, and 5 and substitute the end moments for the members from step 1.

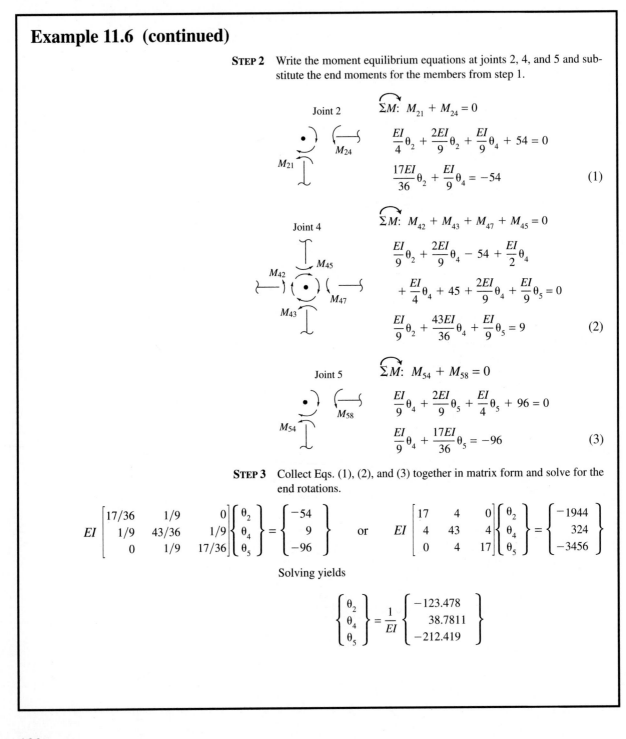

Joint 2 $\Sigma M: \ M_{21} + M_{24} = 0$

$$\frac{EI}{4}\theta_2 + \frac{2EI}{9}\theta_2 + \frac{EI}{9}\theta_4 + 54 = 0$$

$$\frac{17EI}{36}\theta_2 + \frac{EI}{9}\theta_4 = -54 \qquad (1)$$

Joint 4 $\Sigma M: \ M_{42} + M_{43} + M_{47} + M_{45} = 0$

$$\frac{EI}{9}\theta_2 + \frac{2EI}{9}\theta_4 - 54 + \frac{EI}{2}\theta_4$$

$$+ \frac{EI}{4}\theta_4 + 45 + \frac{2EI}{9}\theta_4 + \frac{EI}{9}\theta_5 = 0$$

$$\frac{EI}{9}\theta_2 + \frac{43EI}{36}\theta_4 + \frac{EI}{9}\theta_5 = 9 \qquad (2)$$

Joint 5 $\Sigma M: \ M_{54} + M_{58} = 0$

$$\frac{EI}{9}\theta_4 + \frac{2EI}{9}\theta_5 + \frac{EI}{4}\theta_5 + 96 = 0$$

$$\frac{EI}{9}\theta_4 + \frac{17EI}{36}\theta_5 = -96 \qquad (3)$$

STEP 3 Collect Eqs. (1), (2), and (3) together in matrix form and solve for the end rotations.

$$EI\begin{bmatrix} 17/36 & 1/9 & 0 \\ 1/9 & 43/36 & 1/9 \\ 0 & 1/9 & 17/36 \end{bmatrix}\begin{Bmatrix} \theta_2 \\ \theta_4 \\ \theta_5 \end{Bmatrix} = \begin{Bmatrix} -54 \\ 9 \\ -96 \end{Bmatrix} \quad \text{or} \quad EI\begin{bmatrix} 17 & 4 & 0 \\ 4 & 43 & 4 \\ 0 & 4 & 17 \end{bmatrix}\begin{Bmatrix} \theta_2 \\ \theta_4 \\ \theta_5 \end{Bmatrix} = \begin{Bmatrix} -1944 \\ 324 \\ -3456 \end{Bmatrix}$$

Solving yields

$$\begin{Bmatrix} \theta_2 \\ \theta_4 \\ \theta_5 \end{Bmatrix} = \frac{1}{EI}\begin{Bmatrix} -123.478 \\ 38.7811 \\ -212.419 \end{Bmatrix}$$

Example 11.6 (continued)

STEP 4 Backsubstitute the rotations into the end moment equations established in step 1 to obtain the final member end moments.

$$M_{12} = \frac{EI}{8}\left(-\frac{123.478}{EI}\right) \qquad\qquad = -15.43 \text{ kip-ft}$$

$$M_{21} = \frac{EI}{4}\left(-\frac{123.478}{EI}\right) \qquad\qquad = -30.87 \text{ kip-ft}$$

$$M_{24} = \frac{EI}{9}\frac{2(-123.478) + 38.7811}{EI} + 54 \qquad = 30.87 \text{ kip-ft}$$

$$M_{42} = \frac{EI}{9}\frac{(-123.478) + 2(38.7811)}{EI} - 54 \qquad = -59.10 \text{ kip-ft}$$

$$M_{34} = \frac{EI}{4}\frac{38.7811}{EI} \qquad\qquad = 9.70 \text{ kip-ft}$$

$$M_{43} = \frac{EI}{2}\frac{38.7811}{EI} \qquad\qquad = 19.39 \text{ kip-ft}$$

$$M_{47} = \frac{EI}{4}\frac{38.7811}{EI} + 45 \qquad\qquad = 54.70 \text{ kip-ft}$$

$$M_{45} = \frac{EI}{9}\frac{2(38.7811) + (-212.419)}{EI} \qquad = -14.98 \text{ kip-ft}$$

$$M_{54} = \frac{EI}{9}\frac{(38.7811) + 2(-212.419)}{EI} \qquad = -42.90 \text{ kip-ft}$$

$$M_{58} = \frac{EI}{4}\frac{(-212.419)}{EI} + 96 \qquad\qquad = 42.90 \text{ kip-ft}$$

467

Example 11.6 (continued)

STEP 5   Draw the moment diagram for the structure.

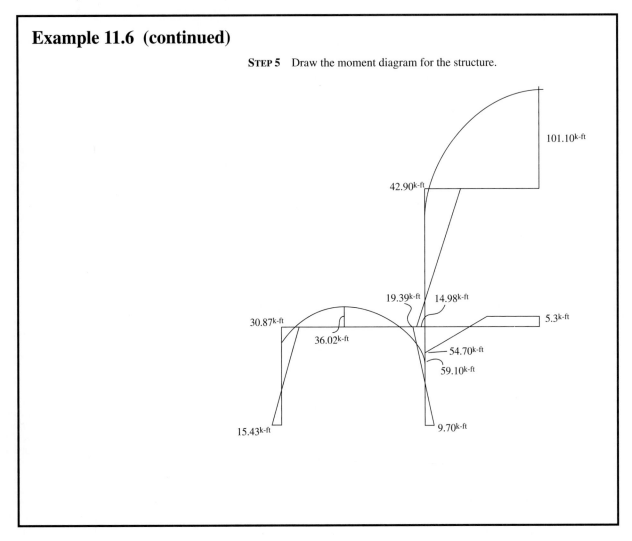

member 2–4, the moment in the center of member 5–8, which is on the axis of symmetry of the structure, is $2 \cdot 24^2/8 - (1/2)(42.90) - (1/2)(42.90) = 101.1$ kip-ft.

Similarly, by symmetry, the end moment for the right end of member 4–7 is -54.70 kip-ft. The moment diagram for the entire member due to the concentrated 10-kip loads alone is obtained as the diagram for the loads acting on a simply supported beam and is a symmetric trapezoid with the ordinates under the loads being 60 kip-ft. The moment in the center then becomes by superposition $60 - (1/2)(54.70) - (1/2)(54.70) = 5.3$ kip-ft. All moment diagrams are plotted on the compression side of the members.

11.6   Summary and Limitations

The slope-deflection method is based on computation of the rotations and displacements of the joints of beams and frames. The method is presented with the assumption that bending behavior is completely independent of the presence of axial forces in the members of a structure. Axial deformations of the members of frames are also neglected. The method is completely independent of the degree of indeterminacy of a structure, and the same sequence of steps is always followed in an analysis by this method of any structure. These characteristics make the method well suited for use with computers, as it is easy to program.

The slope-deflection equations are derived with the use of complementary virtual work to develop the relations between end moments and shear forces and member end rotations and displacements. There are four slope-deflection equations, two [Eq. (11.3)] relating end moments to end displacements and rotations and two [Eq. (11.6)] relating end shear forces to the same end displacements and rotations. The possible presence of a distributed load on a member is accounted for through the inclusion in these equations of fixed end moments and fixed end forces.

The first step in the application of the method for the analysis of beams and frames is to write the equations for the end moments for each and every member of the structure. This is followed by writing the equations for the shear force on the end of each member in the structure that connects to a joint that is free to displace in the direction of the shear force. The result of these activities is a set of equations expressing individual member end moments and end forces in terms of the free rotations and displacements of the joints of the structure. The unknowns in these equations are the joint displacements and rotations.

The second step in the analysis is to proceed joint by joint through the structure and establish moment and force equilibrium equations. At each

469

joint of the structure where a rotation can occur, the individual member end moments of all members that connect to that joint are summed to zero. At each joint of the structure where a displacement can occur, the individual member end shear and axial forces that have components in the direction of that displacement of all members that connect to that joint are summed to zero. The number of equilibrium equations established in these operations is equal to the number of unknown joint displacements and rotations.

The final step in the analysis is to solve the equilibrium equations for the unknown displacements and rotations of the joints. These displacements and rotations can then be used to find the final member end moments and forces by backsubstituting into the individual member slope-deflection equations established in the first step. The analysis is very useful for hand computations of structures that are highly restrained.

Modifications of the slope-deflection equations enable them to be extended to structures that are symmetric. Significant reductions in effort required in a slope-deflection analysis are possible for structures that are symmetric and symmetrically or antisymmetrically loaded. The modified forms of the slope-deflection equations for symmetry or antisymmetry conditions are presented in Eqs. (11.8) and (11.9).

All of the assumptions and limitations discussed in previous chapters are embedded in this method and are important to its intelligent use.

- Because of the use of simple bending theory principles, the slope-deflection equation relations between member end forces and moments and member end displacements and rotations are subject to all of the same limitations and assumptions that are listed at the end of Chapter 5 and in Section 6.2.

- All material deformations must be continuous, without any cracks or gaps.

- All structural deformations must be continuous, without any violation of the conditions of constraint at points of support.

- As presented in this chapter, all axial deformations are assumed to be small and have no significant effect on the displacement behavior of the joints of the structure.

- The distribution and orientation of all loads acting on a structure are not changed by the deformations of the structure.

- The orientation of all reaction components at supports is unchanged by the action of applied loads or any movement of the support.

- If movements occur at supports, they are small and do not affect the initial geometry of the structure.

- The principle of superposition of deformations, with all its limitations, is required for the analysis of symmetric structures under general loading if the analysis is carried out on the symmetric and antisymmetric forms of the structure.

Some questions that are worthy of consideration when using the slope-deflection method of analysis are the following:

- How large can an axial force in a member be before the error caused by its effect on bending becomes significant?
- If a nonlinear moment–curvature relation is defined for a member, what modifications are required for this method of analysis, or is it not possible to use the slope-deflection method in this case?
- How large can the joint displacements and rotations become before a slope-deflection analysis is no longer valid?
- Is the slope-deflection method of analysis a good one to use for statically determinate structures, and why or why not?

PROBLEMS

11.1   Draw the final moment diagram for the structure shown. Use the slope-deflection method (*E* constant).

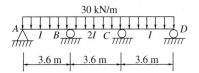

30 kN/m

| 3.6 m | 3.6 m | 3.6 m |

11.2   Solve the plane frame loaded as shown above using the slope-deflection method and draw the final moment diagram. Neglect axial deformations (*E* constant).

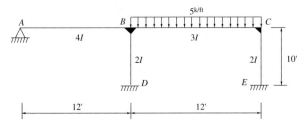

11.3   Draw the final moment diagram for the structure shown below. Neglect axial deformations (*E* constant).

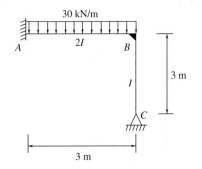

11.4   Using the slope-deflection method, draw the moment diagram for the structure.

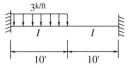

11.5   Solve Problem 10.21 by the slope-deflection method and draw the final moment diagram for all members. Neglect axial deformations.

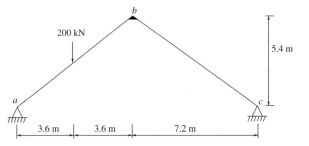

200 kN

| 3.6 m | 3.6 m | 7.2 m |

5.4 m

11.6   Analyze and draw the final moment diagram for the structure shown. Neglect axial deformations.

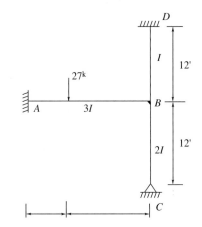

11.7   Solve Problem 10.14 by the slope-deflection method. Neglect axial deformations.

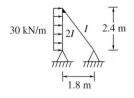

30 kN/m   2*I*   *I*   2.4 m

1.8 m

11.8   Obtain the final moment diagram for the structure shown. Use the slope-deflection method and neglect axial deformations.

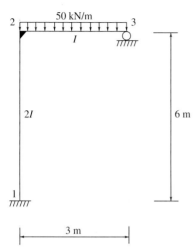

11.9   Using the slope-deflection method, compute the final moments and draw the moment diagram when the structure is subjected to the following support movements:

(Support *B* settles 0.24 in. down;)
(Support *A* rotates 0.005 rad clockwise)
($E = 29,000$ ksi, $I = 400$ in^4).

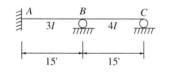

11.10   Draw the final moment diagram for all members of the plane frame shown. Use the slope-deflection method to obtain the final end moments (E constant). Neglect axial deformations.

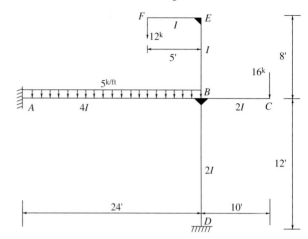

11.11   Draw the final moment diagram for all members of the plane frame shown. Use the slope-deflection method to obtain the end moments (E constant). Neglect axial deformations.

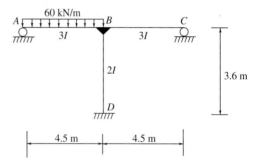

11.12   For the structure shown, obtain the final moment diagram due to the applied loading. Use the slope-deflection method (E constant). Neglect axial deformations.

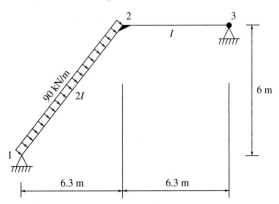

11.13   Analyze and draw the final moment diagram for the structure shown using the slope-deflection method. Neglect axial deformations.

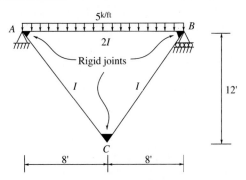

11.14   Solve Problem 10.18 by the slope-deflection method.

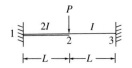

11.15   Solve Problem 10.19 by the slope-deflection method.

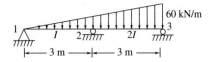

11.16   In the structure shown, draw the final moment diagram due to the applied loading. Use the slope-deflection method. Neglect axial deformations.

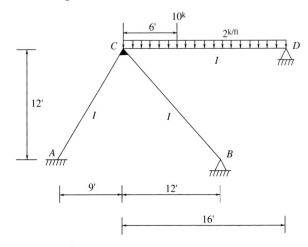

11.17   Using the slope-deflection method, compute the final end moments in the frame loaded as shown (E constant). Neglect axial deformations.

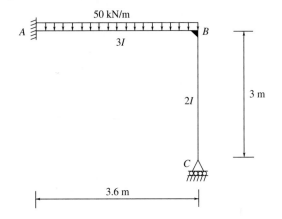

11.18 Draw the moment diagram for members *A–B* and *B–C*. Solve by the slope-deflection method. Neglect axial deformations.

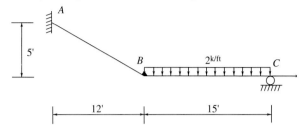

11.19 Solve by the slope-deflection method (*EI* constant).

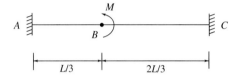

11.20 Use the slope-deflection method. Set up all necessary equations for a complete analysis of this frame. Do *not* solve the resulting equations. Neglect axial deformations.

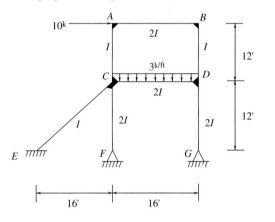

11.21 Solve Problem 10.24 by the slope-deflection method. Neglect axial deformations.

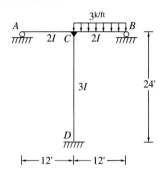

11.22 Solve by the slope-deflection method and draw the final moment diagram for all members (*E* constant). Neglect axial deformations.

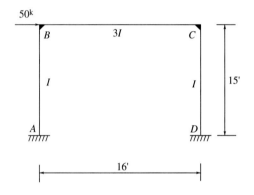

11.23 Solve Problem 10.30 by the slope-deflection method.

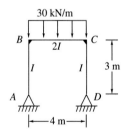

Moment Distribution Method of Analysis

The analysis of statically indeterminate structures has been approached from two different points of view. In Chapter 10 a set of equations equal to the degree of indeterminacy of the structure was derived. These equations established continuity and consistency of deformation of the structure and were called *compatibility equations*. This method of analysis is known as a *force method of analysis*. In Chapter 11 a set of equilibrium equations equal to the number of joint displacements and rotations of the structure were established. These equilibrium equations were based on the slope-deflection equations, and the method of analysis is known as a *displacement method of analysis*.

Both of these methods require the establishment and solution of a set of simultaneous equations. The moment distribution method provides a direct approach to the analysis problem of highly indeterminate frames without the requirement of solving simultaneous equations, or at the worst, only a few simultaneous equations. After its introduction, many extensions and enhancements of the technique were presented, and it evolved into one of the most successful and widely used methods of analysis.

Some authors refer to moment distribution as a displacement method of analysis because it is a very cleverly devised method of solving the slope-deflection equations in an interactive manner although joint displacements and rotations are not obtained explicitly. Other authors refer to it as a force method of analysis since the results obtained at the conclusion of the procedure are the moments on the ends of the members of the structure. The classification of the method is not important, but the elegant simplicity of the method and its enhancement of the structural engineer's understanding of structural behavior is worthy of presentation.

12.1   Moment Distribution Method

The moment distribution method of analysis is an adaptation of the slope-deflection equations which permits the rapid hand calculation of the solution of highly restrained, indeterminate frame structures. The method was presented by Hardy Cross in the early 1930s and was one of the most widely used methods of analysis for over 40 years from that time. The recent wide availability of high-speed computers has reduced the importance of the moment distribution technique as a tool of analysis. However, it still is important to the understanding of structural behavior and is very useful for analysis of small structures and as a tool for checking the consistency of the results of computer analyses.

The technique is developed initially for structures where joint translations (or member chord rotations) cannot occur. The use of the technique is illustrated in several examples, and a modification is made to accommodate members with pin or roller supports at one end. Finally, the technique is extended to structures where joint translation (or member chord rotation) can occur. A thorough understanding of the technique provides a clear illustration of structural behavior which enhances the structural engineer's intuition.

12.2   Moment Distribution Without Joint Translation

For framed structures in which the effects of axial deformation can be neglected, the restraint of joint translation is common. In the slope-deflection method of analysis, structures of this type were studied and presented in Section 11.3. The technique of moment distribution is best presented in the context of an analysis of these types of structures and is introduced with a reexamination of the slope-deflection analysis illustrated in Example 11.2. The essential features of the solution process for the two-member structure are presented in Fig. 12.1, but with the omission of the final moment diagram. The calculations have been condensed to only the essential operations to emphasize the concepts utilized in the solution technique.

Figure 12.2 A moment distribution diagram or work sheet for the structure analyzed in Fig. 12.1 by the slope-deflection method is shown. The numbered steps and operations in Figs. 12.1 and 12.2, respectively, show the sequence of computations for the two processes, many of which involve identical operations. In following the discussion of operations presented in Figs. 12.1 and 12.2, it is recommended that the

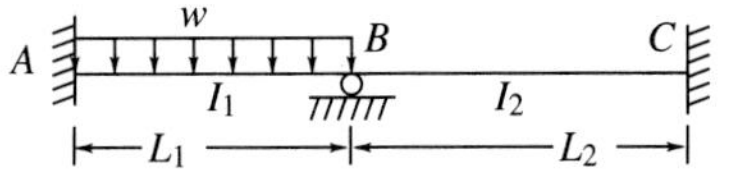

STEP 1. Write slope-deflection equations for end moments for each member in terms of relative stiffness $K = I/L$ and fixed end moments (FEMs) due to uniform load.

$$M_{AB} = 2EK_1\theta_B + \frac{wL_1^2}{12} = 4E\left(\frac{1}{2}\right)K_1\theta_B + \frac{wL_1^2}{12}$$

$$M_{BA} = 4EK_1\theta_B - \frac{wL_1^2}{12}$$

$$M_{BC} = 4EK_2\theta_B$$

$$M_{CB} = 2EK_2\theta_B = 4E\left(\frac{1}{2}\right)K_2\theta_B$$

STEP 2. Write the moment equilibrium equation at B and solve for θ_B.

Moment equilibrium at joint B:

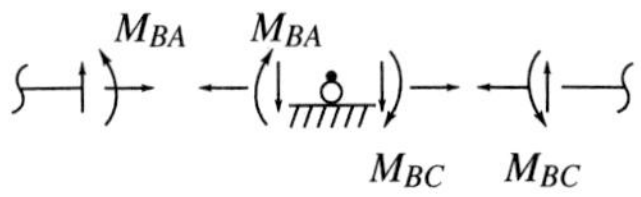

ΣM_B: $M_{BA} + M_{BC} = 0$

$$4EK_1\theta_B - \frac{wL_1^2}{12} + 4EK_2\theta_B = 0$$

On solving for $\theta_B \rightarrow \theta_B = \dfrac{wL_1^2 / 12}{4E\,(K_1 + K_2)}$

STEP 3. Substitute for θ_B in the end moment equations of step 1 and the sum moments due to FEM and θ_B for final values.

$$M_{AB} = \left[\frac{1}{2}\right]\frac{K_1}{K_1 + K_2}\left(\frac{wL_1^2}{12}\right) + \frac{wL_1^2}{12} = \frac{3K_1 + 2K_2}{2\,(K_1 + K_2)}\left(\frac{wL_1^2}{12}\right)$$

$$M_{BA} = \frac{K_1}{K_1 + K_2}\left(\frac{wL_1^2}{12}\right) - \frac{wL_1^2}{12} = \frac{-K_2}{K_1 + K_2}\left(\frac{wL_1^2}{12}\right)$$

$$M_{BC} = \frac{K_2}{K_1 + K_2}\left(\frac{wL_1^2}{12}\right) = \frac{K_2}{K_1 + K_2}\left(\frac{wL_1^2}{12}\right)$$

$$M_{CB} = \left[\frac{1}{2}\right]\frac{K_2}{K_1 + K_2}\left(\frac{wL_1^2}{12}\right) = \frac{K_2}{2\,(K_1 + K_2)}\left(\frac{wL_1^2}{12}\right)$$

Figure 12.1   Slope-deflection solution of Example 11.2.

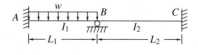

A	B		C	
M_{AB}	M_{BA}	M_{BC}	M_{CB}	(1)
$K_1 = \dfrac{I_1}{L_1}$	$D_{BA} = \dfrac{K_1}{K_1 + K_2}$	$\dfrac{K_2}{K_1 + K_2} = D_{BC}$	$K_2 = \dfrac{I_2}{L_2}$	(2)
$\dfrac{wL_1^2}{12}$	$-\dfrac{wL_1^2}{12}$			(3)
	$\dfrac{K_1}{K_1 + K_2}\left(\dfrac{wL_1^2}{12}\right)$	$\dfrac{K_2}{K_1 + K_2}\left(\dfrac{wL_1^2}{12}\right)$		(4)
$\dfrac{1}{2}\dfrac{K_1}{K_1 + K_2}\left(\dfrac{wL_1^2}{12}\right)$ $\xleftarrow{\;1/2\;}$		$\xrightarrow{\;1/2\;}$ $\dfrac{1}{2}\dfrac{K_2}{K_1 + K_2}\left(\dfrac{wL_1^2}{12}\right)$		(5)
$\dfrac{3K_1 + 2K_2}{2(K_1 + K_2)}\left(\dfrac{wL_1^2}{12}\right)$	$\dfrac{K_2}{K_1 + K_2}\left(\dfrac{wL_1^2}{12}\right)$	$\dfrac{K_2}{K_1 + K_2}\left(\dfrac{wL_1^2}{12}\right)$	$\dfrac{1}{2}\dfrac{K_2}{K_1 + K_2}\left(\dfrac{wL_1^2}{12}\right)$	(6)

<u>Operations:</u>

1. Compute relative stiffness, $K = I / L$.
2. Compute distribution factors.
3. Compute fixed end moments (FEMs).
4. Distribute moment at B.
5. Carry over moments to A and C.
6. Sum columns of moments for final values.

Figure 12.2 Moment distribution solution and work sheet for Example 11.2.

reader take paper and pencil and construct the moment distribution work sheet of Fig. 12.2 as the sequence of steps evolve. A careful study of the operations in the moment distribution diagram reveals the fundamentals of the method.

Step 1 of the slope-deflection solution in Fig. 12.1 involves writing the end moments for the two members of the structure in terms of the fixed end moments for the member AB and the unknown rotation θ_B (all other rotations and displacements in the structure are zero). It is convenient in this process to use the relative stiffness factors, K_1 and K_2, which are the ratios of the moment of inertia of each member to this length. Corresponding to step 1 in Fig. 12.1 are operations 1 and 3 in the moment distribution solution, Fig. 12.2. Here the member relative stiffnesses, K_1 and K_2, and the fixed end moments (FEM) are entered in the work sheet under the ends of member A–B. The second operation in the moment distribution work sheet is to compute the distribution factors at joint B. The distribution factor for a member at a joint is defined as the ratio of the relative stiffness of that member to the sum of the relative stiffnesses of all members that connect to that joint. By definition, the distribution factor for member A–B at B, D_{BA}, is

$$D_{BA} = \frac{K_1}{K_1 + K_2}$$

and the distribution factor for member B–C at B is

$$D_{BC} = \frac{K_2}{K_1 + K_2}$$

The two distribution factors are computed and entered in the boxes on the diagram in operation 2 in Fig. 12.2. The physical significance of the distribution factors will be discussed shortly.

The second step in the slope-deflection solution is to isolate joint B and write the moment equilibrium equation for that joint in terms of M_{BA} and M_{BC}. Replacing the moments with the expressions developed in step 1 for the member end moments in terms of θ_B gives an equation that can be solved for the rotation θ_B that establishes moment equilibrium at joint B. The values of the end moments are obtained when θ_B is substituted back into the expressions for M_{BA} and M_{BC} in step 3. They become $[K_1/(K_1 + K_2)](wL_1^2/12)$ for M_{BA} and $[K_2/(K_1 + K_2)](wL_1^2/12)$ for M_{BC}, respectively, as shown in Fig. 12.1. These same moments are listed in the moment distribution work sheet in the M_{BA} and M_{BC} column at the B end of each member in operation 4 in Fig. 12.2.

In Fig. 12.1 it is seen that these member end moments develop because of the unbalanced moment of $wL_1^2/12$ at joint B before the computation of θ_B in step 2. Rotation of the joint, θ_B, restores moment equilibrium and creates moments at B which are opposite in sign to the existing unbalanced moment. These moments, computed in step 2 in Fig. 12.1, are proportional to the relative rotational stiffness of each member coming into the joint with the assumption that the opposite ends of each member at joint B (A and C in the present example) are fixed against rotation. The relative rotational stiffness for members B–A and B–C are the distribution factors computed in the second operation of the work sheet, Fig. 12.2. Thus the unbalanced moment of $wL_1^2/12$ acting at joint B is reduced to zero when the rotation θ_B occurs. The balancing moments are computed as the product of the distribution factors for the members connected at B and the negative of the unbalanced moment and are referred to as *distributed moments*. These moments appear in the work sheet, Fig. 12.2, as part of operation 4.

The rotation θ_B also creates moments at the ends of the members opposite to B. The moments at A and C are computed in steps 2 and 3 of the slope-deflection solution in Fig. 12.1 when the value of θ_B is substituted into the expressions for the moments M_{AB} and M_{CB}. Note that these

481

moments, due to θ_B, are one half of the distributed moments that occurred at the B end of the same members, and this fact is emphasized by the box drawn around the $\frac{1}{2}$ in step 3 of Fig. 12.1. These moments are entered in the moment distribution work sheet, Fig. 12.2, in the M_{AB} and M_{CB} columns under the ends A and C in operation 5.

Comparing steps 2 and 3 in Fig. 12.1 with operations 4 and 5 in Fig. 12.2 leads to the terminology that the unbalanced moment at B in Fig. 12.2 is *distributed and carried over* to the joints opposite B with the *carryover factor* being $\frac{1}{2}$, boxed as shown in the two figures. Steps 2 and 3 in the slope-deflection solution in Fig. 12.1 are summarized by the distribute and carry-over operations 4 and 5 in the moment distribution work sheet in Fig. 12.2.

The final step in the slope-deflection process is to sum the moments due to the fixed end moments and the joint rotation θ_B. This operation is also accomplished in operation 6 in the work sheet, Fig. 12.2, by drawing two horizontal lines under each column and adding all moments in that column. The moments below the double lines in the work sheet are the final end moments for each member of the structure. The moments are positive when they act counterclockwise on the end of the member just as in the slope-deflection solution.

The simple problem just considered shows the fundamentals of the moment distribution technique. All of the essential details of the slope-deflection solution process are recorded in the work sheet. A great economy of effort is possible because the lengthy operation of establishing the moment equilibrium at joint B, the solution of that equation for the rotation θ_B, and the substitution of the rotation into the individual member slope-deflection equations to obtain the moments created by the rotation is completely replicated by the distribute and carryover process in operations 4 and 5 in the work sheet.

Example 12.1 The application of the moment distribution method to a more interesting problem is shown. The work sheet is set up as it was in Fig. 12.2, except that there are now two joints where rotation can occur. The initial unbalanced moment occurs at B due to the fixed end moment. At first glance it would appear that the distribute and carryover process cannot be used for this problem at joint B because only A is a joint with zero rotation. The process can be started, however, if joint C is temporarily imagined to be fixed or locked against rotation. Under this condition, the problem is identical to that in Fig. 12.2, where an unbalanced moment appears at joint B and the joints opposite B, A and C, are fixed against rotation. The result of the distribute and carryover in operation 4 shows that a moment (of 10 kN-m) develops at C when B is distributed. After distributing and carrying over, a horizontal line is drawn at joint B to show that the joint has been placed in moment equilibrium by the operation.

Example 12.1

Use moment distribution to obtain the final moment diagram.

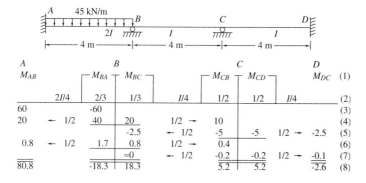

A M_{AB}		B M_{BA}	M_{BC}		C M_{CB}	M_{CD}		D M_{DC}	(1)
	$2I/4$	2/3	1/3	$I/4$	1/2	1/2	$I/4$		(2)
60		-60							(3)
20	← 1/2	40	20	1/2 →	10				(4)
			-2.5	← 1/2	-5	-5	1/2 →	-2.5	(5)
0.8	← 1/2	1.7	0.8	1/2 →	0.4				(6)
			≈0	← 1/2	-0.2	-0.2	1/2 →	-0.1	(7)
$\overline{80.8}$		$\overline{-18.3}$	$\overline{18.3}$		$\overline{5.2}$	$\overline{5.2}$		$\overline{-2.6}$	(8)

Operation

1. Compute the relative stiffnesses (I/L).
2. Compute the distribution factors.

   At B: $D_{BA} = \dfrac{2 \cdot I/4}{2 \cdot I/4 + I/4} = \dfrac{2}{3}$ $D_{BC} = \dfrac{I/4}{3 \cdot I/4} = \dfrac{1}{3}$

   At C: $D_{CB} = \dfrac{I/4}{I/4 + I/4} = \dfrac{1}{2}$ $D_{CD} = \dfrac{I/4}{2 \cdot I/4} = \dfrac{1}{2}$

3. Compute the FEMs: $\pm wL^2/12 = \pm 45 \cdot 4^2/12 = \pm 60$ kN-m.
4. Distribute B and carry over (C fixed).
5. Distribute C and carry over (B fixed).
6. Distribute B and carry over (C fixed).
7. Distribute C and carry over (B fixed).
8. Sum for final moments. For member A–B use the equivalent simple beam technique to obtain the final moment diagram.

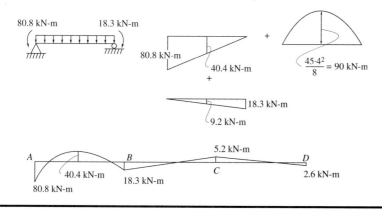

An assessment of the situation at this point shows that joint B is in moment equilibrium but that an unbalanced moment exists at C because it was imagined to be fixed during the distribute and carryover operation. Since joint D is fixed against rotation, the unbalanced moment at C can be reduced to zero by a distribution and carryover process identical to the one just completed at B if B were fixed against rotation, which it is not. But if B is imagined to be fixed temporarily against rotation, the distribute and carryover operation can be used at C. In operation 5, C reaches moment equilibrium by the distribute and carryover operation, so a line is drawn under it, but a new unbalanced moment (of -2.5 kN-m) appears at B from the carryover process.

Now moments appear at the ends of the members at all joints. Joints A and D have moments that are resisted by the fixed condition at the reaction, joint C is in moment equilibrium but joint B is not, a condition of the same nature as that at the start of the computation process at the end of operation 3. Note, though, that this unbalanced moment is about $1/20$ of the initial unbalanced moment at the end of operation 3. The two distribution and carryover processes, first at B and then at C, have produced an unbalanced moment at B that is 20 times smaller. Repeating this two-operation process again will reduce the current unbalanced moment by about $1/20$. Clearly, this process can be repeated as many times as needed to reduce the unbalanced moment at B to an acceptably small value. Repeating the operation 4–5 process again in operations 6 and 7 reduces the unbalanced carryover moment to B in operation 7 to approximately zero. In this problem the calculations are taken to only one decimal place, and when distributed or carryover moments are less than 0.05, they are taken to be zero. The 0.1 carryover moment from C to B in operation 7 is taken to be 0 since the distribution at B of this moment would be less than 0.05. If more significant figures are wanted, the calculations would have to be carried out using more than the three significant figures of the present example and additional cycles of the operation 4–5 process would have to be completed. Such accuracy is not justified in design calculations, as properties of materials, dimensions of cross sections, and magnitudes of loads are not known to more than three significant figures.

At the end of the second distribute and carryover operation at C, the carryover moments are small enough to be called zero. Now, joints B and C are both, within the accuracy of the computation, in equilibrium and no unbalanced moments exist in the structure. A set of double lines is drawn under each column and all moments summed to obtain the final end moments for each member. The final moment diagram is drawn in the usual way. Note that the technique introduced in Chapter 5 of using the equivalent simply supported beam to superimpose the effects of end moments and distributed loads is used to advantage for member A–B.

Example 12.2 In this example it is assumed that axial deformations can be neglected so that joint B cannot deflect vertically or horizontally. The three distribution factors for the three members joining at B are computed in the usual way in operation 2 and joint C, which has only one member, has a distribution factor of 1. The operations proceed in the same manner as in Example 12.1. The unbalanced moments at B are distributed to the three members that connect there with carryover moments going to A, D, and C, which is imagined to be fixed each time the distribution process is executed at B. Each cycle of the distribute and carryover process at B and C reduces the unbalanced moment at B by about $1/15$.

12.3   Reduced Stiffness for Members with Ends Free to Rotate

In Example 12.2 joint C must have a zero moment at the completion of the moment distribution process. This is achieved by constantly distributing and carrying over the unbalanced moments that appear at C when it is imagined to be temporarily fixed against rotation during the distribution process at joint B. A modification in the moment distribution technique can preserve a zero moment condition on the end of a member with a reduced amount of effort.

Figure 12.3 The slope-deflection solution of a structure similar to the one in Fig. 12.1, but with the joint at C being free to rotate and consequently having a zero moment on the C end of the member, is shown. The slope-deflection process and the parallel moment distribution work sheet with the operations for both are shown. It is seen in step 1 of the slope-deflection process that the condition of zero moment at C is enforced in the solution by using the condition that θ_C be equal to $-\theta_B/2$. Because of this relation between θ_C and θ_B, the moment M_{BC} becomes $4E(\frac{3}{4}K_2)\theta_B$, which indicates that the relative stiffness for member B–C has been reduced from K_2 to $\frac{3}{4}K_2$. The solution of the moment equilibrium equation for θ_B at B in step 2 of the slope-deflection solution also is modified by the relative stiffness factor for member B–C being reduced to $\frac{3}{4}K_2$. The reduction in the relative stiffness factor occurs because of the requirement that joint C must undergo a rotation $-\theta_B/2$ simultaneously with any rotation θ_B at B to preserve the zero moment condition at C. Member A–B still has the fixed rotation condition at A during the process and maintains its full relative stiffness K_1.

The moment distribution work sheet has two modifications in it. First, the relative stiffness for member BC is reduced in operation 1 to $\frac{3}{4}K_2$ in the computation of the relative stiffnesses. This produces distribution factors at B that are different from those in Fig. 12.2 because of the lack

485

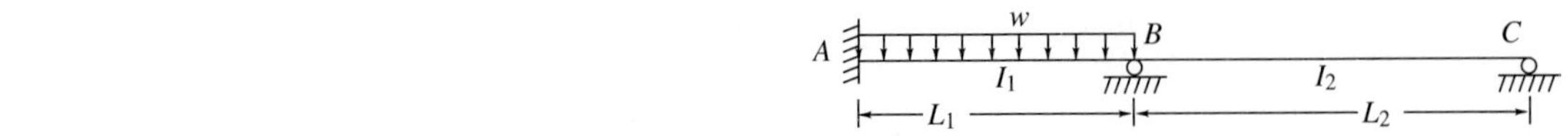

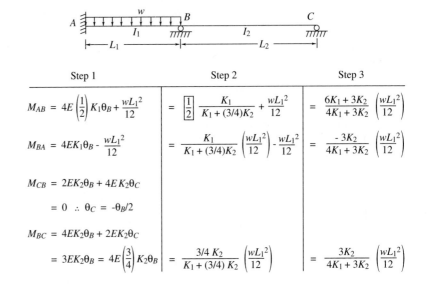

Step 1	Step 2	Step 3
$M_{AB} = 4E\left(\frac{1}{2}\right)K_1\theta_B + \frac{wL_1^2}{12}$	$= \boxed{\frac{1}{2}}\,\frac{K_1}{K_1 + (3/4)K_2} + \frac{wL_1^2}{12}$	$= \frac{6K_1 + 3K_2}{4K_1 + 3K_2}\left(\frac{wL_1^2}{12}\right)$
$M_{BA} = 4EK_1\theta_B - \frac{wL_1^2}{12}$	$= \frac{K_1}{K_1 + (3/4)K_2}\left(\frac{wL_1^2}{12}\right) - \frac{wL_1^2}{12}$	$= \frac{-3K_2}{4K_1 + 3K_2}\left(\frac{wL_1^2}{12}\right)$
$M_{CB} = 2EK_2\theta_B + 4EK_2\theta_C$		
$= 0 \;\therefore\; \theta_C = -\theta_B/2$		
$M_{BC} = 4EK_2\theta_B + 2EK_2\theta_C$		
$= 3EK_2\theta_B = 4E\left(\frac{3}{4}\right)K_2\theta_B$	$= \frac{3/4\,K_2}{K_1 + (3/4)\,K_2}\left(\frac{wL_1^2}{12}\right)$	$= \frac{3K_2}{4K_1 + 3K_2}\left(\frac{wL_1^2}{12}\right)$

STEP 1 Write slope-deflection equations for each member end moment using relative stiffness $K = I/L$ and fixed end moments (FEMs) due to w. For member B–C, use $M_{CB} = 0$ to solve for θ_C in terms of θ_B and then modify M_{BC} to eliminate the rotation θ_C.

STEP 2 Write the moment equilibrium equation at B, and solve for θ_B, and substitute into member end moment equations.

Moment equilibrium at joint B:

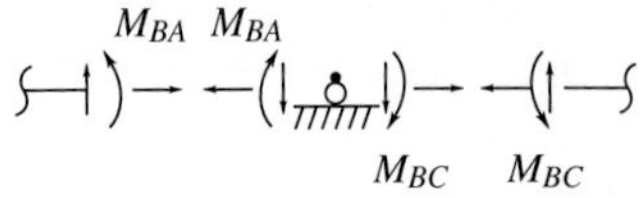

$\curvearrowright \Sigma M_B$: $\quad M_{BA} + M_{BC} = 0$

$$4EK_1\theta_B - \frac{wL_1^2}{12} + 3EK_2\theta_B = 0 \;\therefore\; \theta_B = \frac{wL_1^2/12}{4E\,(K_1 + 3/4\,K_2)}$$

STEP 3 Sum member end moments due to the FEMs and the rotation θ_B for final values.

(a) Slope-deflection solution

Figure 12.3a Slope-deflection solution and moment distribution work sheet solution for structure with free joint rotation.

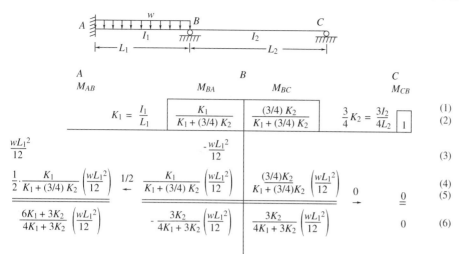

A			B			C		
M_{AB}			M_{BA}		M_{BC}		M_{CB}	
	$K_1 = \dfrac{I_1}{L_1}$		$\dfrac{K_1}{K_1 + (3/4)K_2}$		$\dfrac{(3/4)K_2}{K_1 + (3/4)K_2}$	$\dfrac{3}{4}K_2 = \dfrac{3I_2}{4L_2}$ $\boxed{1}$		(1) (2)
$\dfrac{wL_1^2}{12}$			$-\dfrac{wL_1^2}{12}$					(3)
$\dfrac{1}{2}\cdot\dfrac{K_1}{K_1+(3/4)K_2}\left(\dfrac{wL_1^2}{12}\right)$		$\dfrac{1/2}{\leftarrow}$	$\dfrac{K_1}{K_1+(3/4)K_2}\left(\dfrac{wL_1^2}{12}\right)$		$\dfrac{(3/4)K_2}{K_1+(3/4)K_2}\left(\dfrac{wL_1^2}{12}\right)$	0	$\begin{matrix}0\\ \rightarrow\end{matrix}$	(4) (5)
$\dfrac{6K_1+3K_2}{4K_1+3K_2}\left(\dfrac{wL_1^2}{12}\right)$			$-\dfrac{3K_2}{4K_1+3K_2}\left(\dfrac{wL_1^2}{12}\right)$		$\dfrac{3K_2}{4K_1+3K_2}\left(\dfrac{wL_1^2}{12}\right)$		0	(6)

Operations:

1. Compute relative stiffness, $K = I/L$ and use $(3/4)I/L$ for member B–C
2. Compute distribution factors.
3. Compute fixed end moments (FEMs).
4. Distribute moment at B with C free to rotate.
5. Carry over moments to A only.
6. Sum columns of moments for final values.

(b) Moment distribution solution

Figure 12.3b   Slope-deflection solution and moment distribution work sheet solution for structure with free joint rotation.

of fixity at C. Second, in operation 5 there is no carryover moment to C after the distribution of the unbalanced moments at B in operation 4 since the condition of zero moment at C in enforced by allowing both B and C to rotate during the distribution operation.

The modification of the moment distribution process for joints with free rotation is illustrated in Examples 12.2 and 12.4. Note the significant reduction of work in Example 12.3, which is the same as Example 12.2.

Example 12.3   Using the reduced stiffness factor for B–C eliminates the requirement that moments be carried over to joint C when a distribution is done at joint B. This reduces the distribution and carryover operation for this structure to one cycle and gives the same final moments as those obtained in Example 12.2. The number of distribute and carryover cycles in this case has been cut by a factor of 4 by using the reduced relative stiffness.

487

Example 12.2

Solve for the final moment diagram using moment distribution.

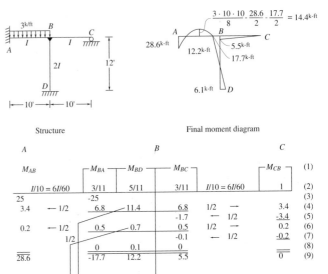

Structure   Final moment diagram

$$\frac{3 \cdot 10 \cdot 10}{8} \cdot \frac{28.6}{2} \cdot \frac{17.7}{2} = 14.4^{\text{k-ft}}$$

	A		B			C	
M_{AB}	M_{BA}	M_{BD}	M_{BC}		M_{CB}		(1)
$I/10 = 6I/60$	3/11	5/11	3/11	$I/10 = 6I/60$	1		(2)
							(3)
25	-25						
3.4 ← 1/2	6.8	11.4	6.8	1/2 →	3.4		(4)
			-1.7	← 1/2	-3.4		(5)
0.2 ← 1/2	0.5	0.7	0.5	1/2 →	0.2		(6)
			-0.1	← 1/2	-0.2		(7)
1/2	0	0.1	0				(8)
28.6	-17.7	12.2	5.5		0		(9)

D — M_{DB} — $I/6 = 10I/60$

5.7

1/2

0.4

6.1

Operation

1. Compute the relative stiffnesses.
2. Compute the distribution factors.

$$D_{CB} = 1 \quad D_{BA} = \frac{I/10}{I/10 + I/10 + I/6} = \frac{6I/10}{6I/60 + 6I/60 + 10I/60}$$

$$= \frac{6}{6 + 6 + 10} = \frac{3}{11}$$

$$D_{BC} = \frac{6I/60}{22I/60} = \frac{3}{11} \qquad D_{BD} = \frac{10I/60}{22I/60} = \frac{5}{11}$$

3. Compute the FEMs: $\pm wL^2/12 = \pm 3 \cdot 10^2/12 = \pm 25$ kip-ft.
4. Distribute B and carry over (C fixed).
5. Distribute C and carry over (B fixed).
6. Distribute B and carry over (C fixed).
7. Distribute C and carry over (B fixed).
8. Distribute B and carry over (C fixed).
9. Sum for final moments.

Example 12.3

Rework Example 12.2 using the reduced relative stiffness factor for member B–C.

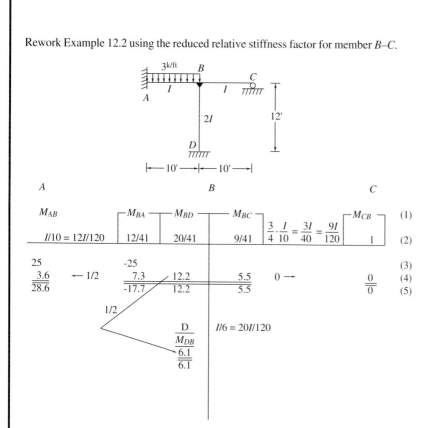

Operation

1. Compute the relative stiffnesses and reduce K_{BC} to $\tfrac{3}{4}K_{BC}$.
2. Compute the distribution factors.

$$D_{CB} = 1$$

$$D_{BA} = \frac{I/10}{I/10 + I/6 + 3/4 \cdot I/10} = \frac{12I/120}{12I/120 + 20I/120 + 9I/120} = \frac{12}{41}$$

$$D_{BC} = \frac{9I/120}{41I/120} = \frac{9}{41} \qquad D_{BD} = \frac{20I/120}{41I/120} = \frac{20}{41}$$

3. Compute the FEMs: $\pm wL^2/12 = \pm 3 \cdot 10^2/12 = \pm 25$ kip-ft.
4. Distribute B and carry over (C free to rotate).
5. Sum for final member end moments and draw the moment diagram.

The final end moments and moment diagram are the same as those in Example 12.2.

The structure shown in Fig. 12.4 is an interstate bridge over the Kishwaukee River, which is a five-span continuous segmented prestresed girder bridge. Example 12.4 uses moment distribution to obtain the moments in the structure due to a particular pattern of live loading.

Example 12.4 The live loading is specified to have magnitude, w, and the spans are length L. The use of moment distribution for a problem without specific magnitudes of the loading of span length is no different than any other, as the fixed end moments of $wL^2/12$ are conveniently taken to have magnitude 100. The final member end moments are recorded as a multiple of wL^2 by dividing those obtained from the moment distribution solution by 1200.

Figure 12.4   Kishwaukee River bridge. (Courtesy Alfred Benesch & Company.)

Example 12.4

Obtain the final moment diagram for uniform loading, w, on spans 1–2, 2–3, and 4–5 of the Kishwaukee River bridge of Fig. 12.4.

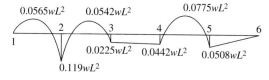

STEP 1   Establish the work sheet, then compute the relative stiffnesses and distribution factors for the joints. Use reduced stiffness in spans 1–2 and 5–6. Compute the fixed end moments of $\pm wL^2/12$ and take, for convenience, $\pm wL^2/12 = 100$ for the the distribution solution.

1		2		3		4		5		6
$\lceil M_{12} \rceil \; \dfrac{3}{4}\cdot\dfrac{I}{L}$	$\lceil M_{21} \; \lceil M_{23} \rceil$	$\lceil M_{32} \; \lceil M_{34} \rceil$		$\lceil M_{43} \; \lceil M_{45} \rceil$		$\lceil M_{54} \; M_{56} \rceil \; \dfrac{3}{4}\cdot\dfrac{I}{L}$		$\lceil M_{6} $		
1	3/7 4/7 I/L	1/2 1/2 I/L		1/2 1/2 I/L		4/7 3/7		1		
100	-100 100	-100				100 -100		0		
-100 1/2 → -50								$\overline{0}$		
0		25 ←1/2 50	50 1/2 → 25							
				28 ←1/2 57	43				
11	14 1/2 → 7									
			-38 ←1/2 -76	-77 1/2 → -38						
				11 ←1/2 22	16					
	8 ←1/2 15	16 1/2 → 8								
			-5 ←1/2 -10	-9 1/2 → -5						
-3	-5 1/2 → -3									
	2 ←1/2 4	4 1/2 → 2								
		$\overline{-27}$ $\overline{27}$				1 ←1/2 3	2			
						$\overline{-61}$ $\overline{61}$				
-1	-1									
-143	143			-2 -1						
				$\overline{-47}$ $\overline{47}$						
$\pm\,0.119wL^2$	$\pm\,0.0225wL^2$		$\pm0.0442wL^2$			$\pm\,0.0508wL^2$				

STEP 2   Complete the moment distribution and sum columns for the final end moments. Each line in the work sheet represents a single distribution and carryover operation. The sequence of operations is as follows: joint 1; 3; 5; 2; 4; 5; 3; 4; 2; 3; 5; 2; and 4; All distribution and carryover operations are executed with values rounded to whole numbers. Since wL^2 was taken as $12 \cdot 100 = 1200$, divide final moments by 1200 to express them in terms of wL^2 and draw the final moment diagram.

$0.0565wL^2$ $0.0542wL^2$ $0.0775wL^2$

$0.0225wL^2$ $0.0442wL^2$

$0.0508wL^2$

$0.119wL^2$

12.4   Structures with Joint Displacements _____

As stated earlier, moment distribution cannot be carried out on a structure where joint displacements (or chord rotations) occur since the process is based on the slope-deflection method with no joint translation.

Figure 12.5 In Fig. 12.5a a structure with two members, *A–B* and *B–C*, is to be analyzed. The structure is indeterminate to the third degree, although an indeterminate analysis would only require the establishment of two compatibility equations due to the lack of any horizontal loads. A slope-deflection analysis requires the solution of two equilibrium equations to determine the unknown joint rotation, θ_B, and displacement, v_B. The possibility of a vertical joint displacement at *B* prevents a direct application of the moment distribution process.

A modification of the moment distribution method can be made that will extend its application to the analysis of structures that can undergo joint displacements. This extension of the moment distribution method is illustrated with the analysis of the structure of Fig. 12.5a in a two-step process. The first step is to carry out a moment distribution for the structure under the action of the uniform loading, which can be done if joint *B* is temporarily prevented from displacing vertically as shown in Fig. 12.5b. The solution that is obtained requires a reactive force at *B* where the temporary restraint against displacement was introduced. As summarized in Fig. 12.5c, the moment distribution process has produced a solution for the uniformly loaded structure of Fig. 12.5a, but with the addition of an upward load, *R*, acting at *B* which prevents any vertical displacement of the joint. To obtain the desired final solution, this load must be eliminated from the structure.

The unwanted load, *R*, is removed in the second step of the process. The following operations are related to the same concept as underlies the start of every moment distribution analysis, that all joints of the structure are initially fixed against displacement and rotation. First, with all joints locked against rotation (only *B* in this problem), introduce a small displacement, *v*, downward at *B* as shown in Fig. 12.5d, and then fix *B* against any further displacement. Second, compute the moments on the ends of members *A–B* and *B–C* using the slope-deflection equations (11.3) under the condition of no joint rotations and the specified or fixed downward displacement, *v*. Because no joint rotation was permitted when the displacement, *v*, was introduced, the end moments developed for no rotations and fixed (or specified) end displacement developed at the ends of the members of the structure are in reality another type of fixed end moment. Since no further displacement of *B* is permitted, these

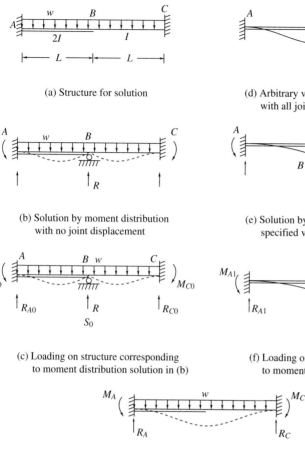

(a) Structure for solution

(d) Arbitrary vertical displacement of B with all joints fixed against rotation

(b) Solution by moment distribution with no joint displacement

(e) Solution by moment distribution with specified vertical displacement at B

(c) Loading on structure corresponding to moment distribution solution in (b)

(f) Loading on structure corresponding to moment distribution solution in (e)

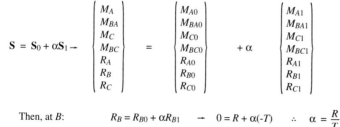

$$\mathbf{S} = \mathbf{S}_0 + \alpha \mathbf{S}_1 \rightarrow \begin{Bmatrix} M_A \\ M_{BA} \\ M_C \\ M_{BC} \\ R_A \\ R_B \\ R_C \end{Bmatrix} = \begin{Bmatrix} M_{A0} \\ M_{BA0} \\ M_{C0} \\ M_{BC0} \\ R_{A0} \\ R_{B0} \\ R_{C0} \end{Bmatrix} + \alpha \begin{Bmatrix} M_{A1} \\ M_{BA1} \\ M_{C1} \\ M_{BC1} \\ R_{A1} \\ R_{B1} \\ R_{C1} \end{Bmatrix}$$

Then, at B: $\qquad R_B = R_{B0} + \alpha R_{B1} \quad \rightarrow \quad 0 = R + \alpha(-T) \quad \therefore \quad \alpha = \dfrac{R}{T}$

(g) Superposition of solutions (c) and (f) to obtain solution (a)

Figure 12.5a–g Moment distribution solution for a structure having a single joint displacement.

moments constitute a set of fixed end moments that can be treated by the moment distribution process. Performing the moment distribution analysis produces the final result, shown in Fig. 12.5e, which gives rise to a reactive force, T, at B. This solution represents the response of the original structure of Fig. 12.5a to a downward concentrated force, T, which yields the displacement, v, at B as shown in Fig. 12.5f.

It is now possible to obtain the final solution to the problem. The desired solution, S, is a matrix of all internal member end moments and external reactions and applied loads for the structure in Fig. 12.5a. It can be obtained by the superposition of the corresponding matrix, S_0, for the solution shown in Fig. 12.5c and some multiple, α, of the matrix, S_1, for the solution shown in Fig. 12.5f. The result can be expressed as

$$S = S_0 + \alpha S_1 \tag{12.1}$$

The terms that appear in these matrices for the structure of Fig. 12.5a are shown in Fig. 12.5g. Selecting from S in Eq. (12.1) the relation for vertical force acting at B, R_B, yields

$$R_B = R_{B0} + \alpha R_{B1} = R + \alpha(-T) \tag{12.2}$$

Since the final value of R_B that acts on the structure must be zero, as seen in Fig. 12.5a, the value of the factor, α, which will yield this result is

$$0 = R - \alpha T \quad \therefore a = \frac{R}{T} \tag{12.3}$$

Figure 12.5g displays the final solution, S, for all external and internal force (or moment) quantities which are obtained by superposition in Eq. (12.1) with the α defined by Eq. (12.3).

The outline to the solution of a problem by moment distribution where a joint displacement can occur has been presented and summarized conceptually in Fig. 12.5. Several details of the solution deserve further explanation. The solution for S_0 shown in Fig. 12.5c is no different than any other moment distribution solution since no joint displacement is permitted. At the end of the moment distribution process for S_0 a final set of internal member end moments is obtained. From these end moments the shear forces on the B ends of members A–B and B–C can be computed from equilibrium considerations of the members and summation of vertical forces at joint B gives the value of the reactive force, R.

Solution S_1 shown in Fig. 12.5f is obtained in the specific two-operation sequence discussed above. In the first operation the magnitude of the end moments for a single member undergoing the arbitrary displacement, v, at the left end (or equivalently, for an arbitrary downward displacement $-v$ at the right end) is shown in Fig. 12.6 and obtained from Eqs. (11.3) as

$$M_{AB} = M_{BA} = M = \frac{2EI}{L}\left\{2 \times 0 + 0 - \frac{3}{L}[0 - (v)]\right\} = \frac{6EIv}{L^2} \quad (12.4)$$

The second operation in the S_1 solution is simply a moment distribution using the moments caused by the arbitrary displacement, v, and calculated by Eq. (12.4) as a set of unbalanced fixed end moments acting on the structure. At the end of this moment distribution process a set of internal member end moments will be obtained. Following the same process as in the solution S_0, the shear forces on the B ends of the two members can be computed, as well as the reactive force, T, required to hold the structure in the vertically displaced position.

Example 12.5 The procedure presented for the structure of Fig. 12.5 is executed for a specific applied load on the structure of 48 kN/m. The step-by-step procedure presents the detailed analysis in the moment distribution solution. The S_0 solution is completed in steps 1 and 2 with the reactive force required to prevent the vertical displacement of B being computed as 144 kN in step 2. Steps 3 to 5 comprise the S_1 solution for an arbitrary vertical displacement at B. In step 3 the value of the moments due to the arbitrary displacement, v, in the S_1 solution is based on a value of v that makes their magnitude a simple or convenient value. In step 6 the value of α is determined from the condition that the magnitude of the vertically applied load at B is zero in the original structure for which solution **S** is desired and the final internal moments for members $A–B$ and $B–C$ are computed using Eq. (12.1).

Example 12.6 The moment distribution technique is used for a more complicated structure that undergoes joint translation. Only a single joint displacement needs to be restrained in this structure if axial deformation effects are neglected. In this example the S_0 solution is particularly simple. Since there are no loads applied on any member between joints, there are no fixed end moments on the ends of the members and hence no moments to distribute. The final end moments in the S_0 solution are all zero, which makes the member shear forces zero and thus the only reactive force that develops is the force, P, at F due to the compressive force $-P$ that is internal to members $D–E$ and $E–F$.

For solution S_1, the value in step 2 of the arbitrary joint displacement U_F is selected to give a magnitude of 1000 to the fixed end moments on the three columns. The selection of the local member coordinate systems for the three columns is identical so that the displacement U_F has the same sign in Eq. (12.4) for each member. After completing the moment distribution in step 3 and computing the three column shear forces and the reactive force at F due to the displacement U_F in step 4, the value of α is computed in step 5 from the condition that the sum of all reactive

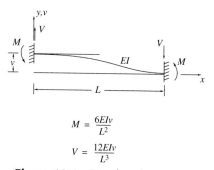

$$M = \frac{6EIv}{L^2}$$

$$V = \frac{12EIv}{L^3}$$

Figure 12.6   Fixed end moments for member with relative end displacement.

Example 12.5

Solve the structure in Fig. 12.5 by moment distribution.

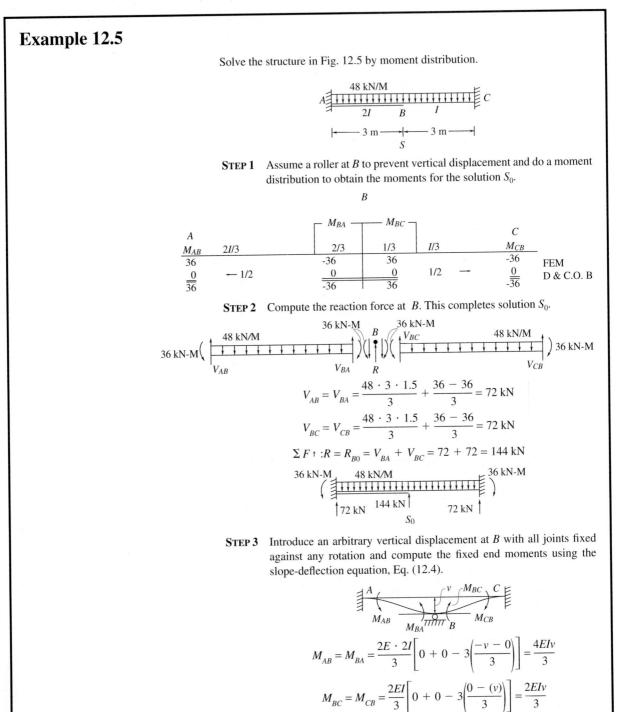

STEP 1   Assume a roller at B to prevent vertical displacement and do a moment distribution to obtain the moments for the solution S_0.

STEP 2   Compute the reaction force at B. This completes solution S_0.

$$V_{AB} = V_{BA} = \frac{48 \cdot 3 \cdot 1.5}{3} + \frac{36 - 36}{3} = 72 \text{ kN}$$

$$V_{BC} = V_{CB} = \frac{48 \cdot 3 \cdot 1.5}{3} + \frac{36 - 36}{3} = 72 \text{ kN}$$

$$\Sigma F \uparrow : R = R_{B0} = V_{BA} + V_{BC} = 72 + 72 = 144 \text{ kN}$$

STEP 3   Introduce an arbitrary vertical displacement at B with all joints fixed against any rotation and compute the fixed end moments using the slope-deflection equation, Eq. (12.4).

$$M_{AB} = M_{BA} = \frac{2E \cdot 2I}{3}\left[0 + 0 - 3\left(\frac{-v - 0}{3}\right)\right] = \frac{4EIv}{3}$$

$$M_{BC} = M_{CB} = \frac{2EI}{3}\left[0 + 0 - 3\left(\frac{0 - (v)}{3}\right)\right] = \frac{2EIv}{3}$$

Example 12.5 (continued)

For convenience take $EIv/3 = 30$; $\therefore M_{AB} = M_{BA} = 120$ kN-m and $M_{BC} = M_{CB} = -60$ kN-m.

STEP 4   Do moment distribution using moments computed in step 3.

		B			
		$\overline{M_{BA}}$	$\overline{M_{BC}}$		
A				C	
M_{AB}		2/3	1/3	M_{CB}	
120		120	-60	-60	FEM
-20	← 1/2	-40	-20	1/2 → -10	D & C.O. B
100		80	-80	-70	

STEP 5   Compute reaction force at B. This completes solution S_1.

$$V_{BA} = -V_{AB} = -\frac{100 + 80}{3} = -60 \text{ kN} \qquad V_{BC} = -V_{CB} - \frac{70 + 80}{3} = -50 \text{ kN}$$

$$\Sigma F \uparrow : R = R_{B1} = -V_{BA} - V_{BC} = 60 + 50 = 110 \text{ kN}$$

100 kN-M   70 kN-M

60 kN   110 kN   50 kN

S_1

STEP 6   Superimpose solutions S_0 and S_1 to obtain the final solution, S. Using the condition of no vertical force at B:

$$R_B = R_{B0} + \alpha R_{B1} \rightarrow 0 = 144 - 110\alpha \qquad \therefore \alpha = \frac{144}{110} = \frac{72}{55}$$

Using this α value yields the final end moments from $\mathbf{S} = \mathbf{S_0} + \alpha\mathbf{S_1}$:

$$\begin{Bmatrix} M_{BA} \\ M_{AB} \\ M_{CB} \\ M_{BC} \end{Bmatrix} = \begin{Bmatrix} -36 \\ 36 \\ -36 \\ 36 \end{Bmatrix} + \frac{72}{55} \begin{Bmatrix} 80 \\ 100 \\ -70 \\ -80 \end{Bmatrix} = \begin{Bmatrix} 68.7 \\ 166.9 \\ -127.6 \\ -68.7 \end{Bmatrix}$$

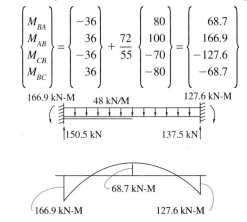

166.9 kN-M   48 kN/M   127.6 kN-M

150.5 kN   137.5 kN

68.7 kN-M

166.9 kN-M   127.6 kN-M

Example 12.6

Obtain the final moments for the frame by moment distribution.

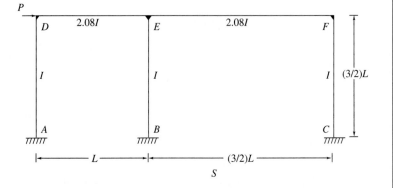

STEP 1 Obtain the solution S_0 by introducing a roller at F to prevent horizontal displacement. By neglecting axial deformations, the joint translations of all joints in the frame are prevented. Since no joint translations or rotations occur, there will be no fixed end moments developed by the applied load P. With all member end moments zero, the reaction components are all zero except at F, which is simply P (i.e., $R_{F0} = P$). This completes the solution S_0.

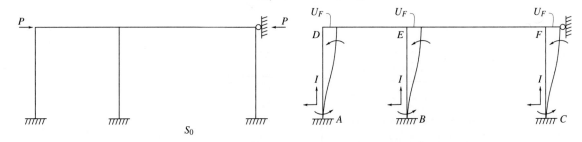

STEP 2 Introduce an arbitrary horizontal displacement at F (and hence also at D and E) with all joints fixed against rotation. Noting the orientation of the individual member coordinate systems, the slope-deflection equations as presented in Eq. (12.4) give fixed end moments that are equal for all columns and of magnitude

$$M = \frac{2EI}{3L/2}\left[0 + 0 \frac{-3}{3L/2}(-U_F - 0)\right] = \frac{8EIU_F}{3L^2}$$

For convenience, define U_F such that $M = 1000$.

Example 12.6 (continued)

STEP 3 Do a moment distribution for the frame locked in the displaced position with horizontal displacement U_F. Distribution factors are computed in the usual way. For example, at E:

$$D_{ED} = \frac{2.08I/L}{2.08\dfrac{I}{L} + \dfrac{2.08I}{(3/2)L} + \dfrac{I}{(3/2)L}} = 0.503$$

$$D_{EB} = \frac{I/\,{}^3\!/_2\,L}{4.133I/L} = 0.061 \qquad D_{EF} = \frac{2.08I/\,{}^3\!/_2\,L}{4.133I/L} = 0.336$$

		D			E				F		
	M_{DA}	M_{DE}		M_{ED}	M_{EB}	M_{EF}		M_{FE}	M_{FC}		
0.243	0.757	2.08I/L	0.503	0.161	0.336	2.08/(3/2)L	0.675	0.325			
1000				1000				1000		FEM
-243	-757	1/2 →	-378		-338	← 1/2	-675	-325	D & C.O.D, F
	-71	← 1/2	-143	-46	-95	1/2 →	-47		D & C.O. E
17	54	1/2 →	27		16	← 1/2	32	15	D & C.O.D, F
	-11	← 1/2	-22	-7	-14	1/2 →	-7		D & C.O E
3	8	1/2 →	4		2	← 1/2	5	2	D & C.O.D, F
777	-777		-3	-1	-2		-692	692	D (only) E
			-515	946	-431				

$$\begin{array}{ccc} \dfrac{I}{(3/2)L} & \dfrac{I}{(3/2)L} & \dfrac{I}{(3/2)L} \\[2mm] A & B & C \\ M_{AD} & M_{BE} & M_{CF} \end{array}$$

1000	1000	1000
-121	-23	-162
8	-3	7
2	-1	1
889	973	846

STEP 4 Compute the shear force at the top of each column and sum to obtain the reaction at F. This reaction and the end moments obtained in the moment distribution comprise the solution S_1.

$$V_1 = \frac{777 + 889}{(3/2)L} = \frac{1110.7}{L} \qquad V_2 = \frac{946 + 973}{(3/2)L} = \frac{1279.3}{L} \qquad V_3 = \frac{692 + 846}{(3/2)L} = \frac{1025.3}{L}$$

AD   BE   CF

$$\Sigma F \rightarrow : R = V_1 + V_2 + V_3 = \frac{3415.3}{L} = R_{F1}$$

Example 12.6 (continued)

STEP 5 Sum S_0 and $\alpha \cdot S_1$ to obtain the final solution S. Note that at F there is no applied force that enables the calculation of α

$$R_F = R_{F0} + \alpha \cdot R_{F1} \rightarrow 0 = P + \alpha\left(-\frac{3415.3}{L}\right) \qquad \therefore \alpha = \frac{PL}{3415.3}$$

STEP 6 Obtain the final moments and draw the moment diagram.

$$\mathbf{S} = \mathbf{S}_0 + \frac{PL}{3415.3} \cdot \mathbf{S}_1$$

$$M_{AD} = 0 + \frac{PL}{3415.3} \cdot 889 = 0.260PL$$

$$M_{DA} = 0 + \frac{PL}{3415.3} \cdot 777 = 0.228PL$$

$$M_{DE} = 0 + \frac{PL}{3415.3} \cdot (-777) = -0.288PL$$

$$M_{ED} = 0 + \frac{PL}{3415.3} \cdot (-515) = -0.151PL$$

$$M_{BE} = 0 + \frac{PL}{3415.3} \cdot 973 = 0.285PL$$

$$M_{EB} = 0 + \frac{PL}{3415.3} \cdot 946 = 0.277PL$$

$$M_{EF} = 0 + \frac{PL}{3415.3} \cdot (-431) = -0.126PL$$

$$M_{FE} = 0 + \frac{PL}{3415.3} \cdot (-692) = -0.203PL$$

$$M_{CF} = 0 + \frac{PL}{3415.3} \cdot 846 = 0.248PL$$

$$M_{FC} = 0 + \frac{PL}{3415.3} \cdot 692 = 0.203PL$$

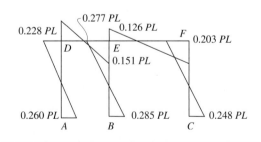

500

forces at F from solutions S_0 and S_1 must be zero. In computing α, the re-active forces are assumed positive to the left. In step 6 the final end moments are computed using Eq. (12.1). Only the member end moments were selected from the $\mathbf{S}$, $\mathbf{S}_0$, and $\mathbf{S}_1$ matrices to illustrate the computations. For numerical calculations with U.S. units, take $L = 20$ ft and $P = 3$ kips; for SI units, take $L = 4$ m and $P = 15$ kN.

The solution of problems where multiple independent joint displacements are possible is outlined conceptually in Fig. 12.7 for a structure where three joint displacements can occur. It is again assumed that axial deformations are negligible.

Figure 12.7 The solution S_0 of Fig. 12.7b is no different from the S_0 of previously worked problems, but the three potential joint displacements must be restrained with three temporary supports. There must be three additional solutions, one for each possible displacement that is re-strained in the S_0 solution. Each solution S_j in Fig. 12.7c to e in computed in the same way as the S_1 solution illustrated in Fig. 12.5 and Examples 12.6 and 12.7. The reactive forces acting at each of the three restraint points must be computed from the set of member end moments for each of the three solutions S_j as shown in Fig. 12.7.

The concept of the superposition of each moment distribution solution to obtain the final desired solution that is expressed in Eq. (12.1) is now extended to the current problem. If the solution S_1 is multiplied by α_1, S_2 by α_2, and S_3 by α_3 and all are added to the solution S_0, the desired final solution $\mathbf{S}$ becomes

$$\mathbf{S} = \mathbf{S}_0 + \alpha_1\mathbf{S}_1 + \alpha_2\mathbf{S}_2 + \alpha_3\mathbf{S}_3 = \mathbf{S}_0 + \mathbf{S}_T\boldsymbol{\alpha} \qquad (12.5)$$

where the matrix $\mathbf{S}_T$ is $m \times 3$ and the matrix $\boldsymbol{\alpha}$ 3×1 for the current problem and m is the sum of the number of member end moments and the three reaction point forces. The latter three equations, which are a subset of the equations in Eq. (12.5), are shown in Fig. 12.7f. The values of the weighting factors, α_j, of the $\boldsymbol{\alpha}$ matrix are obtained by solving these three equations using the values of the three known applied forces in the desired solution $\mathbf{S}$ at the three points where reactive restraints have been introduced and the three reactive forces calculated in the solutions S_0, S_1, S_2, and S_3 to control joint translations. The final solution $\mathbf{S}$ of the problem is obtained from Eq. (12.5) using the values of the α_j obtained from the solution of the equations shown in Fig. 12.7f.

The process depicted in Fig. 12.7 can be generalized to the case of n independent joint translations. The solution $\mathbf{S}$ in Eq. (12.5) will be the sum of $n + 1$ solutions S_j, all but the first one, S_0, being multiplied by a weighting factor, α_j. The size of the matrices $\mathbf{S}_T$ and $\boldsymbol{\alpha}$ will be $m \times n$ and $n \times 1$ respectively. There will be n restraints that have been introduced temporarily in each solution S_j to maintain the specified joint translations in the $n + 1$

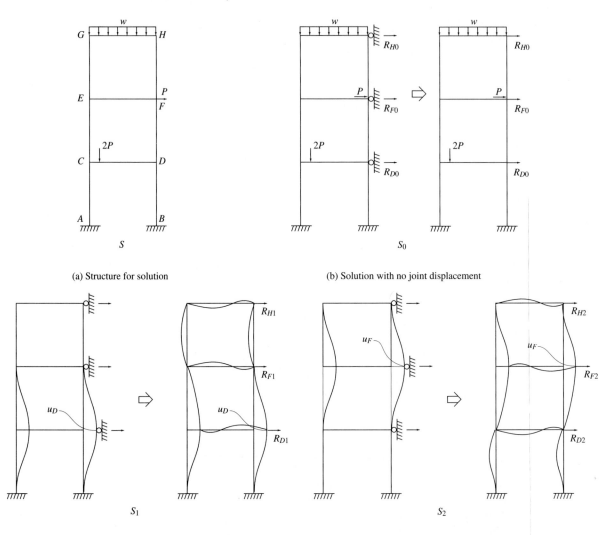

(a) Structure for solution

(b) Solution with no joint displacement

(c) Solution with arbitrary displacement at D

(d) Solution with arbitrary displacement at F

Figure 12.7a–d   Moment distribution solution for structure with multiple joint displacements.

moment distribution solutions. The magnitude of each of the forces at the n points of restraint is known in the final solution **S**, so the n equations of Eq. (12.5) that relate the known force conditions at the temporary reaction points are used to solve for the unknown weighting factors α_j.

From a practical standpoint, solutions to structures by moment distribution where more than one displacement is possible is rarely done. The use of computers obviates the need for these hand computations.

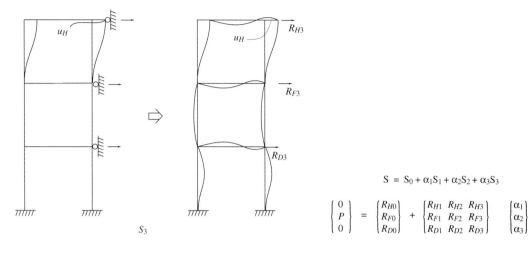

$$S = S_0 + \alpha_1 S_1 + \alpha_2 S_2 + \alpha_3 S_3$$

$$\begin{Bmatrix} 0 \\ P \\ 0 \end{Bmatrix} = \begin{Bmatrix} R_{H0} \\ R_{F0} \\ R_{D0} \end{Bmatrix} + \begin{bmatrix} R_{H1} & R_{H2} & R_{H3} \\ R_{F1} & R_{F2} & R_{F3} \\ R_{D1} & R_{D2} & R_{D3} \end{bmatrix} \begin{Bmatrix} \alpha_1 \\ \alpha_2 \\ \alpha_3 \end{Bmatrix}$$

(e) Solution with arbitrary displacement at H (f) Superposition of solutions and equations for solution for α factors

Figure 12.7e–f Moment distribution solution for structure with multiple joint displacements.

12.5   Modified Relative Stiffness for Symmetric Structures _____

In previous chapters the use of symmetry of a structure has reduced the effort of analysis by about half. It is equally advantageous in moment distribution solutions to use symmetry of the structure and symmetry or antisymmetry of the loading to reduce the computational effort. The division of a structure at the axis of symmetry will often divide members in half as shown in Figs. 12.8 and 12.9, which are taken from Figs. 11.9 and 11.10. The analysis of structures by moment distribution must be modified for these members just as it is in the slope-deflection analyses.

The changes that are necessary in the moment distribution analysis of structures affect only the relative stiffnesses of members that cross the axis of symmetry. Consider first the symmetric structure with symmetric loading as shown in Fig. 12.8a and b. Member D–F is cut in half at M, but the entire member is considered in the analysis by slope-deflection analysis with the end moment, M_{DF}, from Eq. (11.8) being expressed as

$$M_{DF} = 2E\frac{I}{L}\theta_D + \text{FEM}_{DF} = 4E\left(\frac{1}{2}\right)K\theta_D + \text{FEM}_{DF} \qquad (12.6)$$

as is also shown in Fig. 12.8c. Since the uniform loading acts on member D–F, the fixed end moment in this case from Fig. 11.8 is $wL^2/12$.

In Figs. 12.1 and 12.3 the relative stiffness of a member at a joint that undergoes rotation is multiplied by $4E$. In Fig. 12.1 the relative stiffnesses

503

multiplied by $4E$ at the B end of members A–B and B–C are K_1 and K_2, respectively, because the far ends of those members are fixed against rotation. In Fig. 12.3a, the relative stiffnesses multiplied by $4E$ at the B end of members A–B and B–C are K_1 and $\frac{3}{4}K_2$, respectively. Member B–C has a reduced stiffness because the C end of that member is free to rotate an amount $-\theta_B/2$ when a rotation, θ_B, occurs at the B end. These two rotations of the ends of member B–C maintain the condition of zero moment on the C end for any arbitrary rotation θ_B.

In Eq. (12.6) the relative stiffness multiplied by $4E$ of the D end of member D–F is $\frac{1}{2}K$, a relative stiffness that is further reduced from the $\frac{3}{4}K$ of a member that is free to rotate because of a zero moment condition. It is concluded from this analysis that the relative stiffness of a member crossing the axis of symmetry of a symmetrically loaded structure is

$$K_{\text{symmetric}} = \frac{1}{2}\frac{I}{L} = \frac{1}{2}K \tag{12.7}$$

The reduction by one half of the relative stiffness of the member is due to the actions on the ends of the member because of symmetry. When a rotation θ_D occurs at D, symmetry dictates that a rotation $-\theta_D$ occurs at the opposite end, F. These two rotations reduce the effectiveness of the member to resist rotation and the relative stiffness reflects this condition.

Now consider the antisymmetric loading case for the member D–F crossing the axis of symmetry in Fig. 12.9. Again the member D–F is cut in half at M, but the entire member is considered in the slope-deflection analysis with the end moment, M_{DF}, from Eq. (11.9) being expressed as

$$M_{DF} = 6E\frac{I}{L}\theta_D - 12E\frac{I}{L}\frac{v_D}{L} + \text{FEM}_{DF}$$

A moment distribution analysis is always undertaken without joint displacement. With $v_D = 0$, the end moment developed by an antisymmetric set of end rotations θ_D on the D and F ends of member D–F is

$$M_{DF} = 6E\frac{I}{L}\theta_D + \text{FEM}_{DF} = 4E\left(\frac{3}{2}\right)K\theta_D + \text{FEM}_{DF} \tag{12.8}$$

as is also shown in Fig. 12.9c. Since the antisymmetric pair of loads P acts on member D–F, the fixed end moment in this case can be computed with the aid of Fig. 11.8 to be $3PL/32$. Now the relative stiffness multiplied by $4E$ of the D end of member D–F is $\frac{3}{2}K$, a relative stiffness that is greater than the normal $K = I/L$ of a member that fixed against rotation at its opposite end.

It is concluded from this analysis that the relative stiffness of a member crossing the axis of symmetry of an antisymmetrically loaded structure is

$$K_{\text{antisymmetric}} = \frac{3}{2}\frac{I}{L} = \frac{3}{2}K \tag{12.9}$$

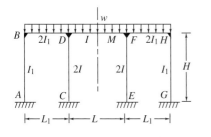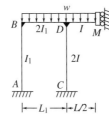

(a) Symmetric structure with symmetric loading

(b) Reduced symmetric structure

$$M_{DF} = \frac{2EI}{L}\,\theta_D + \text{FEM}_{DF} = 4E\left(\frac{1}{2}\right)\frac{I}{L}\,\theta_D + \text{FEM}_{DF} = 4E\left(\frac{1}{2}\right)K\theta_D + \frac{wL^2}{12}$$

$\therefore$ Modified relative stiffness for symmetric loading $= \left(\dfrac{1}{2}\right)K$

(c) End moment for member *DF* from Fig. 11.9 and modified relative stiffness for moment distribution method

Figure 12.8a–c Modified relative stiffness for moment distribution method for a member crossing axis of symmetry of symmetrically loaded structure.

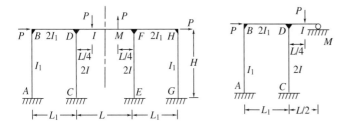

(a) Symmetric structure with antisymmetric loading

(b) Reduced symmetric structure

$$M_{DF} = \frac{6EI}{L}\,\theta_D + \text{FEM}_{DF} = 4E\left(\frac{3}{2}\right)\frac{I}{L}\,\theta_D + \text{FEM}_{DF} = 4E\left(\frac{3}{2}\right)K\theta_D + \frac{3PL}{32}$$

$\therefore$ Modified relative stiffness for antisymmetric loading $= \left(\dfrac{3}{2}\right)K$

(c) End moment for member *DF* from Fig. 11.10 without displacement and modified relative stiffness for moment distribution method

Figure 12.9a–c Modified relative stiffness for moment distribution method for a member crossing axis of symmetry of antisymmetrically loaded structure.

505

The increase by one half of the relative stiffness of the member is due to the actions on the ends of the member because of antisymmetry. When a rotation θ_D occurs at D, antisymmetry dictates that a rotation θ_D also occur at the opposite end, F. These two rotations increase the effectiveness of the member to resist rotation and the relative stiffness reflects this condition.

The relative stiffnesses defined in Eqs. (12.7) and (12.9) for members crossing the axis of symmetry of symmetrically and antisymmetrically loaded structures, respectively, are to be used in moment distribution analyses of such structures. The next example illustrates the analysis of a structure with antisymmetric loadings.

> **Example 12.7** A simple one-bay, one-story frame of arbitrary symmetric geometry and material properties is antisymmetrically loaded by the two forces, P. Axial deformations are again neglected, so that the displacements of joints 2 and 3 are the same and no vertical displacement of 2 can occur. Due to antisymmetry of the loading, the structure is reduced to the left half shown, which can undergo a horizontal displacement of joints 2 and M. The antisymmetric deformations of member 2–3 lead to the conclusion that there is no vertical displacement at M. Because of the displacement of 2, two moment distribution analyses are required. The two solutions are shown in step 1, but the analysis for S_0 is very simple, as there are no loads applied between the joints of the members. For S_0, there are no member end moments and the reaction force, R_0, at the constraint at M is $-P$.
>
> The solution S_1 is a typical specified displacement solution with the magnitude of the specified displacement, U, in step 3 taken to cause end moments at joints 1 and 2 of 1000. Because of the antisymmetry of the loading, member 2–3 has a relative stiffness, as specified in Eq. (12.9), of $\frac{3}{2}(\beta I/L)$. The distribution factors computed in step 2 are expressed in terms of the nondimensional parameter, γ, which is defined as $\gamma = 3\beta H/2L$. Note that the geometric proportions of the frame and the material relations between members are captured in γ.
>
> Once the distribution factors and FEMs for the structure have been established, the solution S_1 in steps 4 and 5 is the same as in any moment distribution problem with joint displacements. The value of α is determined in step 6 and the final member end moments are simply α times the moments found in S_1 since all end moments in the S_0 solution are zero. The final moment diagram is drawn using the antisymmetric conditions in the problem. For numerical calculations using U.S. units, take $L = 30$ ft, $H = 15$ ft, $P = 3$ kips, and $\beta = 4$; for SI units take $L = 10$ m, $H = 5$ m, $P = 15$ kN, and $\beta = 4$.

The solution presented in Example 12.7 demonstrates the value of symmetry in a moment distribution analysis. It illustrates that moment distribution

Example 12.7

Use moment distribution to obtain the final moment diagram for the symmetric structure with the antisymmetric load. Use antisymmetry to simplify the process.

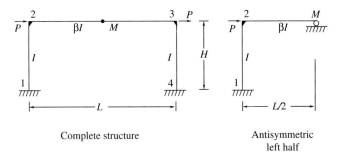

Complete structure   Antisymmetric
left half

STEP 1   Because of the horizontal translation of joints 2 and M, the solution requires an S_0 and an S_1 solution by moment distribution. Obtain solution S_0 by introducing a pin support at M to control horizontal displacement. For the S_0 solution, this prevents the translation of joint 2, and since there are no joint translations or rotations, there will be no fixed end moments developed by the applied load P. With all member end moments zero, the reaction components are all zero except at M, which is simply P (i.e., $R_0 = -P$). This completes solution S_0.

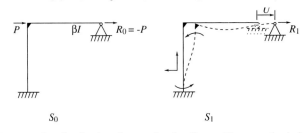

S_0 S_1

STEP 2   Compute the distribution factors for the frame. Because the height, width, moment of inertia of the column, and beam are defined arbitrarily, it is convenient to define the parameter $\gamma = 3\beta H/2L$. Note that the relative stiffness of member 2–M–3 is taken as $\frac{3}{2}\beta I/L$ because of the antisymmetry of the frame response.

$$D_{21} = \frac{I/H}{I/H + 3\beta I/2L} = \frac{1}{1 + \gamma} \quad D_{23} = \frac{3\beta I/2L}{I/H + 3\beta I/2L} = \frac{\gamma}{1 + \gamma} \quad \gamma = \frac{3\beta H}{2L}$$

STEP 3   Introduce an arbitrary horizontal displacement at M (and hence also at 2) with all joints fixed against rotation. Noting the orientation of the coordinate system for member 1–2, Eq. (12.4) gives the fixed end moments:

$$M_{12} = M_{21} = \frac{2EI}{H}\left[0 + 0\frac{-3}{H}(-U - 0)\right] = \frac{6EIU}{H^2}$$

For convenience, define U such that $M_{12} = M_{21} = 1000$.

Example 12.7 (continued)

STEP 4 Set up the moment distribution work sheet and enter the fixed end moments. Only joint 2 has an unbalanced moment, so a single distribution is all that is needed and the sum of the moments in the columns gives the final member end moments.

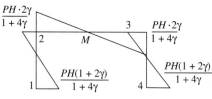

1			2			Operation:
M_{12}			$-M_{21}$	$M_{23}-$		
		I/H	$\dfrac{1}{1+\gamma}$	$\dfrac{\gamma}{1+\gamma}$	$3\beta I/2L$	
1000			1000			FEM
$\dfrac{-500}{1+\gamma}$		$\overleftarrow{\quad}$ $1/2$	$\dfrac{-1000}{1+\gamma}$	$\dfrac{-1000\gamma}{1+\gamma}$		D&C.O. 2
$\dfrac{500\,(1+2\gamma)}{1+\gamma}$			$\dfrac{1000\gamma}{1+\gamma}$	$\dfrac{-1000\gamma}{1+\gamma}$		

STEP 5 Use a free-body diagram of the column and member 2–M to obtain reaction R_1 from the shear in the column.

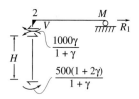

$$\Sigma F\rightarrow : R_1 - V = 0 \qquad \therefore R_1 = V$$

$$\Sigma M:\quad \frac{1000\gamma}{1+\gamma} + \frac{500(1+2\gamma)}{1+\gamma} - VH = 0$$

$$\therefore V = \frac{500(1+4\gamma)}{(1+\gamma)H} = R_1$$

STEP 6 The superposition of solutions S_0 and S_1 gives the final solution S. Using the condition of no applied load at M, the value of α that multiplies the S_1 solution is obtained and the final moment diagram drawn using antisymmetry of the loading.

Final moment diagram

$$\mathbf{S} = \mathbf{S}_0 + \alpha \mathbf{S}_1$$

$$\text{At } M:\ 0 = R + \alpha R_1$$

$$\therefore \alpha = \frac{-R}{R_1} = -(-P)\frac{(1+\gamma)H}{500(1+4\gamma)}$$

$$M_{12} = \frac{PH(1+2\gamma)}{1+4\gamma} \qquad M_{21} = \frac{2PH\gamma}{1+4\gamma} = -M_{23}$$

can be used to solve problems in which one or more parameters of the mathematical model of the structure remains numerically unspecified.

12.6   Summary and Limitations

The moment distribution method of analysis is presented as a means of solving with hand computations highly constrained, and hence highly indeterminate, structures in a manner that is relatively straightforward. The number of computations required in any solution is quite small, and if the number of joint displacements of the structure is of the order of one or possibly two, the moment distribution solution can be executed quite quickly.

The moment distribution method is based on the slope-deflection method of analysis, but without the intermediate step of solving simultaneous equations for unknown joint rotations and displacements. The method is presented in the context of the analysis of the two member structure of Fig. 12.1, but is quickly extended to multiple member structures without joint translations. When members with ends that are free to rotate because of a specified end moment condition are present, a modification of the technique in which the relative stiffnesses of these members are reduced to three quarters of the normal value of I/L speeds its execution.

Every moment distribution analysis requires that no joint displacements be present. If one or more joint displacements can occur in a structure, the analysis is done in stages. In the initial stage all joint displacements are eliminated by the introduction of reactions that prevent any displacement. The moment distribution is executed and the resulting member end moments are used to compute the reaction components at each point where a reaction was introduced. This completes the first stage and yields the solution S_0. A matrix $\mathbf{S}_0$ is created that is a column matrix of all internal member end moments and reaction components at the joints where restraints have been introduced.

Each succeeding stage of the analysis provides a moment distribution solution for a specified or fixed displacement at a different reaction point. A single specified displacement is introduced at a reaction point with all joints in the structure fixed against rotation. This action develops fixed end moments in those members that are connected to the joint with the specified displacement, which are computed using Eq. (12.4). The fixed end moments created in these members present a set of unbalanced moments on the structure and moment distribution is used to eliminate these moments just as is done in the S_0 solution. The resulting end moments are used to compute all reaction components, which completes the solution S_j of the stage. The matrix $\mathbf{S}_j$ is created, which is a column matrix of all internal member end moments and reaction components at the joints where restraints have been introduced for stage j.

509

The final solution, **S**, of the problem is a column matrix of all the final internal member end moments and applied forces (if any) at the joints where restraints have been introduced to prevent joint displacements. In Eq. (12.5) the solution **S** is the superposition of the matrices of the initial solution, S_0, and the solution for each stage, S_j, multiplied by a weighting factor α_j. The α_j are obtained by solving those equations of Eq. (12.5) that relate the known values in **S** of the applied forces at the points where displacement constraints were introduced to the reaction forces obtained in the solutions S_0 and all S_j. This process is straightforward but lengthy and tedious for more than one or two possible joint displacements.

The moment distribution method is extended to symmetric structures that have symmetric and antisymmetric loading. Members that cross an axis of symmetry have modified relative stiffnesses to account for symmetric or antisymmetric behavior. For symmetric loading the relative stiffness is defined as $\frac{1}{2}I/L$ in Eq. (12.7). Antisymmetric loading changes the relative stiffness to $\frac{3}{2}I/L$ as defined in Eq. (12.9). As with other methods of analysis, the use of symmetry can greatly reduce the computational work for a problem.

The moment distribution method is based on the slope-deflection equations, so that all of the assumptions and limitations that are listed at the end of Chapter 11 are applicable to moment distribution as well. Some additional limitations of a practical or procedural nature are also listed below.

- All moment distribution calculations assume that no joint displacements can occur.

- A special staged solution process is required if joint displacements can occur.

- The axial deformations of members of frames are assumed to be zero.

- The moment distribution process can be modified to include axial deformations, but it complicates the process to the point of making it useless.

Some questions that are worthy of consideration when using the moment distribution method of analysis are the following:

- The moment distribution method assumes no axial forces in any of the members. Can the method be extended to include the effects of axial force?

- The moment distribution method can be extended to incorporate nonprismatic members. What equations would form the basis of moment distribution for structures with these members?

- How many significant figures should be used in a moment distribution analysis?

- Will a moment distribution analysis of a continuous beam give the same results as those from an analysis by a computer program that is based on a different formulation of the structure?

- Will a moment distribution analysis of a simple frame give the same results as those from an analysis by a computer program that is based on a different formulation of the structure?

PROBLEMS

12.1 Analyze and draw the final moment diagram for the structure shown. Use moment distribution. Neglect axial deformations.

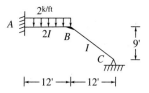

12.2 Using the moment distribution method of analysis, draw the final moment diagram for all members (E constant).

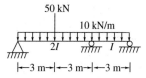

12.3 Analyze and draw the final moment diagram for the structure shown. Use moment distribution. Neglect axial deformations.

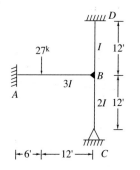

12.4 Draw the final moment diagram for the structure shown using the moment distribution method. Neglect axial deformations.

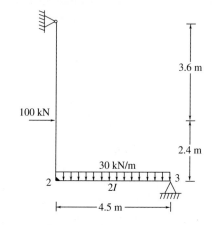

12.5 Draw the final moment diagram for the structure shown using the moment distribution method (E constant). Neglect axial deformations.

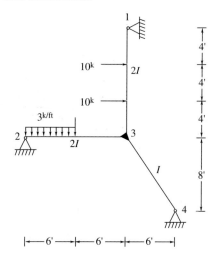

12.6   Draw the final moment diagram for the structure shown using the moment distribution method. Neglect axial deformations (E constant).

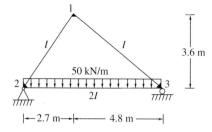

12.7   Obtain the final moment diagram for the structure shown. Use the moment distribution method (E constant).

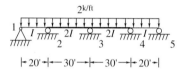

12.8   For the structure shown, obtain the final moment diagram due to the applied loading. Use the moment distribution method (E constant). Neglect axial deformations.

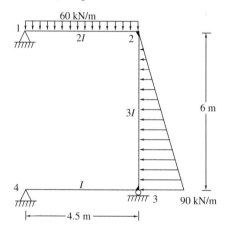

12.9   Draw the final moment diagram for the structure shown. Use the moment distribution method. Neglect axial deformations (E constant).

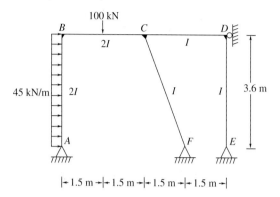

12.10   Draw the final moment diagram for the structure shown. Use the moment distribution method. Neglect axial deformations (E constant).

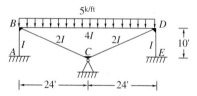

12.11   Using the moment distribution method, draw the final moment diagram for all members, and compute all reaction components (E constant). Neglect axial deformations.

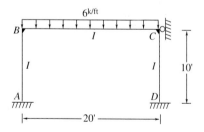

12.12 For the structure shown, draw the final moment diagram. Use the moment distribution method. Neglect axial deformations (*E* constant).

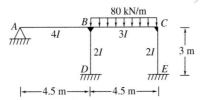

12.13 The cables in the *frame* shown have no bending stiffness. Draw the moment diagram for members *A–B*, *B–C*, and *C–D*. Solve by moment distribution. Neglect axial deformations (*E* constant).

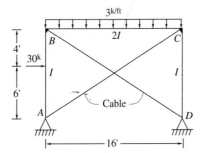

12.14 Draw the final moment diagram for the structure. Neglect axial deformations (*E* constant).

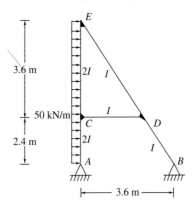

12.15 Draw the final moment diagram for the structure. Neglect axial deformations (*E* constant).

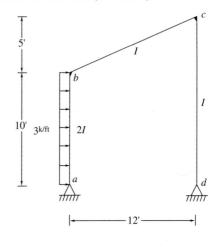

12.16 Obtain the final moment diagram for the frame loaded as shown. Neglect axial deformations (*E* constant).

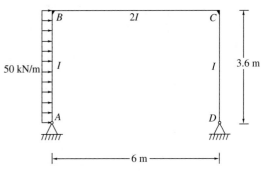

12.17 Draw the final moment diagram for the two-span beam. The material modulus is the same for all members.

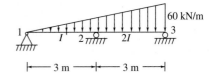

12.18 Draw the final moment diagram for the structure shown. Neglect shortening of members due to axial load (*E* constant).

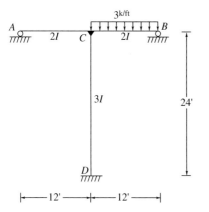

12.19 For the structure shown, discuss how the effects of axial deformations could be included in a moment distribution solution of the structure (*EI* constant).

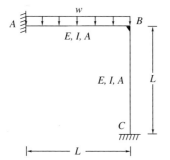

12.20 Solve by moment distribution and draw the moment diagram. Neglect axial deformations (*E* constant).

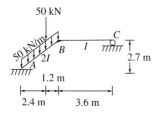

12.21 Using the moment distribution method, draw the final moment diagram for all members (*E* constant). Neglect axial deformations.

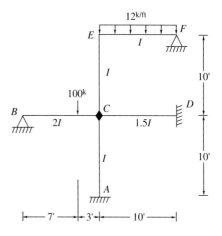

12.22 Using moment distribution, determine and draw the moment diagram for the structure. Neglect axial deformations (*E* constant).

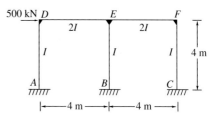

12.23 For the structure shown, obtain the final moment diagram due to the applied loading. Use the moment distribution method (*E* constant) Neglect axial deformations.

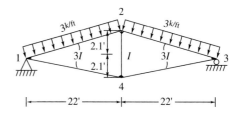

12.24   Analyze and draw the final moment diagram for the structure shown using the moment distribution method. Neglect axial deformations.

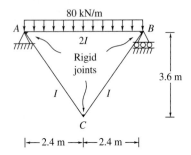

12.25   Draw the final moment diagram for the structure shown. Neglect axial deformations.

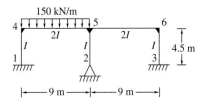

For each of the structures shown, draw the final moment diagram. Use moment distribution as a basis of solution and take axial deformations as negligibly small (*E* constant).

12.26

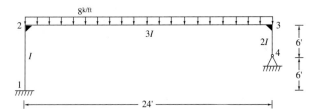

12.27

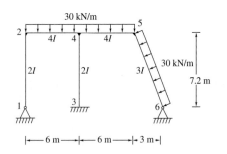

12.28

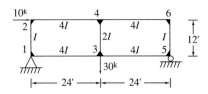

12.29   Solve by moment distribution. Neglect axial deformations (*E* constant).

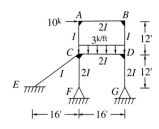

12.30   Solve by moment distribution. Neglect axial deformations (*E* constant).

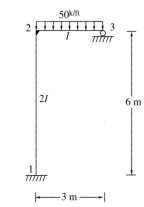

12.31   Solve by moment distribution. Neglect axial deformations (*E* constant).

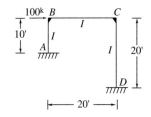

Influence Lines for Statically Indeterminate Structures

The concept of an influence line and the variety of uses for influences lines have been presented in Chapter 8. Only statically determinate structures were considered in that chapter, and the influence lines obtained were all linear, or piecewise linear. Influence lines were used to fulfill two primary functions: They were used to compute the magnitude of some force or moment action in a structure due to a specified loading; and they were used to position a specific configuration of loads on a structure so that the maximum value of a force or moment action is obtained for that loading. Once drawn, the influence lines also presented graphically the variation of a force or moment action with position of a single unit load on the structure which served to illustrate structural behavior under the action of moving loads.

Influence lines for statically determinate structures are easily obtained from the definition of an influence line as shown in Chapter 8. Use of the virtual work principle also provides a simple means of obtaining the shape of an influence line as well as the ordinates for some of the simpler forms. The extension of these methods to indeterminate structures is explored in this chapter. In particular, the use of the Müller–Breslau principle, a principle derived from energy considerations, can reduce significantly the effort to obtain influence lines. More important, when using this principle, influence lines can be sketched from a simple evaluation of structural behavior.

517

The Müller–Breslau principle is based on another energy principle known as Betti's law. A derivation of Betti's law is presented in Section 13.2. A special application of Betti's law called Maxwell's law of reciprocal deflections is derived which establishes the symmetry of the equations used in indeterminate analysis in Chapter 10. Finally, the Müller–Breslau principle is established from yet another special application of Betti's law. The application of this principle enables the representation of influence lines for stress quantities by the deflected shape of a structure when it is deformed in a particular manner.

13.1   Development of Influence Lines by Direct Computation and Their Use

For indeterminate structures the direct computation of the influence line from its definition is very straightforward, although the process tends to be rather tedious. In beams and small frames where joint translation does not occur or where only one joint translation occurs, the moment distribution technique provides a relatively simple means of analysis. The following example for a two-span continuous beam illustrates this approach for the computation of the influence lines for the moment at its center support and the reaction at its left end.

Example 13.1   The analysis of a two-span continuous beam for an arbitrary position of the unit load in the left span is sufficient to establish the required influence lines. This is due to the symmetry of the structure, which establishes that the moment at B for a position of the unit load in the right span is the same as the moment at B for the symmetric position of the load in the left span. Moment distribution is used, and only one distribution at A to eliminate the fixed end moment there and one distribution at B to establish moment equilibrium are required.

The completion of the moment distribution enables the direct computation of the influence line for the moment over the center support. The influence line for the reaction at A requires the use of the free body shown in step 3. The concept of the unit load being in or outside the free body established in Chapter 8 is used to complete the computation. Since the structure is indeterminate to only the first degree, the influence line for the reaction can be obtained from equilibrium once the influence line for the moment at B is known.

For plane trusses the use of the definition requires an indeterminate analysis, with the analysis being carried out with the unit load positioned at each joint along the loading path in the structure. Example 13.2 illustrates the computation for a two-span truss.

Example 13.2 The two-span truss is loaded along its bottom chord by the floor system. The structure is one degree indeterminate and is analyzed by the method of consistent deformations presented in Chapter 10. The structure is symmetric, so that if a load is placed at joint L_1, the forces in the members of span L_4 to L_8 for this position of the load are the same as those developed in the symmetric members of span L_0 to L_4 when the same load is positioned at joint L_7. Because of this symmetry, an analysis of the structure for every possible position of a unit load can be completed by placing the unit load at the L_1, L_2, and L_3 joints of the L_0-to-L_4 span. The three positions of the unit load requires three indeterminate analyses.

The analysis follows the same sequence of steps as those in Example 10.1 except that there are three positions of the unit load so that three deformation systems are established as shown in steps 1 and 2. In step 3 the force system that is to be used to compute the relative displacement u_r in the redundant member U_3–U_4 is created. In steps 4 and 5 the analysis for the value of the redundant force F is completed with the aid of the spreadsheet shown in step 4. In the spreadsheet, note that the deformation system strains and hence also the internal work are all multiplied by E.

The member forces for the three positions of the unit load are given in the last three columns of the spreadsheet. When the unit load is at L_0, L_4, or L_8, the effect of the unit load passes directly into the reactions at those points without causing forces in the members of the structure. This information and the symmetry of the structure provides the forces in all members of the structure for every possible position of the unit load.

The influence line for the reaction at L_0 can be constructed in two different ways. A free body of the left portion of the structure, created by passing a section through the three members U_3–U_4, U_3–L_4, and L_3–L_4, can be used with the summation of moments about L_4 to obtain the reaction in terms of the unit load and the force in U_3–U_4. This is conceptually the same as the procedure used in Example 13.1 to obtain the influence line for the left reaction of the beam. The second way of obtaining the influence line is to isolate joint L_0 by cutting through members U_0–L_0 and L_0–L_1 and summing forces vertically. Except for the unit load positioned at L_0, equilibrium gives the reaction R at L_0 as $R = -F_{U0L0}$. This procedure is followed in the example.

When the unit load is at L_0, all member forces in the structure are zero, and the upward reaction, R, is 1. When the unit load is at L_1, the force in member U_0–L_0 is -0.705, so equilibrium requires that $R = -(-0.705) = 0.705$. The value of R is obtained similarly for the unit load at L_2 and L_3. The value of R when the unit load is at L_4 or L_8 is zero. When the unit load is at L_5, the force in member U_0–L_0 is 0.0626, which is obtained by symmetry as the force in member U_8–L_8 for the unit load at L_3. The reaction R is thus -0.0626. The same procedure is

Example 13.1

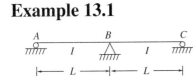

Two-span continuous beam

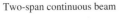

Unit load in span A–B

Obtain the influence line for the moment at B, using moment distribution and the definition of an influence line.

STEP 1 Do moment distribution for a unit lead in span A–B a distance x from A. Note that from symmetry the distribution factors at B are both $\frac{1}{2}$. The fixed end moments are computed from Fig. 11.8. The first distribution is at A with the carryover to B, and the second distribution at B completes the process. The end moments for the members are obtained by summation and are positive when counterclockwise.

A		B			C
$\begin{array}{c}\text{—}M_{AB}\text{—}\\ 1\end{array}$	$(3/4)I/L$	$\begin{array}{c}\text{—}M_{BA}\text{—}\\ 1/2\end{array}$	$\begin{array}{c}M_{BC}\text{—}\\ 1/2\end{array}$	$(3/4)I/L$	$\begin{array}{c}\text{—}M_{CB}\text{—}\\ 1\end{array}$

$$\frac{x(L-x)^2 \cdot 1}{L^2} \qquad\qquad -\frac{x^2(L-x)\cdot 1}{L^2}$$

$$-\frac{x(L-x)^2 \cdot 1}{L^2} \qquad 1/2\rightarrow \quad -\frac{x(L-x)^2 \cdot 1}{2L^2}$$

$$\underline{0}$$

$$\frac{2x^2(L-x)\cdot 1 + x(L-x)^2\cdot 1}{4L^2} \quad\bigg|\quad \frac{2x^2(L-x)\cdot 1 + x(L-x)^2\cdot 1}{4L^2}$$

$$-\frac{x(L^2-x^2)\cdot 1}{4L^2} \qquad\qquad \frac{x(L^2-x^2)\cdot 1}{4L^2} \qquad\qquad \overline{\overline{0}}$$

STEP 2 When the unit load is in B–C, the moment at B can be obtained from the symmetric placement of the load in the span A–B. Therefore, the values of the influence line ordinates are calculated using the end moment expression for M_{BA} from the moment distribution and are given in the table for intervals of $0.1L$, where x is measured from A for A–B and B for B–C. The influence line is drawn below.

STEP 3 The influence line for the reaction R_A is obtained from a free-body diagram of the member A–B. By summing moments about B and recognizing the unit load being in or outside the free body, the influence line is obtained using the ordinates already obtained in step 2 for the influence line for M_B. The ordinates of the influence line for R_A are tabulated in the table for intervals of $0.1L$, where x is measured the same as for M_B. The influence line is drawn below.

Example 13.1 (continued)

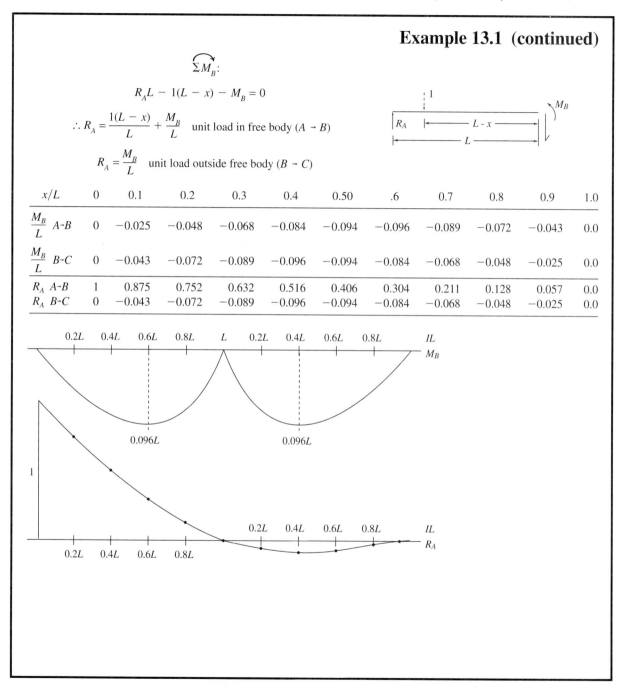

$\overset{\curvearrowright}{\Sigma M_B}:$

$$R_A L - 1(L - x) - M_B = 0$$

$$\therefore R_A = \frac{1(L - x)}{L} + \frac{M_B}{L} \quad \text{unit load in free body } (A \rightarrow B)$$

$$R_A = \frac{M_B}{L} \quad \text{unit load outside free body } (B \rightarrow C)$$

x/L	0	0.1	0.2	0.3	0.4	0.50	.6	0.7	0.8	0.9	1.0
$\frac{M_B}{L}$ A-B	0	−0.025	−0.048	−0.068	−0.084	−0.094	−0.096	−0.089	−0.072	−0.043	0.0
$\frac{M_B}{L}$ B-C	0	−0.043	−0.072	−0.089	−0.096	−0.094	−0.084	−0.068	−0.048	−0.025	0.0
R_A A-B	1	0.875	0.752	0.632	0.516	0.406	0.304	0.211	0.128	0.057	0.0
R_A B-C	0	−0.043	−0.072	−0.089	−0.096	−0.094	−0.084	−0.068	−0.048	−0.025	0.0

Example 13.2

Obtain the influence line for the reaction at L_0 and the force in member U_3–L_4 for the two-span truss shown. Loads enter the truss through a floor system attached to the bottom chord joints. The member areas are: top and bottom chord members, 6 in²; vertical members, 4 in²; and diagonal members, 5 in². ll members have the same material modulus E.

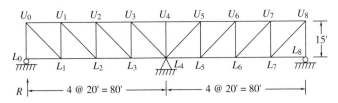

STEP 1 The truss is indeterminate to the first degree, so the method of consistent deformations is used to carry out the analysis. The symmetry of the structure will reduce the analysis to three loading configurations, one for the position of the unit load at each lower chord joint between L_0 and L_4. The redundant selected for the analysis is the force in the top chord member U_3–U_4. The three internal member force distributions for each position of the unit load and for the unknown axial force, F, in member U_3–U_4 are shown below.

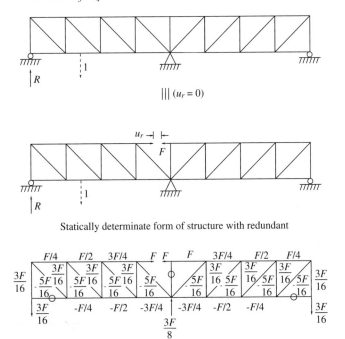

Statically determinate form of structure with redundant

Member forces due to F

Example 13.2 (continued)

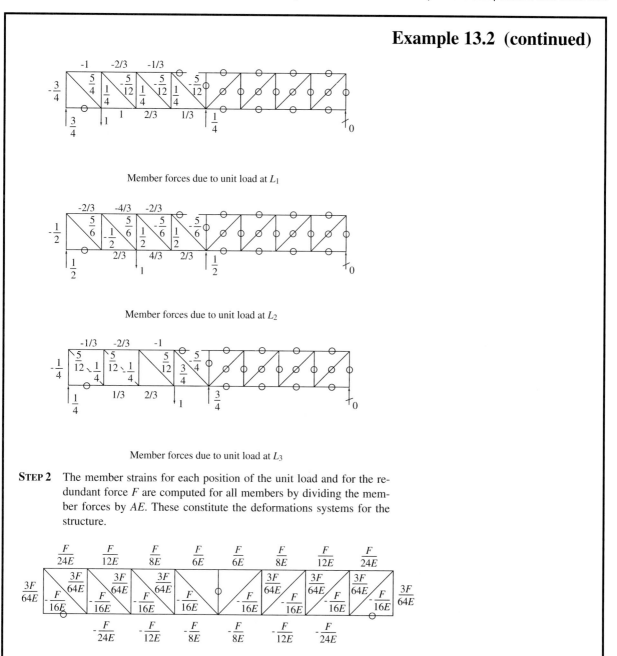

Member forces due to unit load at L_1

Member forces due to unit load at L_2

Member forces due to unit load at L_3

STEP 2 The member strains for each position of the unit load and for the redundant force F are computed for all members by dividing the member forces by AE. These constitute the deformations systems for the
structure.

Strains due to redundant force F

523

Example 13.2 (continued)

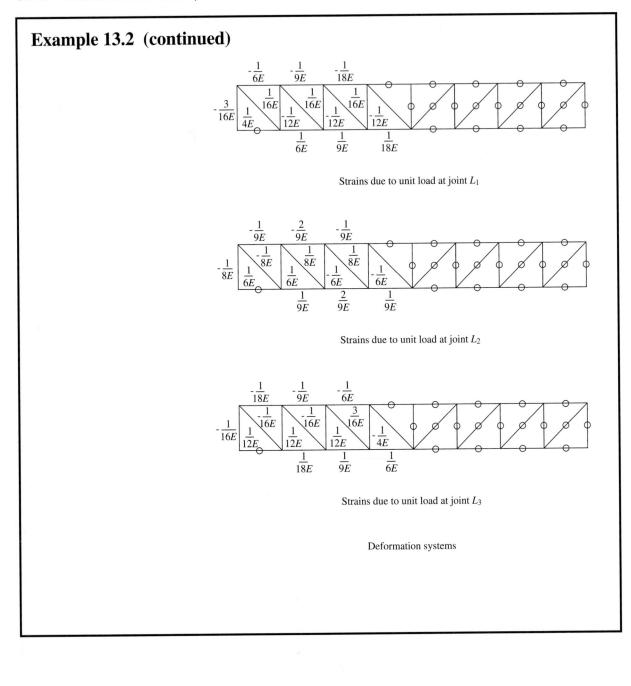

Strains due to unit load at joint L_1

Strains due to unit load at joint L_2

Strains due to unit load at joint L_3

Deformation systems

Example 13.2 (continued)

STEP 3   The virtual force system, δV is established to compute the relative displacement caused by the deformation systems at the cut in member U_3–U_4.

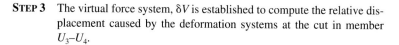

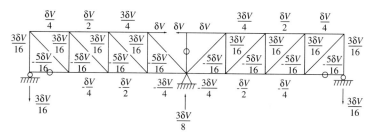

Force system (and member forces)

STEP 4   With the force system acting, the deformations caused by the unit load at L_1 and the redundant force F are applied to the structure and the complementary virtual work principle used to compute the corresponding internal and external work. The same procedure is repeated for the unit load at L_2 and L_3. These operations are summarized in the spreadsheet shown below.

525

Example 13.2 (continued)

MEMBER	LENGTH	AREA SQ.	FORCE SYSTEM	DEFORMATION SYSTEM STRAINS (xE) UNIT LOAD AT:				INTERNAL WORK Wi δFv (Fp/AE) (xE) UNIT LOAD AT:				FINAL FORCES (K) UNIT LOAD AT:		
	feet	IN.	δFv	L1	L2	L3	F	L1	L2	L3	F	L1	L2	L3
U0U1	20	6	0.25 δV	-0.17	-0.11	-0.06	0.042 F	-0.83 δV	-0.56 δV	-0.28 δV	0.208 δV*F	-0.94	-0.571	-0.25
U1U2	20	6	0.5 δV	-0.11	-0.22	-0.11	0.083 F	-1.11 δV	-2.22 δV	-1.11 δV	0.833 δV*F	-0.547	-1.143	-0.5
U2U3	20	6	0.75 δV	-0.06	-0.11	-0.17	0.125 F	-0.83 δV	-1.67 δV	-2.5 δV	1.875 δV*F	-0.154	-0.38	-0.75
U3U4	20	6	1 δV	0	0	0	0.167 F	0	0	0	3.333 δV*F	0.2385	0.3816	0.3339
U4U5	20	6	1 δV	0	0	0	0.167 F	0	0	0	3.333 δV*F	0.2385	0.3816	0.3339
U5U6	20	6	0.75 δV	0	0	0	0.125 F	0	0	0	1.875 δV*F	0.1789	0.2862	0.2504
U6U7	20	6	0.5 δV	0	0	0	0.083 F	0	0	0	0.833 δV*F	0.1192	0.1908	0.1669
U7U8	20	6	0.25 δV	0	0	0	0.042 F	0	0	0	0.208 δV*F	0.0596	0.0954	0.0835
L0L1	20	6	0	0	0	0	0	0	0	0	0	0	0	0
L1L2	20	6	-0.25 δV	0.167	0.111	0.056	-0.04 F	-0.83 δV	-0.56 δV	-0.28 δV	0.208 δV*F	0.9404	0.5713	0.2499
L2L3	20	6	-0.5 δV	0.111	0.222	0.111	-0.08 F	-1.11 δV	-2.22 δV	-1.11 δV	0.833 δV*F	0.5474	1.1425	0.4997
L3L4	20	6	-0.75 δV	0.056	0.111	0.167	-0.13 F	-0.83 δV	-1.67 δV	-2.5 δV	1.875 δV*F	0.1545	0.3805	0.7496
L4L5	20	6	-0.75 δV	0	0	0	-0.13 F	0	0	0	1.875 δV*F	-0.179	-0.286	-0.25
L5L6	20	6	-0.5 δV	0	0	0	-0.08 F	0	0	0	0.833 δV*F	-0.119	-0.191	-0.167
L6L7	20	6	-0.25 δV	0	0	0	-0.04 F	0	0	0	0.208 δV*F	-0.06	-0.095	-0.083
L7L8	20	6	0	0	0	0	0	0	0	0	0	0	0	0
L0U0	15	4	0.188 δV	-0.19	-0.13	-0.06	0.047 F	-0.53 δV	-0.35 δV	-0.18 δV	0.132 δV*F	-0.705	-0.428	-0.187
L1U1	15	4	0.188 δV	0.063	-0.13	-0.06	0.047 F	0.176 δV	-0.35 δV	-0.18 δV	0.132 δV*F	0.2947	-0.428	-0.187
L2U2	15	4	0.188 δV	0.063	0.125	-0.06	0.047 F	0.176 δV	0.352 δV	-0.18 δV	0.132 δV*F	0.2947	0.5715	-0.187
L3U3	15	4	0.188 δV	0.063	0.125	0.188	0.047 F	0.176 δV	0.352 δV	0.527 δV	0.132 δV*F	0.2947	0.5715	0.8126
L4U4	15	4	0	0	0	0	0	0	0	0	0	0	0	0
L5U5	15	4	0.188 δV	0	0	0	0.047 F	0	0	0	0.132 δV*F	0.0447	0.0715	0.0626
L6U6	15	4	0.188 δV	0	0	0	0.047 F	0	0	0	0.132 δV*F	0.0447	0.0715	0.0626
L7U7	15	4	0.188 δV	0	0	0	0.047 F	0	0	0	0.132 δV*F	0.0447	0.0715	0.0626
L8U8	15	4	0.188 δV	0	0	0	0.047 F	0	0	0	0.132 δV*F	0.0447	0.0715	0.0626
U0L1	25	5	-0.31 δV	0.25	0.167	0.083	-0.06 F	-1.95 δV	-1.3 δV	-0.65 δV	0.488 δV*F	1.1755	0.7141	0.3123
U1L2	25	5	-0.31 δV	-0.08	0.167	0.083	-0.06 F	0.651 δV	-1.3 δV	-0.65 δV	0.488 δV*F	-0.491	0.7141	0.3123
U2L3	25	5	-0.31 δV	-0.08	-0.17	0.083	-0.06 F	0.651 δV	1.302 δV	-0.65 δV	0.488 δV*F	-0.491	-0.953	0.3123
U3L4	25	5	-0.31 δV	-0.08	-0.17	-0.25	-0.06 F	0.651 δV	1.302 δV	1.953 δV	0.488 δV*F	-0.491	-0.953	-1.354
U5L4	25	5	-0.31 δV	0	0	0	-0.06 F	0	0	0	0.488 δV*F	-0.075	-0.119	-0.104
U6L5	25	5	-0.31 δV	0	0	0	-0.06 F	0	0	0	0.488 δV*F	-0.075	-0.119	-0.104
U7L6	25	5	-0.31 δV	0	0	0	-0.06 F	0	0	0	0.488 δV*F	-0.075	-0.119	-0.104
U8L7	25	5	-0.31 δV	0	0	0	-0.06 F	0	0	0	0.488 δV*F	-0.075	-0.119	-0.104
							Σ =	-5.56 δV	-8.89 δV	-7.78 δV	23.29 δV*F			

STEP 5 Equate the external work and internal work for the unit load at each joint. The four columns in the spreadsheet for the internal work are summed. Using the method of consistent deformations, the value of the redundant, F, for each position of the unit load can be calculated.

Unit load at L_1: $\delta V u_r = 0 = -5.56 \cdot \dfrac{\delta V}{E} + 23.29 F \dfrac{\delta V}{E}$

$$\therefore F = \frac{5.56}{23.29} = 0.2385$$

Unit load at L_2: $\delta V u_r = 0 = -8.89 \cdot \dfrac{\delta V}{E} + 23.29 F \dfrac{\delta V}{E}$

$$\therefore F = \frac{8.89}{23.29} = 0.3816$$

Example 13.2 (continued)

Unit load at L_3: $\delta V u_r = 0 = -7.78 \cdot \dfrac{\delta V}{E} + 23.29 F \dfrac{\delta V}{E}$

$$\therefore F = \frac{7.78}{23.29} = 0.3339$$

STEP 6 For each position of the unit load the final member forces are computed from the analysis in step 1 as the sum of the forces in the members due to the unit load and the corresponding redundant, F. The results are summarized in the last three columns of the spreadsheet.

STEP 7 All member forces are now known for the unit load at joints L_1, L_2, and L_3. All member forces are zero when the unit load is at the reaction joints, L_0, L_4, and L_8. The influence line can be constructed for the reaction at L_0 by taking the negative of the force in the member $L_0 U_0$ for the unit load at L_1, L_2, and L_3. When the unit load is at L_5, L_6, and L_7, the value of the reaction at L_0 can be obtained by symmetry as the negative of the force in member L_8-U_8 for the unit load at positions L_3, L_2, and L_1, respectively. The influence line for the force in member U_3-L_4 is obtained in a similar manner and the influence lines are plotted below.

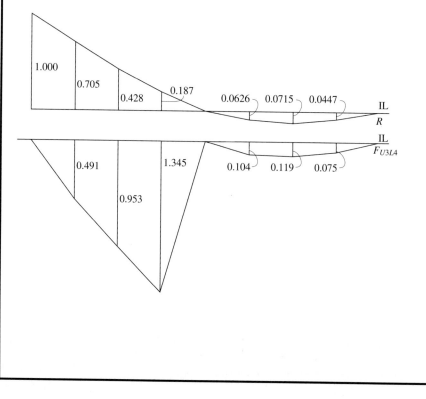

followed for the unit load at positions L_6 and L_7, which yields the ordinates -0.0715 and -0.0447, respectively.

The floor system of the truss distributes the load in proportion to its position between joints, as explained in Chapter 8. Thus the influence line for the reaction, R, at L_0 is completed by drawing straight lines between the values obtained for the unit load at each joint. The influence line for the force in member U_3–L_4 is constructed in a similar fashion using the forces in the last three columns of the spreadsheet for the members U_3–L_4 and U_5–L_4.

The use of the influence lines obtained in Examples 13.1 and 13.2 for the placement of loading on a structure is straightforward for uniform loads. As illustrated in Fig. 8.3 and quantified in Eq. (8.4), the area under the influence line functions times the intensity of the uniform load provides the magnitude of the corresponding force or moment action. The influence lines for the reactions or the forces in members of a trusses are piecewise linear so that the area under the influence line can be obtained as the sum of the areas under a series of trapezoidal figures. The influence lines for the beam of Example 13.1 vary continuously as polynomials of third degree. Direct integration can be used or, more conveniently, the numerical integration expressions given in Eqs. (9.3) to (9.6) can be used to give the same (exact) results. For the truss influence lines, Eq. (9.2) can be used to give exact results. These concepts are illustrated in the next example.

Example 13.3 The influence lines obtained in the first two examples provide the means of computing the magnitude of reactions, internal moments, and axial forces in the two structures due to a uniform load. Numerical integration is used in all the computations because it requires the fewest calculations and still yields the correct result. The influence line ordinates required in Eq. (9.3) are the ordinates at the center and two ends of the span, which are taken from the table in step 3 of Example 13.1. Equation (9.2) is well suited for the calculations using the influence lines for the truss because of their linear variation between panel points. The uniform spacing of the panel points at 20 ft is equal to the value of h in Eq. (9.2). The numerical integration formulas, Eqs. (9.3) to (9.6), *must not be used* for this calculation, as the influence line is piecewise linear and these equations are based on the assumption of curves that are polynomials of second or higher degree for the value of the integral.

The placement of a single concentrated load on a structure to create the maximum magnitude of a force or moment action is straightforward for trusses because the influence lines are piecewise linear and the maximum influence line ordinates can be determined by inspection. For beams and frames, as illustrated by the influence line for the moment M_B in Example 13.1, the maximum ordinate of the influence line must be obtained from differentiation of the influence function.

Example 13.3

Using the influence lines obtained in Examples 13.1 and 13.2, compute the maximum positive and negative values of the reactions at the left end of those structures, the maximum moment at the center of the beam of Example 13.1, and the maximum axial force in member U_3–L_4 of the truss of Example 13.2 due to a uniform load, w.

STEP 1 Compute the maximum positive and negative reactions for the left end of the beam. The influence line is third degree, so the numerical integration with Eq. (9.3) will yield the exact value. For a positive reaction the load is applied in left span only and for the negative reaction in the right span only.

Maximum positive:

$$R_A = \int_0^L wIL(x)\,dx = \frac{w(L-0)}{6} \cdot (1.0 + 4 \cdot 0.406 + 0) = 0.437\ wL$$

Maximum negative:

$$R_A = \int_0^L wIL(x)\,dx = \frac{-w(L-0)}{6} \cdot (0 + 4 \cdot 0.094 + 0) = -0.063\ wL$$

STEP 2 Compute maximum positive and negative reactions for the left end of the truss. The influence line is piecewise linear, so the numerical integration with Eq. (9.2) will yield the exact value. For a positive reaction the load is applied in left span only and for the negative reaction in the right span only.

Maximum positive:

$$R = \int_0^{80} wIL(x)\,dx = w \cdot 20 \left(\frac{1}{2} + 0.705 + 0.428 + 0.187 + \frac{0}{2} \right)$$

$$= 36.40\ w$$

529

Example 13.3 (continued)

Maximum negative:

$$R = \int_0^{80} wIL(x)\, dx = -w \cdot 20 \left(\frac{0}{2} + 0.0626 + 0.0715 + 0.0447 + \frac{0}{2} \right)$$

$$= -3.576\, w$$

STEP 3 Compute the maximum value of moment M_B. Both spans are loaded, but note the symmetry of the influence line and use Eq. (9.3).

$$M_B = 2 \int_0^L wIL(x)\, dx = \frac{-2w(L-0)}{6} L(0 + 4 \cdot 0.094 + 0) = -0.125\, wL^2$$

STEP 4 Compute the maximum value of force in member U_3–L_4. Both spans are loaded with w. Use Eq. (9.2).

$$F_{U3L4} = \int_0^{160} wIL(x)\, dx$$

$$= -w \cdot 20 \cdot \left(\frac{0}{2} + 0.491 + 0.953 + 1.354 + 0 \right.$$

$$\left. + 0.104 + 0.119 + 0.075 + \frac{0}{2} \right) = -61.92\, w$$

The placement of a series of concentrated loads to obtain maximum magnitudes of stress quantities is more difficult. For the truss influence lines, linear interpolation coupled with trial positions of loads can be used, but the computations become very protracted. For beams and frames the problem becomes significantly more difficult. For these structures additional mathematical development is required, which is beyond the scope of this treatment.

As illustrated in Example 13.2, the construction of influence lines for indeterminate trusses is a lengthy process. Also, indeterminate beams and frames require lengthy calculations, the effort escalating with the degree of indeterminacy. Direct use of the definition of the influence line as a means of drawing it requires a considerable amount of effort for indeterminate structures, even if computer programs are used for the analysis. An alternative approach is to use the concept based on the virtual work principle presented in Chapter 8, which is a specialization of the Müller–Breslau principle. The use of this concept for indeterminate structures first requires the development of Betti's law and Maxwell's law of reciprocal deflections.

13.2   Betti's Law and Maxwell's Law of Reciprocal Deflections

The same principle of conservation of work and energy used to obtain the virtual work principle in Chapter 7 is used to obtain Betti's law. Figure 13.1 shows a general structure fabricated from linear members and loaded with two force systems, P and Q. Each force system acting separately on the structure, along with the reactions it develops at restraint points, is in equilibrium. In addition, the force systems do not cause the structure to fail in any way. The P system is comprised of n forces, and the Q system of m forces which can act anywhere on the structure. Although the P and Q force systems are depicted as being composed only of forces, any or all of the forces in either system could be replaced with moments without changing the result, which is obtained below.

The application of each force system to the structure will cause the structure to undergo a corresponding set of displacements. In particular, define as U_{jP} the deflection caused by the P force system in the direction of the force Q_j at the point of application of Q_j. Also, define as U_{jQ} the deflection caused by the Q force system in the direction of P_j at the point of application of P_j. For a structure constructed from linear members the internal stresses are restricted to a single normal stress, σ, parallel to the axis of the member and a shear stress, τ, normal to that axis and the corresponding strains are ε and γ. Designate the stresses σ_Q and τ_Q and corresponding strains ε_Q and γ_Q as those developed by the Q force system and the stresses σ_P and τ_P and corresponding strains ε_P and γ_P as those developed by the P force system.

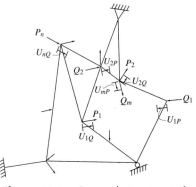

Figure 13.1   General structure of linear members acted upon by two different force systems and the corresponding deformations.

531

Consider, now, two separate loading sequences for the structure. First with the P force system acting on the structure, apply the Q force system, causing the Q system strains ε_Q and γ_Q and the Q system deflections U_{jQ} at the points of action of the P_j. Conservation of energy requires that the external work done by the P force system be equal to the internal work of deformation done by the P system stresses, all of which is due to the deformations caused during application of the Q force system. The external work is the sum of the product of each force P_j times the displacement U_{jQ} and the internal work can be computed from Eq. (7.2). Setting them equal yields the mathematical expression

$$\sum_{j=1}^{n} P_j U_{jQ} = \int_V (\sigma_p \varepsilon_Q + \tau_p \gamma_Q)\, dV \tag{13.1}$$

Since the structure is constructed of linear members, the normal and shear strains due to Q can be expressed in terms of the stress–strain relation for a linear material as

$$\varepsilon_Q = \frac{\sigma_Q}{E}$$

$$\gamma_Q = \frac{\tau_Q}{G} \tag{13.2}$$

and Eq. (13.1) can be rewritten in the equivalent form

$$\sum_{j=1}^{n} P_j U_{jQ} = \int_V \left(\sigma_p \frac{\sigma_Q}{E} + \tau_p \frac{\tau_Q}{G} \right) dV \tag{13.3}$$

For the second loading sequence, consider the Q force system acting on the structure as the P force system is applied, causing deflections U_{jP} at the point of application Q_j and internal strains ε_P and γ_P. Following the same procedure used to obtain Eq. (13.3), the conservation of energy principle requires in this second loading sequence that

$$\sum_{j=1}^{m} Q_j U_{jp} = \int_V \left(\sigma_Q \frac{\sigma_p}{E} + \tau_Q \frac{\tau_p}{G} \right) dV \tag{13.4}$$

An examination of the right-hand sides of Eqs. (13.3) and (13.4) shows them to be equal, leading to the conclusion that the two left-hand sides must also be equal, or

$$\sum_{j=1}^{n} P_j U_{jQ} = \sum_{j=1}^{m} Q_j U_{jp} \tag{13.5}$$

This is the mathematical statement of Betti's law, which may be stated in words as follows:

> In a structure with unyielding supports, constant temperature, and linear elastic material, in which no significant geometry changes occur due to deformations, the external work done by a system of forces, Q, in equilibrium during the application of a force system, P, in equilibrium is equal to the external work done by the P system during application of the Q system.

It is important to note that the displacements and deformations of the structure during application of either the P or Q loading systems are very small, so that the geometry of the deformed and of the undeformed structure is essentially the same.

Betti's law can be specialized to the case of a single load in both the P and Q systems. If the point where P is applied is called point 1 and the point where Q is applied point 2, Eq. (13.5) reduces to

$$Pu_{12} = Qu_{21}$$

or if $P = Q$, this becomes

$$u_{12} = u_{21} \qquad (13.6)$$

which is a mathematical representation of Maxwell's law of reciprocal deflections. This may be stated as follows:

> In a structure having linear elastic material, unyielding supports, and experiencing no temperature changes, the deflection in a specified direction d_1 at point 1 due to a load, P, acting in a specified direction, d_2, at point 2 is numerically equal to the deflection in direction d_2 at point 2 due to the same load, P, acting in the direction d_1 at point 1.

As with Eq. (13.5), the geometry of the structure is assumed to be essentially unchanged by the action of the loads. Figure 13.2a shows the interpretation of Maxwell's law for two points on a simple beam with applied forces. The derivation of Eq. (13.6) is based on the use of forces and displacements, but moment and rotations could have been used as well, as illustrated in Fig. 13.2b with two moments. If a force and a moment have the

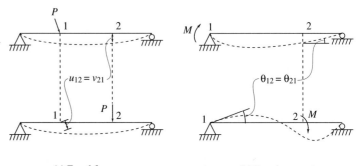

(a) Equal forces   (b) Equal moments

Figure 13.2 Maxwell's law applied to two equal forces or two equal moments.

same magnitude, Eq. (13.6) indicates that the corresponding reciprocal displacement and rotation will be numerically equal in magnitude. This concept is depicted in Fig. 13.3.

13.3 Müller–Breslau Principle

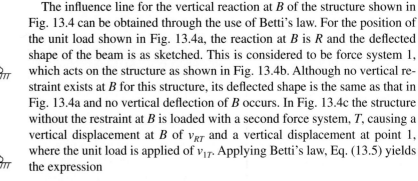

Figure 13.3 Maxwell's law applied to a force and moment having equal magnitudes.

In Chapter 8 the virtual work principle for rigid bodies was used to obtain the influence lines for statically determinate structures. This principle can be broadened to provide a means of obtaining the influence line for a particular force or moment action in an indeterminate structure from the deflected shape of the structure under the action of a loading corresponding to that force or moment action. This more general principle arises from a special application of Betti's law.

The influence line for the vertical reaction at B of the structure shown in Fig. 13.4 can be obtained through the use of Betti's law. For the position of the unit load shown in Fig. 13.4a, the reaction at B is R and the deflected shape of the beam is as sketched. This is considered to be force system 1, which acts on the structure as shown in Fig. 13.4b. Although no vertical restraint exists at B for this structure, its deflected shape is the same as that in Fig. 13.4a and no vertical deflection of B occurs. In Fig. 13.4c the structure without the restraint at B is loaded with a second force system, T, causing a vertical displacement at B of v_{RT} and a vertical displacement at point 1, where the unit load is applied of v_{1T}. Applying Betti's law, Eq. (13.5) yields the expression

$$Rv_{RT} + (-1)v_{1T} = T \cdot 0 \qquad (13.7)$$

The left side of Eq. (13.7) represents the external work done by force system 1 as it rides through the displacement caused by force system T. The

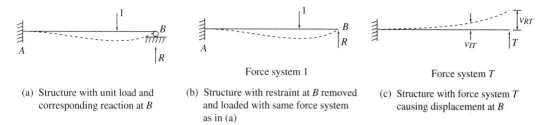

(a) Structure with unit load and corresponding reaction at B

(b) Structure with restraint at B removed and loaded with same force system as in (a)

Force system 1

(c) Structure with force system T causing displacement at B

Force system T

Figure 13.4a–c   Application of Betti's law to obtain an influence line for a reaction force.

negative sign associated with the unit load reflects the fact that the direction of the unit load is opposite the displacement caused by T. The work done by the T force system as it rides through the displacement caused by force system 1 is zero since the displacement at B due to that system is zero. Equation (13.7) can be solved for R to yield

$$R = \frac{(1) \cdot v_{1T}}{v_{RT}} \tag{13.8}$$

The units of the right-hand side of Eq. (13.8) are the force units of the unit load since the length units associated with the displacements disappear in the ratio. If the magnitude of T is selected so that the deflection at B, v_{RT}, is unity, which gives a corresponding deflection at point 1, V_{1T} of V_1, Eq. (13.8) becomes simply

$$R = (1) \cdot v_1 \tag{13.9}$$

where v_1 is the nondimensional deflection at the point of application of the unit load. Since point 1 can be any point between A and B, Eq. (13.9) defines the influence line for the reaction, R, at B and is simply the deflected shape of the structure of Fig. 13.4a when the vertical displacement restraint at B is removed and force T sufficient in magnitude to cause a unit vertical displacement at B is applied.

The result obtained in Eq. (13.9) is a specific application of the Müller–Breslau principle, which can be stated as follows:

> In a structure having linear elastic material, unyielding supports, and constant temperature, the influence line for any force or moment action is the deflected shape of the structure that is obtained by removing the restraint corresponding to that action and introducing a unit deformation in the positive sense of the action.

535

Although the principle has been illustrated in an application to a statically indeterminate structure, the principle applies to determinate structures as well. However, for a statically determinate structure no force or moment is required to introduce the unit deformation, as the release of any restraint in the structure makes it a mechanism. As a result, the Müller–Breslau principle is reduced to the same form as the virtual work principle for rigid bodies and the influence lines for determinate structures are always straight lines, as shown in Chapter 8.

13.4 Influence Lines Using the Müller–Breslau Principle

The greatest value of the Müller–Breslau principle for indeterminate structures is its use to obtain a sketch of the influence line for a particular force or moment action. Unlike determinate structures, the structure obtained when the release corresponding to the desired force or moment action is introduced is indeterminate or at best statically determinate. Introduction of the loading necessary to cause a unit deformation subjects the members of the structure to axial and bending deformations. The deflected shape of the structure depends on the material and geometric properties of the members. The design process is used to define these properties so that each time the design changes a member property, the influence lines for the structure show changes in their ordinates. However, the shape of the influence lines does not change a great deal, so loading patterns that are derived from sketches of influence lines are also essentially unchanged.

The computation of influence lines for indeterminate trusses is long and tedious by hand computation methods as was illustrated in Example 13.2. The Müller–Breslau principle can be employed, but deflection of the truss must be computed using virtual work or specialized techniques such as the bar-chain method (See, e.g., the second edition of Reference 22). Again the hand computations are lengthy. Computer programs, especially those based on the displacement method of analysis discussed in a succeeding chapter, are very useful and represent a practical means of obtaining influence lines for plane trusses.

Quantitative Influence Lines for Beams

Influence line computations by methods of analysis using computers for indeterminate beams and frames present many of the same problems encountered with trusses. Computer programs usually are not designed specifically to output the deflected shapes assumed by individual members in a structure, so that influence lines cannot be conveniently obtained. If the moment variation along a member can be obtained from an analysis, by computer or by hand, the Müller–Breslau principle can be used to obtain the influence line. The following example illustrates this concept.

Example 13.4 The Müller–Breslau principle is used in step 1 to sketch the shape of the influence line for the moment at support B of the three-span continuous beam. Releasing the continuity of slope at B and introducing a unit relative rotation causes the structure to deform in the shape shown, which is a qualitative representation of the influence line. A quantitative representation of the deflected shape or influence line is obtained if the moment and hence curvature variation of the internal moments for the loading causing the relative unit rotation are known.

A moment, M, of undetermined magnitude applied to the hinge at B will cause a relative rotation. The distribution of the member end moments due to M is computed in step 2 by using moment distribution to obtain the final moment diagram for this loading. The moment distribution is needed only for the moment action on the B end of member $B–C$ in the structure $B–C–D$ because the internal moment variation in member $A–B$ due to M acting at B can be obtained from equilibrium. With the completion of the moment distribution for the $B–C–D$ portion of the structure, the moment and curvature diagram for all members of the structure is established.

In step 3 the displacement function for each span is obtained by direct integration of the curvature functions using the zero vertical displacement conditions at each support to evaluate the constants of integration. The slope at the B end of members $A–B$ and $B–C$ is computed from the derivative of the displacement functions. The magnitude of M at support B can be obtained from the condition that the relative rotation, by the Müller–Breslau principle, must be unity. In step 4 the two rotations obtained in step 3 from the member end slopes are added together and the value of M determined. Substituting this value of M into the deflection expressions for each span in step 5 provides the complete quantitative functional variation of the influence line. These functions are evaluated at intervals of $0.1L$ in each span and the results are summarized in the table. The influence line is drawn to scale below the table.

The influence line for the vertical reaction at A for the continuous beam of Example 13.4 can be obtained by taking a free body of span $A–B$ and summing moments about B. The influence line for M_B would provide all of the necessary information to compute the ordinates of the influence line. The structure of Example 13.4 is indeterminate to the second degree, so it would usually be necessary to undertake an additional analysis of the structure and obtain an influence line for a second redundant quantity. However, due to the symmetry of the structure the influence line for this second redundant can be obtained without a second analysis. For example, the influence line for the

Example 13.4

Obtain the influence line for the internal moment at support B.

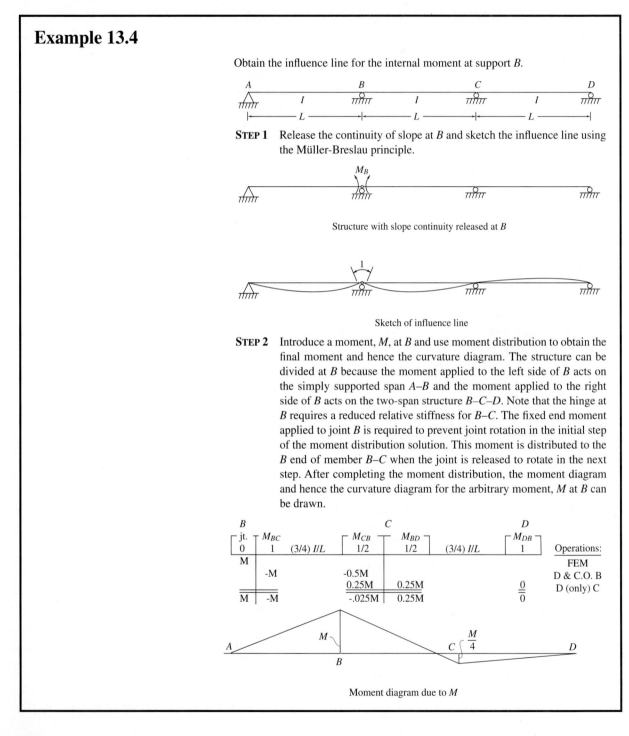

STEP 1 Release the continuity of slope at B and sketch the influence line using the Müller-Breslau principle.

Structure with slope continuity released at B

Sketch of influence line

STEP 2 Introduce a moment, M, at B and use moment distribution to obtain the final moment and hence the curvature diagram. The structure can be divided at B because the moment applied to the left side of B acts on the simply supported span A–B and the moment applied to the right side of B acts on the two-span structure B–C–D. Note that the hinge at B requires a reduced relative stiffness for B–C. The fixed end moment applied to joint B is required to prevent joint rotation in the initial step of the moment distribution solution. This moment is distributed to the B end of member B–C when the joint is released to rotate in the next step. After completing the moment distribution, the moment diagram and hence the curvature diagram for the arbitrary moment, M at B can be drawn.

Moment diagram due to M

Example 13.4 (continued)

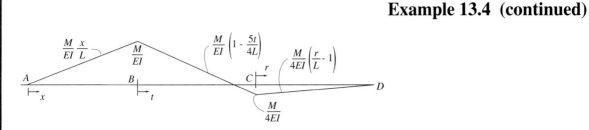

Moment diagram due to M

STEP 3 Using the curvature diagrams defined by functions of x, t, and r for the arbitrary moment, M, at B, the deflection, v, in each span can be obtained by direct integration.

Span AB: v is a function of x.

From Eq. (6.9) using the curvature function for A–B:

$$v'' = \frac{M}{EI}\frac{x}{L} \quad \therefore v' = \frac{M}{EI}\frac{x^2}{2L} + C_1 \quad \text{and} \quad v = \frac{M}{EI}\frac{x^3}{6L} + C_1 x + C_2$$

Evaluate C_1 and C_2 from the zero vertical displacement conditions at A and B.

$$v = 0 \text{ at } x = 0 \quad \therefore C_2 = 0 \quad v = 0 \text{ at } x = L \quad \therefore C_2 = -\frac{M}{EI}\frac{L}{6}$$

Substitute for C_1 and C_2 and obtain the final displacement expression for A–B. Evaluate the slope at B.

$$v = \frac{M x L}{EI\,6}\left(\frac{x^2}{L^2} - 1\right) \quad \text{and} \quad v'\Big|_{x=L} = \frac{M}{EI}\frac{L}{3} \quad \text{(counterclockwise rotation)}$$

Span BC: v is a function of t.

From Eq. (6.9) using the curvature function for B–C:

$$v'' = \frac{M}{EI}\left(1 - \frac{5t}{4L}\right) \quad \therefore v' = \frac{M}{EI}\left(t - \frac{5t^2}{8L}\right) + C_3 \quad \text{and} \quad v = \frac{M}{EI}\left(\frac{t^2}{2} - \frac{5t^3}{24L}\right) + C_3 t + C_4$$

Evaluate C_3 and C_4 from the zero vertical displacement conditions at B and C.

$$v = 0 \text{ at } t = 0 \quad \therefore C_4 = 0 \quad v = 0 \text{ at } t = L \quad \therefore C_3 = -\frac{M}{EI}\frac{7L}{24}$$

Substitute for C_3 and C_4 and obtain the final displacement expression for B–C. Evaluate the slope at B.

$$v = \frac{M t L}{EI\,24}\left(\frac{12t}{L} - \frac{5t^2}{L^2} - 7\right) \quad \text{and} \quad v'\Big|_{t=0} = -\frac{M}{EI}\frac{7L}{24} \quad \text{(clockwise rotation)}$$

539

Example 13.4 (continued)

Span CD: v is a function of r.

From Eq. (6.9) using the curvature function for C–D:

$$v'' = \frac{M}{4EI}\left(\frac{r}{L} - 1\right) \quad \therefore v' = \frac{M}{4EI}\left(\frac{r^2}{2L} - r\right) + C_5 \quad \text{and} \quad v = \frac{M}{4EI}\left(\frac{r^3}{6L} - \frac{r^2}{2}\right) + C_5 r + C_6$$

Evaluate C_5 and C_6 from the zero vertical displacement conditions at C and D.

$$v = 0 \text{ at } r = 0 \quad \therefore C_6 = 0 \qquad v = 0 \text{ at } r = L \quad \therefore C_5 = \frac{M}{EI}\frac{L}{12}$$

Substitute for C_5 and C_6 and obtain the final displacement expression for C–D.

$$v = \frac{M}{EI}\frac{rL}{24}\left(\frac{r^2}{L^2} - \frac{3r}{L} + 2\right)$$

STEP 4 Use the condition that at B the relative rotation must be unity to obtain the magnitude of M.

$$v'_{AB}(x = L) + v'_{BC}(t = 0) = 1 \rightarrow \frac{M}{EI}\frac{L}{3} + \frac{M}{EI}\frac{7L}{24} = 1 \quad \therefore \frac{M}{EI} = \frac{8}{5L}$$

STEP 5 Backsubstitute for M/EI in the deflection expressions for each span, which gives the complete influence line function. These functions are evaluated at intervals of $0.1L$ and are summarized in the table below and the influence line is drawn for the moment at B.

$$v_{AB} = \frac{4x}{15}\left(\frac{x^2}{L^2} - 1\right) \qquad v_{BC} = \frac{t}{15}\left(\frac{12t}{L} - \frac{5t^2}{L^2} - 7\right) \qquad v_{CD} = \frac{r}{15}\left(\frac{r^2}{L^2} - \frac{3r}{L} + 2\right)$$

x, r, t	0	0.1L	0.2L	0.3L	0.4L	0.5L	0.6L	0.7L	0.8L	0.9L	L
$\dfrac{v_{AB}}{L}(x)$	0	-0.026	-0.051	-0.073	-0.090	-0.100	-0.102	-0.095	-0.077	-0.046	0
$\dfrac{v_{BC}}{L}(r)$	0	-0.039	-0.064	-0.077	-0.080	-0.075	-0.064	-0.049	-0.032	-0.015	0
$\dfrac{v_{CD}}{L}(r)$	0	0.011	0.019	0.024	0.026	0.025	0.022	0.018	0.013	0.007	0

moment at support C is the mirror image of the influence line for M_B obtained in the example.

Output from computer programs specify moments and forces on the ends of beams. With this information, the displacement function defining the deflection of any member of a structure between its ends can be established by direct integration as illustrated in Example 13.4. These programs also output the displacements or rotations of joints of the structure, which provide the necessary information for use with the Müller–Breslau principle to obtain influence lines. A more extended presentation of the use of computer programs to obtain influence lines is given in Reference 23.

In Eq. (13.8) the reaction R to the cantilever structure of Fig. 13.4 for a unit load is expressed as the unit load multiplied by the ratio, v_{1T}/v_{RT}, of the displacements at the point of application of the unit load and application of the force T. In Eq. (13.9) the displacement v_{RT} is specified to be 1; that is, the magnitude of T is such that v_{RT} is unity. This yields the result that the displaced shape of the beam for v_{RT} equal to 1 is the influence line for the reaction. If, however, the value of T is different from that which causes v_{RT} to equal 1, the influence line is still given by Eq. (13.8). This provides the basis of using a computer analysis of a structure to obtain the influence line.

Influence Lines for Trusses

In the truss of Example 13.2 the influence line for the vertical reaction at L_0 can be obtained by releasing the vertical reaction constraint at L_0 and introducing a vertical force there of R. Using a computer program with a vertical force corresponding to R, of arbitrary magnitude, acting upward, the result of the analysis will be a vertical displacement at L_0, v_R, and vertical displacements at the other lower chord joints, v_j, where v_1 corresponds to the vertical displacement at L_1, v_2 to the vertical displacement at L_2, and so on. The influence line for the vertical reaction at L_0 can be obtained as the ratio of the computer output displacements v_j/v_R by virtue of Eq. (13.8). This concept is illustrated in Example 13.5.

Construction of Qualitative Influence Lines for Frames

The Müller–Breslau principle also provides the means of obtaining a qualitative representation of the influence lines in highly indeterminate structures. Often the most important information about placement of live loading on a structure can be obtained from a qualitative presentation of one or more influence lines. With the aid of these influence lines computer analyses can be executed with load distributions that cause maximum values of selected force or moment quantities. The following examples show how the influence lines for moments in the members of a multispan bridge and a multistory, multibay plane frame can be sketched.

Example 13.5

Use a computer program to obtain the influence line for the vertical reaction at L_0 of the truss in Example 13.2.

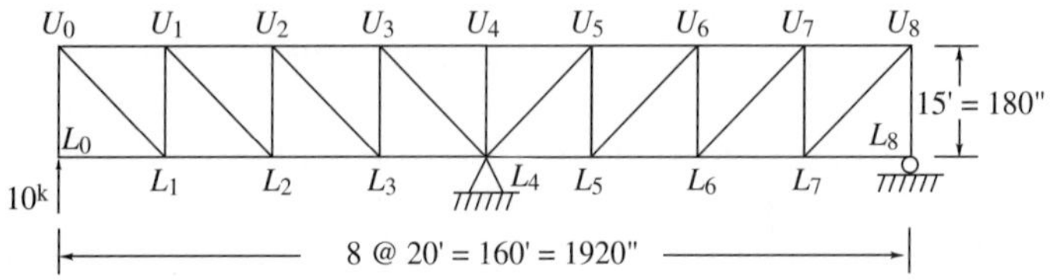

STEP 1 To use the Müller–Breslau principle the vertical reaction at L_0 is removed and a force is applied upward to create an upward displacement. As shown above, a force of 10 kips is used. The input to the computer program uses length units of inches and the modulus of elasticity of the material of the truss is taken as 10,000 ksi. The areas of the members in square inches are the same as in Example 13.2.

STEP 2 The displacements of the joints L_0 to L_8 are obtained from the computer results and are tabulated in the table below. The displacements are positive upward. The last line of the table lists the ordinates of the influence line for the reaction at L_0. These ordinates are obtained by using Eq. (13.8). The displacements in the second line of the table are divided by the displacement at L_0 of 7.951. The influence line matches the one obtained in Example 13.2.

Joint	L_0	L_1	L_2	L_3	L_4	L_5	L_6	L_7	L_8
Displacement	7.951	5.608	3.407	1.490	0.0	-0.4978	-0.5689	-0.3556	0.0
Influence line reaction, L_0	1.00	0.705	0.428	0.187	0.0	-0.0626	-0.0715	-0.0447	0.0

Example 13.6 The influence line for the vertical reaction at joint 3 is obtained through the introduction of a unit vertical displacement at 3 and then estimating the shape of the resulting displaced structure. Recognition of the continuity of slope at reaction points and the diminishing of the vertical displacement at points more distant from the unit displacement aids in sketching the influence line. The influence line for the negative moment at support 2 is obtained by introducing a unit relative rotation in the member at that support. Comparing the sketched deflected shape of the structure, which is the influence line for the moment at 2, with the configuration of the loading on this same structure in Example 12.4, it can be seen that this loading produces the maximum negative moment at support 2.

Example 13.7 The influence lines sketched in this example provide a means of determining what pattern of loading on a rigid frame creates maximum positive and negative moments in beams. The assumption that there is no joint translation associated with deformation of the structure caused by the unit relative rotations is relatively unimportant. Although not shown in the sketched deflected shapes, the magnitude of the displacements rapidly diminishes in a few bays or stories from the point of imposition of the unit rotations.

13.5   Summary and Limitations

The construction of influence lines for indeterminate structures directly from their definition is straightforward but very tedious. Even with the use of computer programs the direct method is lengthy and awkward to use for beam-and-frame structures because the unit load must be placed at discrete points on the structure rather than being allowed to vary in position in a continuous manner. Also, the output of computer programs does not provide for a plot of influence lines. Once constructed, however, influence lines are used in the same manner as illustrated in Chapter 8 for the computation of the magnitude of some internal moment, shear, or axial force for a distributed or concentrated load. Numerical integration using Eq. (9.2) or (9.3) provides a simple means of treating uniform loadings. A series of concentrated loads presents significant difficulties when the loads are to be positioned to create a maximum value of some force or moment action.

The derivation of Betti's law and Maxwell's law of reciprocal deflections is presented in Section 13.2. These two principles depend on considerations of conservation of energy for structures with linear elastic materials undergoing small displacements. With a special application of Betti's law, the

543

Example 13.6

Use the Müller–Breslau principle to sketch the influence lines for the reaction at support 3 and the moment over support 2 in the Kishwaukee River bridge shown in Fig. 12.4.

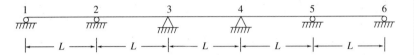

STEP 1 The bridge is a five-span continuous beam structure with the beam being supported on top of high piers. The vertical deflection of the piers due to their axial deformation is very small, and the structure can be taken as a continuous-span structure on reaction supports as shown above.

STEP 2 To obtain the influence line for the reaction at support 3, release the structure from the support and introduce a vertical unit displacement at the support. Sketch the resulting deflected shape of the structure. Note that continuity of the slope of the beam is maintained throughout its length and the magnitude of vertical deflections decreases with distance away from the unit displacement. The influence line is shown below the deflected structure.

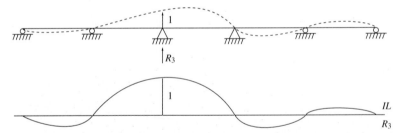

STEP 3 To obtain the influence line for the negative moment over support 2, release continuity of slope in the beam by introducing a hinge over the support and impose a relative unit rotation at the hinge. Sketch the resulting deflected shape of the structure by following the same procedure and guidelines used in step 2. The influence line is shown below the deflected structure.

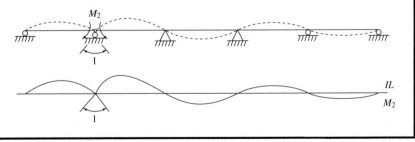

Example 13.7

Use the Müller-Breslau principle to sketch the influence lines for the moment in the center of member 18–25 and the negative moment at the 18 end of member 18–25 in the rigid frame shown. Indicate the pattern of uniform loads required to obtain maximum values of these moments.

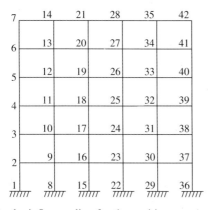

STEP 1   To obtain the influence line for the positive moment in the center of member 18–25, release the continuity of slope in the member by introducing a hinge and impose a relative unit rotation at the hinge. Sketch the resulting deflected shape of the structure. Start by sketching the deflected shape of the horizontal members between joints 4 and 39, being sure to maintain continuity of slope of the members on either side of the hinge. Assume that there are no vertical or horizontal displacements of the joints of the frame. The joints 4, 11, 18, 25, 32, and 39 all undergo a rotation consistent with the deflected shape of the horizontal members. Using the rotations of these joints as a starting point, sketch the deflected shape of each column of the frame, again assuming no joint translations. The rigid joints of the frame must maintain the orthogonal relationship of the members framing into them as the deformations develop. With all of the columns deflected, the deflection of the beams on the other five floors of the frame can be sketched using a continuity slope and maintaining the orthogonality of beams and columns at the joints. The resulting deflected shape represents the influence line for the positive moment in member 18–25.

STEP 2   To create the maximum positive moment in the center of member 18–25, a uniform load is placed on the beams in each bay and on each floor where the influence line is above the undeformed position of the

Example 13.7 (continued)

structure. This gives the pattern of uniform loads indicated by the heavy, dark lines in the sketch.

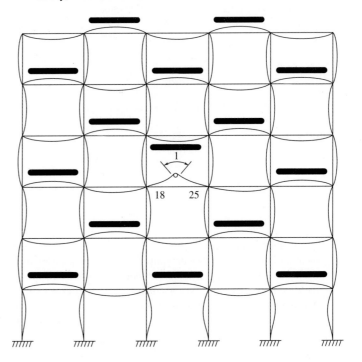

Influence line and loading pattern for positive moment in member 18–25

STEP 3   To obtain the influence line for the negative moment at 18 (or left) end of member 18–25, release the continuity of slope in the member by introducing a hinge and impose a relative unit rotation at the hinge. Sketch the resulting deflected shape of the structure. Follow the sequence of drawing operations of step 1 in sketching the deflected shape. After sketching the deflected shapes of the horizontal members between joints 4 and 39 and the columns, sketch the deflection of the beams on the other five floors of the frame. Note that in the bay

Example 13.7 (continued)

containing the member 18–25 the beams have reversed curvature so that a portion of the deflected shape is above the original position of the member and a portion below it. The resulting deflected shape represents the influence line for the negative moment at the left end of member 18–25.

STEP 4   As in step 2, the maximum negative moment is created when a uniform load is placed on the beams in each bay and on each floor where the influence line is above the undeformed position of the structure. This gives the pattern of uniform loads indicated by the heavy, dark lines in the sketch.

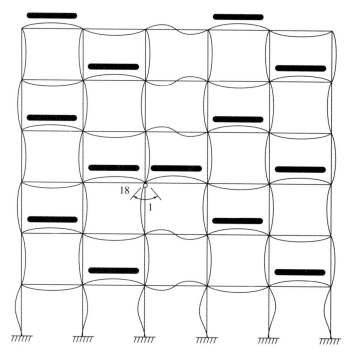

Influence line and loading pattern for the negative moment at the left end of member 18–25

Müller–Breslau principle is derived. This principle states that the influence line for a force or moment in a structure is given by the deflected shape of that structure, which is created by removing the restraint on continuity of deformation corresponding to the force or moment and introducing at the point of release a specified unit displacement or rotation. For statically determinate structures the Müller–Breslau principle reduces to the principle of virtual work used in Chapter 8 to obtain influence lines.

Several applications of the Müller–Breslau principle illustrate how influence lines can be constructed quantitatively. The principle enables computers to be used for this task, which proves to be particularly convenient for trusses. Quantitative influence lines for beams and frames require the use of direct integration in conjunction with an analysis of a structure subjected to a specified unit displacement or rotation. If Eq. (13.8) is used, an analysis using a force or moment action can be substituted for the unit displacement or rotation analysis.

The Müller–Breslau principle is also used as a means of sketching influence lines. The deflected shape of a structure due to the imposition of a unit displacement or rotation at a single point is usually fairly easy to determine. The qualitative representation of an influence line is often sufficient to identify the pattern of uniform loading, which will create the maximum value of some force or moment action. The Müller–Breslau principle provides a powerful means of understanding structural behavior and design of structures for the maximum effects of loading.

The limitations of the analyses presented in this chapter have been stated earlier, but a review of some of the important ones is useful.

- The assumptions and limitations of the analysis of trusses that are listed at the end of Chapter 3 must be observed.

- Because of the use of simple bending theory principles, the deformation analysis of beams and frames is subject to all of the same limitations and assumptions that are listed at the end of Chapter 5 and in Section 6.2 except that members can be nonprismatic.

- All limitations and assumptions that are listed at the end of Chapter 10 must be observed.

- The Müller–Breslau principle can be used for structures with nonlinear materials for the qualitative sketching of influence lines to determine placement of loads, but quantitative influence lines have no meaning and cannot be drawn.

Some questions that are worthy of consideration when using the principles presented in this chapter are the following:

- How can the change of individual member properties make a significant change in the influence line for an indeterminate structure?

- If, as in Example 13.5, a computer program is used to assist the construction of an influence line, under what conditions does the magnitude of the force or moment used in the analysis for the displaced shape of the structure not affect the results?

- Can a computer program analysis such as that in Example 13.5 be used to assist the construction of an influence line for a statically determinate structure?

PROBLEMS

13.1   Use the Müller–Breslau principle to draw the influence line for the vertical reaction at B and compute the ordinates at distances of $0.1L$ in each span (E constant).

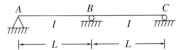

13.2   The influence line, $g(x)$ for the reaction at A of the structure shown is given as:

$$g(x) = \begin{cases} 1 - \dfrac{x}{16} + \dfrac{x^3}{32,000} & 0 \le x \le 20 \\[2mm] 1 - \dfrac{x}{16} + \dfrac{x^3}{32,000} - \dfrac{(x-20)^3}{16,000} & 20 \le x \le 40 \end{cases}$$

Compute the reactions of the structure for (a) a 10-kip load at D, and (b) a uniform load of 2 kips/ft over span A–B–D.

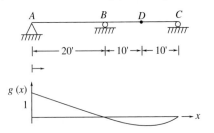

13.3 (a)   Sketch the influence line for the moment at D.

(b)   Calculate the ordinates of the influence line for the moment at D at intervals of $0.25L$ along both spans. Take E constant.

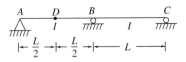

13.4   The influence line, $g(x)$ for the reaction at A of the structure shown is given as:

$$g(x) = \begin{cases} 1 - \dfrac{x}{16} + \dfrac{x^3}{32,000} & 0 \le x \le 20 \\[2mm] 1 - \dfrac{x}{16} + \dfrac{x^3}{32,000} - \dfrac{(x-20)^3}{16,000} & 20 \le x \le 40 \end{cases}$$

(a)   Find the location in B–C where a 100-kN load will cause the maximum upward reaction at A and compute its magnitude.

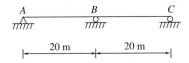

(b)   Compute the reaction at A for a uniform load of 50 kN/m in A–B.

13.5   For the structure shown, the loads all will act on the horizontal portion of the structure (E constant).

(a)   Sketch the influence line for the horizontal reaction at 3.

(b)   Sketch the influence line for the moment just to the left of 2.

(c)   Obtain the ordinates for the influence line for the horizontal reaction at 3 at intervals of $0.1L$.

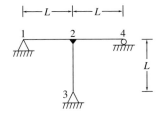

13.6   The two-span continuous beam shown is symmetric. Using the Müller–Breslau principle, obtain the influence line for the moment at 2 and the reaction at 1. Plot the two influence lines and show the value of the ordinates at intervals of $L/6$ (E constant).

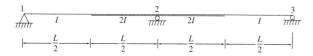

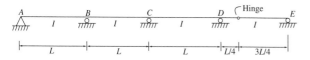

13.7   The four-span beam below has a hinge in span D–E as shown. Using the results presented in Example 13.4, obtain quantitative expressions for the influence line for **(a)** the moment at B, and **(b)** the vertical reaction at A (E constant).

13.8 (a) *Sketch* the influence line for the *negative* moment just to the left of joint 3 for unit loads acting on span 2–3–4.
  (b) Obtain the ordinates of the influence line for intervals along 2–3–4 of $L/10$.

(*E* constant)

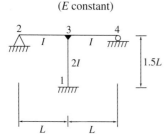

13.9   Use the Müller–Breslau principle to sketch the influence lines for the moment at support 4 and the shear to the right of support 3.

13.10   Using the results presented in Example 13.2, obtain the influence line for the vertical reaction at (a) L_8, and (b) L_4.

13.11   Using the results presented in Example 13.2, obtain the following influence lines.
  (a) Force in member U_2–L_3
  (b) Force in member U_5–L_5
  (c) Force in member L_4–L_5
  (d) Force in member U_2–U_3

13.12   Using the results presented in Example 13.2, compute the following for a concentrated load of 20 kips and a uniform load of 4 kips/ft of variable length.
  (a) maximum force in member L_4–L_5
  (b) Maximum force in member U_3–L_3

13.13   Obtain the influence lines for the vertical reaction at L_0 and the force in member U_1–U_2. (Take advantage of symmetry; E constant. Areas: chords 30 cm², verticals 15 cm², diagonals 35 cm².)

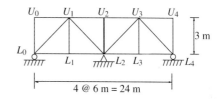

13.14   Use the Müller–Breslau principle to sketch the influence line for:
  (a) Positive moment in the center of span 10–11
  (b) Negative moment just to the left of joint 6
  (c) Shear just to the right of joint 6
  (*Note*: Base of structure is hinged.)

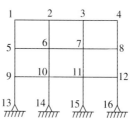

551

13.15   Use the Müller–Breslau principle to sketch the influence lines for the positive moment in the center of member 14–15, negative moment just to the right of joint 14, positive shear just to the right of joint 20, and vertical upward reaction at joint 33.

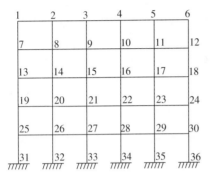

13.16   Sketch the influence line using the Müller–Breslau principle for the upward reaction at L_0, the force in member U_1–U_2, the force in member U_1–L_2, and the force in member U_3–L_3. Assume that the unit load is applied to the bottom chord of the truss.

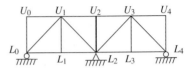

13.17   Use the Müller–Breslau principle to sketch the influence lines for the force in members U_4–U_5, L_6–L_7, and U_5–L_6. Assume that the unit load acts along the bottom chord of the truss.

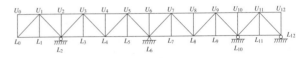

Approximate Analysis Using Force-Based Assumptions

In the preceding chapters the analysis of structures has been carried out using a mathematical model of the structure that represents its physical response to applied loads in a reasonably accurate manner. The analysis of these mathematical models has yielded a solution that is designated as "exact." These solutions require a considerable amount of computational effort that may, for preliminary design of structures, not be warranted.

In this chapter some very simple methods of hand analysis are presented as a means of obtaining a reasonable estimate of the distribution of internal forces and moments in a structure without having to go through an extensive analysis or even knowing the properties of the members of the structure. Even though the structure may be highly indeterminate, the methods can still be used, which is a great help in the preliminary design stage. Although these methods have been developed for hand analysis, they can with a little effort be programmed for use on a computer.

14.1 Need for Approximate Methods of Analysis

The structural engineer must design structures that are safe and economical. In statically determinate structures the distribution of forces and moments due to applied loadings can be calculated without knowing member properties.

The design process is greatly simplified in this case since a single analysis provides the complete set of information needed for stress calculations. Deflection calculations are also greatly simplified.

The design and construction of indeterminate structures is very common in present practice. The distribution of forces and moments in indeterminate structures depends on the geometric and material properties of the members of the structures; hence some knowledge of these properties is needed before any analysis can be performed. In particular, computer programs for analysis need specific member properties as input. To initiate the design–analysis cycles, the structural engineer must provide an initial set of member properties based on previous designs of similar structures, his or her own experience in designing structures, or some systematic analysis not requiring specific member properties. The convergence to a final economical design in the design–analysis iteration is significantly influenced by the initial values of member properties. The better the initial estimates of member properties, the fewer cycles are required for convergence.

In this chapter several methods of approximate analysis are presented. In all the methods, assumptions are made about the internal distribution of forces or moments in the structure. These assumptions together with the equations of equilibrium provide a set of equations that will yield a distribution of forces and moments in the structure which can be used for an initial design. The number of assumptions required to reduce an indeterminate structure to a determinate one is just equal to the degree of indeterminacy of the structure.

The first few sections of this chapter are devoted to frames subjected to vertical and lateral loads. The final section is devoted to trusses. No specific treatment of continuous beams is given since this type of structure can be analyzed approximately by the same methods as those discussed for vertical loads on beams in plane frame structures.

14.2 Vertical Loading of Beams in a Frame

Figure 14.1 shows the distribution of moments in a single uniformly loaded beam with various types of end rotation restraints. For all three conditions of support the moment at the center of the span must be $wL^2/8$ relative to the moment at the ends of the span, a condition dictated by equilibrium. Points of zero moment or inflection points exist in all three beams. The location of these points depends on the degree of rotational restraint that exists at the ends of the beams. The simply supported beam has the points of zero moments at the end because of the free rotation condition there, while the fully fixed beam has zero moments at a distance of $0.211L$ in from the ends.

As indicated in Fig. 14.1b, any restraint of rotation at the ends of the beam is reflected by moments M, which become $wL^2/12$ for the case of full end rotational restraint. The end restraining moments causes the points of

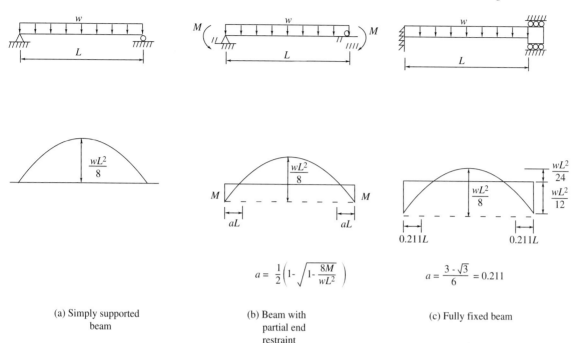

Figure 14.1a–c Effect of end rotation restraint on moments in a uniformly loaded beam.

zero moment or inflection points to move toward the middle of the span as it increases from zero of the simply supported condition. The distance the inflection points move in is expressed by the factor a defined in Fig. 14.1b. As shown in Fig. 14.2, a typical member such as i–j in a frame and under vertical load has moments develop at the ends due to the restraining effects of the columns and other beams. A value of the factor a between 0.00 and 0.211 is appropriate in characterizing the restraint on the member ends. If a is taken as 0.1, the two inflection points in the member are located and the beam can be analyzed by statics. For $a = 0.1$, the moments M are $0.045wL^2$ at the ends and $0.08wL^2$ in the center.

The rigidly connected member i–j in Fig. 14.2 is indeterminate to the third degree. Assuming the location of the two inflection points as in Fig. 14.1b defines the configuration of the moment diagram based on static equilibrium. In addition, it is observed that under vertical loading alone, the axial force in girders is small and can be assumed to be zero. If applied to every horizontal member of the frame in Fig. 14.1, these three assumptions provide a number of assumptions equal to the degree of indeterminacy of the frame. For girders that are interior to the frame, the moments developed for a preliminary design may be taken as those for a fully fixed beam ($0.083wL^2$) since the

555

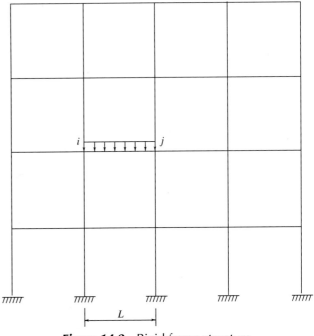

Figure 14.2   Rigid frame structure.

loading causing maximum moment at the ends of the beam will essentially prevent any joint rotation. For girders connected to an exterior column, the moment ($0.045wL^2$) suggested above is a reasonable approximation.

Axial forces and moments in columns for the vertical loading conditions can easily be computed by the equations of equilibrium. Typically, interior columns have axial loads that are equal to vertical loading acting on one bay and little or no moment. Exterior columns have axial loading equal to the vertical loading acting on one-half of the exterior bay. The moments in these columns can be taken approximately as $0.045wL^2$ at each story, divided approximately in half ($0.0225wL^2$) to each column framing into the story. At the base of the exterior columns the moment may be taken as one-half of the moment on the first story end of column, or approximately $0.012wL^2$.

14.3   Moments and Forces in Simple Frames with Lateral Loads

The approximate analysis for vertical loads provides only one-half of the necessary information for the design of a building. Lateral loads must also be resisted, and the behavior of a laterally loaded rigid frame, such as that shown in Fig. 14.2, is quite different from its vertical loading behavior. A study of lateral loading behavior is most easily introduced with the analysis of portal frame behavior.

Figure 14.3 A simple portal frame with pinned supports subjected to a lateral load is shown. An exact analysis that neglects axial shortening effects in the girder produces the final moment diagram shown. An inflection point occurs in the center of the girder which is due to the symmetry of the structure and the antisymmetry of the loading. The value of α has no affect on the location of the inflection point.

Figure 14.4 shows a simple portal frame with fixed bases. This frame is indeterminate to the third degree, so any analysis using static equilibrium equations will require three assumptions. The pinned base frame of Fig. 14.3 required only one assumption, which yielded exact results when the inflection point was assumed in the center of the beam. The assumption of an inflection point in the center of the beam is still the best choice and in fact corresponds to the exact solution for the frame of Fig. 14.4. If the beam were rigid, the effect of the load would be to displace the frame laterally and deform the columns into shapes having double curvature, such as shown in Fig. 12.6, and with inflection points at their centers, as shown in Fig. 14.4b. The resulting moment diagram in Fig. 14.4c would have points of zero moment or inflection points at the center of the columns and beams.

If the three assumptions are made for the structure of Fig. 14.4 that inflection points occur at the center of the beams and the columns, the structure is reduced to a statically determinate one that is simply analyzed. The inflection point in the center of the beam is exactly correct as long as the columns have symmetric stiffness. The inflection points at the center of the

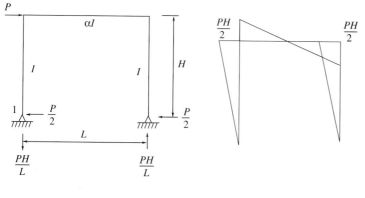

(a) Structure with load   (b) Final moment diagram

Figure 14.3a–b   Exact analysis of simple portal frame.

557

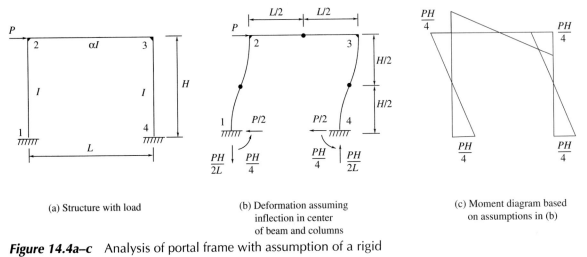

(a) Structure with load

(b) Deformation assuming inflection in center of beam and columns

(c) Moment diagram based on assumptions in (b)

Figure 14.4a–c   Analysis of portal frame with assumption of a rigid beam.

columns are correct only if the beam is rigid. Bending deformations of the beam will cause the inflection points in the columns to move upward from the midpoint.

Figure 14.5 The distance to the inflection point in the columns from the base of the columns of the structure for the various ratios of the relative stiffness of the beam to the relative stiffness of the column is shown. For relative stiffness ratios of beam to column greater than 3, the error in assuming the inflection point at the midpoint of the column is just over 5% and the error in the moments in the column is less than 6%. Typically, the relative stiffness ratio of beams to columns in frames is 2 or more, so that the assumption of inflection points at the midpoint of columns is a good one for an initial analysis.

The assumptions made in Fig. 14.4 lead to the moment diagram shown in Fig. 14.4c. The moment diagram yields a distribution of moments in the frame that satisfies equilibrium, but compatibility of deformation is not satisfied. That is, the deformations of the frame are such that the rotation and relative horizontal and vertical displacement of joint 4 with respect to joint 1 are not zero. This can easily be verified if an analysis similar to the one illustrated and discussed in Fig. 10.21 is carried out using the curvature diagram for the frame obtained by dividing the moment diagram of Fig. 14.4c by the EI of the members. Compatibility is not satisfied, because the distribution of the moments in Fig. 14.4c is not the exact one, and hence the curvatures are also not exact. The consequence of making assumptions about the distribution of axial forces or moments in a structure in order to use equilibrium equations for its analysis is to create deformations that are not compatible.

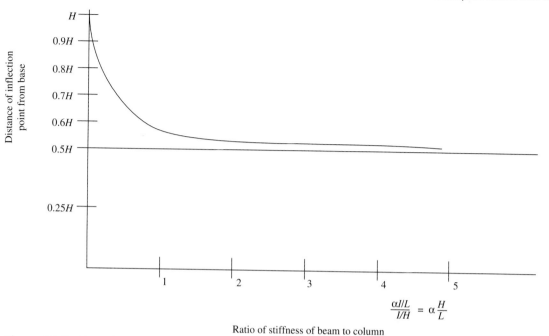

Figure 14.5   Exact location of inflection point in columns of frame in Fig. 14.4.

Another type of portal frame, one in which the two columns are not restrained in the same way at their bases, is shown in Fig. 14.6. The exact analysis of this structure depends on the relation of the height to the width of the frame as well as the relative bending stiffness of the three members.

Figure 14.6 To make an approximate analysis of this frame, two assumptions must be made about the distribution of forces or moments because the frame is indeterminate to the second degree. An exact analysis can be made if the proportions and relative stiffnesses of members are known. In the absence of this information, a reasonable set of assumptions is that the moment is zero in the center of the beam and left column as shown in Fig. 14.6b. These two assumptions render the structure statically determinate, and the equations of equilibrium may be used to obtain the shear forces in the two column and the final moment diagram. The reader can easily verify that the column shears in Fig. 14.6b and moment diagrams in Fig. 14.6c are correct.

The three simple frames shown in Figs. 14.3 to 14.6 provide reasonable approximations of the distribution of shears to the columns of single-bay frames with different support conditions and subjected to a lateral load.

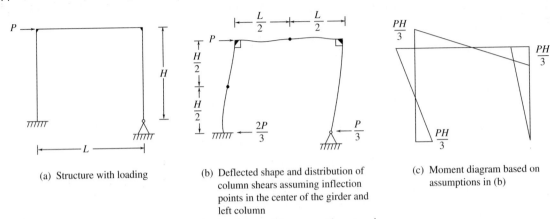

(a) Structure with loading

(b) Deflected shape and distribution of column shears assuming inflection points in the center of the girder and left column

(c) Moment diagram based on assumptions in (b)

Figure 14.6a–c Approximate analysis of portal frame with mixed base constraint conditions.

14.4 Portal Method of Analysis

For structures having multiple bays and stories the assumption of inflection points in the center of beams and columns does not provide a sufficient number of conditions to reduce the structure to a statically determinate form. This is easily seen in the structure of Fig. 14.7.

Figure 14.7 The structure is indeterminate to the ninth degree, but assuming an inflection point at the center of each column and each beam gives only seven conditions on the distribution of moments in the frame. The two additional assumptions come from the manner in which the lateral load, $3P$, is resisted by shear in the columns of the structure.

The assumption is made in the portal method for this structure, which has identical bays, that the lateral load is resisted equally by the frame representing each bay of the structures, as indicated in Fig. 14.7b. This is equivalent to making two assumptions; the first is that one-third of the lateral load is resisted in the first bay of the structure and the second that one-third of the lateral load is resisted in the second bay of the structure. Horizontal equilibrium then requires that the remaining one-third of the load be resisted in the third bay of the structure. From these two assumptions it is clear from the analysis in Fig. 14.4 that the shear force shown in Fig. 14.7b for each column of the portal frame that represents a single bay is $P/2$. Adding the shear forces in the columns in each bay yields the result that the shear force in the two exterior columns is $P/2$, while it is P in the two interior columns. The two assumptions relating to the distribution of shears in the bays, along with the seven assumed in-

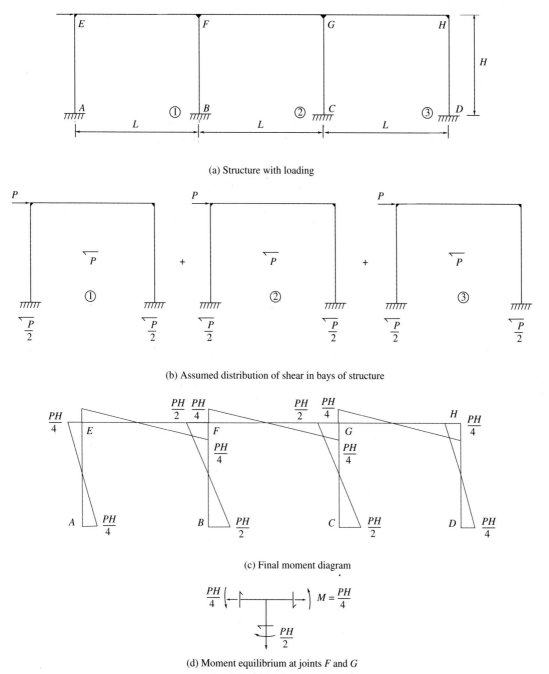

(a) Structure with loading

(b) Assumed distribution of shear in bays of structure

(c) Final moment diagram

(d) Moment equilibrium at joints F and G

Figure 14.7a–d Assumed distribution of shear forces in a structure based on the portal method.

flection points, enable the analysis of the structure to be completed by statics alone.

The analysis for the moments in the structure starts by obtaining the moments in the columns. Since the moment is zero at the center of the columns, the lower half of each column is a cantilever beam of length $H/2$ with a load applied at the "tip" (the inflection point) equal to the shear in the column. This yields the moments at the base of the columns as the product of the shear in the column and half-height of the column. There are no intermediate loads on the column, so the variation of the moment is linear, which makes the moment at the top of the column equal in magnitude to the moment at the bottom by virtue of the zero moment at mid-height. Moment equilibrium at E and H fixes the moment in beams E–F and G–H since points of zero moment are assumed at the centers of the beam. Finally, the moment diagram for beam F–G is obtained by calculating the moment of F from moment equilibrium in the free-body diagram at joint F shown in Fig. 14.7d. The completed moment diagram is shown in Fig. 14.7c.

When an irregular structure such as shown in Fig. 14.8 is considered, a variation on the portal method is required.

Figure 14.8 The structure in Fig. 14.8a is indeterminate to the seventh degree. The assumption of an inflection point in the center of each beam is still appropriate, and an inflection point can be assumed to occur in the center of the two left columns. This yields five assumptions; two more are required to complete the analysis.

The two assumptions about the shear developed in two of the bays being $3P$ and $3P$ in the third bay by horizontal equilibrium are shown in Fig. 14.8b. The column shears for each bay are then obtained by distributing the bay shear as $3P/2$ to each column in the left bay, as is done in Fig. 14.4; $3P/2$ to each column in the right bay, as is done in Fig. 14.3; and $2P/3$ to the left column and $P/3$ to the right column of the center bay, as is done in Fig. 14.6. This completes the assumptions, and the moment diagram for the structure is obtained by equilibrium in a manner similar to that discussed in Fig. 14.7.

The equal distribution of shear to the bays in Fig. 14.8b is not the only assumption that can be used. Perhaps a more accurate or appropriate distribution of the shear would be in proportion to the apparent resistance each bay can provide to the lateral load. The fixity at the base of the columns will affect the resistance. The left bay, with both columns fixed, will have a greater resistance to the shear than the center bay, where only one column is fixed, and the center bay will have greater resistance than the right bay, where neither column is fixed at the base. One might expect the left bay to have twice the resistance to shear of the right bay from the standpoint of column fixity, and the center bay $1\frac{1}{2}$ times the re-

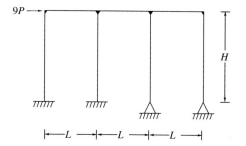

(a) Structure with lateral load

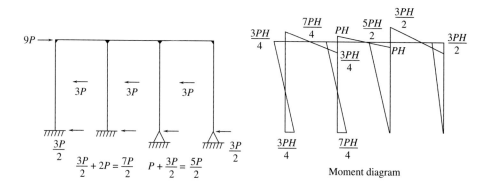

(b) Shear distributed equally to each bay of structure

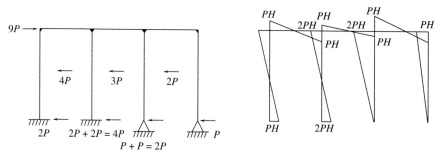

(c) Shear distributed in proportion to the lateral stiffness of each bay

Figure 14.8a–c Assumed distribution of shear forces in irregular structure based on the portal method.

sistance of the right bay. If this observation is used, the shear in the bays of the frame in Fig. 14.8 will be distributed as shown in Fig. 14.8c. Distributing the bay shears to the columns as is done in Figs. 14.3 to 14.6, the column shears and moments become those shown in Fig. 14.8c.

In Fig. 14.8 the distribution of the shears to the bays and columns in both Fig. 14.8b and c is not exact but is based on a particular set of two assumptions. In the absence of information and for the purpose of a preliminary analysis, the assumptions in Fig. 14.8c are perhaps the most accurate and useful.

The use of the portal method for multiple-story frames is a simple extension of the analysis just completed. It is assumed that there is an inflection point in the center of each column and each beam and that the shear in each story is distributed to the bays in that story in relation to the column fixities. Bay shears are distributed to columns by the proportions shown in Figs. 14.3 to 14.6. For most frames the bays have column fixities like those for the frame of Fig. 14.4, so that interior columns tend to resist twice the shear of exterior columns.

The first step in the process is to calculate the total shear in each story. This is simply the sum of all lateral loads acting on the structure above the story of interest. The shear in each story is then distributed in an appropriate manner to the bays and then to the columns. Nearly always this results in the interior columns having twice the shear force of exterior columns. When this is the case, the interior column shears can be computed as the total story shear divided by the number of bays in the story, and the exterior column shears are one-half of this. The next step is to compute the moments in all columns in the structure. Finally, the moments in the beams of the structure are computed by using the conditions of moment equilibrium of the joints of the frame and proceeding from the exterior to the interior bays. Shear forces in beams are calculated as the end moment divided by the half-length of the beam. Axial forces in beams and columns are obtained from force equilibrium in free bodies cut from the structure through the midpoints of beams and columns. Example 14.1 illustrates a complete analysis of a frame by the portal method.

Example 14.1 The analysis by the portal method is very structured and easy to execute. The column shears are computed first as indicated in step 1. Entering the shears on a diagram of the structure aids the succeeding analysis. In steps 2 and 3 a series of free-body diagrams for each of the two left columns in every story of the frame is used to obtain column end moments as the column shear times the story half-height $(L/2)$. The circled 2 is placed next to this computation. The beam end moments are calculated from the series of free-body diagrams of the joints along the two left columns. The moment at the left end of each beam is computed first by starting with the left exterior joint of each story and summing joint moment equilibrium. The moment on the right end of each

Example 14.1

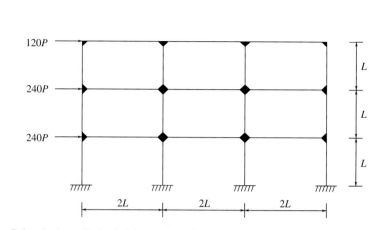

Solve the laterally loaded frame above by the portal method.

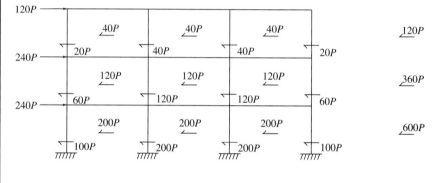

Column shears   Story shears

STEP 1   Obtain story shears by summing the applied load at each story for all stories above. Since all columns are rigidly connected at both ends, divide the story shear by 3 (three bays) to obtain the interior column shears for that story. Divide the interior column shears by 2 to obtain exterior column shears. Note that circled numbers in the diagrams correspond to the operations defined in the step numbers below.

STEP 2   Obtain the moments at the top and bottom of each column by multiplying the shear in that column by half the story height. This is illustrated by the free bodies of the left two columns.

STEP 3   Obtain the beam moments at each joint using moment equilibrium and summing moments clockwise. Since all column moments are known from step 2, the beam moments in the left bay are obtained. Because inflection points are assumed in the center of each beam, the moment

565

Example 14.1 (continued)

at the opposite end of the beam is equal and opposite. Proceed from left to right to obtain beam moments.

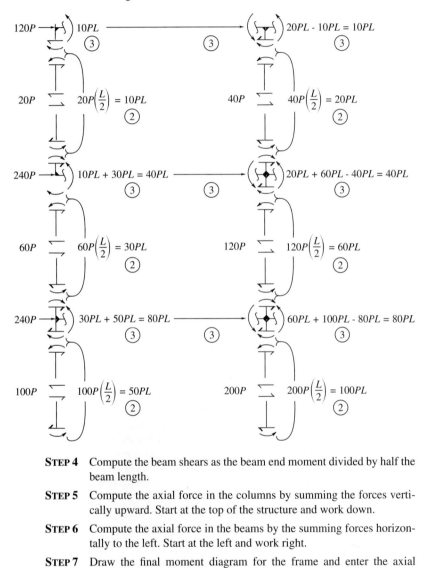

STEP 4 Compute the beam shears as the beam end moment divided by half the beam length.

STEP 5 Compute the axial force in the columns by summing the forces vertically upward. Start at the top of the structure and work down.

STEP 6 Compute the axial force in the beams by the summing forces horizontally to the left. Start at the left and work right.

STEP 7 Draw the final moment diagram for the frame and enter the axial forces for all members.

Example 14.1 (continued)

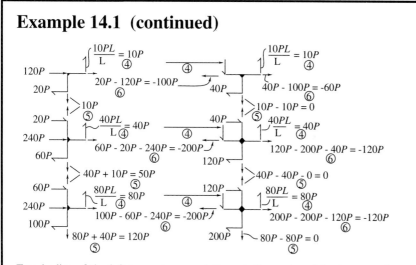

Free bodies of the left two columns cut through the center of the beams and columns.

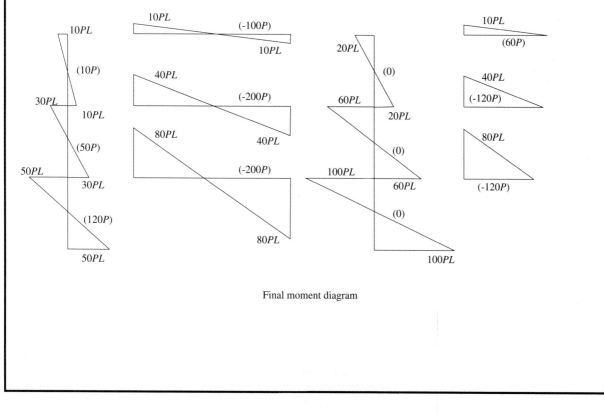

Final moment diagram

beam is equal in magnitude but opposite in sense and the process proceeds from the left toward the center of the structure. Calculations of beam end moments are designated by the circled 3.

The next series of free-body diagrams for the left two column stacks are used in steps 4 to 6 to calculate the beam shears (circled 4), the column axial loads (circled 5), and the beam axial loads (circled 6). These free bodies are obtained by taking a section through the center of each beam and the center of each column so that a joint is completely isolated, with one-half of the beam and column members attached to it. All beam shears are computed as the beam end moment divided by the beam half-length. The column axial forces are computed next by using vertical force equilibrium and proceeding from the top story downward.

Finally, the beam axial forces are computed from horizontal force equilibrium in each free body by starting at the left exterior joint and proceeding to the right in each story. In step 7 the final moment diagram is assembled from the computations of the preceding steps. For numerical calculations with U.S. units, take $L = 15$ ft and $P = 0.1$ kips for SI units, take $L = 5$ m and $P = 0.5$ kN.

The structure is symmetric, so only the moments, shears, and axial forces for the left half of the structure are displayed. For clarity of presentation, the columns with their moment diagrams and the beams with their moment diagrams are separated from one another. It should be noted that shears and moments are symmetric, but the axial forces in the beams are not. For the right bay the beam axial forces from top to bottom are $-20P$, $-40P$, and $-40P$ respectively.

14.5   Cantilever Method of Analysis

An alternative to the portal method of analysis is to make assumptions about the distribution of axial forces in columns rather than shear forces. It is reasonable to expect that a tall, slender building would act like a *cantilever beam*. The cross section of the beam in any story is a horizontal cross section made up of the areas of the columns in that story. The assumption that plane sections remain plane in simple beam theory leads to a presumption that the centroid of the column areas in any story is the centroid of the *beam cross section*, as shown in Fig. 14.9.

Using the assumptions of inflection points in the center of beams and columns as in the portal method, the free body obtained in Fig. 14.9b when the centers of the columns have been cut by a horizontal section will satisfy moment equilibrium when the moment of the axial forces in the columns is equal to the story moment, M_S. This result is dependent on the condition that all the column shear forces are collinear. From beam theory the strain

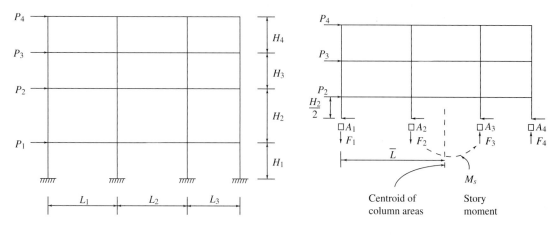

(a) Structure with loading

(b) Free body of second story showing column axial loads, centroid of column areas, and story moment

$$\overline{L} = \frac{A_2 L_1 + A_3(L_1 + L_2) + A_4(L_1 + L_2 + L_3)}{A_1 + A_2 + A_3 + A_4}$$

Applied story moment:

$$M_s = P_2\left(\frac{H_2}{2}\right) + P_3\left(\frac{H_2}{2} + H_3\right) + P_4\left(\frac{H_2}{2} + H_3 + H_4\right)$$

Resisting story moment:

$$M_S = F_1\overline{L} + F_2(\overline{L} - L_1) + F_3(L_1 + L_2 - \overline{L}) + F_4(L_1 + L_2 + L_3 - \overline{L})$$

(c) Computations of story centroid and the applied and resisting story moments

Figure 14.9a–c   Principal features of the cantilever method.

and hence the stress will vary linearly across the story. If the stress in the left column in Fig. 14.9b is F_1/A_1, the stress in all other columns can be calculated from this stress and the location of the centroid of the columns of the story. The stress F_2/A_2 in the second column can be obtained by proportion to the stress in the outside column as $(\overline{L} - L_1)(F_1/A_1)/\overline{L}$. The force in each column is the stress in each column times the area of the column. Thus the outside column has force F_1, the next column F_2, or in terms of F_1, it is $(A_2/A_1)(\overline{L} - L_1)F_1/\overline{L}$. If moments are summed about the centroid of the

569

column areas in Fig. 14.9b, recognizing that columns to the left of the centroid are in tension and those to the right are in compression, the result is

$$M_s = F_1 \overline{L} + \frac{A_2}{A_1}(\overline{L} - L_1)^2 \frac{F_1}{\overline{L}} + \frac{A_3}{A_1}(L_1 + L_2 - \overline{L})^2 \frac{F_1}{\overline{L}}$$

$$+ \frac{A_4}{A_1}(L_1 + L_2 + L_3 - \overline{L})^2 \frac{F_1}{\overline{L}}$$

$$= \frac{F_1}{A_1 \overline{L}}[A_1 \overline{L}^2 + A_2(\overline{L} - L_1)^2 + A_3(L_1 + L_2 - \overline{L})^2$$

$$+ A_4(L_1 + L_2 + L_3 - \overline{L})^2] \tag{14.1}$$

or

$$M_s = \frac{F_1}{A_1 \overline{L}} I_s$$

where I_s is the story moment of inertia with respect to the story centroid. The values of the axial forces in each column can be obtained through use of Eq. (14.1) and the linear distribution of stress assumption. For the structures of Fig. 14.9, the axial forces become (in magnitude)

$$F_1 = \frac{A_1 \overline{L}}{I_s} M_s$$

$$F_2 = \frac{A_2(\overline{L} - L_1)}{I_s} M_s$$

$$F_3 = \frac{A_3(\overline{L} - L_1 - L_2)}{I_s} M_s \tag{14.2}$$

$$F_4 = \frac{A_4(\overline{L} - L_1 - L_2 - L_3)}{I_s} M_s$$

The axial forces in the columns of each story depend on the location of the centroid of the column areas in the story and the area of each column. If the column areas are unknown, it is common practice to assume them all to be equal.

The analysis just described provides a number of assumptions equal to two less than the number of columns in each story. This follows from the fact that once the centroid of the column areas in a story is obtained and the force, F_1, in the first exterior column is calculated in Eq. (14.2), the remaining column forces are assumed to be given by the linear variation of the stress horizontally across the frame. For the structure shown in Fig. 14.7, the inflection point at the center of each column and beam assumptions provide only seven of the nine needed to perform the analysis. Locating the centroid and computing the force in the exterior column fixes the force in the other exterior column. The assumption of linear variation of the stress provides the two interior column forces. Note that structures such as those in Figs. 14.6 and 14.8 cannot be analyzed by the cantilever method because there is no single horizontal plane that can be passed through the inflection points in the columns.

The method of analysis that is followed in the cantilever method is three-step. First, the axial forces are obtained in all columns using Eqs. (14.1) and (14.2). The second step is to compute the shear forces in the beams. This process is started at the top story and the exterior columns. A free body taken by cutting through the columns and beams at their centers and summing forces vertically will yield the beam shears. The sequence of computation of beam shear can proceed either vertically along a column stack or horizontally across a floor. The final step is to obtain the moments in the beams and columns. The beam moments are the shear in each beam times the half-beam length. Column moments are obtained by moment equilibrium at the frame joints starting at the top-story exterior columns and proceeding horizontally or vertically through the structure. The method is illustrated in Example 14.2, which is the same structure as that used in Example 14.1.

Example 14.2   The step-by-step process for the cantilever method is similar to the portal method, but with a slightly different sequencing of the equilibrium computations. First the column axial forces are computed from the story moments in step 2. The beam shears are computed next (circled 3) by using vertical equilibrium in the column stacks for the two left columns and proceeding from the top floor downward. The free bodies in steps 2 and 3 for the left exterior column are obtained by making cuts at the center of the column in each story and the center of the beams in the left bay. The free bodies for the first interior column are obtained similarly by making cuts at the center of the columns and the beams in the two bays. With beam shears now known, the beam end moments are computed and, as in the portal method, are the product of the shear and beam half-length. Column end moments are computed from joint moment equilibrium by proceeding across each story from left to right. The column shears are computed by dividing the column end moments by the half-height (circled 6). If they are wanted, beam axial forces can now be computed from horizontal equilibrium in the free bodies (circled 7). The final moment diagram is drawn in step 8. For numerical calculations use the values in Example 14.1.

The structure is symmetric, so only the moments, shears, and axial forces for the left half of the structure are displayed. For clarity of presentation, the columns with their moment diagrams and the beams with their moment diagrams are separated from one another. As in Example 14.1, the beam

571

Example 14.2

Solve the frame of Example 14.1 by the cantilever method.

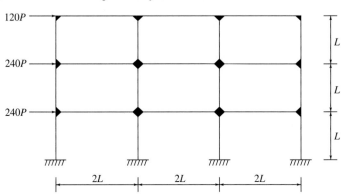

Note that circled numbers on the diagrams correspond to step numbers in the solution procedure.

STEP 1 Compute the centroid of the column areas for each story. Column areas are not given in this problem, so assume that all are equal. This gives the centroid of all stories at the center of the structure. From Eq. (14.1)

$$I_s = A(3L)^2 + AL^2 + AL^2 + A(3L)^2 = 20L^2A$$

STEP 2 Compute the axial forces in the columns. From Eq. (14.2):

$$F_1 = \frac{A(3L)M_s}{20L^2A} = \frac{3M_s}{20L} = -F_4 \quad F_2 = \frac{ALM_s}{20L^2A} = \frac{M_s}{20L} = -F_3$$

The story moments are computed at the center of each story.

Top story: $M_s = 120P\dfrac{L}{2} = 60PL$

Second story: $M_s = 120P\left(L + \dfrac{L}{2}\right) + 240P\dfrac{L}{2} = 300PL$

Bottom story:

$$M_s = 120P\left(L + L + \frac{L}{2}\right) + 240P\left(L + \frac{L}{2}\right) + 240P\left(\frac{L}{2}\right) = 780PL$$

Example 14.2 (continued)

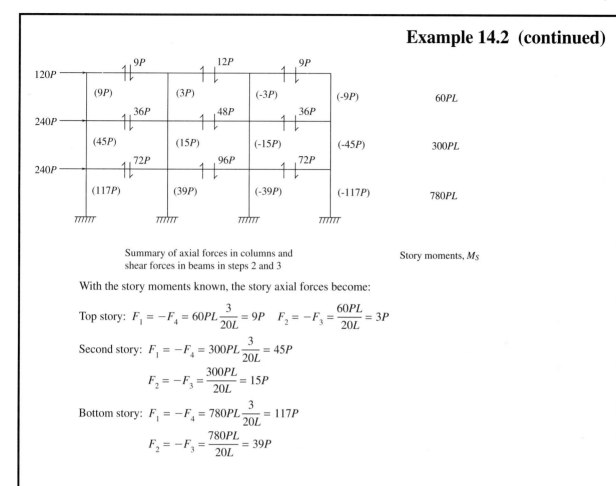

Summary of axial forces in columns and
shear forces in beams in steps 2 and 3

Story moments, M_S

With the story moments known, the story axial forces become:

Top story: $F_1 = -F_4 = 60PL\dfrac{3}{20L} = 9P$ $F_2 = -F_3 = \dfrac{60PL}{20L} = 3P$

Second story: $F_1 = -F_4 = 300PL\dfrac{3}{20L} = 45P$

$F_2 = -F_3 = \dfrac{300PL}{20L} = 15P$

Bottom story: $F_1 = -F_4 = 780PL\dfrac{3}{20L} = 117P$

$F_2 = -F_3 = \dfrac{780PL}{20L} = 39P$

573

Example 14.2 (continued)

STEP 3 Compute the beam shears from the free-body diagram of the column stacks. Free bodies are created from cuts at the center of beams and columns.

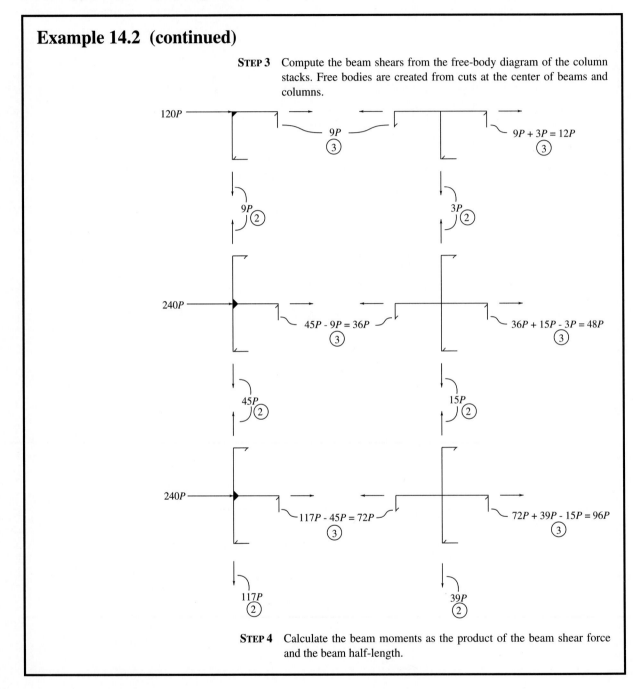

120P

$$9P \over 3$$

$$9P + 3P = 12P \over 3$$

$$9P \over 2$$

$$3P \over 2$$

240P

$$45P - 9P = 36P \over 3$$

$$36P + 15P - 3P = 48P \over 3$$

$$45P \over 2$$

$$15P \over 2$$

240P

$$117P - 45P = 72P \over 3$$

$$72P + 39P - 15P = 96P \over 3$$

$$117P \over 2$$

$$39P \over 2$$

STEP 4 Calculate the beam moments as the product of the beam shear force and the beam half-length.

Example 14.2 (continued)

STEP 5 Calculate the column moments from moment equilibrium at each joint. The calculation can proceed either horizontally or vertically through the frame.

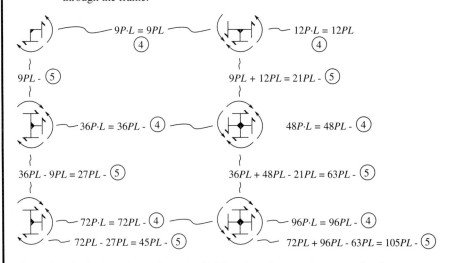

$$9P \cdot L = 9PL \quad ④$$

$$12P \cdot L = 12PL \quad ④$$

$$9PL - ⑤$$

$$9PL + 12PL = 21PL - ⑤$$

$$36P \cdot L = 36PL - ④$$

$$48P \cdot L = 48PL - ④$$

$$36PL - 9PL = 27PL - ⑤$$

$$36PL + 48PL - 21PL = 63PL - ⑤$$

$$72P \cdot L = 72PL - ④$$

$$96P \cdot L = 96PL - ④$$

$$72PL - 27PL = 45PL - ⑤$$

$$72PL + 96PL - 63PL = 105PL - ⑤$$

STEP 6 Obtain the column shears by dividing the column end moments by the column half-height.

Example 14.2 (continued)

STEP 7 Obtain the beam axial forces by using the horizontal equilibrium equation in the free bodies below with the column shear forces acting.

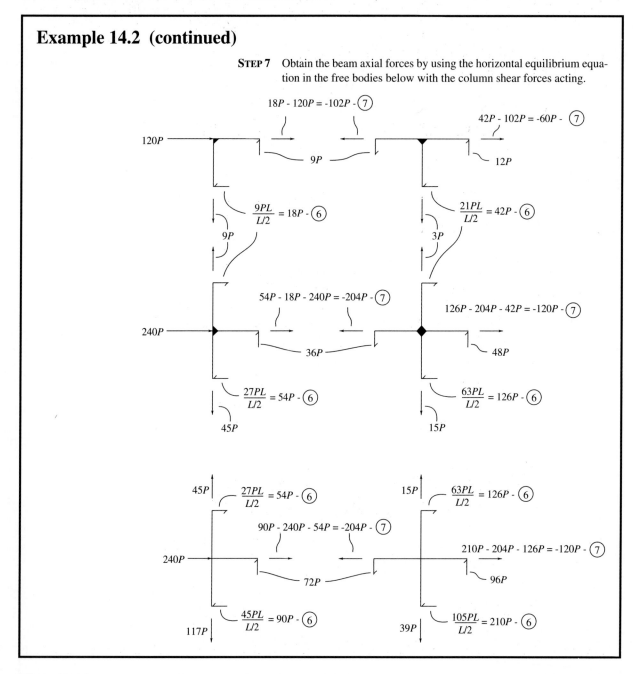

Example 14.2 (continued)

STEP 8   Draw the final moment diagram and enter the axial forces.

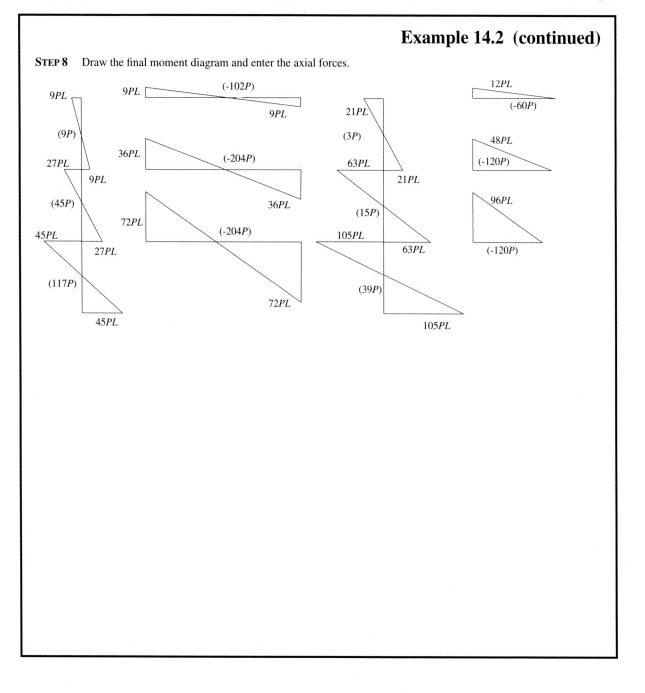

577

axial forces are not symmetric. For the right bay the beam axial forces from top to bottom are $-18P$, $-36P$, and $-36P$.

The moments obtained in the portal and cantilever methods for this frame are different, but not by large amounts. Either method is adequate for the estimation of moments in the example frame used. In general, the cantilever method gives better results for frames that are tall and slender and the portal method for frames that are wider than they are tall. However, either method can be used on most frames for a reasonable estimate of internal moments due to lateral loading.

14.6 Estimating Lateral Displacements Due to Lateral Loads

The calculation of displacements requires a knowledge of member properties. In the preliminary design stage, the need to estimate the horizontal deflection or drift of a structure is important. With the use of the portal or cantilever methods, some estimate of the size of the column and beam members of a frame can be obtained with little effort. In what follows the deflection of a frame based on the assumptions of zero moments at the center of beam and column members is presented. It is assumed that the calculation can be based on the deformations of any column, exterior or interior, and that the member sizes and beam and column moments are known.

In Fig. 14.10 the displaced position of the joint k relative to the floor below of a frame is shown to an exaggerated scale. The displacement from floor $k-1$ to k is designated u_k and is caused by deformations of the column of the story, u_{ck}, and the rotations of joints $k-1$ and k, θ_{k-1} and θ_k, due to the deformations of the beams to the right of the column, designated u_{bk-1} and u_{bk} in the figure. The beams to the left of the column (if they exist) also experience deformations consistent with the rotations of joints $k-1$ and k, but the beams to the right of the column will be used in the displacement calculations. The displacements between floors caused by the rotation of the beam are, for small rotations,

$$u_{bk-1} = \theta_{k-1}\frac{L_k}{2} \quad \text{and} \quad u_{bk} = \theta_k\frac{L_k}{2}$$

which becomes

$$u_b = (\theta_{k-1} + \theta_k)\frac{L_k}{2}$$

The column displacements are the deflections from the rotated tangent to the joints at $k-1$ and k, and are the deflection of the end of a cantilever

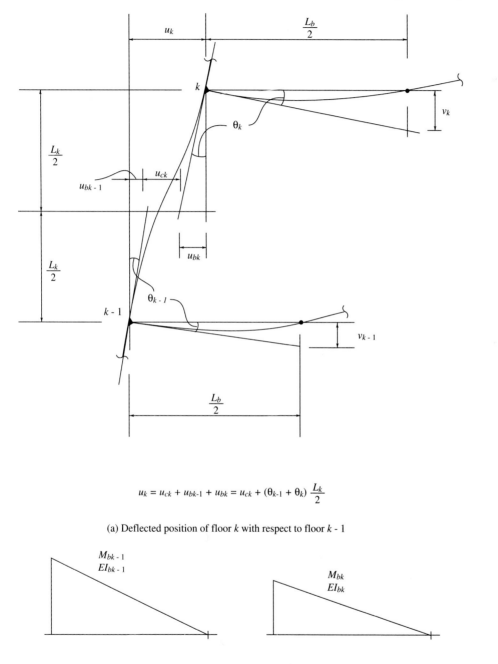

$$u_k = u_{ck} + u_{bk\text{-}1} + u_{bk} = u_{ck} + (\theta_{k\text{-}1} + \theta_k)\frac{L_k}{2}$$

(a) Deflected position of floor k with respect to floor k - 1

(b) Curvature diagrams for left half of beams at joints k - 1 and k

Figure 14.10a–b   Displacement of story k.

beam of length l with a concentrated load P acting on the end $Pl^3/3EI$. The columns are of length $l = L_k/2$ and assumed to have a shear V_k applied at their midpoints. The displacement u_{ck} is computed as

$$u_{ck} = 2\frac{V_k(L_k/2)^3}{3EI_k} = V_k\frac{L_k^3}{12EI_k} = M_K\frac{L_K^2}{6EI_k}$$

since the column moments in story k are $M_k = V_k L_k/2$.

The rotations θ_{k-1} and θ_k are due to the beam deformations as shown in Fig. 14.10. The vertical deflections of the center of the beams are zero because of the antisymmetry of the bending of the beams, so for small deflections the rotations can be computed as

$$\theta_k = \frac{v_k}{L_b/2} = 2\frac{v_k}{L_b}$$

$$\theta_{k-1} = \frac{v_{k-1}}{L_b/2} = 2\frac{v_{k-1}}{L_b}$$

The deformation of each beam is due to its linearly varying curvature diagram, which has the maximum ordinate M_{bk-1}/EI_{bk-1} and M_{bk}/EI_{bk}, where the moments M_{bk-1} and M_{bk} are the moments at the joints at the left ends of the beams and can be expressed as

$$M_{bk-1} = V_{bk-1}\frac{L_b}{2}$$

$$M_{bk} = V_{bk}\frac{L_b}{2}$$

so the shear forces in the beams are

$$V_{bk-1} = \frac{M_{bk-1}}{L_b/2} = 2\frac{M_{bk-1}}{L_b}$$

$$V_{bk} = \frac{M_{bk}}{L_b/2} = 2\frac{M_{bk}}{L_b}$$

The vertical displacements v_{k-1} and v_k are the vertical displacements of the rotated beams, cantilevered from joints $k-1$ and k, respectively. For these cantilever segments the vertical displacements are

$$v_{k-1} = \frac{M_{bk-1}L_b^2}{12EI_{bk-1}} \quad \text{and} \quad v_k = \frac{M_{bk}L_b^2}{12EI_{bk}}$$

From Fig. 14.10 the story displacement is

$$u_k = u_{ck} + (\theta_{k-1} + \theta_k)\frac{L_k}{2}$$

which upon substitution of the individual displacements from the preceding expressions becomes

$$u_k = \frac{M_k L_k^2}{6EI_k} + \frac{L_k L_b}{12E}\left(\frac{M_{bk-1}}{I_{bk-1}} + \frac{M_{bk}}{I_{bk}}\right) \qquad (14.3)$$

which gives an estimate of the displacement in story k.

The value of the displacement at any story in the structure can be obtained by summing from the base of the structure each story displacement computed by Eq. (14.3). The column and beam moments in any story are available from either a portal or cantilever analysis, and the member properties of the columns and beams are assumed to have been selected by some simple design procedure. A more extensive discussion of computing lateral displacements can be found in Reference 38.

Figure 14.11 The two-bay three-story frame shown in Fig. 10.17 is analyzed for the displacements caused by laterally applied loads. The properties of the members of the frame are given in Table 10.1. The results of the analysis of the frame (the analysis includes the deformations from axial forces) are shown in Fig. 14.11b and c. The moment diagrams of beams and columns are again separated for clarity of presentation.

Figure 14.12 The moments that result from an analysis of the laterally loaded frame of Fig. 14.11 by the portal method are shown in Fig. 14.12b. The axial forces and shears in the final results have not been entered on the diagram. If the structure is analyzed using the cantilever method, the exact same distribution of moments is obtained.

The use of Eq. (14.3) to calculate the lateral deflections of the frame of Fig. 14.11 using the moments obtained by the approximate analysis in Fig. 14.12 is illustrated in Example 14.3.

Example 14.3 The exterior column of the frame is used to compute the lateral displacements. The rotation of the column at its base is zero due to the full fixity. The errors in the displacements when compared with the exact results in Fig. 14.11 are around 10%.

14.7   Approximate Analysis of Indeterminate Trusses _____

Trusses that are indeterminate can also be analyzed approximately with assumptions that relate to the manner in which they are constructed. A common type of truss structure that is often used for lateral (or wind) bracing in truss bridges is shown in Fig. 14.13.

Figure 14.13 The truss is indeterminate to the sixth degree because of the six extra diagonal members in the panels of the truss. For this structure to

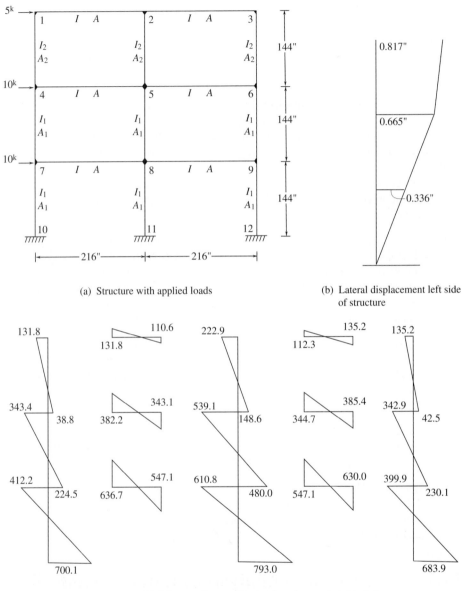

(a) Structure with applied loads

(b) Lateral displacement left side of structure

(c) Moment diagrams for members (moments in kip-in.)

Figure 14.11a–c   Exact analysis of structure of Fig. 10.17 due to lateral loads only.

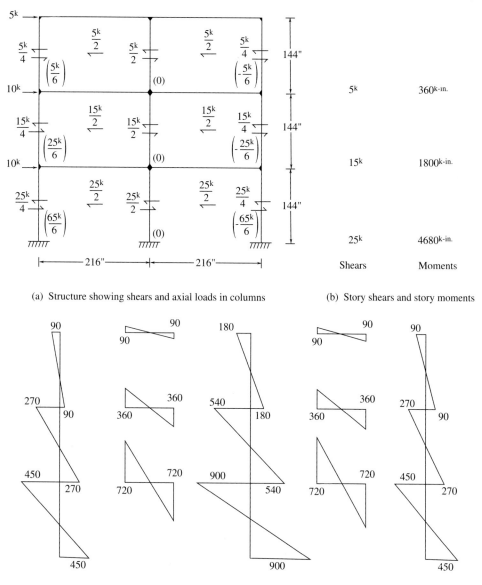

(a) Structure showing shears and axial loads in columns

(b) Story shears and story moments

(c) Moment diagrams for members (moments in kip-in.)

Figure 14.12a–c   Portal and cantilever method analysis of structure of Fig. 14.11.

583

Example 14.3

Obtain an estimate of the lateral displacements of the frame of Fig. 14.11 using the approximate analysis of Fig. 14.12.

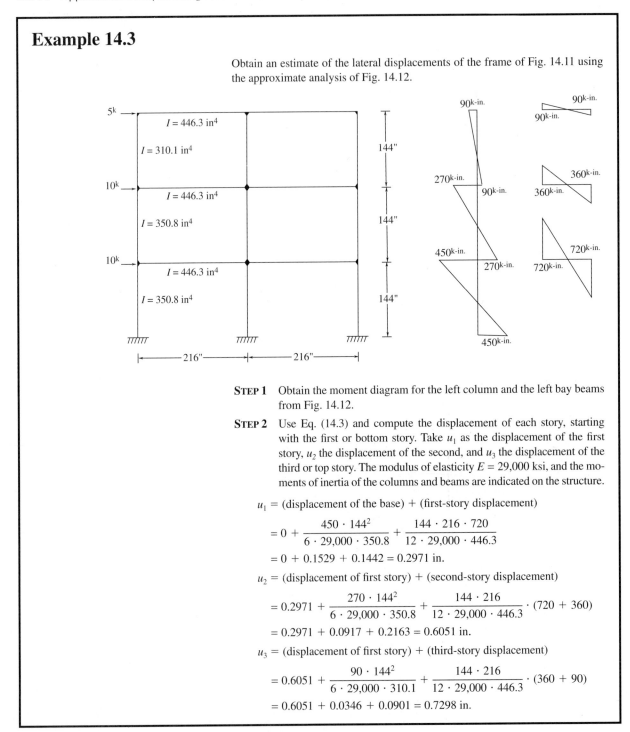

STEP 1   Obtain the moment diagram for the left column and the left bay beams from Fig. 14.12.

STEP 2   Use Eq. (14.3) and compute the displacement of each story, starting with the first or bottom story. Take u_1 as the displacement of the first story, u_2 the displacement of the second, and u_3 the displacement of the third or top story. The modulus of elasticity $E = 29,000$ ksi, and the moments of inertia of the columns and beams are indicated on the structure.

$u_1 = $ (displacement of the base) $+$ (first-story displacement)

$$= 0 + \frac{450 \cdot 144^2}{6 \cdot 29,000 \cdot 350.8} + \frac{144 \cdot 216 \cdot 720}{12 \cdot 29,000 \cdot 446.3}$$

$$= 0 + 0.1529 + 0.1442 = 0.2971 \text{ in.}$$

$u_2 = $ (displacement of first story) $+$ (second-story displacement)

$$= 0.2971 + \frac{270 \cdot 144^2}{6 \cdot 29,000 \cdot 350.8} + \frac{144 \cdot 216}{12 \cdot 29,000 \cdot 446.3} \cdot (720 + 360)$$

$$= 0.2971 + 0.0917 + 0.2163 = 0.6051 \text{ in.}$$

$u_3 = $ (displacement of first story) $+$ (third-story displacement)

$$= 0.6051 + \frac{90 \cdot 144^2}{6 \cdot 29,000 \cdot 310.1} + \frac{144 \cdot 216}{12 \cdot 29,000 \cdot 446.3} \cdot (360 + 90)$$

$$= 0.6051 + 0.0346 + 0.0901 = 0.7298 \text{ in.}$$

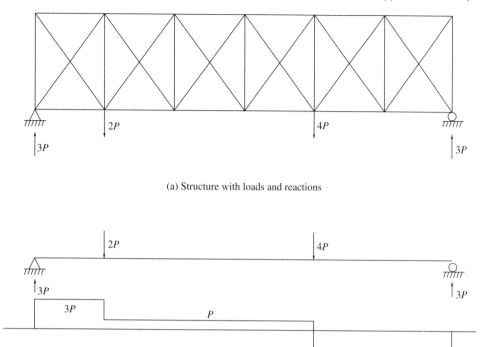

(a) Structure with loads and reactions

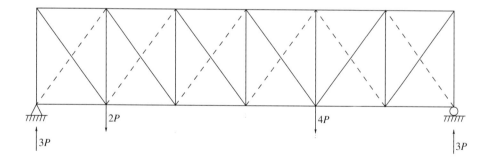

(b) Shear diagram for a beam loaded like a truss

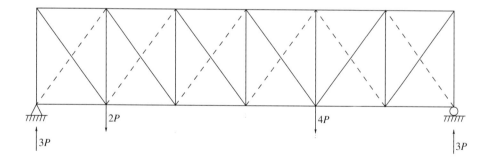

(c) Strucure with only the diagonals in tension active
in resisting the shear

Figure 14.13a–c   Indeterminate plane truss and analysis based on
assumption of tension-only diagonals.

585

be analyzed using static equilibrium equations, six assumptions must be made with respect to the internal member force distribution. Treating the truss as a simply supported beam, it must provide the resistance to the shear shown in the shear diagram of Fig. 14.13b. Each panel of the truss has two diagonals, although only one diagonal is needed to resist the shear force in the panel.

Trusses such as these usually are constructed in one of two ways. The truss diagonals may be designed to carry tension only and are thus very long and slender. Alternatively, the diagonals may be designed to carry both tension and compression. If the diagonals are designed to carry tension only, the diagonal that is subjected to compression will buckle under the load and may be assumed to have zero force. An analysis of the shear that exists in each panel of the truss will indicate which diagonal is in compression and hence not carrying any significant axial force. For the six panels, six zero-force members can be identified which yield the six necessary assumptions, as shown in Fig. 14.13c. The truss analysis is then carried out in the usual manner.

If the diagonals are designed to carry both tension and compression, it is common to assume that the forces in the diagonals of each panel are equal and opposite. The six assumptions constitute the necessary number to make the analysis using only the equilibrium equations.

Figure 14.14 The industrial building shown in Fig. 14.14a is constructed with X-type long, slender steel rods in the two center bays to provide resistance to lateral applied loads. Except for the diagonal rods, the structure is constructed of members capable of carrying bending but has connections detailed to act as pins, as shown in the idealization of the structure in Fig. 14.14b. Only the two center bays can be active in resisting laterally applied loads, and for the applied load P, the members in compression in the X-bracing can be assumed to be ineffective in providing resistance. This leaves two members in tension, one in each bay, to resist the lateral load, which can be assumed to be distributed equally between them. For the proportions of the structure the horizontal component of the member tension force, T, is equal to

$$\frac{37}{(37^2 + 40^2)^{1/2}}T = \frac{37}{\sqrt{2969}}T$$

For the lateral load, P, the tension in each member of each bay is approximately $(\sqrt{2969}/37)(P/2) = 0.74P$.

Another type of structure that can be analyzed effectively by an assumed force distribution is the truss of Fig. 14.15.

Figure 14.14a Industrial building with X-bracing in two bays. (Courtesy Bruce W. Abbott.)

Figure 14.15 The truss is indeterminate to the first degree. The type of diagonal web members shown in this truss is common in three-dimensional structures. The loads acting on the three-dimensional structure can be resolved into the planes of the trusses that make up the structure, and a plane truss analysis carried out. The assumption that can be made in this case is to divide the truss into two trusses, each carrying the loading shown. Each of these two trusses is statically determinate and can be analyzed in the usual manner. When a force acts on a joint that is common to both of the divided structures, it is assumed to be divided equally between the two, as shown by the load P_1. The final member forces in the approximate analysis are assumed to be the sum of the member forces obtained from the analysis of each structure.

587

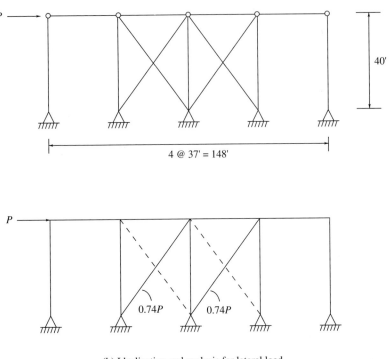

(b) Idealization and analysis for lateral load

Figure 14.14b Industrial building with X-bracing in two bays.

14.8   **Summary and Limitations**

The approximate analysis of highly indeterminate structures by assuming distributions of internal forces or moments can be carried out through the use of equilibrium equations. For an indeterminate structure, the introduction of a number of assumptions equal to the degree of indeterminacy of the structure reduces it to a statically determinate structure. For preliminary design, such an analysis provides a starting point or can be used as a means of comparing several alternative design concepts without having to go through sophisticated and time-consuming analyses.

A discussion of the location of zero-moment points in vertically loaded members of a large frame provides a means of calculating internal moments. The internal distribution of forces and moments due to lateral loads on a highly indeterminate rigid frame is obtained by approximate analysis using the portal or cantilever method. Both methods assume that there are points of zero moment in the center of the beams and column of the frame. The portal method makes the additional assumption that the shear forces in any story of the frame are distributed equally in the bays. The cantilever

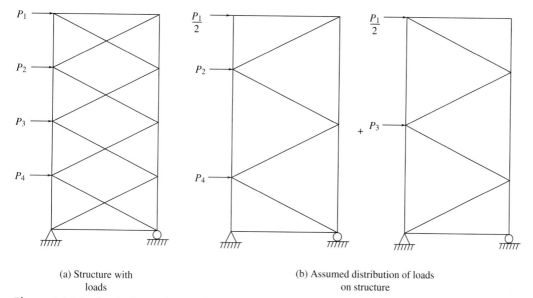

(a) Structure with loads

(b) Assumed distribution of loads on structure

Figure 14.15a–b   Indeterminate plane truss with assumed division by load path for applied loads.

method makes the additional assumption that the axial forces in the columns are distributed so as to create a linear variation of axial stress in the columns across a horizontal plane passing through the zero-moment points in the center of the columns.

Estimation of the lateral displacements of a structure based on the results of an approximate analysis for shears and moments is discussed in Section 14.6. The analysis is based on obtaining the lateral displacement of each story and summing these from the base of the structure to the story where the displacement is desired. Approximate member properties are assumed to be obtained from a simplified design based on the forces and moments available from an analysis by the portal or cantilever method.

The final section of the chapter provides a means of obtaining the internal forces in indeterminate trusses. The nature of the physical behavior and the method of construction provide the basis for the assumptions of the analysis.

In using approximate methods of analysis the structural engineer should be aware of several assumptions in their use.

- The deformations of the structure are very small, so that the equilibrium equations can be established using the dimensions of the original undeformed structural geometry.

- The deformations of the structure under the assumed internal distribution of forces and moments do not satisfy compatibility.

- The estimation of lateral displacements using Eq. (14.3) assumes that the member properties of the column and the beams used in the calculation are representative of the properties and behavior of the remaining members in the structure.

The use of approximate methods provides a means for hand computations to be effective in the preliminary design stage. Some questions about these methods that are worthy of consideration are the following:

- Some of these methods can be programmed for computer use, but how does one evaluate the correctness of the computer-generated solution?
- Explain how these methods can, or cannot, be extended to be used with structures that are being constructed of nonlinear materials.
- What is the effect of making more assumptions about the internal distribution of forces and moments than the degree of indeterminacy of the structure?

PROBLEMS

14.1 Obtain the moments for the laterally loaded frame of Fig. 14.11 by the portal and cantilever methods and show that they both give the same result.

14.2 Draw the final moment diagram for the structure shown using the portal method.

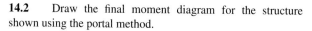
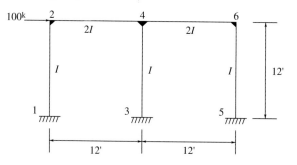

14.3 Analyze and draw the final moment diagram for the frame shown using both the portal and cantilever methods.

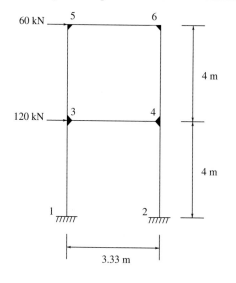

14.4 Analyze and draw the final moment diagram for the frame shown using the portal method.

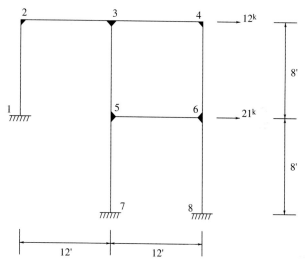

14.5 Use the portal method to draw the final moment diagram for the structure.

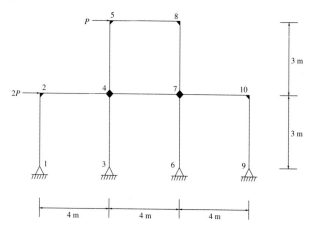

591

14.6   Analyze and draw the final moment diagram for the frame shown using the portal method.

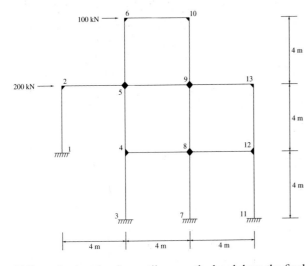

14.8   Analyze by the cantilever method and draw the final moment diagram.

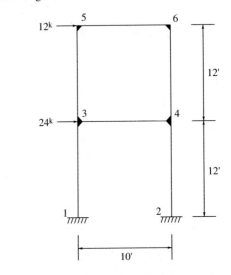

14.7   Analyze by the cantilever method and draw the final moment diagram.

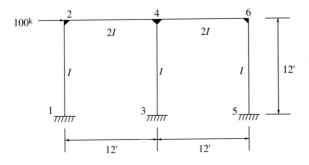

14.9   Using the portal method, draw the final moment diagram for the structure shown.

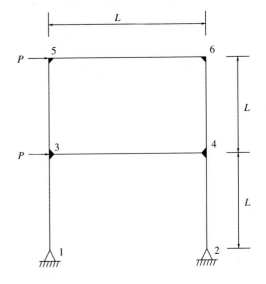

14.10   Use a modification of the cantilever method to obtain the final moment diagram.

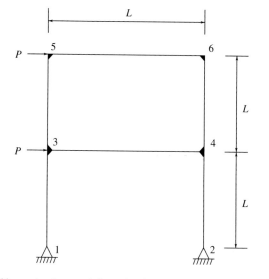

14.11   Analyze and draw the final moment diagram for the frame shown using the portal method.

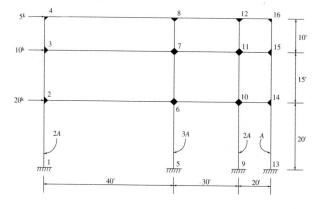

14.12   Analyze and draw the final moment diagram for the frame shown using the cantilever method.

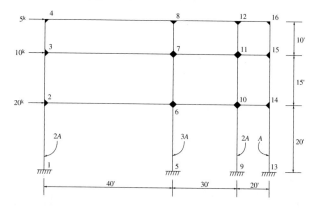

14.13   For the rigidly connected Vierendeel truss shown, use an appropriate approximate method of analysis to obtain the axial forces and moments in all members. Draw the moment diagram for the analysis.

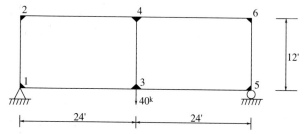

14.14   For the rigidly connected Vierendeel truss shown, use an appropriate approximate method of analysis to obtain the axial forces and moments in all members. Draw the moment diagram for the analysis.

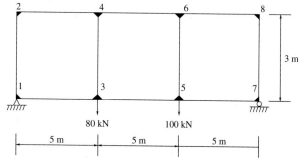

593

14.15(a) Using the principles of the portal method, obtain the moment diagram for the laterally loaded frame.

(b) Explain why the cantilever method cannot be used for this structure.

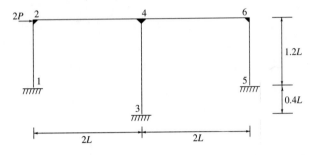

14.17 Analyze and draw the final moment diagram for the frame using the cantilever method.

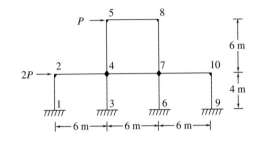

14.16 Use an appropriate modification of the portal method to obtain the final moment diagram for the frame shown.

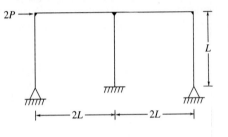

Approximate Analysis Using Assumed Deformations

In Chapter 14 highly indeterminate structures were analyzed by making assumptions about the distribution of moments, axial forces, or shears. The number of assumptions made was equal to the degree of indeterminacy of the structure, and the equations of equilibrium were then utilized to obtain a complete analysis of the structure. Deformations were largely ignored in this process or were calculated on the basis of the assumed internal distribution of forces and moments. The procedures were well suited for preliminary design and relatively simple hand calculations.

In this chapter the deformation behavior of structures is used as a basis for making approximate analyses of structural elements as well as complete structures. Equilibrium equations are always available when assumptions are made about the distribution of force actions in structures. When assumptions are made about deformation behavior of structures, the internal forces and moments are obtained from the force–deformation characteristics of the structural elements and equilibrium conditions are derived from energy considerations. An example of this for rigid bodies is the use of the virtual work principle in Chapter 7 to obtain the reactions of structures based on an imagined (or assumed) deformation system.

A review of the concept of strain energy, complementary strain energy, potential energy, and complementary potential energy is presented in Section 15.1. These energy considerations are developed in the context of linear members, that is, members in which the length of a member along its axis is several times its width and depth. Although strain energy will be presented in a general form, the use of strain energy in all subsequent treatments will be limited to materials with linear stress–strain relations.

The development and use of Castigliano's theorems is presented in Section 15.2. Their use in the computation of the deformations of structures and their relation to virtual work and complementary virtual work is presented. The use of Castigliano's first theorem for the approximate analysis of nonprismatic members is presented.

The chapter concludes with the analysis in Sections 15.4 to 15.6 of laterally loaded plane frame structures. The use of energy concepts is shown to be the basis of the widely used shear building approximation of lateral displacement behavior of rigid frames. Energy concepts also are shown to be a means of developing simplified mathematical models of structural behavior that are easily analyzed with computer programs based on assumed displacement analysis.

15.1   Strain, Complementary Strain, Potential, and Complementary Potential Energy

Strain and Complementary Strain Energy

In a structure made from an elastic material the members deform under the action of externally applied forces and moments, but when these actions are removed the members return to their original undeformed configurations. The recovery of these members is due to internal stored energy, more commonly called *strain energy*, that is created during the application of the loads. In Fig. 15.1 the internal strain energy stored in an element of material is defined as the area under the elastic stress–strain curve and is expressed as

$$u = \int_0^\varepsilon \sigma \, d\varepsilon = \int_0^\varepsilon f(\varepsilon) \, d\varepsilon \tag{15.1}$$

where u is the strain energy density, that is, the strain energy per unit volume. The total strain energy, U, stored in a structural member is obtained by summing the strain energy in all of the elements of the member or simply by the integration over the volume of the member of the strain energy density, which yields

$$U = \int_{\text{vol}} u \, dV \tag{15.2}$$

Once the stress–strain relation, $\sigma = f(\varepsilon)$, is defined, the strain energy density, u, can be obtained for the material from Eq. (15.1).

Also shown in Fig. 15.1 is the area between the stress–strain relation and the σ axis. This area is called the *complementary strain energy density*, or the complementary strain energy per unit volume of the material, u_c. From the figure, this area is

$$u_c = \int_0^\sigma \varepsilon \, d\sigma = \int_0^\sigma g(\sigma) \, d\sigma \tag{15.3}$$

where the stress–strain relation is inverted and expressed in the alternative form as a strain–stress relation, $\varepsilon = g(\sigma)$. The total complementary strain energy associated with a member is obtained by summing or integrating u_c over the volume of the member, which yields

$$U_c = \int_{vol} u_c \, dV \tag{15.4}$$

Again, the specification of the strain–stress relation $\varepsilon = g(\sigma)$ enables the complementary strain energy density u_c to be obtained from Eq. (15.3).

The complementary strain energy does not have a physical interpretation as does the strain energy for a member given by Eq. (15.2), but it has a very important role in the mathematical analysis of structural behavior. The computation of deflections by the complementary virtual work principle in Chapter 7 is closely related to complementary strain energy, and the development of the analysis of indeterminate structures is also possible from complementary strain energy considerations. These ideas are explored later in this section.

Equations (15.2) and (15.4) are valid for any stress–strain (or strain–stress) relation, but only a linear relation is considered in this chapter. For linear members fabricated with a linear elastic material having

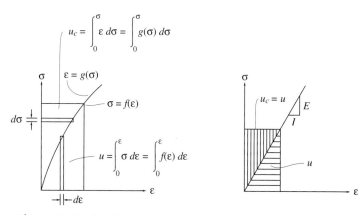

Figure 15.1 Strain energy and complementary strain energy for an elastic material.

modulus of elasticity E, the stress–strain relation is $\sigma = E\varepsilon$ and the strain–stress relation is $\varepsilon = \sigma/E$. Substituting these relations into Eqs. (15.1) and (15.3), the strain energy and complementary strain energy densities become

$$u = \int_0^\varepsilon E\varepsilon \, d\varepsilon = \frac{1}{2}E\varepsilon^2 \tag{15.5a}$$

$$u_c = \int_0^\varepsilon \frac{\sigma}{E} \, d\sigma = \frac{1}{2}\frac{\sigma^2}{E} \tag{15.5b}$$

In a linear member of length L with cross-sectional area A, which may vary along its length, the strain and complementary strain energy associated with the member become, from Eqs. (15.2) and (15.4), respectively,

$$U = \int_0^L \int_A \frac{1}{2}E\varepsilon^2 \, dA \, dx \tag{15.6a}$$

$$U = \int_0^L \int_A \frac{1}{2}\frac{\sigma^2}{E} \, dA \, dx \tag{15.6b}$$

The concepts developed in Eqs. (15.1) to (15.6) have been based on the normal strains and stresses that exist in members due to bending and axial force. Associated with bending action and torsional action are shear strains, γ, and stresses, τ. A completely parallel development of strain and complementary strain energy for shear action will yield a series of equations of exactly the same form as Eqs. (15.1) to (15.6), but with τ replacing σ, γ replacing ε, and the shear modulus G replacing E. The resulting pair of equations for linear members with linear elastic materials that correspond to Eq. (15.6) are

$$U = \int_0^L \int_A \frac{1}{2}G\gamma^2 \, dA \, dx \tag{15.7a}$$

$$U = \int_0^L \int_A \frac{1}{2}\frac{\tau^2}{G} \, dA \, dx \tag{15.7b}$$

In linear members the stresses and strains arise from the actions of axial force, torsion, bending, and shear due to bending. From considerations of mechanics of materials [see also Table 1.1 and Eqs. (6.6) and (6.9) and Reference 7, 17, or 26] the stresses and strains associated with these actions are:

Axial force: $\qquad\qquad\qquad\qquad\qquad \sigma = \dfrac{P}{A} \qquad \varepsilon = \varepsilon = \dfrac{du}{dx}$

Torsion:
$$\tau = \frac{T\rho}{J} \qquad \gamma = \rho\left(\frac{d\theta}{dx}\right)^2$$

Bending:
$$\sigma = \frac{-My}{I} \qquad \varepsilon = \frac{-y}{R} = -y\phi = \frac{-yd^2v}{dx^2}$$

Shear from bending:
$$\tau = \frac{VQ}{Ib} \qquad \gamma = \frac{VQ}{GIb}$$

In these expressions internal force actions are axial force, P, torque, T, bending moment, M, and shear due to bending, V.

The stress, σ, and strain, ε, associated with axial action are constant on the cross section, but the stresses and strains associated with the other actions vary with position in the cross section. The position coordinate for torsion is the radius ρ, for bending the vertical distance y and for shear due to bending the coordinate is implicit in the variable Q, which is the first moment of a portion of the area of the cross section about the centroidal axis. All of these position coordinates use the centroidal axis of the member as the reference axis.

The constants, area A, polar moment of inertia J, and moment of inertia I, are geometric properties of the cross section and are defined as

$$A = \int_A dA$$

$$I = \int_A y^2\, dA$$

$$J = \int_A \rho^2\, dA$$

The displacement functions u for axial action, θ for torsion action, and v for vertical displacement in bending action are all functions of position along the length of the member.

Using the definitions in the preceding paragraph, the strain energy in a member of length L becomes, from Eqs. (15.6a) and (15.7a),

$$U = \frac{1}{2}\int_0^L\int_A E\varepsilon^2\, dA\, dx = \frac{1}{2}\int_0^L AE\varepsilon^2\, dx = \frac{1}{2}\int_0^L AE\left(\frac{du}{dx}\right)^2 dx \quad (15.8a)$$

due to axial action,

$$U = \frac{1}{2}\int_0^L\int_A G\left(\rho\frac{d\theta}{dx}\right)^2 dA\, dx = \frac{1}{2}\int_0^L GJ\left(\frac{d\theta}{dx}\right)^2 dx \qquad (15.8b)$$

599

total complementary energy. Again, only stable equilibrium states will be considered in this book, so the title *minimum total complementary energy* will be used. Note that the change in the complementary strain energy, δU_c, becomes equal to the change in complementary potential energy of the applied loads, δV_c, at the stationary point. In the complementary virtual work principle, Eq. (7.4), the change in the external work, or simply the external work, corresponds to δV_c since it is the product of *specified displacements and rotations* with the arbitrary forces and moments. The change in internal work, or simply the internal work, corresponds to δU_c since it is the product of the *known or specified strains* with the arbitrary stresses. The complementary virtual work principle and the principle of minimum total complementary energy are equivalent at the point where the structure is in stable equilibrium and the displacements and deformations are compatible.

Equations (15.12) and (15.13) thus form the basis of the analysis in the following sections of structures by energy methods. They represent alternative forms of the two virtual work principles, Eqs. (7.3) and (7.4). These mathematically derived methods are extremely flexible in their application and provide the foundation of many of the methods that have been programmed for computer analysis of structures. Further discussions of these energy methods, as well as their relation to the method discussed in the next section, can be found in Reference 24.

15.2 Castigliano's and Engesser's Theorems

The development of Castigliano's theorems was originally based on detailed analysis of the strain energy of structures with linear elastic materials. These theorems are specializations of the principles of minimum total potential and complementary energy and will be presented in that context. The second theorem is limited in application to structures having linear elastic materials, as is discussed later.

Consider a structure in equilibrium in a displaced state. From Eq. (15.12), any change from that state which is occasioned by a small change in some displacement, $d\Delta_i$, and in the direction of a force, P_i, acting at that point will change δV by an amount $P_i d\Delta_i$. The change in internal strain energy δU will be $(\partial U/\partial \Delta_i)d\Delta_i$. Because $\delta U = \delta V$ in Eq. (15.12) this results in

$$\frac{\partial U}{\partial \Delta_i} d\Delta_i = P_i d\Delta_i \quad \text{or} \quad \frac{\partial U}{\partial \Delta_i} = P_i \qquad (15.14)$$

Equation (15.14) is the mathematical presentation of Castigliano's first theorem, which can be stated as follows:

The displacement function for a pri
from a linearly elastic material is de
member is constant and therefore it m
displacements divided by L. This yield
the figure, which can be used as the
nonprismatic member axial deformatic
ber subjected to a constant axial force i
two examples.

Example 15.1 The displacement
under constant axial force shown in
formation of the member. Th
Castigliano's first theorem is in error
$0.69315PL/AE$.

Example 15.2 An improved resul
Example 15.1 is divided into two seg
center of the member being an unkno
The strain energy is computed in the
is expressed in terms of two displacer
applied on the member, the force P
Castigliano's first theorem is used twi
are solved for u_3 and u_2, which is the
is now reduced to 1%.

The deformation behavior of nonpris
treated in a similar way as the axial force
ber in the previous two examples. The dev
tion that describes bending of prismatic n

Figure 15.3 The assumption is made
tween its ends but that both a displace
the ends of the member. The curvature
be taken directly from the curvatures s
being zero due to the absence of the
$\phi_A = \phi_{AB} = M_{AB}/EI$ and $\phi_B = \phi_{BA} = M$
tion presented in Fig. 15.3. The direct i
tion using Eq. (6.9) yields a displaceme
constants C_1 and C_2 are evaluated using
θ_A at the left end at $x = 0$.

The curvatures ϕ_A and ϕ_B can be eli
placement expression, by using the s
(11.3), in which the FEMs are zero si
$q(x)$. Dividing both sides of those equa
M_{BA}/EI on the left-hand sides, which a
ϕ_{AB} ($= \phi_A$) and ϕ_{BA} ($= \phi_B$), respecti

> For a structure in equilibrium and in which displacements are small, the temperature is constant, and the supports are unyielding, the first partial derivative of the strain energy for that structure with respect to a displacement is equal to the force in the direction of that displacement.

The theorem defines the equilibrium condition for the force and provides a means of solving problems in which displacements are unknown. The displacement, Δ_i and force, P_i, in Eq. (15.14) can be replaced with a rotation, θ_i, and moment, M_i, respectively, and the equation provides the equilibrium condition for the moment. As will be seen later, Castigliano's first theorem will provide a powerful means of obtaining approximate solutions to structural problems based on assumed displacement functions.

Again consider a structure in equilibrium in a displaced state. From Eq. (15.13), any change from that state which is occasioned by a small change in some force, dP_i, and in the direction of a specified displacement, Δ_i, acting at that point will change δV_c by an amount $\Delta_i dP_i$. The change in internal complementary strain energy δU_c will be $(\partial U_C/\partial P_i)dP_i$. Because $\delta U_c = \delta V_c$ in Eq. (15.13), this results in

$$\frac{\partial U_c}{\partial P_i} dP_i = \Delta_i \, dP_i \quad \text{or} \quad \frac{\partial U_c}{\partial P_i} = \Delta_i \tag{15.15}$$

Equation (15.15) is the mathematical presentation of Engesser's theorem, which can be stated as follows:

> For a structure in equilibrium and in which displacements are small, the temperature is constant, and the supports are unyielding, the first partial derivative of the complementary strain energy for that structure with respect to a force is equal to the displacement of the structure in the direction of that force.

The theorem defines the compatibility condition for the force and provides a means of solving problems in which applied loads are unknown. The analysis by the method of consistent deformations of indeterminate structures in which the magnitude of the redundant forces or moments are unknown can be obtained from Engesser's theorem. The displacement, Δ_i, and force, P_i, in Eq. (15.15) can be replaced with a rotation, θ_i, and moment,

605

Example 15.1

For the nonprismatic member shown under axial load, obtain the end displacement using Castigliano's first theorem and a prismatic member displacement relation.

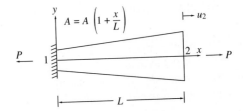

STEP 1   Assume the displacement function for the prismatic member given in Fig. 15.2, and use Eq. (15.8a) to obtain the strain energy for the nonprismatic member.

$$u = u_1\left(1 - \frac{x}{L}\right) + u_2\frac{x}{L} = u_2\frac{x}{L}$$

$$\frac{du}{dx} = \varepsilon = \frac{u_2}{L}$$

$$U = \frac{1}{2}\int_0^L AE\left(\frac{du}{dx}\right)^2 dx = \frac{1}{2}\int_0^L A\left(1 + \frac{x}{L}\right)E\left(\frac{u_2}{L}\right)^2 dx = \frac{3}{4}\left(\frac{AE}{L}\right)u_2^2$$

STEP 2   Apply Castigliano's first theorem, Eq. (15.14), to obtain the end displacement u_2.

$$\frac{\partial U}{\partial u_2} = \frac{3}{4}\left(\frac{AE}{L}\right)2u_2 = \frac{3AE}{2L}u_2 = P \qquad \therefore u_2 = \frac{2PL}{3AE} = 0.667\frac{PL}{AE}$$

15.3   Approximate

Example 15.2

For the nonprismatic member of Example 15.1, obtain the end displacement u_2 by dividing the member into two segments and using the prismatic member displacement function in each segment.

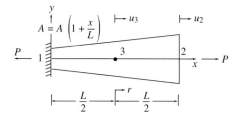

STEP 1 Establish the displacement function for each segment from Fig. 15.2. Take the displacement at the center of the member as u_3 and use r as the coordinate defining the displacement function in the right half of the member.

$$1\text{–}3: \quad u = u_1\left(1 - \frac{2x}{L}\right) + u_3\left(\frac{2x}{L}\right) = 2u_3\frac{x}{L} \quad A = A\left(1 + \frac{x}{L}\right)$$

$$3\text{–}2: \quad u = u_3\left(1 - \frac{2r}{L}\right) + u_2\frac{2r}{L} \quad A = A\left(\frac{3}{2} + \frac{r}{L}\right)$$

$$\varepsilon_{13} = \frac{du}{dx} = \frac{2u_3}{L} \quad \varepsilon_{32} = \frac{du}{dr} = \frac{2(u_2 - u_3)}{L}$$

STEP 2 Obtain the strain energy from Eq. (15.8a) for the whole member.

$$U = \frac{1}{2}\int_0^{L/2} AE\left(1 + \frac{x}{L}\right)\frac{4}{L^2}u_3^2 \, dx + \frac{1}{2}\int_0^{L/2} AE\left(\frac{3}{2} + \frac{r}{L}\right)\frac{4}{L^2}(u_2 - u_3)^2 \, dr$$

$$U = \frac{5AE}{4L}u_3^2 - \frac{7AE}{4L}(u_2 - u_3)^2$$

Example 15.2 (continued)

STEP 3　Apply Castigliano's first theorem, Eq. (15.14), twice, once at the center for U_3, where there is no applied load, and once at the right end, where the applied load is P.

$$\frac{\partial U}{\partial u_3} = 0 = \frac{5AE}{2L}u_3 - \frac{7AE}{2L}(u_2 - u_3)$$

$$\frac{\partial U}{\partial u_2} = P = \frac{7AE}{2L}(u_2 - u_3)$$

STEP 4　Solve the two equations for the displacements u_2 and u_3.

$$\begin{Bmatrix} 0 \\ P \end{Bmatrix} = \frac{AE}{L}\begin{bmatrix} 6 & -7/2 \\ -7/2 & 7/2 \end{bmatrix}\begin{Bmatrix} u_3 \\ u_2 \end{Bmatrix} \quad \therefore \quad \begin{Bmatrix} u_3 \\ u_2 \end{Bmatrix} = \frac{PL}{AE}\begin{Bmatrix} 2/5 \\ 24/35 \end{Bmatrix}$$

Thus

$$u_2 = \frac{24}{35}\frac{PL}{AE} = 0.686\frac{PL}{AE}$$

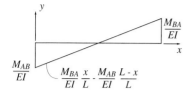

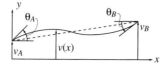

Curvature from Fig. 11.4   Displacement function $v(x)$

Define curvatures ϕ_A and ϕ_B: $\phi_A = \dfrac{M_{AB}}{EI}$ $\phi_B = \dfrac{M_{BA}}{EI}$

The curvature function becomes: $\phi = \dfrac{M}{EI} = \phi_B \dfrac{x}{L} - \phi_A \dfrac{L-x}{L}$

Use Eq. (6.9) and integrate twice to obtain $v(x)$. Use conditions at $x = 0$ that $v(0) = v_A$ and $dv/dx\,(0) = \theta_A$ to evaluate the constant on integration.

$$\frac{d^2v}{dx^2} = \phi = \phi_B \frac{x}{L} - \phi_A \frac{L-x}{L}$$

$$\frac{dv}{dx} = \phi_B \frac{x^2}{2L} + \phi_A \frac{(L-x)^2}{2L} + C_1$$

$$v = \phi_B \frac{x^3}{6L} - \phi_A \frac{(L-x)^3}{6L} + C_1 x + C_2$$

$$\frac{dv}{dx}(0) = \theta_A = \phi_A \frac{L}{2} + C_1 \qquad v(0) = v_A = -\phi_A \frac{L^2}{6} + C_2$$

$$\therefore C_1 = \theta_A - \phi_A \frac{L}{2} \qquad\qquad \therefore C_2 = v_A + \phi_A \frac{L^2}{6}$$

$$\therefore v(x) = \phi_B \frac{x^3}{6L} + \phi_A \left[\frac{L^2}{6} - \frac{Lx}{2} - \frac{(L-x)^3}{6L} \right] + v_A + \theta_A x$$

Substitute for ϕ_A and ϕ_B from Eq. (11.3), noting that the FEMs are zero, and simplify the result.

$$\phi_B = \frac{2[\theta_A + 2\theta_B - 3(v_B - v_A)/L]}{L} \qquad \phi_A = \frac{2[2\theta_A + \theta_B - 3(v_B - v_A)/L]}{L}$$

$$v(x) = \frac{x^3}{3L^2}\left(\theta_A + 2\theta_B - \frac{3(v_B - v_A)}{L}\right) + \frac{2}{L}\left(2\theta_A + \theta_B - \frac{3(v_B - v_A)}{L}\right)\left(\frac{L^2}{6} - \frac{Lx}{2} + \frac{(L-x)^3}{6L}\right) + v_A + \theta_A x$$

Collecting terms yields

$$v(x) = v_A\left(1 + \frac{2x^3}{L^3} - \frac{3x^2}{L^2}\right) + \theta_A x\left(1 - \frac{2x}{L} + \frac{x^2}{L^2}\right) + v_B \frac{x^2}{L^2}\left(3 - \frac{2x}{L}\right) + \theta_B \frac{x^2}{L}\left(\frac{x}{L} - 1\right)$$

Displacement function $v(x)$ in terms of end displacements v_A and v_B and end rotations θ_A and θ_B

Figure 15.3   Displacement function $v(x)$ for a prismatic member undergoing deformations from a linear variation of curvature.

Eq. (11.3) divided by EI are the curvature functions for ϕ_B and ϕ_A presented in Fig. 15.3. Substituting these expressions, ϕ_A and ϕ_B into the expression for $v(x)$ and simplifying yields the desired cubic relation in terms of the end displacements, v_A and v_B, and end rotations, θ_A and θ_B.

The use of the cubic displacement function for a prismatic member presented in Fig. 15.3 to approximate the behavior of a nonprismatic member is illustrated by Example 15.3.

Example 15.3 The cantilever beam has a moment of inertia that varies linearly. The strain energy is established using the boxed cubic displacement function from Fig. 15.3 for a prismatic member as an approximation of the displaced shape of the member. The curvature function required for the strain energy defined in Eq. (15.8c) is obtained from the second derivative of the assumed cubic function. The integral for the strain energy in Eq. (15.8c) becomes a cubic polynomial which is the product of three linear functions, the $I(x)$ function and the square of the curvature function. There are two displacements that can occur, so that Castigliano's first theorem is used at the left end, where the vertical load (P) and the applied end moment ($= 0$) are known. The resulting two expressions have three different integrals of cubic functions, which can be evaluated by direct integration or by numerical integration with the use of Eq. (9.3). This later option is chosen with the three linear functions g_1, g_2, and g_3 defined as shown. The solution for θ_1 and v_1 gives values that are in error by no more than 0.5%.

The error in the displacements is very small in Example 15.3 because the variation of $I(x)$ is very gradual, but other variations of the moment of inertia of the beam can have much larger errors. The accuracy of the approximate analysis in these situations can be improved if the nonprismatic beam is divided into two segments and the cubic displacement function for a prismatic member used in each segment. This is analogous to the procedure followed in Example 15.2 for the axially loaded member. For the bending action of a cantilever beam like that in Example 15.3, the use of two segments would require the application of Castigliano's first theorem four times to obtain the four equilibrium equations, since there are two unknown displacements and two unknown rotations.

Castigliano's first theorem was used in the preceding examples to obtain the equilibrium equations for solution. As explained in Section 15.2, Castigliano's first theorem is identical in concept to minimizing the total potential energy of the structure. Thus the same results would be obtained in these examples if the total potential energy, Eq. (15.10), were minimized with respect to the joint displacements and rotations, which is equivalent to using Eq. (15.12). The procedures followed in Examples 15.1 to 15.3 are special applications of a more general method of approximate analysis

Example 15.3

For the tapered cantilever beam shown, obtain the displacement and rotation of the left end using the prismatic member displacement function of Fig. 15.3.

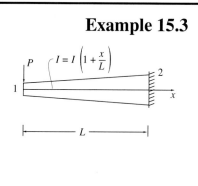

STEP 1 Obtain the displacement function from Fig. 15.3 and differentiate it twice to obtain the curvature expression. Note that $A \rightarrow 1$ and $B \rightarrow 2$.

$$\theta_2 = 0 \text{ and } v_2 = 0 \quad \therefore v(x) = v_1\left(1 + \frac{2x^3}{L^3} - \frac{3x^2}{L^2}\right) + \theta_1 x\left(1 - \frac{2x}{L} + \frac{x^2}{L^2}\right)$$

$$\frac{d^2x}{dx^2} = v_1\left(\frac{12x}{L^3} - \frac{6}{L^2}\right) + \theta_1\left(\frac{-4}{L} + \frac{6x}{L^2}\right) = \frac{6}{L^2}\left(\frac{2x}{L} - 1\right)v_1 + \frac{2}{L}\left(\frac{3x}{L} - 2\right)\theta_1$$

STEP 2 Compute the strain energy in the member using Eq. (15.8c).

$$U = \frac{1}{2}\int_0^L EI\left(\frac{d^2v}{dx^2}\right)^2 dx = \int_0^L \frac{1}{2}EI\left(1 + \frac{x}{L}\right)\left[\frac{6}{L^2}\left(\frac{2x}{L} - 1\right)v_1 + \frac{2}{L}\left(\frac{3x}{L} - 2\right)\theta_1\right]^2 dx$$

STEP 3 Use Castigliano's first theorem, Eq. (15.14), twice and differentiate before performing the integration. Note that the applied load P acts in the negative coordinate direction and there is no moment acting on the left end of the beam. Express the integrals as the product of defined functions.

$$\frac{\partial U}{\partial v_1} = -P = EI\int_0^L\left(1 + \frac{x}{L}\right)\left[\frac{6}{L^2}\left(\frac{2x}{L} - 1\right)v_1 + \frac{2}{L}\left(\frac{3x}{L} - 2\right)\theta_1\right]\frac{6}{L^2}\left(\frac{2x}{L} - 1\right) dx$$

$$\frac{\partial U}{\partial \theta_1} = 0 = EI\int_0^L\left(1 + \frac{x}{L}\right)\left[\frac{6}{L^2}\left(\frac{2x}{L} - 1\right)v_1 + \frac{2}{L}\left(\frac{3x}{L} - 2\right)\theta_1\right]\frac{2}{L}\left(\frac{3x}{L} - 2\right) dx$$

Define $g_1 = (1 + x/L)$, $g_2 = (2x/L - 1)$, and $g_3 = (3x/L - 2)$. Then

$$\frac{\partial U}{\partial v_1} = -P = EI\int_0^L g_1\left(\frac{6}{L^2}g_2 v_1 + \frac{2}{L}g_3\theta_1\right)\frac{6}{L^2}g_2 \, dx = EI(J_1 v_1 + J_2\theta_1)$$

$$\frac{\partial U}{\partial v_1} = 0 = EI\int_0^L g_1\left(\frac{6}{L^2}g_2 v_1 + \frac{2}{L}g_3\theta_1\right)\frac{2}{L}g_3 \, dx = EI(J_2 v_1 + J_3\theta_1)$$

Example 15.3 (continued)

where

$$J_1 = \int_0^L \frac{36}{L^4} g_1 g_2^2 \, dx \quad J_2 = \int_0^L \frac{12}{L^3} g_1 g_2 g_3 \, dx \quad \text{and} \quad J_3 = \int_0^L \frac{4}{L^2} g_1 g_3^2 \, dx$$

STEP 4   The integrals can be evaluated conveniently by numerical integration. Note that Eq. (9.3), Simpson's rule, will give the exact results.

x/L	g_1	g_2	g_3	$g_1 g_2^2$	$g_1 g_2 g_3$	$g_1 g_3^2$
0	1	-1	-2	1	2	4
$\frac{1}{2}$	$\frac{3}{2}$	0	$-\frac{1}{2}$	0	0	$\frac{3}{8}$
1	2	1	1	2	2	2

$$J_1 = \frac{36L}{L^4 6}(1 + 4 \cdot 0 + 2) = \frac{18}{L^3}$$

$$J_2 = \frac{12L}{L^3 6}(2 + 4 \cdot 0 + 2) = \frac{8}{L^2}$$

$$J_3 = \frac{4L}{L^2 6}(4 + 4 \cdot \frac{3}{8} + 2) = \frac{5}{L}$$

STEP 5   Collect and solve the equations from step 3 using the values of the integrals, J_i, from step 4. Compare the approximate solution with the exact values, $\theta_1 = 0.3068PL^2/EI$ and $v_1 = -0.1931PL^3/EI$.

$$EI \begin{bmatrix} 18/L^3 & 8/L^2 \\ 8/L^2 & 5/L \end{bmatrix} \begin{Bmatrix} v_1 \\ \theta_1 \end{Bmatrix} = -P \begin{Bmatrix} 1 \\ 0 \end{Bmatrix}$$

Solving yields

$$\begin{Bmatrix} v_1 \\ \theta_1 \end{Bmatrix} = \frac{PL^2}{EI} \begin{Bmatrix} -5L/26 \\ 8/26 \end{Bmatrix} = \frac{PL^2}{EI} \begin{Bmatrix} -0.192L \\ 0.308 \end{Bmatrix}$$

known as the finite element method. More details about this method are given in References 5 and 11.

15.4 Approximate Analysis of Single-Story Frames

In Chapter 14 the approximate analysis of single-story frames was based on assumptions about where the moment in a member was equal to zero. Another approach is to make an assumption about the deformation behavior of the frame and apply minimum total potential energy, Eq. (15.12), to obtain the approximate result. Castigliano's first theorem can also be used in many cases, but the minimum total potential energy approach is more general and can easily treat multiple applied loads. This approach uses the strain energy stored in a structure, so to aid in the computation, the strain energy stored in a single member undergoing bending action is derived in Fig. 15.4.

> **Figure 15.4** The linear curvature variation in the member is taken from Fig. 15.3, which, in turn, is taken from the slope-deflection equations with no applied loads acting transverse to the member. Substitution for the curvatures ϕ_A and ϕ_B is made from Eq. (11.3) divided by EI, as illustrated in Fig. 15.3. After simplification, the resulting expression for the strain energy in a member is exact for any prismatic member made with linear elastic material undergoing small arbitrary end displacements and rotations. Note that the positive sense of displacements is in the direction of the individual member coordinate axis and rotations are positive counterclockwise.

The process of making an approximate analysis of a laterally loaded frame is to assume some type of deformation behavior of the structure. In Section 15.3 this involved assuming a displacement function along the member and using Castigliano's first theorem to establish the equilibrium equations for solution. For laterally loaded frames with prismatic members and using the boxed expression in Fig. 15.4 for the strain energy in each member, the assumption of a pattern of displacements and rotations of the joints of the structure is all that is required. Under lateral loading a single-story frame undergoes a single horizontal joint displacement, since axial deformations are so small they may be neglected. This reduces the approximation procedure to assuming a pattern of joint rotations in the frame. The next example illustrates this concept.

> **Example 15.4** The frame analyzed by moment distribution in Example 12.6 is analyzed assuming that all joint rotations are equal. The two unknown displacements in the structure become the joint displacement at D ($=$ to the joint displacements at E and F) and a rotation that is assumed

615

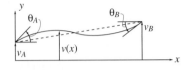

Member with displacements and rotations

Strain energy from Eq. (15.8c): $\quad U = \dfrac{1}{2} \displaystyle\int_0^L EI \left(\dfrac{d^2v}{dx^2} \right)^2 dx = \dfrac{1}{2} \displaystyle\int_0^L EI \, \phi^2 \, dx$

Curvature from Fig. 11.4 or 15.3: $\quad \phi = \phi_B \dfrac{x}{L} - \phi_A \dfrac{L-x}{L}$

Substitute into Eq. (15.8c) and integrate:

$$U = \frac{1}{2} \int_0^L EI \, \phi^2 \, dx = \frac{1}{2} \int_0^L EI \left(\phi_B^2 \frac{x^2}{L^2} - 2 \, \phi_A \, \phi_B \, \frac{x}{L} \, (L-x) + \phi_A^2 \, \frac{(L-x)^2}{L^2} \right) dx$$

$$U = \frac{EIL}{6} \left(\phi_B^2 - \phi_A \, \phi_B + \phi_A^2 \right)$$

Curvatures ϕ_A and ϕ_B from Eq. (11.3) or Fig. 15.3:

$$\phi_B = \frac{2[\theta_A + 2\theta_B - 3(v_B - v_A)/L]}{L} \qquad \phi_A = \frac{2[2\theta_A + \theta_B - 3(v_B - v_A)/L]}{L}$$

Compute ϕ_A^2, ϕ_B^2, and $\phi_A\phi_B$:

$$\phi_A^2 = \frac{4}{L^2} \left[4\theta_A^2 + \theta_B^2 + \frac{9(v_B - v_A)^2}{L^2} + 4\theta_A \theta_B - 12\theta_A \, \frac{v_B - v_A}{L} - 6\theta_B \, \frac{v_B - v_A}{L} \right]$$

$$\phi_B^2 = \frac{4}{L^2} \left[\theta_A^2 + 4\theta_B^2 + \frac{9(v_B - v_A)^2}{L^2} + 4\theta_A \theta_B - 6\theta_A \, \frac{v_B - v_A}{L} - 12\theta_B \frac{v_B - v_A}{L} \right]$$

$$\phi_A\phi_B = \frac{4}{L^2} \left[2\theta_A^2 + 2\theta_B^2 + \frac{9(v_B - v_A)^2}{L^2} + 5\theta_A \theta_B - 6\theta_A \, \frac{v_B - v_A}{L} - 3\theta_A \, \frac{v_B - v_A}{L} \right.$$

$$\left. - 6\theta_B \, \frac{v_B - v_A}{L} - 3\theta_B \, \frac{v_B - v_A}{L} \right]$$

Collecting terms in the expression for U yields

$$\boxed{U = B \left[\frac{6(v_B - v_A)^2}{L^2} - 6(\theta_A + \theta_B) \, \frac{v_B - v_A}{L} + 2(\theta_A^2 + \theta_A \theta_B + \theta_B^2) \right]}$$

where $B = EI/L$.

Figure 15.4 Strain energy of prismatic member with arbitrary end displacements v_A and v_B and end rotations θ_A and θ_B.

to be the same at joints *D, E,* and *F.* The strain energy is obtained for each member of the structure and summed. The potential energy of the applied load, *P,* is just *Pu.* The use of the minimum total potential energy principle yields the two equations of solution. Once the displacement and rotation have been determined, the moments in the members of the frame are obtained from the slope-deflection equations. The approximate deflection is very close to the exact value, but the rotation varies considerably from the exact values, showing that the assumption of constant rotation in this frame is not a good one. Except for member *EF* the maximum moment in each member has less than 9% error.

The analysis of the frame in Example 15.4 gave very good results for the displacement of the structure. An examination of the moment diagram for the frame shows that moment equilibrium is not satisfied at the joints. This is a consequence of the assumption of a single rotation for all joints of the frame, when, in fact, there are three quite different rotations. The analysis being performed in the example is an approximate analysis that is appropriate for preliminary design and estimates of lateral displacements. If the frame had been analyzed approximately by the portal method of Chapter 14, moment equilibrium would have been satisfied at all of the joints, but displacements and rotations would not have been compatible. If a portal method analysis is made on this frame, it will be found that the errors in the moments from that analysis are significantly larger than those in Example 15.4.

15.5   Shear Building Approximation

In the preliminary design of buildings an estimate of the lateral deflection or drift is needed and a simple means of obtaining it is through the use of the shear building approximation. Often this same approximation is used in the dynamic analysis of buildings subject to earthquake excitation or as a means of comparing the dynamic characteristics of several preliminary designs. The shear building approximation is based on energy considerations, as is shown in the following discussion using the structure in Fig. 15.5.

Figure 15.5 The *n*-story rigid frame under lateral load is to be approximated by a single "cantilever" member called a *shear building.* The shear building has no rotations at each of the joints, which represents a floor in the original building. The deflection behavior of these two structures can be made equivalent through energy considerations.

The total potential energy of the original structure is taken as *W* and that of the shear building as *W**. Assume in the original building that all of the vertical displacements and rotations of the joints are zero, just as they are in the shear building. In addition, assume that all of the displacements of each floor of the shear building, u_k^*, are equal to the

Example 15.4

Obtain an approximate solution to the laterally loaded single-story frame of Example 12.6. Assume that all rotations at the top of the frame are the same and equal to θ. Use the minimum total potential energy principle.

STEP 1   Assume displacement of u to the right at joint D and counterclockwise rotations θ at joints D, E, and F. Note that in the coordinate systems for the columns the u displacement is negative. Compute the strain energy of all members of the frame using the expression in Fig. 15.4.

Strain energy beams (no displacements, all rotations $= \theta$)

$$U_{DE} = B_{DE}[2(\theta^2 + \theta^2 + \theta^2)] = 6B_{DE}\theta^2 = 6 \cdot 2.08\frac{EI}{L}\theta^2 = 12.48B\theta^2$$

$$U_{EF} = 6B_{EF}\theta^2 = 6 \cdot 2.08\frac{EI}{(3L/2)}\theta^2 = 8.32B\theta^2$$

Strain energy columns (all the same, and $v_B = -u$ and $\theta_B = \theta$ in the expression of Fig. 15.4.):

$$U_{AD} = U_{BE} = U_{CF} = B_{AD}\left[6\frac{(-u-0)^2}{(3L/2)^2} - 60\frac{-u-0}{3L/2} + 2\theta^2\right]$$

$$= \frac{EI}{3L/2}\left[\frac{8}{3}\left(\frac{u}{L}\right)^2 + 4u\frac{\theta}{L} + 2\theta^2\right]$$

$$= B\left[\frac{16}{9}\left(\frac{u}{L}\right)^2 + \frac{8}{3}\theta\frac{u}{L} + \frac{4}{3}\theta^2\right]$$

STEP 2   Compute the potential energy of the loads and the total potential energy of the structure.

$$V = Pu$$

$$U = U_{DE} + U_{EF} + 3U_{AD}$$

$$= 12.48B\theta^2 + 8.32B\theta^2 + 3B\left[\frac{16}{9}\left(\frac{u}{L}\right)^2 + \frac{8}{3}\theta\frac{u}{L} + \frac{4}{3}\theta^2\right]$$

$$= B\left[\frac{16}{3}\left(\frac{u}{L}\right)^2 + 8\theta\frac{u}{L} + 24.8\,\theta^2\right]$$

$$W = U - V = B\left[\frac{16}{3}\left(\frac{u}{L}\right)^2 + 8\theta\left(\frac{u}{L}\right) + 24.8\theta^2\right] - Pu$$

STEP 3   Minimize the total potential energy with respect to u and θ. Solve the resulting equations. Exact value of u is $0.1095PL^3/EI$.

Example 15.4 (continued)

$$\frac{\partial W}{\partial u} = \frac{32}{3}B\frac{u}{L^2} + 8B\frac{\theta}{L} - P = 0$$

$$\frac{\partial W}{\partial \theta} = 8B\frac{u}{L} + 49.6B\theta = 0$$

$$B\begin{bmatrix} 32/(3L^2) & 8/L \\ 8/L & 49.6 \end{bmatrix}\begin{Bmatrix} u \\ \theta \end{Bmatrix} = \begin{Bmatrix} P \\ 0 \end{Bmatrix} \quad \text{Solving:} \quad \begin{Bmatrix} u \\ \theta \end{Bmatrix} = \frac{P}{B}\begin{Bmatrix} 0.10665L^2 \\ -0.17202L \end{Bmatrix}$$

STEP 4   Use the slope-deflection equations, Eqs. (11.3), to obtain the end moments from u and θ.

Columns: $M_{AD} = M_{BE} = M_{CF} = \dfrac{2EI}{3L/2}\left[0 + \theta - \dfrac{3}{3L/2}(-u - 0)\right] = \dfrac{4B}{3}\left(\theta + \dfrac{2u}{L}\right)$

$$= \frac{4}{3}P(-0.017202L + 2 \cdot 0.10665L) = 0.261PL$$

$$M_{DA} = M_{EB} = M_{FC} = \frac{4}{3}B\left(2 \cdot \theta + 0 + 2 \cdot \frac{u}{L}\right)$$

$$= \frac{4}{3}P(-2 \cdot 0.017202L + 2 \cdot 0.10665L) = 0.239PL$$

Beams: $M_{DE} = M_{ED} = 2E\dfrac{2.08I}{L}(2\theta + \theta - 0) = 4.16B(3\theta)$

$$= 12.48B\theta = -0.215PL$$

$$M_{EF} = M_{FE} = 2E\frac{2.08I}{3L/2} \cdot 3\theta = 8.32B\theta = -0.143PL$$

STEP 5   Draw the final moment diagram for the frame.

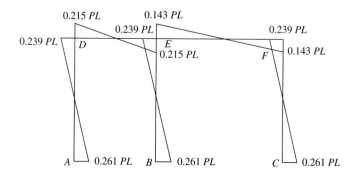

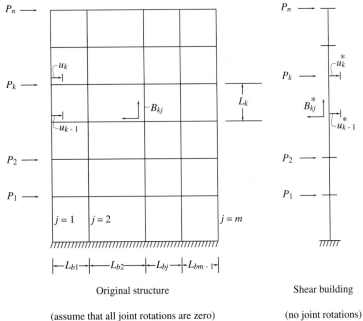

Original structure

(assume that all joint rotations are zero)

Shear building

(no joint rotations)

$$W = U - V$$

$$V = \sum_{k=1}^{n} P_k u_k$$

$$W^* = U^* - V^*$$

$$V^* = \sum_{k=1}^{n} P_k u_k^*$$

For $W = W^*$ assume that $u_k^* = u_k$.

Then $V^* = V$ and $W = W^*$ if $U^* = U$.

Figure 15.5 Equivalence of deformation behavior between a shear building and a rigid frame with certain restrictions on frame deflection behavior.

corresponding displacements of the floors of the original building, u_k, and that the lateral displacements of all joints of floor k in the original structure are the same u_k. As shown in Fig. 15.5, this will make the potential energies V and V^* of the applied loads in the two structures the same. If the strain energy, U^*, in the shear building is the same as the strain energy, U, in the original structure under the assumption of no joint rotations, the total potential energy of the two structures will be the same.

The strain energy developed in the original structure is limited to that in the columns. The beams do not develop strain energy because axial deformations and transverse end displacements and rotations of the joints on the ends of these members are all assumed to be zero. Thus the strain energy in the two structures can be made the same if the strain energy in the column of each story of the shear building is made equal to the strain energy in all the columns in the corresponding story of the original building.

Referring to Fig. 15.5, the strain energy in the column of the kth story of the shear building is obtained from the boxed expression in Fig. 15.4, as, since all rotations are zero,

$$U_k^* = \frac{6B_k^*(u_k - u_{k-1})^2}{L_k^2} \tag{15.17}$$

where $B_k^* = EI_k^*/L_k$. Similarly, the strain energy in a single column in the jth column line of the kth story of the original building will be

$$U_{kj} = \frac{6B_{kj}(u_k - u_{k-1})^2}{L_k^2}$$

in which $B_{kj} = EI_{kj}/L_k$. The strain energy developed in the m columns of the kth story of the original structure is simply the sum of the U_{kj}, with the result

$$U_k = \sum_{j=1}^{m} U_{kj} = 6\frac{(u_k - u_{k-1})^2}{L_k^2} \sum_{j=1}^{m} B_{kj} \tag{15.18}$$

Equating Eqs. (15.17) and (15.18) yields the result that U_k^* will be equal to U_k if for the modulus of elasticity of the materials in the two structures being the same, the moment of inertia I_k^* is equal to the sum of the moments of inertia I_{kj} of the m columns in the original structure, or

$$I_k^* = \sum_{j=1}^{m} I_{kj} \tag{15.19}$$

Using Eq. (15.19) on each floor of the two structures makes the strain energy of them equal and hence the total potential energy of them as well. Equation (15.19) provides the definition of the member properties for the shear building model.

The deflections of the shear building due to the applied loads P_k can easily be determined from slope-deflection equations (11.6) for shear, by starting at the base of the structure and working upward. Since the rotations are zero on each end of the members of the shear building, the displacement of story k can be obtained in terms of the shear in story k, $V_{k-1 \cdot k}$, by inverting Eq. (11.6) to give

$$u_k = u_{k-1} - \frac{V_{k-1 \cdot k} L_k^3}{12EI_k^*} \tag{15.20}$$

621

Since the displacement at the base of the structure is zero, the displacement u_1 can be determined, and then u_2, and so on, sequentially applying Eq. (15.20). Although this model is somewhat crude, it will provide reasonable estimates of the nature of the variation of the lateral displacements of a rigid frame as illustrated in Example 15.5.

Example 15.5 The three-story two-bay frame of Fig. 10.17 analyzed exactly in Fig. 14.11 is approximated as a shear building. The displacements due to the lateral loading are obtained from Eq. (15.20) and are in error by 39 to 54%. That the displacements are too small is not surprising since the top and bottom of the columns are prevented from rotating. The pattern of the displacements matches that of the exact solution fairly well.

The moments in the columns of each story are obtained directly from slope-deflection equation (11.3). They are the same for all columns in each story because the moments of inertia are the same for all columns in a story. The beam moments are obtained from moment equilibrium at the joints. The assumption is made that the beam moments at the center column joints are equal. The moments on the columns are seen to be average values in each story, which give order-of-magnitude approximations of the exact values.

15.6   Single-Bay Building Approximation

The contribution to the deflections of the rigid frame structure in Fig. 15.5 of the beams is completely neglected since the rotations of the joints make the beams "rigid." A more sophisticated mathematical model that accounts for their contribution is the one-bay model. This idealization is developed in the same manner as that of the shear building, except that the strain energy associated with the beams in the original structure is included in the model shown in Fig. 15.6.

Figure 15.6 The single-bay building model shown has a beam with possible joint rotations. The one-bay building is assumed to have equal joint rotations, θ_k^*, at each end of the horizontal member in story k. The original structure is assumed to deform so that all of the rotations θ_k in any story, k, will also be equal. Again the procedure is to make the total potential energy in the original and the one-bay building the same.

Take the total potential energy of the original structure as W and that of the one-bay building as W^*. Assume in the original building that all of the vertical displacements of the joints are zero, just as they are in the one-bay building. In addition, assume that all the displacements of each floor of the one-bay building, u_k^*, are equal to the corresponding displacements of the floors of the original building, u_k, and also assume that the lateral displacements of all joints of floor k in the original structure are the same u_k. As shown in Fig. 15.6, this will make the potential energy of the applied loads in the two structures, V and V^*, the same. If the strain energy, U^*, in the one-bay building is the same as the strain energy, U, in the original structure, the total potential energy of the two structures will be the same.

The strain energy developed in the original structure now has a contribution from the beams as well as the columns. Referring to Fig. 15.6, each beam in story k of the original building will develop strain energy due to the two equal end rotations, θ_k, which can be expressed using the boxed relation of Fig. 15.4 as

$$U_{bkj} = 6G_{kj}\theta_k^2 \qquad (15.21)$$

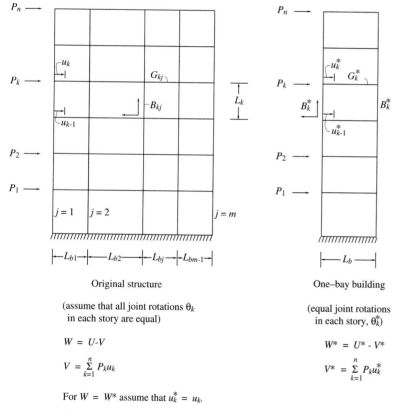

Original structure

(assume that all joint rotations θ_k
in each story are equal)

$W = U - V$

$V = \sum\limits_{k=1}^{n} P_k u_k$

For $W = W^*$ assume that $u_k^* = u_k$.

Then $V^* = V$ and $W = W^*$ if $U^* = U$.

One–bay building

(equal joint rotations
in each story, θ_k^*)

$W^* = U^* - V^*$

$V^* = \sum\limits_{k=1}^{n} P_k u_k^*$

Figure 15.6 Equivalence of deformation behavior between a one-bay building and a rigid frame with certain restrictions on frame deflection behavior.

Example 15.5

Use the shear building approximation to obtain the displacements of the laterally loaded frame of Fig. 14.11. Also obtain the member moments from the analysis.

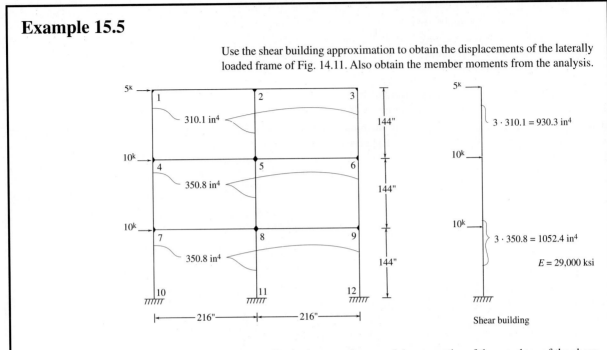

STEP 1 Obtain the story shears and the properties of the members of the shear building using Eq. (15.19).

STEP 2 Use Eq. (15.20) to obtain the displacements at each story. Start at the base and work upward. Displacements to the right are negative because of local coordinate system of the columns.

$$u_1 = u_0 - \frac{V_{01}L_1^3}{12EI_1} = 0 - 25 \cdot \frac{144^3}{12 \cdot 29{,}000 \cdot 1052.4} = -0.2038 \text{ in}$$

$$u_2 = -0.2038 - 15 \cdot \frac{144^3}{12 \cdot 29{,}000 \cdot 1052.4} = -0.2038 - 0.1223 = -0.3261 \text{ in}$$

$$u_3 = -0.3261 - 5 \cdot \frac{144^3}{12 \cdot 29{,}000 \cdot 930.3} = -0.3261 - 0.0461 = -0.3722 \text{ in}$$

STEP 3 Compute column moments in each story using slope-deflections equation (11.3) and use joint moment equilibrium to obtain beam moments. Divide moments equally between beams at joints having more than one.

$$\text{Story 1: } M = \frac{-6EI(u_1 - 0)}{L_1^2} = 6 \cdot 29{,}000 \cdot \frac{350.8(-0.2038)}{144^2} = 600 \text{ kip-in.}$$

$$\text{Story 2: } M = \frac{6EI(u_2 - u_1)}{L_2^2} = 6 \cdot 29{,}000 \cdot \frac{350.8(-0.1223)}{144^2} = 350 \text{ kip-in.}$$

Example 15.5 (continued)

Story 3: $M = \dfrac{6EI(u_3 - u_2)}{L_3^2} = 6 \cdot 29{,}000 \cdot \dfrac{310.1(-0.0461)}{144^2} = 120$ kip-in.

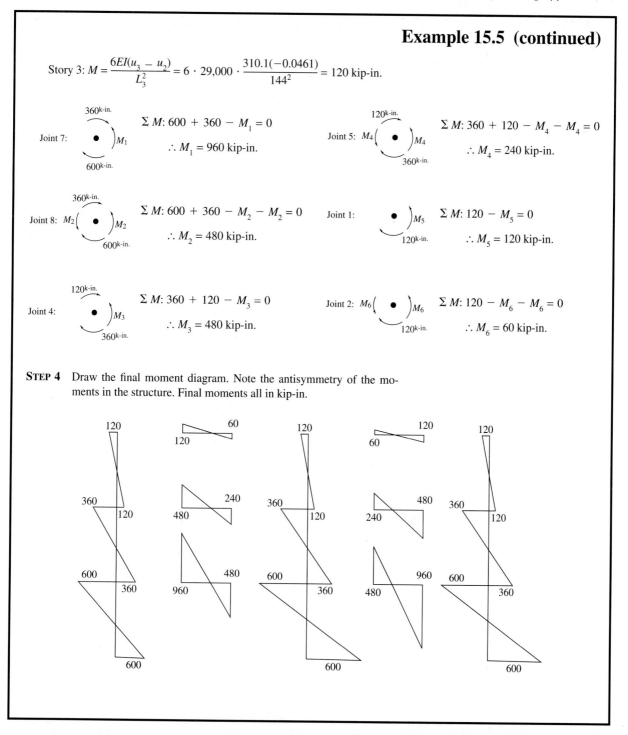

Joint 7:   360^{k-in.} $\bigodot M_1$ 600^{k-in.}

ΣM: $600 + 360 - M_1 = 0$

$\therefore M_1 = 960$ kip-in.

Joint 5: $M_4 \bigodot M_4$ 120^{k-in.}   360^{k-in.}

ΣM: $360 + 120 - M_4 - M_4 = 0$

$\therefore M_4 = 240$ kip-in.

Joint 8: $M_2 \bigodot M_2$ 360^{k-in.}   600^{k-in.}

ΣM: $600 + 360 - M_2 - M_2 = 0$

$\therefore M_2 = 480$ kip-in.

Joint 1: $\bigodot M_5$ 120^{k-in.}

ΣM: $120 - M_5 = 0$

$\therefore M_5 = 120$ kip-in.

Joint 4:   120^{k-in.} $\bigodot M_3$ 360^{k-in.}

ΣM: $360 + 120 - M_3 = 0$

$\therefore M_3 = 480$ kip-in.

Joint 2: $M_6 \bigodot M_6$ 120^{k-in.}

ΣM: $120 - M_6 - M_6 = 0$

$\therefore M_6 = 60$ kip-in.

STEP 4   Draw the final moment diagram. Note the antisymmetry of the moments in the structure. Final moments all in kip-in.

where $G_{kj} = EI_{bkj}/L_{bj}$. This constitutes the strain energy for the beam in bay j because there are no axial deformations and no end vertical displacements. The total strain energy for all beams in the kth story, U_{bk}, is obtained by summing to yield

$$U_{bk} = \sum_{j=1}^{m-1} U_{bkj} = \sum_{j=1}^{m-1} 6G_{kj}\theta_k^2 \qquad (15.22)$$

The strain energy in the beam of the kth story of the one-bay structure is

$$U_{bk}^* = 6G_k^* \theta_k^{*2} \qquad (15.23)$$

where $G_k^* = EI_{bk}^*/L_b^*$. If θ_k^* is taken equal to θ_k, the strain energy in the beams of the original and one-bay structure will be equal if

$$G_k^* = \sum_{j=1}^{m-1} G_{kj} \qquad (15.24a)$$

or for the same material modulus E in the two structures, if

$$\frac{I_{bk}^*}{L_b} = \sum_{j=1}^{m-1} \frac{I_{bkj}}{L_{bj}} \qquad (15.24b)$$

Now the strain energy in the two structures can be made the same if the strain energy in the columns of each story of the one bay building is made equal to the strain energy in all of the columns in the corresponding story of the original building. Again referring to Fig. 15.6, the strain energy in the two identical columns of the kth story of the one bay building is obtained from the boxed expression of Fig. 15.4, as

$$U_k^* = 2B_k^* \frac{[6(u_k - u_{k-1})^2}{L_k^2} - \frac{6(\theta_k + \theta_{k-1})(u_k - u_{k-1})}{L_k}$$
$$+ 2(\theta_k^2 + \theta_k\theta_{k-1} + \theta_{k-1}^2)] \qquad (15.25)$$

where $B_k^* = EI_k^*/L_k$. Similarly, the strain energy in a single column in the column line of the kth story of the original building will develop strain energy

$$U_{kj} = B_{kj} \left[\frac{6(u_k - u_{k-1})^2}{L_k^2} - \frac{6(\theta_k + \theta_{k-1})(u_k - u_{k-1})}{L_k} \right.$$
$$\left. + 2(\theta_k^2 + \theta_k\theta_{k-1} + \theta_{k-1}^2) \right] \qquad (15.26)$$

in which $B_{kj} = EI_{kj}/L_k$. The strain energy developed in the m columns of the kth story of the original structure is simply the sum of the U_{kj} with the result

$$U_k = \left[6\frac{(u_k - u_{k-1})^2}{L_k^2} - 6(\theta_k + \theta_{k-1})\frac{u_k - u_{k-1}}{L_k} \right.$$
$$\left. + 2(\theta_k^2 + \theta_k\theta_{k-1} + \theta_{k-1}^2) \right] \sum_{j=1}^{m} B_{kj} \qquad (15.27)$$

Equating Eqs. (15.25) and (15.27) yields the result that U_k^* will be equal to U_k if for the modulus of elasticity of the materials in the two structures being the same, the moment of inertia I_k^* is equal to one-half the sum of the moments of inertia I_{kj} of the m columns in the original structure, or

$$I_k^* = \frac{1}{2} \sum_{j=1}^{m} I_{kj} \tag{15.28}$$

Using Eqs. (15.24) and (15.28) on each floor of the two structures makes both their strain energy and their total potential energy equal. These two equations provide the definition of the member properties for the one-bay building model.

The one-bay building model will provide improved displacement results for lateral loading as well as improved dynamic characteristics. Solution of the one-bay building requires the use of a computer, as there are now an unknown displacement and rotation at each floor of the structure. There is no simple hand calculation that can obtain the desired results, although the reduced number of equations that need to be solved for the one-bay building make it easy to program and run on the smallest computers.

Example 15.6 The two-bay, three-story building of Example 15.6 is analyzed as a one-bay building. The displacements of the building are vastly improved over the previous example and have a maximum error of just over 2%. The column moments calculated from the slope-deflection equations are also improved over the previous example, but again are average values. The beam moments can now be calculated from Eq. (11.3) and show average values of moments in relation to the exact ones.

15.7   Summary and Limitations

The use of energy methods of analysis provides an alternative means of obtaining approximate solutions of structures. These displacement-based methods are easily programmed and existing computer programs can be used for the analysis. The methods are introduced in the first section with a review of strain and complementary strain energy. This is followed by a presentation of the total potential and the total complementary energy of a structure.

The principle of minimum total potential energy and the principle of minimum complementary energy for stable structures are presented and the relation of these principles to the virtual work and the complementary virtual work principles is discussed. In the next section the two theorems of Castigliano and Engesser are presented and related to the potential and complementary energy principles. Castigliano's first theorem

627

Example 15.6

Use the single-bay building approximation for the analysis of the two-bay three-story building of Fig. 14.11.

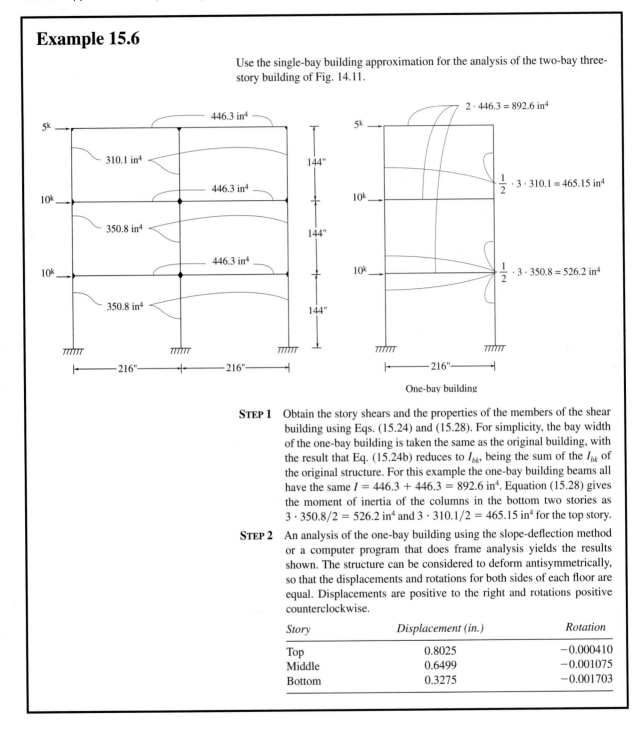

One-bay building

STEP 1 Obtain the story shears and the properties of the members of the shear building using Eqs. (15.24) and (15.28). For simplicity, the bay width of the one-bay building is taken the same as the original building, with the result that Eq. (15.24b) reduces to I_{bk}, being the sum of the I_{bk} of the original structure. For this example the one-bay building beams all have the same $I = 446.3 + 446.3 = 892.6$ in^4. Equation (15.28) gives the moment of inertia of the columns in the bottom two stories as $3 \cdot 350.8/2 = 526.2$ in^4 and $3 \cdot 310.1/2 = 465.15$ in^4 for the top story.

STEP 2 An analysis of the one-bay building using the slope-deflection method or a computer program that does frame analysis yields the results shown. The structure can be considered to deform antisymmetrically, so that the displacements and rotations for both sides of each floor are equal. Displacements are positive to the right and rotations positive counterclockwise.

Story	Displacement (in.)	Rotation
Top	0.8025	−0.000410
Middle	0.6499	−0.001075
Bottom	0.3275	−0.001703

Example 15.6 (continued)

STEP 3 To obtain the moments, use the slope-deflection equations, Eq. (11.3). There is no applied loading between joints, so the fixed end moments are all zero. For all the beams in one floor, the member end moments are all equal since the moments of inertia are equal and there is only one end rotation per floor. All columns in any story have equal end moments at each end because of equal moments of inertia and a single set of displacements and rotations per floor. The local coordinate for columns is taken with the x axis vertical.

Bottom-story columns:

$$M_{bot} = \frac{2 \cdot 29{,}000 \cdot 350.8}{144} \cdot \left[2 \cdot 0 + (-0.001703)\right.$$
$$\left. - \frac{3}{144}(-0.3275 - 0)\right] = 723 \text{ kip-in.}$$

$$M_{top} = \frac{2 \cdot 29{,}000 \cdot 350.8}{144} \cdot \left[2(-0.001703) + 0\right.$$
$$\left. - \frac{3}{144}(-0.3275 - 0)\right] = 483 \text{ kip-in.}$$

Beams: $M = \frac{2 \cdot 29{,}000 \cdot 446.3}{216} \cdot \left[2(-0.001703)\right.$
$$\left. + (-0.001703) - 0\right] = -612 \text{ kip-in.}$$

Middle-story columns:

$$M_{bot} = \frac{2 \cdot 29{,}000 \cdot 350.8}{144}\left[2(-0.001703) - 0.001075\right.$$
$$\left. - \frac{3}{144}(0.6499 - (-0.3275))\right] = 316 \text{ kip-in.}$$

$$M_{top} = \frac{2 \cdot 29{,}000 \cdot 350.8}{144}\left[2(-0.001075) - 0.001703\right.$$
$$\left. - \frac{3}{144}(0.6499 - (-0.3275))\right] = 405 \text{ kip-in.}$$

629

Example 15.6 (continued)

Beams: $M = \dfrac{2 \cdot 29{,}000 \cdot 446.3}{216} \cdot [2(-0.001705)$

$+ (0.001075 - 0] = -386$ kip-in.

Top-story columns:

$$M_{bot} = \frac{2 \cdot 29{,}000 \cdot 310.1}{144}\left[2(-0.001705) - 0.00041\right.$$

$$\left. - \frac{3}{144}(-0.8025 - (-0.6499))\right] = 77.3 \text{ kip-in.}$$

$$M_{top} = \frac{2 \cdot 29{,}000 \cdot 310.1}{144}\left[2(-0.00041) - 0.001075\right.$$

$$\left. - \frac{3}{144}(-0.8025 - (-0.6499))\right] = 160 \text{ kip-in.}$$

Beams: $M = \dfrac{2 \cdot 29{,}000 \cdot 446.3}{216}[2 \cdot (-0.000410)$

$$\left. + (0.000410) - 0\right] = -147 \text{ kip-in.}$$

STEP 4   Draw the final moment diagram (moments in kip-in.)

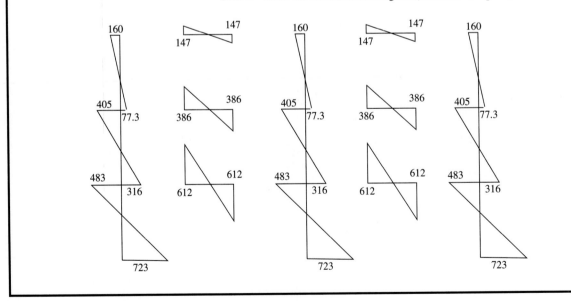

is used as a means of approximating the deformation behavior of non-prismatic members.

The use of minimum total potential energy as a means of approximating the behavior of single-story frames is presented in Section 15.4. Using the strain energy for a prismatic member with arbitrary end displacements and rotations derived in Fig. 15.4, the approximate analysis of frames based on assumed displacement and rotation patterns of the joints of the frame is developed. The extraction of member end moments from the slope-deflection equations is illustrated.

The last two sections of the chapter are devoted to the approximation of the deformation of rigid frames under the action of lateral loads. The development of two simple models of frame behavior, the shear building and the one-bay building, is shown to be based on the equivalence of the total potential energy of the simple structures and the rigid frame they model. The rigid frame is assumed to undergo certain displacement and rotation patterns that approximate its deformation behavior.

The energy methods presented are based on the concepts of axial force behavior and simple bending theory for beams. The assumptions and limitations of energy techniques are listed below.

- When energy techniques in this chapter are used for approximate analysis of trusses, they are subject to the same limitations and assumptions as are listed at the end of Chapter 3 except that the material can be nonlinear.

- Because of the use of simple bending theory principles, the deformation analysis of beams and frames is subject to the same limitations and assumptions as are listed at the end of Chapter 5 and in Section 6.2 except that members can be nonprismatic and have materials with nonlinear stress–strain relations.

- The use of Castigliano's second theorem is limited to structures that have linear elastic materials.

- All material deformations must be continuous, without any cracks or gaps.

- All structural deformations must be continuous, without any violation of the conditions of constraint at points of support.

- The distribution and orientation of all loads acting on a structure are not changed by the deformations of the structure.

- The orientation of all reaction components at supports is unchanged by the action of applied loads or any movement of the support.

- If movements occur at supports, they are small and do not affect the initial geometry of the structure.

631

- Sudden changes in cross-sectional properties in nonprismatic members are not so large that the assumed simple deformation behavior from mechanics of materials is greatly in error.

The use of energy methods is important in current methods of structural analysis, particularly those that are programmed for digital computers. These energy methods are the basis of the widely used finite element methods of structural analysis. Since the use of computer programs based on energy methods is extensive, some questions that should be considered by the structural engineer are:

- Can energy methods account for the interaction between axial force and bending action, and does the computer program being used attempt to model these effects?
- How can the error of an approximate solution based on energy methods be determined?
- Will a computer program based on the virtual work principle give the same results as a program based on the minimum total potential energy principle, and why or why not?
- If the strain and complementary strain energy for a linear elastic material are the same, why don't the minimum total potential energy and minimum complementary energy principles give the same equations?

PROBLEMS

15.1 The plane truss member has an area that varies as

$$A = A_0\left(1 + \frac{3x}{L}\right)$$

Use the displacement function of Fig. 15.2 and an energy method to obtain the displacement, u_2, of the right end (E constant).

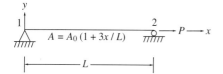

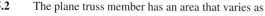

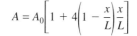

15.2 The plane truss member has an area that varies as

$$A = A_0\left[1 + 4\left(1 - \frac{x}{L}\right)\frac{x}{L}\right]$$

Use the displacement function of Fig. 15.2 and an energy method to obtain the displacement, u_2, of the right end (E constant).

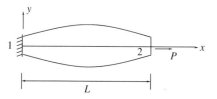

15.3 Repeat Problem 15.2, but divide the member in half and use the displacement function of Fig. 15.2 in each half. Take the displacement of the center of the member as u_3.

15.4 (a) Obtain the exact displacement, u_2, of the right end of the member as a function of α, (E constant).

(b) Repeat part (a), but use the displacement relation of Fig. 15.2 and an energy technique to obtain an approximation of u_2 as a function of α. Compare with part (a).

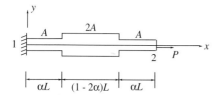

15.5 Using the assumed cubic expression of Fig. 15.3 for deflection v for a prismatic member, compute approximate values of the vertical displacement v and end rotation of the tapered cantilever beam [(E constant), $I = I_0(1 + 7x/l)$]. Compare with the exact values:

$$v_1 = -0.057083\frac{Pl^3}{EI_0}$$

$$\theta_1 = +0.10042\frac{Pl^2}{EI_0}$$

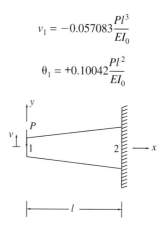

15.6 Obtain an approximate value of the deflection of the center of the beam. Use an energy technique and the displacement function of Fig. 15.3. Take advantage of symmetry (E constant).

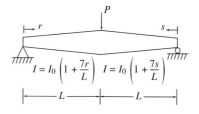

15.7 For the tapered cantilever beam with the concentrated end moment shown, use the displacement function of Fig. 15.3 and an energy technique to obtain an approximate value of the vertical deflection at A (E constant).

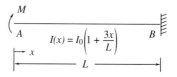

633

15.8   Repeat Problem 15.5, but divide the member in half and use the displacement function of Fig. 15.3 in each half. Take the displacement and rotation of the center of the member as v_3 and θ_3, respectively. Compare the result with the exact values.

15.9   Use the prismatic member displacement function of Fig. 15.3 and an energy technique to obtain an approximation of the deflection of the right end of the beam, v_z (E constant).

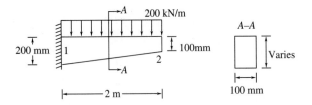

15.10   Repeat Problem 15.9, but use the displacement function defined in Example 6.1 with the $wL^4/24EI$ being replaced by $v_2/3$. Will the approximate value of v_2 obtained with this function be better or worse than the v_2 obtained in Problem 15.9?

15.11   Use the prismatic member displacement function of Fig. 15.3 and an energy technique to obtain the deflection of the left end of the beam (E constant).

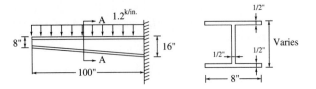

15.12   Repeat Problem 15.10 but with the beam of Problem 15.11.

15.13   Use the shear building approximation to make an approximate analysis of the structure of Example 12.7. Obtain the final moments and compare them with those in Examples 15.4 and 12.6.

15.14   Show that the strain energy for each column of the frame shown below is

$$U = \frac{3v^2 B_j}{L^2}$$

where v is the displacement of the top of the frame and no rotations occur at the joints at the top of the columns. Use this result to show that the column of the equivalent shear building has a B^* defined as

$$B^* = \tfrac{1}{4} \sum_j B_j$$

[*Hint:* Use Eq. (11.3) to solve for the rotation of the base of the column in terms of the lateral displacement, v, of the top of the column for the condition of no moment at the base and no rotation at the top. Then use the energy expression of Fig. 15.4.]

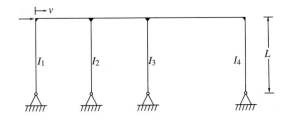

15.15   Use the shear building approach and the results of Problem 15.14 to make an approximate analysis of the frame shown (E constant). Draw the final moment diagram.

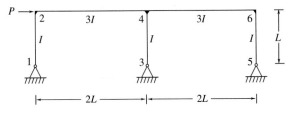

15.16   Use the shear building approach and the results of Problem 15.14 to make an approximate analysis of the frame shown (E constant). Draw the final moment diagram.

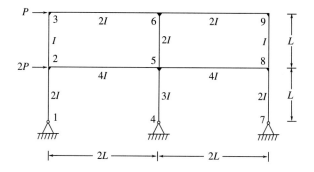

15.17   Show that Eqs. (15.24) and (15.28) can be used for the one-bay building approximation of the frame of Problem 15.14 if the base of the one-bay building in pinned.

15.18   Using the observation of Problem 15.17 and the one-bay building approach, obtain an approximate analysis of the frame of Problem 15.15. Draw the final moment diagram.

15.19   Using the observation of Problem 15.17 and the one-bay building approach, obtain an approximate analysis of the frame of Problem 15.16. Draw the final moment diagram.

16

Analysis of Structures by the Stiffness Matrix Method

With the exception of the slope-deflection analysis described in Chapter 11 and the displacement-based approximate methods presented in Chapter 15, the methods of analysis of structures that have been studied to this point are based on the direct computation of the forces or moments that develop in loaded structures. These techniques are known as *force methods of analysis* since the results that are first obtained are the forces or moments in the structure. The determination of displacements or rotations requires additional computations.

The force techniques use the equations of equilibrium to obtain moments and forces in the structure. When the structures are indeterminate, one or more equations describing the deformation behavior of the structure are developed and deformation conditions imposed that reflected the manner in which the structure deforms. These equations, called *compatibility equations,* are written in terms of all forces and moments acting in and on the structure, including the unknown force or moment redundants. The solution of these equations yields the value of the unknown redundant forces and moments, and equilibrium is used to obtain the final distribution of forces and moments in the structure.

In this chapter a method of analysis is presented that is based on establishing the force–displacement relations for truss members and using these relations in a series of force equilibrium equations at each joint of the truss.

The equilibrium equations are developed with the displacements of the joints being unknown. The equations can be solved for the unknown displacements and the individual member forces obtained from the force–displacement relations. This method is a *displacement method of analysis* since the model is based on the unknown joint displacements. It is also often referred to as a *matrix method* or *matrix analysis* since its implementation involves extensive operations with matrices. In this chapter the development of the displacement method for plane trusses is described. It is assumed that the student is familiar with the use of matrices. A brief review of determinants and matrices is included in the Appendix.

16.1 Displacement Method of Analysis for Plane Trusses

The solution of the mathematical equations established for any problem in structural engineering must identically satisfy the three following conditions: (1) all equilibrium equations for structure are satisfied, (2) the stress–strain (force–deformation) equation for the material of the structure is followed exactly in every member, and (3) the deformations of the structure are compatible (the structure does not come apart and it deforms in a manner consistent with the restraints at supports). If these three conditions are satisfied in a structure having linear elastic material, the solution of the equations constituting its mathematical model is unique as long as the structure is stable. In the equations constituting the mathematical model in matrix methods, these three conditions are always satisfied. For articulated structures (trusses and frames), the solutions for stable structures are exact within the limitations of the assumptions made in establishing the equations of the mathematical models.

The analysis of plane trusses by the stiffness matrix method is based on the formulation of the relation between the axial force in a member and the displacements of the ends of that member. The axial forces that develop in the members of a plane truss are summed at each joint, which establishes two equilibrium equations per joint. The assumptions used in the analysis of plane trusses by the stiffness matrix method are:

- The materials of the truss are all linear elastic.
- All strains and displacements are small, so that the geometry of the structure remains essentially unchanged.
- Individual members of the structure do not become unstable under the action of the applied loads.
- The structure as a whole, or any portion of it, does not become unstable under the action of the applied loads.
- The magnitude and distribution of the loading applied to the truss is independent of the deformations of the structure.

638

- The centroidal axes of all members that come together at each joint of the structure intersect at a common point.

The stiffness matrix method is ideally suited for analysis using computer programs. The configuration of the structure, the applied loads, and the support conditions are input to a computer program, which then solves for the joint displacements and the member forces. The solution technique used by the program is derived and described in the remainder of this section.

A general truss is shown in Fig. 16.1. If the analysis of the structure is to be done using a computer program, the geometry of the structure, the connectivity of its members, the properties of its members, the loading of the structure, and the manner in which it is supported must be established in a form that can be input into the computer program. The necessary information can be most conveniently defined if the joints and members of the structure are numbered consecutively. The location of each numbered joint is given by the coordinates of the joint in a global coordinate system X–Y. Each numbered member of the structure has a cross-sectional area and material modulus of elasticity associated with it and is located in the structure by the two joints to which it connects. The member e, for example, has area, A, and modulus of elasticity, E, and connects to joints i and j.

Loads on the structure and reaction constraints are identified by the joint at which they act. The magnitude and direction of the loads are defined in terms of force components in the positive coordinate directions. The restraint of displacements at reaction joints is identified by the joint number and the direction $(X, Y,$ or both X and $Y)$ of the restraint. Displacements in the global coordinate X direction are called u, while displacements in the global coordinate Y direction are called v.

The development of the deformation behavior of any member of the structure can be characterized by the displacements of the ends of the member in the global coordinate system. The member, e, between joints i and j of the structure in Fig. 16.1 is shown in Fig. 16.2. The orientation of the member is defined in terms of the angles θ and ϕ between the member and the global X axis and the global Y axis, respectively. If member e is considered to be a vector with its line of action going from joint i to joint j, the cosine of these angles are referred to as the direction cosines of the member. Also shown in Fig. 16.2 are the internal forces, $F_{xi}, F_{yi}, F_{xj},$ and F_{yj}, that act in the global coordinate directions on the ends of the member which are components of the internal axial force in the member, S, caused by the end displacements. The force components are shown to act in their positive sense. The member axial force, S, is also shown to be positive when the member is in tension. From the figure it is clear that

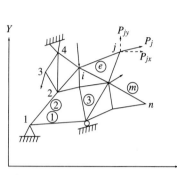

Figure 16.1 General plane truss under load.

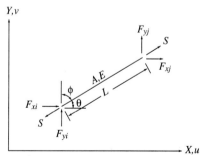

Figure 16.2 Member e between joints i and j in the structure of Fig. 16.1.

$$F_{xj} = -F_{xi} = S \cos \theta = S\alpha$$

$$F_{yj} = -F_{yi} = S \cos \phi = S\beta \qquad (16.1)$$

639

where

$$\alpha = \cos\theta = \frac{X_j - X_i}{L}$$

$$\beta = \cos\phi = \frac{Y_j - Y_i}{L} \qquad (16.2)$$

and

$$L = \sqrt{(X_j - X_i)^2 + (Y_j - Y_i)^2}$$

In Eq. (16.2), α and β are the direction cosines of the member.

The force, S, in the member develops when the end displacements of the member cause an elongation, ΔL. The force–deformation relation for an axially loaded member of linear elastic material undergoing small strains is

$$S = \frac{AE}{L}\Delta L \qquad (16.3)$$

in which A is the cross-sectional area of the member and E the modulus of elasticity of the material. It is also assumed in Eq. (16.3) that the member is prismatic.

The displacements of the two ends of the member are shown to an exaggerated scale in Fig. 16.3a and are referenced in the global coordinate system. The subscripts on u and v identify the joint of the structure at which the displacements occur. Figure 16.3b and c show the relative end displacements of the member (i.e., those displacements that cause the member to elongate an amount, ΔL). The displacements shown in Fig. 16.3 are assumed to be small in relation to the length of the member, so that the orientation of the member before and after the occurrence of the end displacements is very nearly the same. The relative end displacements in Fig. 16.3c

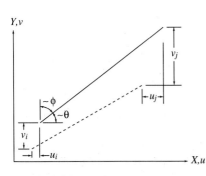

(a) Displaced position of member e

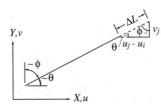

(b) Change in length ΔL of member e

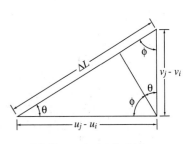

(c) Geometric detail of ΔL

Figure 16.3a–c   Displacements and relative displacements of the ends of member e.

have been further exaggerated so that the geometric relations are more clearly defined. From Fig. 16.3c, the change in length, ΔL, of member is computed simply as

$$\Delta L = (u_j - u_i)\cos\theta + (v_j - v_i)\cos\phi$$
$$= (u_j - u_i)\alpha + (v_j - v_i)\beta \tag{16.4}$$

The change in length of the member, ΔL, is an approximate relation since the angles θ and ϕ have in fact changed a small amount. However, the approximation is valid for the assumed small strain and displacement behavior of structures. Knowing ΔL, the axial force, S, in the member becomes, from Eqs. (16.3), and (16.4)

$$S = \frac{AE}{L}[(u_j - u_i)\alpha^2 + (v_j - v_i)\alpha\beta] \tag{16.5}$$

Equation (16.5) may now be substituted into the four equations for member end forces, Eq. (16.1). These equations, after substitution of Eq. (16.5), can be written as a system of four equations using matrix notation. The results are

$$F_{xj} = -F_{xi} = S\alpha = \frac{AE}{L}[(u_j - u_i)\alpha^2 + (v_j - v_i)\alpha\beta]$$
$$F_{yj} = -F_{yi} = S\beta = \frac{AE}{L}[(u_j - u_i)\alpha\beta + (v_j - v_i)\beta^2] \tag{16.6a}$$

or

$$\begin{Bmatrix} F_{xi} \\ F_{yi} \\ F_{xj} \\ F_{yj} \end{Bmatrix} = \frac{AE}{L} \begin{bmatrix} \alpha^2 & \alpha\beta & -\alpha^2 & -\alpha\beta \\ \alpha\beta & \beta^2 & -\alpha\beta & -\beta^2 \\ -\alpha^2 & -\alpha\beta & \alpha^2 & \alpha\beta \\ -\alpha\beta & -\beta^2 & \alpha\beta & \beta^2 \end{bmatrix} \begin{Bmatrix} u_i \\ v_i \\ u_j \\ v_j \end{Bmatrix} \tag{16.6b}$$

which can be condensed to

$$\begin{Bmatrix} \mathbf{F}_{ei} \\ \mathbf{F}_{ej} \end{Bmatrix} = \begin{bmatrix} \mathbf{K}_e & -\mathbf{K}_e \\ -\mathbf{K}_e & \mathbf{K}_e \end{bmatrix} \begin{Bmatrix} \mathbf{U}_i \\ \mathbf{U}_j \end{Bmatrix} \tag{16.6c}$$

in which

$$\mathbf{K}_e = \frac{AE}{L} \begin{bmatrix} \alpha^2 & \alpha\beta \\ \alpha\beta & \beta^2 \end{bmatrix}$$

$$\mathbf{F}_{ei} = \begin{Bmatrix} F_{exi} \\ F_{eyi} \end{Bmatrix} \qquad \mathbf{F}_{ej} = \begin{Bmatrix} F_{exj} \\ F_{eyj} \end{Bmatrix} \tag{16.6d}$$

$$\mathbf{U}_i = \begin{Bmatrix} u_i \\ v_i \end{Bmatrix} \qquad \mathbf{U}_j = \begin{Bmatrix} u_j \\ v_j \end{Bmatrix}$$

641

or finally, in the compact matrix form

$$\mathbf{F}_e = \mathbf{K}_e \mathbf{U}_e \qquad (16.6e)$$

Equation (16.6e) defines the relation between the forces developed on the ends of a plane truss member by the displacements of the ends of the member. These forces and displacements are all referred to global coordinate system.

Equation (16.6) is a force-deformation relation for the member i–j. It has exactly the same form as the familiar relation between the force in a spring, F, and the elongation of the spring, u, which is $F = ku$, where k is the stiffness of the spring. The matrix $\mathbf{K}_e$ in Eq. (16.6e) performs the same role as k in the spring relation and is referred to as the *stiffness matrix* of the member. The units of $\mathbf{K}_e$ are force per length, which are the units of stiffness. Note that the matrix $\mathbf{K}_e$ is symmetric, which will lead to the important result that the final set of equilibrium equations for the structure is also symmetric. Finally, it is also important to recognize that the formulation of the force–deformation relation for member i–j in Eq. (16.6) is dependent on the coordinates of joints i and j and the member properties. The process is independent of the loading on the truss, the manner of support of the truss, and whether or not the truss is statically determinate.

16.2 Formulation and Solution of Equations of Equilibrium

In Fig. 16.4 the complete solution process for the displacement method of analysis for a simple two-bar truss is presented. The displacement method is not the method one would select for hand computations in analyzing this statically determinate truss. However, for this very small and simple structure, the presentation of all steps of the displacement method is easy to follow and can be illustrated clearly. Although Fig. 16.4 has many parts, the complete concept of the displacement method of analysis can be understood with some explanatory comments about the several steps of the process.

Figure 16.4 The construction of the member force–deformation relations in the form of the stiffness matrices in Fig. 16.4b is taken from the definitions and concepts presented in Eqs. (16.1), (16.2), (16.5), and (16.6). The use of these equations specifically involves the stress–strain relation for the material which guarantees that the action of the members of the truss satisfies it. The known zero displacement conditions at the reaction joints is ignored in this stage of the process, but are imposed at a later step in the analysis. The form of the relations given in Eq. (16.6c) is particularly useful in the solution process.

In Fig. 16.4c the two equilibrium equations for each joint of the truss are established from the free-body diagram showing the internal and

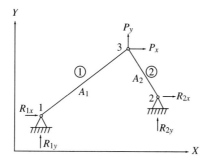

(a) Structure and loading

From Eqs. (16.2):

$$L_1 = \sqrt{(x_3 - x_1)^2 + (y_3 - y_1)^2}$$

$$L_2 = \sqrt{(x_3 - x_2)^2 + (y_3 - y_2)^2}$$

$$\alpha_1 = \frac{x_3 - x_1}{L_1} \quad \beta_1 = \frac{y_3 - y_1}{L_1} \quad k_1 = \frac{A_1 E}{L_1}$$

$$\alpha_2 = \frac{x_3 - x_2}{L_2} \quad \beta_2 = \frac{y_3 - y_2}{L_2} \quad k_2 = \frac{A_2 E}{L_2}$$

From Eqs. (16.5) and (16.6):

$$S_1 = k_1[(u_3 - u_1)\alpha_1 + (v_3 - v_1)\beta_1]$$

$$S_2 = k_2[(u_3 - u_2)\alpha_2 + (v_3 - v_2)\beta_2]$$

$$
\begin{Bmatrix} F_{1x1} \\ F_{1y1} \\ \text{--} \\ F_{1x3} \\ F_{1y3} \end{Bmatrix}
= k_1
\begin{bmatrix}
a_1^2 & a_1\beta_1 & \vdots & -a_1^2 & -a_1\beta_1 \\
a_1\beta_1 & \beta_1^2 & \vdots & -a_1\beta_1 & -\beta_1^2 \\
\cdots & \cdots & \vdots & \cdots & \cdots \\
-a_1^2 & -a_1\beta_1 & \vdots & a_1^2 & a_1\beta_1 \\
-a_1\beta_1 & -\beta_1^2 & \vdots & a_1\beta_1 & \beta_1^2
\end{bmatrix}
\begin{Bmatrix} u_1 \\ v_1 \\ \text{--} \\ u_3 \\ v_3 \end{Bmatrix}
$$

$$
\begin{Bmatrix} F_{2x2} \\ F_{2y2} \\ \text{--} \\ F_{2x3} \\ F_{2y3} \end{Bmatrix}
= k_2
\begin{bmatrix}
a_2^2 & a_2\beta_2 & \vdots & -a_2^2 & -a_2\beta_2 \\
a_2\beta_2 & \beta_2^2 & \vdots & -a_2\beta_2 & -\beta_2^2 \\
\cdots & \cdots & \vdots & \cdots & \cdots \\
-a_2^2 & -a_2\beta_2 & \vdots & a_2^2 & a_2\beta_2 \\
-a_2\beta_2 & -\beta_2^2 & \vdots & a_2\beta_2 & \beta_2^2
\end{bmatrix}
\begin{Bmatrix} u_2 \\ v_2 \\ \text{--} \\ u_3 \\ v_3 \end{Bmatrix}
$$

or, in compact matrix notation:

$$
\begin{Bmatrix} \mathbf{F}_{11} \\ \mathbf{F}_{13} \end{Bmatrix}
=
\begin{bmatrix} \mathbf{K}_1 & -\mathbf{K}_1 \\ -\mathbf{K}_1 & \mathbf{K}_1 \end{bmatrix}
\begin{Bmatrix} \mathbf{U}_1 \\ \mathbf{U}_3 \end{Bmatrix}
$$

$$
\begin{Bmatrix} \mathbf{F}_{22} \\ \mathbf{F}_{23} \end{Bmatrix}
=
\begin{bmatrix} \mathbf{K}_2 & -\mathbf{K}_2 \\ -\mathbf{K}_2 & \mathbf{K}_2 \end{bmatrix}
\begin{Bmatrix} \mathbf{U}_2 \\ \mathbf{U}_3 \end{Bmatrix}
$$

Member 1

Member 2

(b) Member force–deformation relations

Figure 16.4a–b Displacement solution for two-bar plane truss.

external forces that act at each joint. A small piece of each member that connects at a joint is also shown to indicate the direction of the global coordinate system force components of the internal axial force of that member. The opposite of these force components act on the joint. The

643

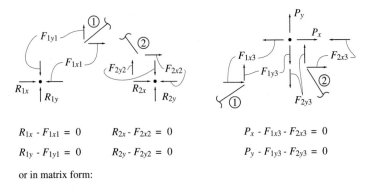

$$R_{1x} - F_{1x1} = 0 \qquad R_{2x} - F_{2x2} = 0 \qquad P_x - F_{1x3} - F_{2x3} = 0$$

$$R_{1y} - F_{1y1} = 0 \qquad R_{2y} - F_{2y2} = 0 \qquad P_y - F_{1y3} - F_{2y3} = 0$$

or in matrix form:

$$\mathbf{P}_1 = \mathbf{F}_{11} \qquad\qquad \mathbf{P}_2 = \mathbf{F}_{22} \qquad\qquad \mathbf{P}_3 = \mathbf{F}_{13} + \mathbf{F}_{23}$$

Joint 1   Joint 2   Joint 3

(c) Joint equilibrium equations

Collecting the joint equilibrium equations from (c):

$$\mathbf{P}_1 = \mathbf{F}_{11} = \mathbf{K}_1\mathbf{U}_1 + \mathbf{0}\mathbf{U}_2 - \mathbf{K}_1\mathbf{U}_3$$

$$\mathbf{P}_2 = \mathbf{F}_{22} = \mathbf{0}\mathbf{U}_1 + \mathbf{K}_2\mathbf{U}_2 - \mathbf{K}_2\mathbf{U}_3$$

$$\mathbf{P}_3 = \mathbf{F}_{13} + \mathbf{F}_{23} = -\mathbf{K}_1\mathbf{U}_1 + \mathbf{K}_1\mathbf{U}_3 + (-\mathbf{K}_2\mathbf{U}_2 + \mathbf{K}_2\mathbf{U}_3)$$

$$= -\mathbf{K}_1\mathbf{U}_1 + \mathbf{K}_2\mathbf{U}_2 + (\mathbf{K}_1 + \mathbf{K}_2)\,\mathbf{U}_3$$

or in assembled matrix form:

$$\begin{Bmatrix} \mathbf{P}_1 \\ \mathbf{P}_2 \\ \mathbf{P}_3 \end{Bmatrix} = \begin{bmatrix} \mathbf{K}_1 & \mathbf{0} & -\mathbf{K}_1 \\ \mathbf{0} & \mathbf{K}_2 & -\mathbf{K}_2 \\ -\mathbf{K}_1 & -\mathbf{K}_2 & \mathbf{K}_1 + \mathbf{K}_2 \end{bmatrix} \begin{Bmatrix} \mathbf{U}_1 \\ \mathbf{U}_2 \\ \mathbf{U}_3 \end{Bmatrix}$$

(d) Assemble global stiffness matrix from member
force–deformation relations in (b) and equilibrium
relations in (c)

Figure 16.4c–d   Displacement solution for two-bar plane truss.

equilibrium equations for the free bodies in Fig. 16.4c of each joint show
that the sum of the internal member force components resist or balance
the external applied forces on that joint. If no external force is applied at
a joint, the internal force components simply sum to zero. The reaction
forces at joints 1 and 2 are also treated as external forces acting on the
structure even though they are unknown.

To establish a final form for the six equilibrium equations presented in
Fig. 16.4c, it is convenient to express the two equations at each joint as a
matrix relation. The matrix equation at each joint is then expanded in

$$
\begin{Bmatrix} R_{1x} \\ R_{1y} \\ R_{2x} \\ R_{2y} \\ \hdashline P_x \\ P_y \end{Bmatrix}
=
\left[\begin{array}{cccc:cc}
k_1\alpha_1^2 & k_1\alpha_1\beta_1 & 0 & 0 & -k_1\alpha_1^2 & -k_1\alpha_1\beta_1 \\
k_1\alpha_1\beta_1 & k_1\beta_1^2 & 0 & 0 & -k_1\alpha_1\beta_1 & -k_1\beta_1^2 \\
0 & 0 & k_2\alpha_2^2 & k_2\alpha_2\beta_2 & -k_2\alpha_2^2 & -k_2\alpha_2\beta_2 \\
0 & 0 & k_2\alpha_2\beta_2 & k_2\beta_2^2 & -k_2\alpha_2\beta_2 & -k_2\beta_2^2 \\
\hdashline
-k_1\alpha_1^2 & -k_1\alpha_1\beta_1 & -k_2\alpha_2^2 & -k_2\alpha_2\beta_2 & k_1\alpha_1^2+k_2\alpha_2^2 & k_1\alpha_1\beta_1+k_2\alpha_2\beta_2 \\
-k_1\alpha_1\beta_1 & -k_1\beta_1^2 & -k_2\alpha_2\beta_2 & -k_2\beta_2^2 & k_1\alpha_1\beta_1+k_2\alpha_2\beta_2 & k_1\beta_1^2+k_2\beta_2^2
\end{array}\right]
\begin{Bmatrix} u_1 \\ v_1 \\ u_2 \\ v_2 \\ \hdashline u_3 \\ v_3 \end{Bmatrix}
$$

(e) Expanded stiffness matrix relation for entire structure from (d)

$$u_1 = v_1 = u_2 = v_2 = 0$$

$$
\begin{Bmatrix} R_{1x} \\ R_{1y} \\ R_{2x} \\ R_{2y} \\ \hdashline P_x \\ P_y \end{Bmatrix}
=
\left[\begin{array}{cc}
-k_1\alpha_1^2 & -k_1\alpha_1\beta_1 \\
-k_1\alpha_1\beta_1 & -k_1\beta_1^2 \\
-k_2\alpha_2^2 & -k_2\alpha_2\beta_2 \\
-k_2\alpha_2\beta_2 & -k_2\beta_2^2 \\
\hdashline
k_1\alpha_1^2+k_2\alpha_2^2 & k_1\alpha_1\beta_1+k_2\alpha_2\beta_2 \\
k_1\alpha_1\beta_1+k_2\alpha_2\beta_2 & k_1\beta_1^2+k_2\beta_2^2
\end{array}\right]
\begin{Bmatrix} u_3 \\ v_3 \end{Bmatrix}
$$

(f) Impose zero displacement conditions at 1 and 2

$$
\begin{Bmatrix} P_x \\ P_y \end{Bmatrix}
=
\begin{bmatrix} K_1 & K_2 \\ K_2 & K_3 \end{bmatrix}
\begin{Bmatrix} u_3 \\ v_3 \end{Bmatrix}
\qquad \therefore \qquad
\begin{array}{l}
u_3 = (K_3 \cdot P_x - K_2 \cdot P_y)/D \\
v_3 = (K_1 \cdot P_y - K_2 \cdot P_x)/D
\end{array}
$$

$$K_1 = k_1\alpha_1^2 + k_2\alpha_2^2 \qquad K_2 = k_1\alpha_1\beta_1 + k_2\alpha_2\beta_2 \qquad K_3 = k_1\beta_1^2 + k_2\beta_2^2 \qquad D = K_1 \cdot K_3 - K_2^2$$

(g) Solve the last two equations in (f) for u_3 and v_3

$$R_{1x} = -k_1\alpha_1^2 \cdot u_3 - k_1\alpha_1\beta_1 \cdot v_3 \qquad\qquad R_{2x} = -k_2\alpha_2^2 \cdot u_3 - k_2\alpha_2\beta_2 \cdot v_3$$

$$R_{1y} = -k_1\alpha_1\beta_1 \cdot u_3 - k_1\beta_1^2 \cdot v_3 \qquad\qquad R_{2y} = -k_2\alpha_2\beta_2 \cdot u_3 - k_2\beta_2^2 \cdot v_3$$

(h) Obtain reactions from the first four equations in (f)

$$S_1 = k_1\alpha_1 u_3 + k_1\beta_1 v_3 \qquad\qquad S_2 = k_2\alpha_2 u_3 + k_2\beta_2 v_3$$

(i) Obtain member forces from relations in (a)

Figure 16.4e–i Displacement solution for two-bar plane truss.

Fig. 16.4d so that it is expressed as the product of a series of 2×2 matrices and all of the joint displacements. For example, at joint 1 the equilibrium equation in Fig. 16.4c is $\mathbf{P}_1 = \mathbf{F}_{11}$ and from Fig. 16.4b, $\mathbf{F}_{11} = \mathbf{K}_1\mathbf{U}_1 - \mathbf{K}_1\mathbf{U}_3$ and hence $\mathbf{P}_1 = \mathbf{K}_1\mathbf{U}_1 - \mathbf{K}_1\mathbf{U}_3$. In Fig. 16.4d this equation is expanded to include the product of the null matrix, $\mathbf{0}$, and $\mathbf{U}_2$ so that the equation now contains 2×2 matrices that multiply all of the 2×1 joint displacement matrices. The matrix equations at joints 2 and 3 are treated in a similar manner to yield the three matrix equations shown. The collected set of three equations in Fig. 16.6d constitutes all of the equilibrium equations for the structure. Each row of 2×2 matrices in the 3×3 array corresponds to the two equations for forces acting in the global X and Y directions at one joint of the structure, and each column of 2×2 matrices corresponds to the displacement in the global X and Y directions at a joint of the structure.

An examination of the 3×3 array of 2×2 matrices in Fig. 16.4d shows how the joint equilibrium equations of the structure are constructed. Because pairs of rows represent the X and Y forces at a joint and pairs of columns represent the u and v displacements at a joint, the terms in the individual member 4×4 stiffness matrices are defined as the 2×2 blocks $\pm\mathbf{K}_1$ or $\pm\mathbf{K}_2$ shown in Fig. 16.4b. These terms are moved in the assembly process into the 3×3 array and are placed in the row and column positions of that array that correspond to the forces and displacements of the two joints to which the member is connected. For example, the four 2×2 $\mathbf{K}_1$ matrices in the stiffness matrix for member 1 which is connected to joints 1 and 3 appear in the first and third rows and the first and third columns of the 3×3 array of Fig. 16.4d.

The six equilibrium equations for the entire structure in the three rows of 2×2 matrices of Fig. 16.4d are presented in the completely expanded form of Eq. (16.6b) in Fig. 16.4e. The individual member end force components appearing in each of the six equilibrium equations in Fig. 16.4c can be identified in that 6×6 array. The F_{1x1} and F_{1y1} forces in the two equations in Fig. 16.4c for joint 1 appear in the first two equations in Fig. 16.4e and the F_{2x2} and F_{2y2} forces in the two equations in Fig. 16.4c and joint 2 appear as the third and fourth equations in Fig. 16.4c. The last two equilibrium equations for joint 3 in Fig. 16.4c require the sum of the force components for both members 1 and 2 at joint 3. Consequently, the last two equations in Fig. 16.4c contains the sum of the contributions of members 1 and 2. The 6×6 stiffness matrix of Fig. 16.4c for the entire structure relates all joint displacements in the global coordinate system with all the corresponding applied joint forces.

The matrix equation in Fig. 16.4e involves six unknowns. There are four unknown reaction components that act as applied loads on joints 1 and 2 and two unknown displacements at joint 3. Since there are six unknowns and six equations, the problem can be solved directly. The four displacements at joints 1 and 2 are known to be zero because they occur at reaction joints where displacements are restrained and the two applied loads at joint 3 are known as the forces acting on the structure.

In Fig. 16.4f the known zero displacements at the reaction joints are imposed. This causes the first four columns of the stiffness matrix in Fig. 16.4e to be multiplied by the zero displacements at joints 1 and 2, and hence these columns along with the four corresponding zero displacements can be dropped from the matrix equation. The six reduced equations shown in Fig. 16.4f still involve six unknowns, the two unknown joint displacements u_3 and v_3 and the four unknown reaction components $R_{1x} \cdots R_{2y}$ at joints 1 and 2.

When expanded, the first four equations in Fig. 16.4f each involve one unknown reaction component and both unknown joint displacements, while the last two equations involve only the unknown joint displacements. As shown in Fig. 16.4g, these two equations can be solved directly for the unknown displacements. With the displacements known, all other unknowns in the structure can be obtained. Figure 16.4h shows the computation of the reaction components from the first four equations in Fig. 16.4f. The calculation of member forces using Eq. (16.5) is shown in Fig. 16.4i.

The process illustrated in Fig. 16.4 exactly parallels the slope-deflection method of solution for plane frames presented in Chapter 11 because all displacement methods of analysis always follow the same prescribed set of general steps. First the relation between each member's end forces or moments and member's end displacements or rotations is established. This corresponds to Eqs. (16.6) for truss members and Eqs. (11.3) and (11.6) for frame members. The second step is to establish the equilibrium equations at each joint of the structure. These equations involve at each joint the unknown displacements for trusses and the unknown displacements and rotation for frames. The equilibrium equations are solved for the unknown displacements and rotations. The final step is to substitute the newly computed displacements and rotations into the member force–displacement relations to obtain the member end forces and moments. Reactions are also computed in this step.

The process illustrated in Fig. 16.4 can be summarized in a formal manner using matrix notation. The individual force–deformation relations in Fig. 16.4b are taken from Eqs. (16.6). The stiffness relation for the entire structure, obtained from the joint equilibrium equations in Fig. 16.4c, can be written as

$$\mathbf{P} = \mathbf{KU} \qquad (16.7)$$

and is shown in Fig. 16.4d and e, in which $\mathbf{K}$ is a 6×6 matrix. Equation (16.7) gives the stiffness relation for the entire structure. The rows of the matrix $\mathbf{K}$ can be subdivided or partitioned so that the equilibrium equations for the unknown reaction components, $\mathbf{R}$, are separated from the equilibrium equations for the known applied joint loads, $\mathbf{P}_a$. The columns of the matrix $\mathbf{K}$ can be partitioned so that the known or specified displacements, $\mathbf{U}_r$, in the direction of reaction components are separated from the unknown displacements, $\mathbf{U}_u$, in the direction of the known applied loads. This partitioning divides $\mathbf{P}$ and $\mathbf{U}$ in Eq. (16.7) into two submatrices and $\mathbf{K}$ into four submatrices. This partitioning is shown by the dashed lines in the matrices in Fig. 16.4e and is formalized by partitioning the matrices in Eq. (16.7) and causing the equation to take the form

$$\left\{ \begin{array}{c} \mathbf{R} \\ \mathbf{P}_a \end{array} \right\} = \left[\begin{array}{cc} \mathbf{K}_{rr} & \mathbf{K}_{ru} \\ \mathbf{K}_{ur} & \mathbf{K}_{uu} \end{array} \right] \left\{ \begin{array}{c} \mathbf{U}_r \\ \mathbf{U}_u \end{array} \right\} \tag{16.8}$$

The partitions of the matrices in Fig. 16.4e correspond with the subscripted $\mathbf{K}$ matrices in Eq. (16.8). As seen in Fig. 16.4e, $\mathbf{K}_{rr}$ is a 4×4, $\mathbf{K}_{ru}$ a 4×2, $\mathbf{K}_{ur}$ a 2×4, and $\mathbf{K}_{uu}$ a 2×2 matrix.

The solution process is continued by expanding Fig. (16.8) into the two matrix relations

$$\mathbf{R} = \mathbf{K}_{rr}\mathbf{U}_r + \mathbf{K}_{ru}\mathbf{U}_u$$
$$\mathbf{P}_a = \mathbf{K}_{ur}\mathbf{U}_r + \mathbf{K}_{uu}\mathbf{U}_u \tag{16.9}$$

The $\mathbf{U}_r$ matrix contains all of the known or specified displacements at the reaction joints of the structure. Usually, the displacements at reaction joints are zero, as is shown in Fig. 16.4f, so that only the $\mathbf{K}_{ru}$ and $\mathbf{K}_{uu}$ terms in Eq. (16.9) are present. For the general case where it is presumed that the displacements at the reaction joints, $\mathbf{U}_r$, are not all zero, Eqs. (16.9) are solved by rewriting the second of these as

$$\mathbf{K}_{uu}\mathbf{U}_u = \mathbf{P}_a - \mathbf{K}_{ur}\mathbf{U}_r \tag{16.10}$$

In Eq. (16.10), only the displacements in $\mathbf{U}_u$ are unknown and they can be obtained by any simultaneous equation solution scheme such as the Gauss elimination technique shown in the Appendix. The structure in Fig. 16.4 has a $\mathbf{K}_{uu}$ matrix which is 2×2, and the solution for the $\mathbf{U}_u$ is shown in Fig. 16.4g. Having solved for $\mathbf{U}_u$, all displacements of the structure are known; hence the reactions are computed from the first of Eqs. (16.9) as

$$\mathbf{R} = \mathbf{K}_{rr}\mathbf{U}_r + \mathbf{K}_{ru}\mathbf{U}_u \tag{16.11}$$

as shown in Fig. 16.4h (note that $\mathbf{K}_{rr}\mathbf{U}_r$ is zero in that figure since $\mathbf{U}_r = 0$). Finally, the member forces can be computed from Eq. (16.5) using the known member end displacements. This is illustrated in Fig. 16.4i.

The complete solution process for any structure is the same regardless of whether the structure is determinate or indeterminate. Only the sizes of the matrices in Eqs. (16.7) to (16.11) change, according to the number of joints in the structure and the number of reaction components. This makes the stiffness matrix method of analysis a technique that is easy to program since it is independent of the complexity of the structure. Once the joints of the structure have been numbered, the structure and sequence of the equations in the method is completely defined. The solution of the linear simultaneous equations (16.10) is a simple task for present-day computers. Many refinements of this process have improved the efficiency and capacity of even small computers to carry out a displacement analysis. The examples discussed below illustrate some of these points.

Example 16.1 The simple two-member truss of Fig. 16.4 is analyzed using the numerical values shown. The steps in the process exactly parallel the sequence of operations in Fig. 16.4. The reader may wish to review the discussion of Fig. 16.4 above while proceeding through the operations in each of the steps. The displacements obtained for joint 3 in step 6 are very small, and the assumption of no essential change in geometry of the structure is indeed satisfied. The 144-kN force computed in member 1 (1–3) in step 7 is about the maximum that can be sustained by mild steel in tension.

A more substantial example is presented below. The sequence of operations is essentially the same as Example 16.1. The use of the 2×2 $\mathbf{K}_e$ submatrices of Eqs. (16.6c) greatly aids in forming the equations.

Example 16.2 The five-joint, seven-member, simply supported plane truss is analyzed using the stiffness matrix method approach. The analysis follows the same sequence of steps presented in Fig. 16.4. The table shown in step 1 summarizes all of the important geometric and material data associated with the truss. The values of L, α, and β are calculated using Eqs. (16.2). The individual member stiffness matrices are obtained from Eq. (16.6a), but are expressed in the form of Eq. (16.6c) with the 2×2 matrices, $\mathbf{K}_i$ defined in Eq. (16.6d).

In step 2 the equilibrium equations at each of the five joints are established in matrix form, where each $\mathbf{P}_i$ and $\mathbf{U}_i$ matrix is 2×1 and each $\mathbf{K}_i$ 2×2. The internal member forces are summed and set equal to the externally applied loads, which are also defined as 2×1 matrices in the step. The collection of five matrix equilibrium equations is assembled into the final 10×10 stiffness matrix of the entire structure. This matrix is shown as five rows and columns of 2×2 blocks created from 2×2 null matrices, $\mathbf{0}$, and the $\mathbf{K}_i$ of the individual member stiffness matrices. The final structure of the equilibrium equations shows how the individual member stiffness matrices are assembled.

In step 3 the three reaction equations, which are the first two and the last equation of the structural stiffness matrix, are identified and extracted to form the 3×7 $\mathbf{K}_{ru}$ matrix of Eq. (16.9). The zero displacements at the reaction joints reduce the number of terms in these equations to seven since they multiply the terms in the first two and last columns of the original equations. The seven equations making up the $\mathbf{K}_{uu}$ matrix are also established, and are the third to the ninth equations and third to ninth columns in the original 10×10 matrix, since the terms in the first two and last columns of these equations are deleted due to the zero displacements at the reaction joints.

Example 16.1

Using the stiffness method of analysis, compute the displacements, reactions, and member forces for the plane truss shown. For both members, $A = 900$ mm^2 and $G = 200$ GPa.

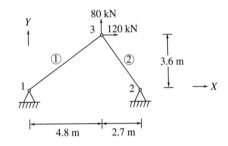

STEP 1 Obtain the geometric parameters from Eqs. (16.2) (see Fig. 16.4b).

Member 1: $L_1 = [(4.8 - 0)^2 + (3.6 - 0)^2]^{1/2} = 6$ m

$$\alpha_1 = \frac{4.8}{6} = 0.8 \quad \beta_1 = \frac{3.6}{6} = 0.6 \quad k_1 = 900 \cdot 10^{-6} \cdot 200 \cdot \frac{10^9}{6} = 30{,}000 \text{ kN/m}$$

Member 2: $L_2 = [(4.8 - 7.5)^2 + (3.6 - 0)^2]^{1/2} = 4.5$ m

$$\alpha_2 = \frac{-2.7}{4.5} = -.6 \quad \beta_2 = \frac{3.6}{4.5} = .8$$

$$k_2 = 900 \cdot 10^{-6} \cdot 200 \cdot \frac{10^9}{4.5} = 40{,}000 \text{ kN/m}$$

STEP 2 Establish relation for member forces and stiffness matrices form Eqs. (16.5) and (16.6) (see Fig. 16.4b).

$$S_1 = 30{,}000[(u_3 - u_1) \cdot 0.8 + (v_3 - v_1) \cdot 0.6]$$
$$= 24{,}000(u_3 - u_1) + 18{,}000(v_3 - v_1)$$

$$S_2 = 40{,}000[(u_3 - u_2) \cdot (-0.6) + (v_3 - v_2) \cdot 0.8]$$
$$= -24{,}000(u_3 - u_2) + 32{,}000(v_3 - v_2)$$

$$\begin{Bmatrix} F_{11} \\ F_{13} \end{Bmatrix} = \begin{bmatrix} K_1 & -K_1 \\ -K_1 & K_1 \end{bmatrix} \begin{Bmatrix} U_1 \\ U_3 \end{Bmatrix} \quad \begin{Bmatrix} F_{22} \\ F_{23} \end{Bmatrix} = \begin{bmatrix} K_2 & -K_2 \\ -K_2 & K_2 \end{bmatrix} \begin{Bmatrix} U_2 \\ U_3 \end{Bmatrix}$$

$$K_1 = 30{,}000 \begin{bmatrix} (0.8)^2 & (0.6)(0.8) \\ (0.6)(0.8) & (0.6)^2 \end{bmatrix}$$

$$= \begin{bmatrix} 19{,}200 & 14{,}400 \\ 14{,}400 & 10{,}800 \end{bmatrix}$$

Example 16.1 (continued)

$$K_2 = 40,000 \begin{bmatrix} (-0.6)^2 & (-0.6)(0.8) \\ (-0.6)(0.8) & (0.8)^2 \end{bmatrix}$$

$$= \begin{bmatrix} 14,400 & -19,200 \\ -19,200 & 25,600 \end{bmatrix}$$

STEP 3 Write the equilibrium equations at each joint. Each equation involves forces in the x and y direction, so they are written in matrix form (see Fig. 16.4c and d).

Joint 1: $P_1 = F_{11} = K_1U_1 - K_1U_3$ where $P_1 = \begin{Bmatrix} R_{1x} \\ R_{1y} \end{Bmatrix}$

Joint 2: $P_2 = F_{22} = K_2U_2 - K_2U_3$ where $P_2 = \begin{Bmatrix} R_{2x} \\ R_{2y} \end{Bmatrix}$

Joint 3: $P_3 = F_{13} + F_{23} = -K_1U_1 + K_1U_3 + (-K_2U_2) + K_2U_3$

$$= -K_1U_1 - K_2U_2 + (K_1 + K_2)U_3 \text{ where } P_3 = \begin{Bmatrix} 120 \\ 80 \end{Bmatrix}$$

or in numerical form:

$$\begin{Bmatrix} R_{1x} \\ R_{1y} \end{Bmatrix} = \begin{bmatrix} 19,200 & 14,400 \\ 14,400 & 10,800 \end{bmatrix} \begin{Bmatrix} u_1 \\ v_1 \end{Bmatrix} - \begin{bmatrix} 19,200 & 14,400 \\ 14,400 & 10,800 \end{bmatrix} \begin{Bmatrix} u_3 \\ v_3 \end{Bmatrix}$$

$$\begin{Bmatrix} R_{2x} \\ R_{2y} \end{Bmatrix} = \begin{bmatrix} 14,400 & -19,200 \\ -19,200 & 25,600 \end{bmatrix} \begin{Bmatrix} u_2 \\ v_2 \end{Bmatrix} - \begin{bmatrix} 14,400 & -19,200 \\ -19,200 & 25,600 \end{bmatrix} \begin{Bmatrix} u_3 \\ v_3 \end{Bmatrix}$$

$$\begin{Bmatrix} 120 \\ 80 \end{Bmatrix} = \begin{bmatrix} 19,200 & 14,400 \\ 14,400 & 10,800 \end{bmatrix} \begin{Bmatrix} u_1 \\ v_1 \end{Bmatrix} - \begin{bmatrix} 14,400 & -19,200 \\ -19,200 & 25,600 \end{bmatrix} \begin{Bmatrix} u_2 \\ v_3 \end{Bmatrix}$$

$$+ \begin{bmatrix} 33,600 & -4,800 \\ -4,800 & 36,400 \end{bmatrix} \begin{Bmatrix} u_3 \\ v_3 \end{Bmatrix}$$

collecting in matrix form yields (see Fig. 16.4d)

$$\begin{Bmatrix} P_1 \\ P_2 \\ P_3 \end{Bmatrix} = \begin{bmatrix} K_1 & 0 & -K_1 \\ 0 & K_2 & -K_2 \\ -K_1 & -K_2 & K_1 + K_2 \end{bmatrix} \begin{Bmatrix} U_1 \\ U_2 \\ U_3 \end{Bmatrix}$$

or, from Eq. (16.7), $P = KU$

Example 16.1 (continued)

which becomes in numerical form, where the symmetry of the matrix is shown (see Fig. 16.4c):

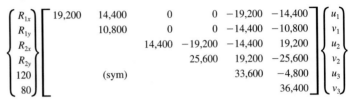

$$\begin{Bmatrix} R_{1x} \\ R_{1y} \\ R_{2x} \\ R_{2y} \\ 120 \\ 80 \end{Bmatrix} = \begin{bmatrix} 19{,}200 & 14{,}400 & 0 & 0 & -19{,}200 & -14{,}400 \\ & 10{,}800 & 0 & 0 & -14{,}400 & -10{,}800 \\ & & 14{,}400 & -19{,}200 & -14{,}400 & 19{,}200 \\ & & & 25{,}600 & 19{,}200 & -25{,}600 \\ & \text{(sym)} & & & 33{,}600 & -4{,}800 \\ & & & & & 36{,}400 \end{bmatrix} \begin{Bmatrix} u_1 \\ v_1 \\ u_2 \\ v_2 \\ u_3 \\ v_3 \end{Bmatrix}$$

STEP 4 Impose the specified zero-displacement conditions at the reaction joints 1 and 2 ($U_r = 0$). This reduces the equations to [from Eq. (16.11)] (see Fig. 16.4f)

$$\begin{Bmatrix} \mathbf{R} \\ \mathbf{P}_a \end{Bmatrix} = \begin{bmatrix} \mathbf{K}_{ru} \\ \mathbf{K}_{uu} \end{bmatrix} \mathbf{U}_u$$

or in numerical form:

$$\begin{Bmatrix} R_{1x} \\ R_{1y} \\ R_{2x} \\ R_{2y} \\ 120 \\ 80 \end{Bmatrix} = \begin{bmatrix} -19{,}200 & -14{,}400 \\ -14{,}400 & -10{,}800 \\ -14{,}400 & 19{,}200 \\ 19{,}200 & -25{,}600 \\ 33{,}600 & -4{,}800 \\ -4{,}800 & 36{,}400 \end{bmatrix} \begin{Bmatrix} u_1 \\ v_1 \end{Bmatrix}$$

STEP 5 Solve the last two equations in the matrix in step 4 for the displacements of joint 3 as indicated in Eq. (16.10) with $\mathbf{U}_r = 0$ (see Fig. 16.4g).

$$\begin{Bmatrix} 120 \\ 80 \end{Bmatrix} = \begin{bmatrix} 33{,}600 & -4{,}800 \\ -4{,}800 & 36{,}400 \end{bmatrix} \begin{Bmatrix} u_3 \\ v_3 \end{Bmatrix} \quad \text{solving} \quad \begin{Bmatrix} u_3 \\ v_3 \end{Bmatrix} = \begin{Bmatrix} 0.00396 \\ 0.00272 \end{Bmatrix} \text{m} = \begin{Bmatrix} 3.96 \\ 2.72 \end{Bmatrix} \text{mm}$$

STEP 6 Compute the reactions from the first four equations of in step 4 using Eq. (16.11) with $\mathbf{U}_{rr} = 0$ (see Fig. 16.4h).

$$\mathbf{R} = \mathbf{K}_{ru}\mathbf{U}_u$$

$$\begin{Bmatrix} R_{1x} \\ R_{1y} \\ R_{2x} \\ R_{2y} \end{Bmatrix} = \begin{bmatrix} -19{,}200 & -14{,}400 \\ -14{,}400 & -10{,}800 \\ -14{,}400 & 19{,}200 \\ 19{,}200 & -25{,}600 \end{bmatrix} \begin{Bmatrix} 0.00396 \\ 0.00272 \end{Bmatrix} = \begin{Bmatrix} -115.2 \\ -86.4 \\ -4.8 \\ 6.4 \end{Bmatrix} \text{kN}$$

Example 16.1 (continued)

STEP 7 Compute the member forces from expressions in step 2 and summarize the final force results on a sketch of the structure (see Fig. 16.4i).

$$S_1 = 24{,}000(0.00396 - 0) + 18{,}000(0.00272 - 0) = 144 \text{ kN}$$

$$S_2 = -24{,}000(0.00396 - 0) + 32{,}000(0.00272 - 0) = -8 \text{ kN}$$

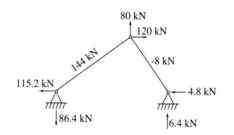

Example 16.2

Using the stiffness method of analysis, compute the displacements, reactions, and member forces for the plane truss shown.

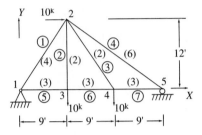

STEP 1 Obtain the individual member stiffness matrices for the seven members of the truss. Establish the origin of the global coordinate system at joint 1 and summarize the parameters needed to establish the matrices defined by Eq. (16.6) in the table below. The member stiffness matrices can be represented as an array of 2×2 $\mathbf{K}_i$'s (see Fig. 16.4b).

Joint	Coordinates X	Y	Member	Area, A	Length, L	α	β	AE/L*α^2	AE/L*β^2	AE/L*α*β
1	0	0	1 1–2	4	15	0.6	0.8	2784	4949.333	3712
2	9	12	2 3–2	2	12	0	1	0	4833.333	0
3	9	0	3 4–2	2	15	−0.6	0.8	1392	2474.667	−1856
4	18	0	4 5–2	6	21.633	−0.832	0.5547	5568.337	2474.816	−3712.22
5	27	0	5 1–3	3	9	1	0	9666.667	0	0
			6 3–4	3	9	1	0	9666.667	0	0
E = 29,000			7 4–5	3	9	1	0	9666.667	0	0

Define the following joint displacement matrices:

$$\mathbf{U}_1 = \begin{Bmatrix} u_1 \\ v_1 \end{Bmatrix} \quad \mathbf{U}_2 = \begin{Bmatrix} u_2 \\ v_2 \end{Bmatrix} \quad \mathbf{U}_3 = \begin{Bmatrix} u_3 \\ v_3 \end{Bmatrix} \quad \mathbf{U}_4 = \begin{Bmatrix} u_4 \\ v_4 \end{Bmatrix} \quad \mathbf{U}_5 = \begin{Bmatrix} u_5 \\ v_5 \end{Bmatrix}$$

Member 1:

$$\begin{Bmatrix} \mathbf{F}_{11} \\ \mathbf{F}_{12} \end{Bmatrix} = \begin{bmatrix} \mathbf{K}_1 & -\mathbf{K}_1 \\ -\mathbf{K}_1 & \mathbf{K}_1 \end{bmatrix} \begin{Bmatrix} \mathbf{U}_1 \\ \mathbf{U}_2 \end{Bmatrix} \quad \mathbf{F}_{11} = \begin{Bmatrix} F_{1x1} \\ F_{1y1} \end{Bmatrix} \quad \mathbf{F}_{12} = \begin{Bmatrix} F_{1x2} \\ F_{1y2} \end{Bmatrix}$$

$$\mathbf{K}_1 = \begin{bmatrix} 2784.0 & 3712.0 \\ 3712.0 & 4949.3 \end{bmatrix}$$

Example 16.2 (continued)

Member 2:

$$\begin{Bmatrix} \mathbf{F}_{23} \\ \mathbf{F}_{22} \end{Bmatrix} = \begin{bmatrix} \mathbf{K}_2 & -\mathbf{K}_2 \\ -\mathbf{K}_2 & \mathbf{K}_2 \end{bmatrix} \begin{Bmatrix} \mathbf{U}_3 \\ \mathbf{U}_2 \end{Bmatrix} \quad \mathbf{F}_{23} = \begin{Bmatrix} F_{2x3} \\ F_{2y3} \end{Bmatrix} \quad \mathbf{F}_{22} = \begin{Bmatrix} F_{2x2} \\ F_{2y2} \end{Bmatrix}$$

$$\mathbf{K}_2 = \begin{bmatrix} 0.0 & 0.0 \\ 0.0 & 4833.3 \end{bmatrix}$$

Member 3:

$$\begin{Bmatrix} \mathbf{F}_{34} \\ \mathbf{F}_{32} \end{Bmatrix} = \begin{bmatrix} \mathbf{K}_3 & -\mathbf{K}_3 \\ -\mathbf{K}_3 & \mathbf{K}_3 \end{bmatrix} \begin{Bmatrix} \mathbf{U}_4 \\ \mathbf{U}_2 \end{Bmatrix} \quad \mathbf{F}_{34} = \begin{Bmatrix} F_{3x4} \\ F_{3y4} \end{Bmatrix} \quad \mathbf{F}_{32} = \begin{Bmatrix} F_{3x2} \\ F_{3y2} \end{Bmatrix}$$

$$\mathbf{K}_3 = \begin{bmatrix} 1392.0 & -1856.0 \\ -1856.0 & 2474.7 \end{bmatrix}$$

Member 4:

$$\begin{Bmatrix} \mathbf{F}_{45} \\ \mathbf{F}_{42} \end{Bmatrix} = \begin{bmatrix} \mathbf{K}_4 & -\mathbf{K}_4 \\ -\mathbf{K}_4 & \mathbf{K}_4 \end{bmatrix} \begin{Bmatrix} \mathbf{U}_5 \\ \mathbf{U}_2 \end{Bmatrix} \quad \mathbf{F}_{45} = \begin{Bmatrix} F_{4x5} \\ F_{4y5} \end{Bmatrix} \quad \mathbf{F}_{42} = \begin{Bmatrix} F_{4x2} \\ F_{4y2} \end{Bmatrix}$$

$$\mathbf{K}_4 = \begin{bmatrix} 5568.3 & -3712.2 \\ -3712.2 & 2474.8 \end{bmatrix}$$

Member 5:

$$\begin{Bmatrix} \mathbf{F}_{51} \\ \mathbf{F}_{53} \end{Bmatrix} = \begin{bmatrix} \mathbf{K}_5 & -\mathbf{K}_5 \\ -\mathbf{K}_5 & \mathbf{K}_5 \end{bmatrix} \begin{Bmatrix} \mathbf{U}_1 \\ \mathbf{U}_3 \end{Bmatrix} \quad \mathbf{F}_{51} = \begin{Bmatrix} F_{5x1} \\ F_{5y1} \end{Bmatrix} \quad \mathbf{F}_{53} = \begin{Bmatrix} F_{5x3} \\ F_{5y3} \end{Bmatrix}$$

$$\mathbf{K}_5 = \begin{bmatrix} 9666.7 & 0.0 \\ 0.0 & 0.0 \end{bmatrix}$$

Member 6:

$$\begin{Bmatrix} \mathbf{F}_{63} \\ \mathbf{F}_{64} \end{Bmatrix} = \begin{bmatrix} \mathbf{K}_6 & -\mathbf{K}_6 \\ -\mathbf{K}_6 & \mathbf{K}_6 \end{bmatrix} \begin{Bmatrix} \mathbf{U}_3 \\ \mathbf{U}_4 \end{Bmatrix} \quad \mathbf{F}_{63} = \begin{Bmatrix} F_{6x3} \\ F_{6y3} \end{Bmatrix} \quad \mathbf{F}_{64} = \begin{Bmatrix} F_{6x4} \\ F_{6y4} \end{Bmatrix}$$

$$\mathbf{K}_6 = \begin{bmatrix} 9666.7 & 0.0 \\ 0.0 & 0.0 \end{bmatrix}$$

Member 7:

$$\begin{Bmatrix} \mathbf{F}_{74} \\ \mathbf{F}_{75} \end{Bmatrix} = \begin{bmatrix} \mathbf{K}_7 & -\mathbf{K}_7 \\ -\mathbf{K}_7 & \mathbf{K}_7 \end{bmatrix} \begin{Bmatrix} \mathbf{U}_4 \\ \mathbf{U}_5 \end{Bmatrix} \quad \mathbf{F}_{74} = \begin{Bmatrix} F_{7x4} \\ F_{7y4} \end{Bmatrix} \quad \mathbf{F}_{75} = \begin{Bmatrix} F_{7x5} \\ F_{7y5} \end{Bmatrix}$$

$$\mathbf{K}_7 = \begin{bmatrix} 9666.7 & 0.0 \\ 0.0 & 0.0 \end{bmatrix}$$

Example 16.2 (continued)

STEP 2   Write the equilibrium equations at each of the five joints. These equations can be assembled into the final matrix for the entire structure with those equations being obtained by the summation of the appropriate 2×2 matrices, $\mathbf{K}_i$, established in step 1. Define the following applied joint load matrices:

$$\mathbf{P}_1 = \left\{ \begin{matrix} R_{1x} \\ R_{1y} \end{matrix} \right\} \quad \mathbf{P}_2 = \left\{ \begin{matrix} 10 \\ 0 \end{matrix} \right\} \quad \mathbf{P}_3 = \left\{ \begin{matrix} 0 \\ -10 \end{matrix} \right\} \quad \mathbf{P}_4 = \left\{ \begin{matrix} 0 \\ -10 \end{matrix} \right\} \quad \mathbf{P}_5 = \left\{ \begin{matrix} 0 \\ R_{5y} \end{matrix} \right\}$$

The equilibrium equations at the five joints become (see Fig. 16.4d)

Joint 1: $\mathbf{P}_1 = \mathbf{F}_{11} + \mathbf{F}_{51} = \mathbf{K}_1(\mathbf{U}_1 - \mathbf{U}_2) + \mathbf{K}_5(\mathbf{U}_1 - \mathbf{U}_3)$

Joint 2: $\mathbf{P}_2 = \mathbf{F}_{12} + \mathbf{F}_{22} + \mathbf{F}_{32} + \mathbf{F}_{42} = \mathbf{K}_1(-\mathbf{U}_1 + \mathbf{U}_2)$
$\qquad\qquad + \mathbf{K}_2(-\mathbf{U}_3 + \mathbf{U}_2) + \mathbf{K}_3(-\mathbf{U}_4 + \mathbf{U}_2) + \mathbf{K}_4(-\mathbf{U}_5 + \mathbf{U}_2)$

Joint 3: $\mathbf{P}_3 = \mathbf{F}_{23} + \mathbf{F}_{53} + \mathbf{F}_{63} = \mathbf{K}_2(\mathbf{U}_3 - \mathbf{U}_2) + \mathbf{K}_5(-\mathbf{U}_1 + \mathbf{U}_3)$
$\qquad\qquad + \mathbf{K}_6(\mathbf{U}_3 - \mathbf{U}_4)$

Joint 4: $\mathbf{P}_4 = \mathbf{F}_{34} + \mathbf{F}_{64} + \mathbf{F}_{74} = \mathbf{K}_3(\mathbf{U}_4 - \mathbf{U}_2) + \mathbf{K}_6(-\mathbf{U}_3 + \mathbf{U}_4)$
$\qquad\qquad + \mathbf{K}_7(\mathbf{U}_4 - \mathbf{U}_5)$

Joint 5: $\mathbf{P}_5 = \mathbf{F}_{45} + \mathbf{F}_{75} = \mathbf{K}_4(\mathbf{U}_5 - \mathbf{U}_2) + \mathbf{K}_7(-\mathbf{U}_4 + \mathbf{U}_5)$

Collecting these equations together gives the final set of equations for the whole structure. Using the matrix definitions already established, the equations become (see Fig. 16.4e)

$$\mathbf{P} = \mathbf{KU}$$

or

$$\left\{ \begin{matrix} \mathbf{P}_1 \\ \mathbf{P}_2 \\ \mathbf{P}_3 \\ \mathbf{P}_4 \\ \mathbf{P}_5 \end{matrix} \right\} = \left[\begin{matrix} \mathbf{K}_1 + \mathbf{K}_5 & -\mathbf{K}_1 & -\mathbf{K}_5 & 0 & 0 \\ -\mathbf{K}_1 & \mathbf{K}_1 + \mathbf{K}_2 + \mathbf{K}_3 + \mathbf{K}_4 & -\mathbf{K}_2 & -\mathbf{K}_3 & -\mathbf{K}_4 \\ -\mathbf{K}_5 & -\mathbf{K}_2 & \mathbf{K}_2 + \mathbf{K}_5 + \mathbf{K}_6 & -\mathbf{K}_6 & 0 \\ 0 & -\mathbf{K}_3 & -\mathbf{K}_6 & \mathbf{K}_3 + \mathbf{K}_6 + \mathbf{K}_7 & -\mathbf{K}_7 \\ 0 & -\mathbf{K}_4 & 0 & -\mathbf{K}_7 & \mathbf{K}_4 + \mathbf{K}_7 \end{matrix} \right] \left\{ \begin{matrix} \mathbf{U}_1 \\ \mathbf{U}_2 \\ \mathbf{U}_3 \\ \mathbf{U}_4 \\ \mathbf{U}_5 \end{matrix} \right\}$$

Example 16.2 (continued)

and the full numerical values in the **K** matrix in units of kips/ft are

$$
\mathbf{K} = \begin{bmatrix}
12450.67 & 3712.00 & -2784.00 & -3712.00 & -9666.67 & 0.00 & 0.00 & 0.00 & 0.00 & 0.00 \\
3712.00 & 4949.33 & -3712.00 & -4949.33 & 0.00 & 0.00 & 0.00 & 0.00 & 0.00 & 0.00 \\
-2784.00 & -3712.00 & 9744.34 & -1856.33 & 0.00 & 0.00 & -1392.00 & 1856.00 & -5568.34 & 3712.22 \\
-3712.00 & -4949.33 & -1856.22 & 14732.15 & 0.00 & -4833.33 & 1856.00 & -2474.67 & 3712.22 & -2474.82 \\
-9666.67 & 0.00 & 0.00 & 0.00 & 19333.33 & 0.00 & -9666.67 & 0.00 & 0.00 & 0.00 \\
0.00 & 0.00 & 0.00 & -4833.33 & 0.00 & 4833.33 & 0.00 & 0.00 & 0.00 & 0.00 \\
0.00 & 0.00 & -1392.00 & 1856.00 & -9666.67 & 0.00 & 20725.33 & -1856.00 & -9666.67 & 0.00 \\
0.00 & 0.00 & 1856.00 & -2474.67 & 0.00 & 0.00 & -1856.00 & 2474.67 & 0.00 & 0.00 \\
0.00 & 0.00 & -5568.34 & 3712.22 & 0.00 & 0.00 & -9666.67 & 0.00 & 15235.00 & -3712.22 \\
0.00 & 0.00 & 3712.22 & -2474.82 & 0.00 & 0.00 & 0.00 & 0.00 & -3712.22 & 2474.82
\end{bmatrix}
$$

STEP 3   Separate the three equations associated with the reaction components from the seven equations associated with the unknown joint displacements. Using the conditions that $u_1 = v_1 = v_5 = 0$, the first two and last columns of the **K** matrix are multiplied by zero, so those columns can be dropped from the equations. Following the notation of Eq. (16.9) gives (see Fig. 16.4f)

$$\mathbf{R} = \mathbf{K}_{ru}\mathbf{U}_u \qquad \text{and} \qquad \mathbf{P}_a = \mathbf{K}_{uu}\mathbf{U}_u$$

where

$$
\mathbf{R} = \begin{Bmatrix} R_{1x} \\ R_{1y} \\ R_{5y} \end{Bmatrix}
\qquad
\mathbf{U}_u = \begin{Bmatrix} u_2 \\ v_2 \\ u_3 \\ v_3 \\ u_4 \\ v_4 \\ u_5 \end{Bmatrix}
\qquad
\mathbf{P}_a = \begin{Bmatrix} 10.00 \\ 0.00 \\ 0.00 \\ -10.00 \\ 0.00 \\ -10.00 \\ 0.00 \end{Bmatrix}
$$

$$
\mathbf{K}_{uu} = \begin{bmatrix}
9744.34 & -1856.22 & 0.00 & 0.00 & -1392.00 & 1856.00 & -5568.34 \\
-1856.22 & 14732.15 & 0.00 & -4833.33 & 1856.00 & -2474.67 & 3712.22 \\
0.00 & 0.00 & 19333.33 & 0.00 & -9666.67 & 0.00 & 0.00 \\
0.00 & -4833.33 & 0.00 & 4833.33 & 0.00 & 0.00 & 0.00 \\
-1392.00 & 1856.00 & -9666.67 & 0.00 & 20725.33 & -1856.00 & -9666.67 \\
1856.00 & -2474.97 & 0.00 & 0.00 & -1856.00 & 2474.67 & 0.00 \\
-5568.34 & 3712.22 & 0.00 & 0.00 & -9666.67 & 0.00 & 15235.00
\end{bmatrix}
$$

$$
\mathbf{K}_{ru} = \begin{bmatrix}
-2784.00 & -3712.00 & -9666.67 & 0.00 & 0.00 & 0.00 & 0.00 \\
-3712.00 & -4949.33 & 0.00 & 0.00 & 0.00 & 0.00 & 0.00 \\
3712.22 & -2474.82 & 0.00 & 0.00 & 0.00 & 0.00 & -3712.22
\end{bmatrix}
$$

The units of the terms in $\mathbf{P}_a$ are kips, and those in the $\mathbf{K}_{ru}$ and $\mathbf{K}_{uu}$ matrices are kips/ft.

Example 16.2 (continued)

STEP 4 Solve for the displacements and backsubstitute for the reactions from Eq. (16.9) and the member axial forces from Eq. (16.5). The displacements have units of feet and the axial forces, kips (see Fig. 16.4g to i).

$$\mathbf{U}_u = \begin{Bmatrix} 0.005543 \\ 0.00528 \\ 0.001466 \\ -0.00735 \\ 0.002931 \\ -0.01128 \\ 0.005172 \end{Bmatrix} \qquad \mathbf{R} = \mathbf{K}_{ru}\mathbf{U}_u = \begin{Bmatrix} -10.00 \\ 5.56 \\ 14.44 \end{Bmatrix}$$

The member forces become

$$S_1 = \left(\frac{AE}{L}\right)_1 [(u_2 - u_1)\alpha_1 + (v_2 - v_1)\beta_1] = -6.944 \text{ kips}$$

$$S_2 = \left(\frac{AE}{L}\right)_2 [(u_2 - u_3)\alpha_2 + (v_2 - v_3)\beta_2] = 10.000 \text{ kips}$$

$$S_3 = \left(\frac{AE}{L}\right)_3 [(u_2 - u_4)\alpha_3 + (v_2 - v_4)\beta_3] = 12.500 \text{ kips}$$

$$S_4 = \left(\frac{AE}{L}\right)_4 [(u_2 - u_5)\alpha_4 + (v_2 - v_5)\beta_4] = -26.040 \text{ kips}$$

$$S_5 = \left(\frac{AE}{L}\right)_5 [(u_3 - u_1)\alpha_5 + (v_3 - v_1)\beta_5] = 14.167 \text{ kips}$$

$$S_6 = \left(\frac{AE}{L}\right)_6 [(u_4 - u_3)\alpha_6 + (v_4 - v_3)\beta_6] = 14.167 \text{ kips}$$

$$S_7 = \left(\frac{AE}{L}\right)_7 [(u_5 - u_4)\alpha_7 + (v_5 - v_4)\beta_7] = 21.167 \text{ kips}$$

Finally, in step 4 the $\mathbf{K}_{uu}$ equations are solved for the unknown joint displacements. The now known and complete set of joint displacements of the entire structure is substituted into Eq. (16.11) to obtain the reaction components in the $\mathbf{R}$ matrix. The member axial forces are computed using Eq. (16.5). The final reaction components and member axial forces are shown on the sketch at the end of the example. The truss in this example is statically determinate, but this fact was not used in the solution process. The displacements of the joints of the truss are again very small and the assumption of no significant change in geometry due to the loading is satisfied. Although member stresses are low in this structure, increasing the load by a factor of 3 will raise member stresses to the limits of safe design with mild steel but will not affect the assumptions of no significant geometry change.

16.3   Local and Global Coordinate Systems and Transformations Between Them

The development of the individual member stiffness matrices for plane truss elements presented in Eq. (16.6) has been done in the global coordinate system. This system provides the common reference system for forces and displacements needed to establish the final equilibrium equations for the entire structure expressed in Eq. (16.7). In the next two sections the development of individual member stiffness matrices will be undertaken in what is called a *local* or *member coordinate system* that is defined by the member axis. The member stiffness matrix must still be referenced to the global coordinate system as in Eq. (16.6) to establish Eq. (16.7). This means that the stiffness matrix referenced to the local coordinate system must be transformed to the global coordinate system. Since both of these systems are orthogonal systems, the transformation between them will be referred to as an *orthogonal transformation*.

As shown in Section 16.4, the transformation of individual member stiffness relations between the local and global coordinate system is in fact a transformation of the displacement and force components that occur on the ends of a member. If in a member such as e shown in Fig. 16.2, the local coordinate system x axis is defined to be the member centroidal axis, the local system makes an angle θ with the global x axis. The transformation between local and global is simply a rotational transformation of force and displacement components and will be developed in terms of displacements.

In Fig. 16.5 a displacement is defined by components u and v in the global coordinate system. The same displacement is defined by the components u'

and v' in the local or member coordinate system. From the geometry of the figure, u' and v' can be related to u and v by the expressions

$$u' = u \cos \theta + v \cos \phi = u\alpha + v\beta$$

$$v' = -u \cos \phi + v \cos \theta = -u\beta + v\alpha \tag{16.12}$$

in which the α and β are defined in the same manner as in Eqs. (16.1) and (16.2). Equation (16.12) can be written in matrix form as

$$\mathbf{u}_l = \mathbf{Q}\mathbf{u}_g \tag{16.13a}$$

in which

$$\mathbf{Q} = \begin{bmatrix} \alpha & \beta \\ -\beta & \alpha \end{bmatrix} \tag{16.13b}$$

and the $\mathbf{u}_l$ and the $\mathbf{u}_g$ are 2×1 matrices with terms u' and v' and u and v, respectively. If the u and v displacement components in the global system were replaced with F_x and F_y force components in the global system and the u' and v' displacement components in the local system with f_x and f_y force components in the local system in Fig. 16.5, the relation between local and global force components would be the same as Eq. (16.13):

$$\mathbf{f}_l = \mathbf{Q}\mathbf{F}_g \tag{16.14}$$

in which the $\mathbf{f}_l$ and $\mathbf{F}_g$ are 2×1 matrices with terms f_x and f_y and F_x and F_y, respectively, and the $\mathbf{Q}$ matrix is the one defined in Eq. (16.13).

The transformations between the local and global systems can be inverted to yield global displacements or forces in terms of the corresponding local displacements or forces. Premultiplying both sides of Eqs. (16.13) and (16.14) by $\mathbf{Q}^{-1}$ yields the global quantities in terms of the corresponding local quantities. Because the $\mathbf{Q}$ matrix is the transformation matrix for an orthogonal transformation, it has the property that

$$\mathbf{Q}^{-1} = \mathbf{Q}^T \tag{16.15}$$

which is easily verified by expanding out the product $\mathbf{Q}^T \cdot \mathbf{Q}$ and showing the $\mathbf{Q}^T \cdot \mathbf{Q} = \mathbf{Q} \cdot \mathbf{Q}^T = \mathbf{I}$. This product requires the use of the identity $\alpha^2 + \beta^2 = \cos^2 \theta + \sin^2 \theta = 1$. With the aid of Eq. (16.15), the inverses of the relations in Eqs. (16.13) and (16.14) become

$$\mathbf{u}_g = \mathbf{Q}^T \mathbf{u}_l$$

$$\mathbf{F}_g = \mathbf{Q}^T \mathbf{f}_l \tag{16.16}$$

These relations will be used to transform stiffness matrices for individual members established in the local member coordinate system to the global coordinate system.

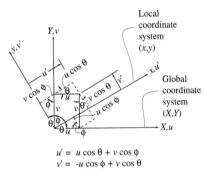

$u' = u \cos \theta + v \cos \phi$
$v' = -u \cos \phi + v \cos \theta$

Figure 16.5   Relation between displacements in the local and global coordinate systems.

16.4 Formulation of the Stiffness Matrix for a Plane Truss Member in Its Local Coordinate System

The stiffness matrix force–deformation relation for member e of Fig. 16.2, Eq. (16.6), was obtained directly from geometric and equilibrium considerations of the member in the global coordinate system in Section 16.1. This same relation can be obtained from an analysis of the member force–deformation behavior referenced to a local member coordinate system. Consider the local coordinate system definition for the member e shown in Fig. 16.6a. As in Fig. 16.2, the orientation of the member and its local coordinate system is at an angle θ with respect to the global coordinate system x axis.

The member end displacements are shown in Fig. 16.6b and the relative member end displacements in Fig. 16.6c, all referenced to the local coordinate

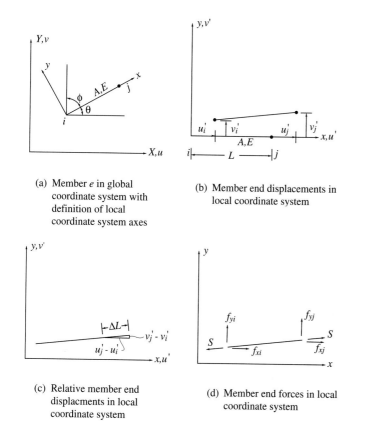

(a) Member e in global coordinate system with definition of local coordinate system axes

(b) Member end displacements in local coordinate system

(c) Relative member end displacments in local coordinate system

(d) Member end forces in local coordinate system

Figure 16.6a–d Displacements, relative displacements, and end forces for member e as referenced in the local coordinate system.

system. The change in length of the member ΔL shown in Fig. 16.6c is due to both the relative x displacement $u_j' - u_i'$ and the relative y displacement $v_j' - v_i'$. However, the contribution to the change in length ΔL from the relative y displacements is extremely small and can be neglected, which is a consequence of the assumption of small strains and small displacements, which is used in the development of this method. The change in length becomes

$$\Delta L = u_j' - u_i' \tag{16.17}$$

The error in neglecting the contribution of the relative displacements in the y direction to ΔL in Eq. (16.17) can be shown to be less than $(v_j' - v_i')^2/L$, which is clearly small for the assumed small displacement behavior of the member. The axial force developed in the member from the relative end displacements is, from Eq. (16.3),

$$S = \frac{AE}{L}\Delta L = \frac{AE}{L}(u_j' - u_i') \tag{16.18}$$

The four components of the internal axial force on the two ends of the member are shown in Fig. 16.d, where use of the lowercase f's indicate that the forces are referenced to the local coordinate system axes. Again, rotation of the member that is related to the relative y displacements is assumed to be so small that it can be neglected. From Fig. 16.6d and the assumption of small displacements, the member end forces become

$$f_{xj} = -f_{xi} = S = \frac{AE}{L}(u_j' - u_i')$$

$$f_{yj} = -f_{yi} = 0 \tag{16.19a}$$

which can be written in matrix form as

$$\begin{Bmatrix} f_{xi} \\ f_{yi} \\ f_{xj} \\ f_{yj} \end{Bmatrix} = \frac{AE}{L} \begin{bmatrix} 1 & 0 & -1 & 0 \\ 0 & 0 & 0 & 0 \\ -1 & 0 & 1 & 0 \\ 0 & 0 & 0 & 0 \end{bmatrix} \begin{Bmatrix} u_i' \\ v_i' \\ u_j' \\ v_j' \end{Bmatrix} \tag{16.19b}$$

or in the compact matrix form

$$\mathbf{f}_e = \mathbf{k}_e \mathbf{u}_e \tag{16.19c}$$

The lowercase $\mathbf{f}$, $\mathbf{k}$, and $\mathbf{u}$ matrices indicate that the relation is referenced to the local coordinate system. This relation is the same as Eq. (16.6), which is referenced to the global coordinate system.

The force–deformation relation in Eq. (16.19) needs to be transformed to the global coordinate system in order to assemble the final stiffness matrix for the entire structure, Eq. (16.7). The displacement and force components in the local system at the i and j joints of the member are related to the corresponding displacement and force components in the global system by Eqs. (16.13) and (16.14). The matrices $\mathbf{u}_e$ and $\mathbf{f}_e$ in Eq. (16.19c) are related to their global system counterparts $\mathbf{U}_e$ and $\mathbf{F}_e$ by the relations

$$\mathbf{u}_e = \mathbf{T}\mathbf{U}_e$$

$$\mathbf{f}_e = \mathbf{T}\mathbf{F}_e \tag{16.20a}$$

in which the 4×4 transformation matrix T is defined as

$$\mathbf{T} = \begin{bmatrix} \mathbf{Q} & \mathbf{0} \\ \mathbf{0} & \mathbf{Q} \end{bmatrix} \tag{16.20b}$$

where $\mathbf{0}$ is a 2×2 null matrix and $\mathbf{Q}$ is defined in Eq. (16.13b). It is easy to show by the expansion of Eq. (16.20a) that the relations between the local and global system displacements and forces defined in Eqs. (16.13) and (16.14) are obtained for joints i and j. It is also easy to verify by direct expansion using the relation in Eq. (16.15) that

$$\mathbf{T}^{-1} = \mathbf{T}^T \tag{16.21}$$

The substitution of Eq. (16.20) into Eq. (16.19) and premultiplying both sides of the resulting equation by $\mathbf{T}^T$ yields the result

$$\mathbf{F}_e = \mathbf{K}_e\mathbf{U}_e \tag{16.22a}$$

in which

$$\mathbf{K}_e = \mathbf{T}^T\mathbf{k}_e\mathbf{T} \tag{16.22b}$$

Equation (16.22a) is identical to Eq. (16.6b) and the complete expansion of Eq. (16.22b) using the definitions of Eqs. (16.13b) and (16.20b) shows it to be identical to the stiffness matrix in Eq. (16.6a). The development of Eq. (16.22), which is identical to Eq. (16.6), shows that the use of a local coordinate system to define the relations between end forces and end displacements of a member and then transform that relation to the global coordinate system provides an alternative means of obtaining the stiffness matrix for that member. The local coordinate system development is important to the stiffness matrix formulation for plane frame members in Chapter 17.

16.5   Summary and Limitations

The procedures presented in this chapter are based on the computation of the displacements of the joints of plane trusses. The procedures are relatively easy to program for execution on a digital computer. They involve the establishment and solution of a system of linear simultaneous algebraic equations. The number of equations that must be solved is large, even for a modest-size structure, so the methods are not suitable for hand computation unless a structure is very small and is highly restrained against displacement.

The stiffness matrix method is introduced in Section 16.1 for a plane truss structure and presented in its entirety for a simple two-member truss in Fig. 16.4. The equations of solution, which are the two force equilibrium equations at each joint of the truss, are ordered by the sequence of numbering of the joints of the structure. The displacements are also numbered in the same sequence so that the equations of solution have coefficient matrices that are symmetric. The coefficient matrices are called stiffness matrices because the individual terms of the matrices have the units of force over length, or stiffness.

In the next section the solution procedure is generalized by presenting it in abstract matrix form. The relation of each matrix in the procedure to the equations and matrices established for the solution of the simple truss in Fig. 16.4 is documented. Each of the matrix equations is related to physical behavior of the truss, and the separation of the equations between those associated with unknown reaction components and those associated with unknown joint displacements is identified.

An alternative development of the stiffness matrix relation for plane truss members is presented in Section 16.4. It is based on establishing the force–deformation relation of a plane truss member in a local coordinate system that uses the member axis as the x axis. The matrix relation in the local coordinate system is transformed to the global coordinate system using the transformation operations and matrices derived in the preceding section. The development of the stiffness matrix of the member in the local coordinate system and transforming it to the global coordinate system simplifies the geometric considerations of the process.

The most important observation to be recognized in the procedures presented in this chapter is that they are based on the classical analysis of prismatic member behavior. The capability of computer programs to produce analyses of very large and very complicated structures underscores the need to understand the limitations on both individual members and overall structural behavior. It is imperative that the assumptions and limitations listed below are understood by the structural engineer when computer analyses are evaluated.

- The assumptions of axial force behavior and the limitations listed at the end of Chapter 3 are also applicable here.
- All material deformations must be continuous, without any cracks or gaps.
- All structural deformations must be continuous, without any violation of the conditions of constraint at points of support or at points (joints) of attachment.

- All axial deformations are assumed to be small, and axial forces have no significant effect on the displacement behavior of the joints of the structure.

- The distribution and orientation of all loads acting on a structure are not changed by the deformations of the structure.

- The orientation of all reaction components at supports is unchanged by the action of applied loads or any movement of the support.

- Any movements that occur at supports are assumed to be small and do not affect the initial geometry of the structure.

- The principle of superposition of deformations with all its limitations is required for the analysis of symmetric structures under general loading if the analysis is carried out on the symmetric and antisymmetric forms of the structure.

- All members are assumed to be prismatic so that simple deformation behavior from mechanics of materials is valid.

The use of computer programs for the analysis and design of structures places a heavy burden on the structural engineer because a complete check of the analysis by hand computations is not possible. Nevertheless, the structural engineer is totally and completely responsible for the design and analysis of a structure. Some questions that should be considered in evaluating a computer-generated analysis are:

- Has the analysis generated by the computer program incorporated all loading conditions that are critical to the structure?

- Has the critical loading configuration for each and every member of the structure been considered in the analysis?

- If two "different" computer programs give the same result in the analysis of a structure, does this prove the correctness of the analysis or validate the computer programs?

- How large must displacements be before there is a need for a geometrically nonlinear analysis?

- If one (or more) of the assumptions or limitations of the stiffness matrix method of analysis listed above is not met by the computer program being used for a particular analysis, is it possible to approximate the effect of the violated limitation or assumption by further (or a more in-depth) analysis with the existing computer program?

- Even though the input to a computer-generated analysis of a structure is absolutely correct, who is responsible if the analysis generated by the computer is incorrect?

- How valid can the results of a computer-generated analysis be if a computer program that generates analyses of structures with linear materials is used with structures with "nearly" linear materials?

- Even though the analysis generated by a computer program is correct, how does one check that the input to that analysis is correct?

- If a hand-generated analysis of a structure by indeterminate analysis is correctly executed and the analysis for the same structure generated by a computer program based on the stiffness matrix method presented in this chapter is different, which analysis is the correct one?

PROBLEMS

16.1 For the pin-connected truss shown below (E constant):

(a) Establish the final matrix of all equations required for solution by the stiffness matrix method and show symmetry of the equations.

(b) Write the reaction equations for the system in matrix form with each matrix explicitly defined.

(c) Write the equations of solution for the joint displacements in matrix form and show the symmetry of the equations.

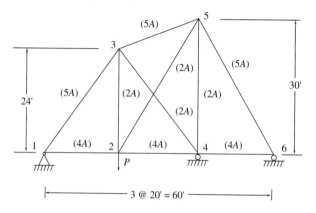

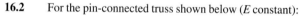

16.2 For the pin-connected truss shown below (E constant):

(a) Establish the final matrix of all equations required for solution by the stiffness matrix method and show symmetry of the equations.

(b) Write the reaction equations for the system in matrix form with each matrix explicitly defined.

(c) Write the equations of solution for the joint displacements in matrix form and show the symmetry of the equations.

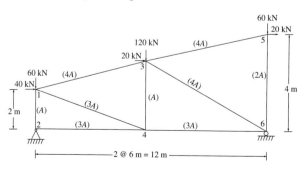

16.3 Obtain the solution for the displacements, member forces, and the reactions for the plane truss shown. Use the stiffness matrix method ($E = 29,000$ ksi).

Verticals: $A = 4$ in^2
Member 3–4: $A = 2$ in^2

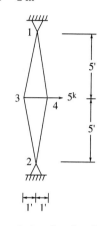

16.4 Obtain the solution for the displacements, member forces, and the reactions for the plane truss shown. Use the stiffness matrix method ($E = 200$ GPa).

Member 3–4: $A = 15$ cm^2
All other members: $A = 40$ cm^2

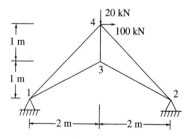

667

16.5   Obtain the solution for the displacements, member forces, and the reactions for the plane truss shown. Use the stiffness matrix method (all members: $A = 5$ in^2; $E = 29,000$ ksi).

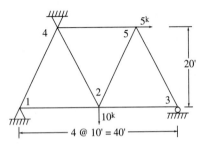

16.6   Obtain the solution for the displacements, member forces, and the reactions for the plane truss shown. Use the stiffness matrix method [Area $(\cdot)$ in cm^2, $E = 200$ GPa].

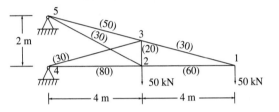

16.7   Solve by the stiffness matrix method for all displacements, reactions, and member forces. Take $L = \sqrt{a^2 + b^2}$ for members 1–3 and 2–3.

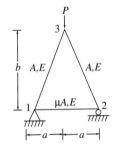

16.8   Repeat Problem 16.5 with the added support movement of joint 4 of 0.02 ft to the right.

16.9   Repeat Problem 16.6 with the added support movement of joint 4 of 6 mm to the left.

COMPUTER PROBLEMS

The following problems should be solved using a computer program that does plane truss analysis by the stiffness matrix method.

C16.1   Analyze the truss of Example 10.1.

C16.2   Analyze the truss of Problem 16.1.

C16.3   Analyze the truss of Problem 16.2.

C16.4   Analyze the truss of Problem 16.5.

C16.5   Analyze the truss of Problem 16.6.

C16.6   Obtain the influence line for the left support reaction of the truss of Example 13.5.

C16.7   Obtain the influence line for the center support reaction of the truss of Example 13.5.

C16.8   Obtain the influence line for the force in member U_3–U_4 of the truss of Example 13.5.

Analysis of Plane Frames by the Stiffness Matrix Method

In Chapter 16 the stiffness matrix method for plane trusses was developed as a displacement method of analysis. By combining the concepts of Chapter 16 with the slope-deflection method presented in Chapter 11, a stiffness matrix analysis for plane frames is established. In this technique the moments and forces in a frame are all expressed in terms of the rotations and displacements of the frame. Three equilibrium equations are established at each joint of the frame. These equations contain displacements, rotations, or reaction components as unknowns. To obtain moments and forces in the members of the frame, additional computations involving the slope-deflection equations are required.

The slope-deflection technique is a displacement method of analysis. For hand computation purposes as presented in Chapter 11, the technique is attractive when the number of joint translations and rotations in a frame is less than the degree of indeterminacy of the frame. For automatic computation using digital computers, the slope-deflection method and other displacement methods are nearly always superior. The development of the method in this chapter will parallel the development of the stiffness matrix method of Chapter 16 and will relate closely to concepts introduced there.

17.1   Local Coordinate System Stiffness Matrices _____

The analysis of plane frames by the stiffness matrix method is an extension of the slope-deflection method of analysis presented in Chapter 11 with the incorporation of the plane truss techniques described in the preceding chapter. The development of the stiffness relation for an individual plane frame member is most easily accomplished by using the local member coordinate system. An expression similar to Eqs. (16.19) will be developed and then, through the use of an appropriately defined transformation matrix, transformed into a relation in the global coordinate system.

The assumptions that underlie the stiffness matrix method for plane frames are the same as those summarized in Chapter 3 for plane truss analysis, in Chapter 5 for plane frame analysis, and in Chapter 6 for simple bending analysis. The presence of axial forces in plane frames underscores the importance of the assumption of no interaction of the axial force action and bending actions. The error of neglecting this interaction for linear elastic materials is less than about 6% as long as the magnitude of the axial load (compression or tension) in a member is less than EI/L^2.

Figure 17.1 shows a plane frame member connected to joints i and j, member i–j, or simply member e, with all of the displacement and force actions referenced to the local x–y coordinate system. The actions are similar

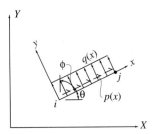

(a) Plane frame member in global
coordinate system

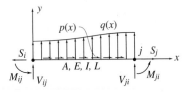
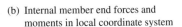

(b) Internal member end forces and
moments in local coordinate system

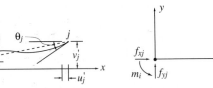

(c) Member end displacements and
rotations in local coordinate system

(d) Local coordinate system member
end forces and moments

Figure 17.1a–d   Forces and displacements on the ends of a plane frame member, e, connected to joints i and j of a structure.

to those shown in Figs. 11.1 and 11.2, but the additional presence of axial force action requires that the axial forces, S_i and S_j, the distributed axial force per unit length, $p(x)$, and the member end displacements u_i' and u_j' be shown. As shown in Fig. 17.1a, member e is understood to be part of a plane frame structure that is defined in a global coordinate system X–Y and is oriented such that the local X axis makes an angle θ with the global X axis and an angle ϕ with the global Y axis.

In Fig. 17.1b the complete set of force actions that develop on the member due to the applied transverse load $q(x)$, the distributed axial load $p(x)$ and the member end displacements and rotations are shown. All forces are shown to act in their positive sense following the sign conventions established in earlier chapters. The internal member end forces are the member end axial forces, S_i and S_j, the member end shear forces, V_{ij} and V_{ji}, and the member end moments, M_{ij} and M_{ji}. The member end displacements, u_i', v_i', θ_i, u_j', v_j', and θ_j are shown in Fig. 17.1c. Finally, Fig. 17.1d shows the six member force and moment actions referenced to the positive local coordinate axes. The lowercase symbols for these actions emphasize the local coordinate reference.

The relations between the member end force actions and end displacement actions have already been defined in preceding sections. With the assumption of no interaction between axial force and bending action, the axial force S is related to the u' displacements, by Eq. (16.18). In the presence of the distributed axial load $p(x)$, the axial end forces become

$$S_i = \frac{AE}{L}(u_j' - u_i') - fef_{xi} \qquad (17.1a)$$

$$S_j = \frac{AE}{L}(u_j' - u_i') + fef_{xj} \qquad (17.1b)$$

where

$$fef_{xi} = -\frac{1}{L}\int_0^L F(x)\,dx \qquad (17.1c)$$

$$fef_{xj} = \frac{1}{L}\int_0^L F(x)\,dx - \int_0^L p(x)\,dx = -fef_{xi} - \int_0^L p(x)\,dx \quad (17.1d)$$

and

$$F(x) = \int_0^x p(x)\,dx \qquad (17.1e)$$

The fef_{xi} and fef_{xj} are the member fixed-end axial forces in the positive x direction. They are calculated from the condition that the displacements u_j of the ends of the member are zero under the action of $p(x)$. The deformation analysis follows from the techniques presented in Section 6.6.

The relations between the member end shears, moments, displacements, and rotations are taken directly from Eqs. (11.3) to (11.7) and become, using the notation of Fig. 17.1,

$$M_{ij} = \frac{2EI}{L}\left(2\theta_i + \theta_j - 3\frac{v_j' - v_i'}{L}\right) + fem_{ij}$$

$$M_{ji} = \frac{2EI}{L}\left(2\theta_j + \theta_i - 3\frac{v_j' - v_i'}{L}\right) + fem_{ji}$$

(17.2a)

$$V_{ij} = \frac{6EI}{L^2}\left(\theta_i + \theta_j - 2\frac{v_j' - v_i'}{L}\right) + fef_{yi}$$

$$V_{ji} = -\frac{6EI}{L^2}\left(\theta_i + \theta_j - 2\frac{v_j' - v_i'}{L}\right) + fef_{yj}$$

(17.2b)

in which

$$fem_{ij} = \frac{2EI}{L}(2M_{00} - M_0)$$

$$fem_{ji} = \frac{2EI}{L}(M_{00} - 2M_0)$$

(17.2c)

$$fef_{yi} = \frac{6EI}{L^2}(M_{00} - M_0) + R_{iq}$$

$$fef_{yj} = \frac{6EI}{L^2}(M_{00} - M_0) + R_{jq}$$

(17.2d)

$$M_0 = \int_0^L \frac{x}{L}\frac{M(x)}{EI}\,dx$$

(17.2e)

$$M_{00} = \int_0^L \left(\frac{L - x}{L}\right)\frac{M(x)}{EI}\,dx$$

(17.2f)

In Eqs. (17.2), $M(x)$ is the moment diagram obtained when the applied distributed load $q(x)$ acts on a simply supported beam, and the R_{iq} and R_{jq} are the reaction forces on the i and j ends of that simply supported beam.

The local coordinate system end forces and moments shown in Fig. 17.1d are easily related to the member internal end forces and moments as

$$f_{xi} = -S_i$$

(17.3a)

$$f_{xj} = S_j$$

(17.3b)

$$f_{yi} = V_{ij} \tag{17.3c}$$

$$m_i = M_{ij} \tag{17.3d}$$

$$f_{yj} = V_{ji} \tag{17.3e}$$

$$m_j = M_{ji} \tag{17.3f}$$

With the definition of the internal force actions in terms of the applied loads and the displacement actions for the member established in Eqs. (17.1) and (17.2), Eqs. (17.3) can be expressed in matrix form as

$$\mathbf{f}_e = \mathbf{k}_e \mathbf{u}_e + \mathbf{f}_{fe} \tag{17.4a}$$

where

$$\mathbf{f}_e = \begin{Bmatrix} f_{xi} \\ f_{yi} \\ m_i \\ f_{xj} \\ f_{yj} \\ m_j \end{Bmatrix} \qquad \mathbf{u}_e = \begin{Bmatrix} u'_i \\ v'_i \\ \theta_i \\ u'_j \\ v'_j \\ \theta_j \end{Bmatrix} \qquad \mathbf{f}_{fe} = \begin{Bmatrix} fef_{xi} \\ fef_{yi} \\ fem_{ij} \\ fef_{xj} \\ fef_{yj} \\ fem_{ij} \end{Bmatrix} \tag{17.4b}$$

and using the definitions $k = AE/L$ and $B = EI/L$, the 6×6 $\mathbf{k}_e$ becomes

$$\mathbf{k}_e = \begin{bmatrix} k & 0 & 0 & -k & 0 & 0 \\ 0 & 12\frac{B}{L^2} & 6\frac{B}{L} & 0 & -12\frac{B}{L^2} & 6\frac{B}{L} \\ 0 & 6\frac{B}{L} & 4B & 0 & -6\frac{B}{L} & 2B \\ -k & 0 & 0 & k & 0 & 0 \\ 0 & -12\frac{B}{L^2} & -6\frac{B}{L} & 0 & 12\frac{B}{L^2} & -6\frac{B}{L} \\ 0 & 6\frac{B}{L} & 2B & 0 & -6\frac{B}{L} & 4 \end{bmatrix} \tag{17.4c}$$

The $\mathbf{k}_e$ can be written in the convenient partitioned form shown by the dashed lines in Eq. (17.4c) as

$$\mathbf{k}_e = \begin{bmatrix} \mathbf{k}_{e1} & \mathbf{k}_{e2} \\ \mathbf{k}_{e2}^T & \mathbf{k}_{e3} \end{bmatrix} \tag{17.4d}$$

where the $\mathbf{k}_{ei}$ are all 3×3 matrices.

Equation (17.4) is the stiffness matrix relation for a plane frame member in the local coordinate system. It corresponds with the combined formulation of the slope-deflection equations (11.3) for member end moments and (11.6) for member end shears in Chapter 11 and the axial force-deformation relation for the member. The solution of a problem using the formulation is illustrated below for the structure of Example 11.4.

Example 17.1 The structure with members *A–B* and *B–C* in Example 11.4 is relabeled, making those members 1 (1–2) and 2 (2–3), respectively, in the current example. Both members 1 and 2 have local coordinate systems with the same orientation. As a consequence, the primes on the displacements and rotations in Eqs. (17.4) have been dropped for convenience. The formation of the equations of solution follows the procedure set forth in Fig. 16.4, but with the submatrices being 3 × 3 rather than the 2 × 2 in plane truss problems.

The final equations for solution presented in step 3 are not the same as the equations in Example 11.4 because the rotation θ_3 (θ_C in Example 11.4) and the displacements u_2 and u_3 are included. However, the five equations will reduce to the same equations as in Example 11.4 if u_2 and u_3 (both of which are zero as shown in step 4) and θ_3 are eliminated from the solution by solving the first and the fourth and fifth equations in terms of v_2 and θ_2. The verification of this observation is left to the reader.

The solution obtained in step 4 gives the same rotation and displacement at joint 2 as was obtained at joint *B* in Example 11.4. The calculation of the reactions and member end forces completes the analysis. The final moment diagram due to the loading is not shown but would be the same as the one in Example 11.4.

The solution by the stiffness matrix method in Example 17.1 is much more lengthy than the slope-deflection solution of Example 11.4, although the results from the two analyses are identical. However, the stiffness matrix method approach is more general and can accommodate the solution of any plane frame problem, including the effects of axial deformations. Extending the slope-deflection method to frames where axial deformation effects are included becomes difficult at best and is not well suited to problems in which members are inclined rather than horizontal or vertical. A generalization of the stiffness matrix method which merely requires the transformation of the stiffness matrix equations to an arbitrary orientation provides a straightforward treatment of these problems.

17.2   Global Coordinate System Stiffness Matrices for Plane Frame Problems

The stiffness matrix relation of Eq. (17.4) can be transformed to the global coordinate system by following the same procedure that was used for the plane truss member. At each end of the member there are two displacements and a rotation as well as two forces and a moment. The displacements and

Example 17.1

Use the stiffness matrix method of analysis to obtain the displacements, rotations, reactions, and member end forces and moments for the structure of Example 11.4 (E constant).

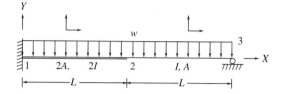

STEP 1   Obtain the individual member stiffness matrices and fixed end forces and moments for the two members. Use the local coordinate systems for the two members, which have the same orientation. Since displacements and rotations in the two systems have the same positive sense, drop the primes from the displacements and rotations in Eqs. (17.4) and (17.4d) and represent the member stiffness matrices as the 3×3 arrays $\mathbf{k}_{ei}$. Define the stiffness parameters in the $\mathbf{k}_{ei}$ in terms of $k = AE/L$ and $B = EI/L$.

Member 1: $k_1 = 2AE/L = 2k;\quad B_1 = E \cdot 2I/L = 2B.$
Member 2: $k_2 = AE/L = k;\quad B_2 = EI/L = B.$
$\left.\rule{0pt}{20pt}\right\}$ Define: $G = k/B.$

Define the following joint displacement matrices:

$$\mathbf{U}_1 = \left\{\begin{matrix} u_1 \\ v_1 \\ \theta_1 \end{matrix}\right\} \quad \mathbf{U}_2 = \left\{\begin{matrix} u_2 \\ v_2 \\ \theta_2 \end{matrix}\right\} \quad \mathbf{U}_3 = \left\{\begin{matrix} u_3 \\ v_3 \\ \theta_3 \end{matrix}\right\}$$

Member 1:

$$\left\{\begin{matrix} \mathbf{f}_{11} \\ \mathbf{f}_{12} \end{matrix}\right\} = \begin{bmatrix} \mathbf{k}_{11} & \mathbf{k}_{12} \\ \mathbf{k}_{12}^T & \mathbf{k}_{13} \end{bmatrix} \left\{\begin{matrix} \mathbf{U}_1 \\ \mathbf{U}_2 \end{matrix}\right\} + \left\{\begin{matrix} \mathbf{f}_{f11} \\ \mathbf{f}_{f12} \end{matrix}\right\} \quad \mathbf{f}_{11} = \left\{\begin{matrix} f_{1x1} \\ f_{1y1} \\ m_{11} \end{matrix}\right\} \quad \mathbf{f}_{12} = \left\{\begin{matrix} f_{1x2} \\ f_{1y2} \\ m_{12} \end{matrix}\right\}$$

$$\mathbf{k}_{11} = 2B \begin{bmatrix} G & 0 & 0 \\ 0 & 12/L^2 & 6/L \\ 0 & 6/L & 4 \end{bmatrix} \quad \mathbf{k}_{12} = 2B \begin{bmatrix} -G & 0 & 0 \\ 0 & -12/L^2 & 6/L \\ 0 & -6/L & 2 \end{bmatrix}$$

$$\mathbf{k}_{13} = 2B \begin{bmatrix} G & 0 & 0 \\ 0 & 12/L^2 & -6/L \\ 0 & -6/L & 4 \end{bmatrix} \quad \mathbf{f}_{f11} = \frac{wL}{12}\left\{\begin{matrix} 0 \\ 6 \\ L \end{matrix}\right\} \quad \mathbf{f}_{f12} = \frac{wL}{12}\left\{\begin{matrix} 0 \\ 6 \\ -L \end{matrix}\right\}$$

Example 17.1 (continued)

Member 2:

$$
\left\{ \begin{matrix} \mathbf{f}_{22} \\ \mathbf{f}_{23} \end{matrix} \right\} = \begin{bmatrix} \mathbf{k}_{21} & \mathbf{k}_{22} \\ \mathbf{k}_{22}^T & \mathbf{k}_{23} \end{bmatrix} \left\{ \begin{matrix} \mathbf{U}_2 \\ \mathbf{U}_3 \end{matrix} \right\} + \left\{ \begin{matrix} \mathbf{f}_{f22} \\ \mathbf{f}_{f23} \end{matrix} \right\} \quad \mathbf{f}_{22} = \left\{ \begin{matrix} f_{2x2} \\ f_{2y2} \\ m_{22} \end{matrix} \right\} \quad \mathbf{f}_{23} = \left\{ \begin{matrix} f_{2x3} \\ f_{2y3} \\ m_{23} \end{matrix} \right\}
$$

$$
\mathbf{k}_{21} = B \begin{bmatrix} G & 0 & 0 \\ 0 & 12/L^2 & 6/L \\ 0 & 6/L & 4 \end{bmatrix} \quad \mathbf{k}_{22} = B \begin{bmatrix} -G & 0 & 0 \\ 0 & -12/L^2 & 6/L \\ 0 & -6/L & 2 \end{bmatrix}
$$

$$
\mathbf{k}_{23} = B \begin{bmatrix} G & 0 & 0 \\ 0 & 12/L^2 & -6/L \\ 0 & -6/L & 4 \end{bmatrix} \quad \mathbf{f}_{f22} = \frac{wL}{12} \left\{ \begin{matrix} 0 \\ 6 \\ L \end{matrix} \right\} \quad \mathbf{f}_{f23} = \frac{wL}{12} \left\{ \begin{matrix} 0 \\ 6 \\ -L \end{matrix} \right\}
$$

STEP 2   Write the equilibrium equations at each of the three joints. These equations can be assembled into the final matrix for the entire structure with those equations being obtained by the summation of the appropriate 3×3 matrices, $\mathbf{k}_{ei}$, established in step 1.

Define the following applied joint load matrices:

$$
\mathbf{P}_1 = \left\{ \begin{matrix} R_{1x} \\ R_{1y} \\ R_{1m} \end{matrix} \right\} \quad \mathbf{P}_2 = \left\{ \begin{matrix} 0 \\ 0 \\ 0 \end{matrix} \right\} \quad \mathbf{P}_3 = \left\{ \begin{matrix} 0 \\ R_{3y} \\ 0 \end{matrix} \right\}
$$

Joint 1

$\Sigma F_x: R_{1x} = f_{1x1}$

$\Sigma F_y: R_{1y} = f_{1y1}$ or $\mathbf{P}_1 = \mathbf{f}_{11}$

$\Sigma M: R_{1m} = m_{11}$

Joint 2

$\Sigma F_x: 0 = f_{1x2} + f_{2x2}$

$\Sigma F_y: 0 = f_{1y2} + f_{2y2}$ or $\mathbf{P}_2 = \mathbf{f}_{12} + \mathbf{f}_{22}$

$\Sigma F_m: 0 = m_{12} + m_{22}$

Joint 3

$\Sigma F_x: 0 = f_{2x3}$

$\Sigma F_y: R_{3y} = f_{2y3}$ or $\mathbf{P}_3 = \mathbf{f}_{23}$

$\Sigma F_m: 0 = m_{23}$

Example 17.1 (continued)

Collecting, the equilibrium equations at the joints become

Joint 1: $\mathbf{P}_1 = \mathbf{f}_{11} = \mathbf{k}_{11}\mathbf{U}_1 + \mathbf{k}_{12}\mathbf{U}_2 + \mathbf{f}_{f11}$

Joint 2: $\mathbf{P}_2 = \mathbf{f}_{12} + \mathbf{f}_{22} = \mathbf{k}_{12}^T\mathbf{U}_1 + \mathbf{k}_{13}\mathbf{U}_2 + \mathbf{f}_{f12}$

$\qquad\qquad + \mathbf{k}_{21}\mathbf{U}_2 + \mathbf{k}_{22}\mathbf{U}_3 + \mathbf{f}_{f22}$

Joint 3: $\mathbf{P}_3 = \mathbf{f}_{23} = \mathbf{k}_{22}^T\mathbf{U}_2 + \mathbf{k}_{23}\mathbf{U}_3 + \mathbf{f}_{f23}$

Collecting these equations together gives the final set of equations for the whole structure. Using the matrix definitions already established, the equations become

$$\mathbf{P} = \mathbf{KU} + \mathbf{F}_f$$

or

$$\left\{\begin{matrix} \mathbf{P}_1 \\ \mathbf{P}_2 \\ \mathbf{P}_3 \end{matrix}\right\} \begin{bmatrix} \mathbf{k}_{11} & \mathbf{k}_{12} & \mathbf{0} \\ \mathbf{k}_{12}^T & \mathbf{k}_{13}+\mathbf{k}_{21} & \mathbf{k}_{22} \\ \mathbf{0} & \mathbf{k}_{22}^T & \mathbf{k}_{23} \end{bmatrix} \begin{bmatrix} \mathbf{U}_1 \\ \mathbf{U}_2 \\ \mathbf{U}_3 \end{bmatrix} + \left\{\begin{matrix} \mathbf{f}_{f11} \\ \mathbf{f}_{f12}+\mathbf{f}_{f22} \\ \mathbf{f}_{f23} \end{matrix}\right\}$$

and the individual terms in the $\mathbf{K}$ and $\mathbf{F}_f$ matrices are

$$\mathbf{K} = B\begin{bmatrix} 2G & 0 & 0 & -2G & 0 & 0 & 0 & 0 & 0 \\ & 24/L^2 & 12/L & 0 & -24/L^2 & 12/L & 0 & 0 & 0 \\ & & 8 & 0 & -12/L & 4 & 0 & 0 & 0 \\ & & & 3G & 0 & 0 & -G & 0 & 0 \\ & & & & 36/L^2 & -6/L & 0 & -12/L^2 & 6/L \\ & & & & & 12 & 0 & -6/L & 2 \\ & & (\text{sym}) & & & & G & 0 & 0 \\ & & & & & & & 12/L^2 & -6/L \\ & & & & & & & & 4 \end{bmatrix} \qquad \mathbf{F}_f = \frac{wL}{12}\left\{\begin{matrix} 0 \\ 6 \\ L \\ 0 \\ 12 \\ 0 \\ 0 \\ 6 \\ -L \end{matrix}\right\}$$

STEP 3   Select the four equations associated with the reaction components and the five equations associated with the unknown joint displacements and rotations. Using the conditions that $u_1 = v_1 = \theta_1 = v_3 = 0$, the first three and the eighth columns of the $\mathbf{K}$ matrix are multiplied by zero, so those columns can be dropped from the equations. Following the notation of Eq. (16.9) and incorporating the fixed end force matrices $\mathbf{F}_{fr}$ and $\mathbf{F}_{fa}$ for the fixed end forces at the reaction joints and at the joints with the unknown displacements gives

$$\mathbf{R} = \mathbf{K}_{ru}\mathbf{U}_u + \mathbf{F}_{fr} \qquad \mathbf{P}_a = \mathbf{K}_{uu}\mathbf{U}_u + \mathbf{F}_{fa}$$

Example 17.1 (continued)

where

$$
\mathbf{R} = \begin{Bmatrix} R_{1x} \\ R_{1y} \\ R_{1m} \\ R_{3y} \end{Bmatrix} \quad \mathbf{F}_{fr} = \frac{wL}{12} \begin{Bmatrix} 0 \\ 6 \\ L \\ 6 \end{Bmatrix} \quad \mathbf{P}_a = \begin{Bmatrix} 0 \\ 0 \\ 0 \\ 0 \\ 0 \end{Bmatrix} \quad \mathbf{F}_{fa} = \frac{wL}{12} \begin{Bmatrix} 0 \\ 12 \\ 0 \\ 0 \\ -L \end{Bmatrix} \quad \mathbf{U}_u = \begin{Bmatrix} u_2 \\ v_2 \\ \theta_2 \\ u_3 \\ \theta_3 \end{Bmatrix}
$$

$$
\mathbf{K}_{ru} = B \begin{bmatrix} -2G & 0 & 0 & 0 & 0 \\ 0 & -24/L^2 & 12/L & 0 & 0 \\ 0 & -12/L & 4 & 0 & 0 \\ 0 & -12/L^2 & -6/L & 0 & -6/L \end{bmatrix}
$$

$$
\mathbf{K}_{uu} = B \begin{bmatrix} 3G & 0 & 0 & -G & 0 \\ & 36/L^2 & -6/L & 0 & 6/L \\ & & 12 & 0 & 2 \\ & \text{(sym)} & & G & 0 \\ & & & & 4 \end{bmatrix}
$$

STEP 4   Solve the equation $\mathbf{P}_a = \mathbf{K}_{uu}\mathbf{U}_u + \mathbf{F}_{fa}$ for the displacements, $\mathbf{U}_u$ and backsubstitute for the reactions from Eq. (16.9) with the addition of the $\mathbf{F}_{fr}$ matrices. The member end forces and moments are obtained from Eq. (17.4) or $\mathbf{f}_e = \mathbf{k}_e \mathbf{u}_e + \mathbf{f}_{fe}$ which is referenced to the local co-ordinate system of each member, with the matrices being defined in step 1. The displacements $\mathbf{U}_u$ become

$$
\mathbf{U}_u = \begin{Bmatrix} u_2 \\ v_2 \\ \theta_2 \\ u_3 \\ \theta_3 \end{Bmatrix} = \frac{wL^2}{96B} \begin{Bmatrix} 0 \\ -17L/3 \\ -5 \\ 0 \\ 13 \end{Bmatrix}
$$

With the displacements known, the reactions and member end forces can be calculated.

$$
\mathbf{R} = \mathbf{K}_{ru}\mathbf{U}_u + \mathbf{F}_{fr} = \frac{wL}{24} \begin{Bmatrix} 0 \\ 19 \\ 12L \\ 5 \end{Bmatrix} + \frac{wL}{24} \begin{Bmatrix} 0 \\ 6 \\ L \\ 6 \end{Bmatrix} = \frac{wL}{24} \begin{Bmatrix} 0 \\ 31 \\ 14L \\ 17 \end{Bmatrix}
$$

Example 17.1 (continued)

Member 1:

$$\begin{Bmatrix} f_{1x1} \\ f_{1y1} \\ m_{11} \\ f_{1x2} \\ f_{1y2} \\ m_{12} \end{Bmatrix} = \frac{wL}{24} \begin{Bmatrix} 0 \\ 19 \\ 12L \\ 0 \\ -19 \\ 7L \end{Bmatrix} + \frac{wL}{12} \begin{Bmatrix} 0 \\ 6 \\ L \\ 0 \\ 6 \\ -L \end{Bmatrix} \frac{wL}{24} \begin{Bmatrix} 0 \\ 31 \\ 14L \\ 0 \\ -7 \\ 5L \end{Bmatrix}$$

Member 2:

$$\begin{Bmatrix} f_{2x2} \\ f_{2y2} \\ m_{12} \\ f_{2x3} \\ f_{2y3} \\ m_{23} \end{Bmatrix} \frac{wL}{24} \begin{Bmatrix} 0 \\ -5 \\ -7L \\ 0 \\ 5 \\ 2L \end{Bmatrix} + \frac{wL}{12} \begin{Bmatrix} 0 \\ 6 \\ L \\ 0 \\ 6 \\ -L \end{Bmatrix} \frac{wL}{24} \begin{Bmatrix} 0 \\ 7 \\ -5L \\ 0 \\ 17 \\ 0 \end{Bmatrix}$$

forces are transformed the same way as for trusses using the matrix of Eq. (16.3). The rotations and the moments are measured in the two coordinate systems with respect to the local z and global Z axes, respectively, which are parallel and in the same direction. As a consequence, for a rotational transformation between the local and global coordinate systems in the x–y and X–Y planes, the moments and rotations in both systems are unchanged. The $\mathbf{Q}$ matrix in Eq. (16.13) can be enlarged to 3×3 for the complete transformation and is defined as

$$\mathbf{Q} = \begin{bmatrix} \alpha & \beta & 0 \\ -\beta & \alpha & 0 \\ 0 & 0 & 1 \end{bmatrix} \tag{17.5}$$

The transformation of displacements and forces between local and global becomes

$$\mathbf{u}_l = \mathbf{Q}\mathbf{u}_g \tag{17.6}$$

$$\mathbf{f}_l = \mathbf{Q}\mathbf{F}_g$$

where

$$\mathbf{u}_l = \begin{Bmatrix} u' \\ v' \\ \theta \end{Bmatrix} \quad \mathbf{u}_g = \begin{Bmatrix} u \\ v \\ \theta \end{Bmatrix} \quad \mathbf{f}_l = \begin{Bmatrix} f_x \\ f_y \\ m \end{Bmatrix} \quad \mathbf{F}_g = \begin{Bmatrix} F_x \\ F_y \\ M \end{Bmatrix} \tag{17.7}$$

The 3×3 $\mathbf{Q}$ transformation matrix defined in Eq. (17.5) has the same property as the $\mathbf{Q}$ matrix defined in Eq. (16.15), because any orthogonal transformation matrix has the property that

$$\mathbf{Q}^{-1} = \mathbf{Q}^T \tag{17.8}$$

The transformation of the displacements, rotation, forces, and moment at each end of the member becomes, similar to Eq. (16.20),

$$\mathbf{u}_e = \mathbf{T}\mathbf{U}_e \tag{17.9a}$$

$$\mathbf{f}_e = \mathbf{T}\mathbf{F}_e$$

in which the 6×6 transformation matrix $\mathbf{T}$ is defined as

$$\mathbf{T} = \begin{bmatrix} \mathbf{Q} & \mathbf{0} \\ \mathbf{0} & \mathbf{Q} \end{bmatrix} \tag{17.9b}$$

where $\mathbf{0}$ is a 3×3 null matrix, $\mathbf{Q}$ is defined in Eq. (17.5), $\mathbf{u}_e$ and $\mathbf{f}_e$ are defined in Eq. (17.4b), and

$$\mathbf{F}_e = \begin{Bmatrix} F_{xi} \\ F_{yi} \\ M_i \\ F_{xj} \\ F_{yj} \\ M_j \end{Bmatrix} \qquad \mathbf{U}_e = \begin{Bmatrix} u_i \\ v_i \\ \theta_i \\ u_j \\ v_j \\ \theta_j \end{Bmatrix} \qquad (17.9c)$$

It is easy to show by the expansion of Eq. (17.9a) that the relations between the local and global system displacements and forces defined in Eqs. (17.6) are obtained for joints i and j. It is also easy to verify by direct expansion using the relation in Eq. (17.8) that

$$\mathbf{T}^{-1} = \mathbf{T}^T \qquad (17.10c)$$

The substitution of Eq. (17.9a) into Eq. (17.4a) and premultiplying both sides of the resulting equation by $\mathbf{T}^T$ yields the result

$$\mathbf{F}_e = \mathbf{K}_e \mathbf{U}_e + \mathbf{F}_{fe} \qquad (17.11a)$$

in which

$$\mathbf{K}_e = \mathbf{T}^T \mathbf{k}_e \mathbf{T} \qquad (17.11b)$$

and

$$\mathbf{F}_{fe} = \mathbf{T}^T \mathbf{f}_{fe} \qquad (17.11c)$$

The expanded forms of $\mathbf{K}_e$ and $\mathbf{F}_{fe}$ are

$$\mathbf{K}_e = \begin{bmatrix}
k\alpha^2 + 12\dfrac{B}{L^2}\beta^2 & \left(k - 12\dfrac{B}{L^2}\right)\alpha\beta & -6\dfrac{B}{L}\beta & -k\alpha^2 - 12\dfrac{B}{L^2}\beta^2 & -\left(k - 12\dfrac{B}{L^2}\right)\alpha\beta & -6\dfrac{B}{L}\beta \\
 & k\beta^2 + 12\dfrac{B}{L^2}\alpha^2 & 6\dfrac{B}{L}\alpha & -\left(k - 12\dfrac{B}{L^2}\right)\alpha\beta & -k\beta^2 - 12\dfrac{B}{L^2}\alpha^2 & 6\dfrac{B}{L}\alpha \\
 & & 4B & 6\dfrac{B}{L}\beta & -6\dfrac{B}{L}\alpha & 2B \\
 & & & k\alpha^2 + 12\dfrac{B}{L^2}\beta^2 & \left(k - 12\dfrac{B}{L^2}\right)\alpha\beta & 6\dfrac{B}{L}\beta \\
 & \text{(sym)} & & & k\beta^2 + 12\dfrac{B}{L^2}\alpha^2 & -6\dfrac{B}{L}\alpha \\
 & & & & & 4B
\end{bmatrix}$$

$$(17.12a)$$

$$\mathbf{F}_{fe} = \begin{Bmatrix} FEF_{xi} \\ FEF_{yi} \\ FEM_{ij} \\ FEF_{xj} \\ FEF_{yj} \\ FEM_{ji} \end{Bmatrix} = \begin{Bmatrix} fef_{xi}\alpha - fef_{yi}\beta \\ fef_{xi}\beta + fef_{yi}\alpha \\ fem_{ij} \\ fef_{xj}\alpha - fef_{yj}\beta \\ fef_{xj}\beta + fef_{yj}\alpha \\ fem_{ij} \end{Bmatrix} \qquad (17.12b)$$

683

where $\mathbf{K}_e$ can be written in the partitioned form

$$\mathbf{K}_e \begin{bmatrix} \mathbf{K}_{e1} & \mathbf{K}_{e2} \\ \mathbf{K}_{e2}^T & \mathbf{K}_{e3} \end{bmatrix} \tag{17.12c}$$

and the $\mathbf{K}_{ei}$ are all 3×3 matrices. Note that

$$\mathbf{K}_{ei} = \mathbf{Q}^T \mathbf{k}_{ei} \mathbf{Q} \tag{17.12d}$$

Equations (17.11) and (17.12) define the stiffness relation for a plane frame member in the global coordinate system.

17.3 Application of the Stiffness Matrix Method to Plane Frames

The solution of a general plane frame problem follows the same sequence of steps that is illustrated in Fig. 16.4 for plane truss structures. The differences between plane truss and plane frame analysis are due to the presence of moments and rotations in plane frame structures. As a consequence, the equilibrium equations in Fig. 16.4c must include a summation of moments equation in addition to the two summation of forces equations at each joint. The final stiffness matrix for the entire frame structure that corresponds to the one for a truss in Fig. 16.4e will have three equilibrium equations and three columns at each joint, the columns being associated with the two displacements and one rotation. These ideas are illustrated in Example 17.2.

Example 17.2 The two-member plane frame shown is loaded with both a uniform and a concentrated load. The joints and members are numbered, with the member numbers appearing in the circles. The global form of the stiffness matrix for an individual member is given in Eq. (17.12a), which can be represented by the four 3×3 matrices $\mathbf{K}_{e1}$, $\mathbf{K}_{e2}$, $\mathbf{K}_{e3}$, and $\mathbf{K}_{e2}^T$ of Eq. (17.12c). The values in the table in step 1 all have units of kips for force and inches for length. The tabular values for $k = AE/L$, $B = EI/L$, α and β are used to establish the two 6×6 stiffness matrices of the members. The applied distributed load of 3 kips/ft for member 1 gives rise to the fixed-end forces and moments of 30 kips and ± 1200 kip-in., respectively, which are computed using the expressions shown in Fig. 11.8a. For member 2 the fixed-end forces of 5 kips and moments of ± 180 kip-in., respectively, are taken from Fig. 11.8b. These fixed-end forces and moments appear in the $\mathbf{F}_{fe}$ matrices for the two members.

In step 2 the equilibrium equations at each of the three joints are established in matrix form. There are three equilibrium equations at each joint for a plane frame, so the final stiffness matrix of the entire structure

Example 17.2

Use the stiffness matrix method of analysis to obtain the displacements, rotations, reactions, and member forces and moments for the plane frame shown

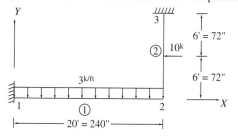

STEP 1 Obtain the individual member stiffness matrices and fixed end forces and moments for the two members of the frame. Establish the origin of the global coordinate system at joint 1 and summarize the parameters needed to establish the matrices defined by Eqs. (17.11) and (17.12) in the table below. The member stiffness matrices can be represented as arrays of 3×3 $\mathbf{K}_{ei}$'s.

| Joint | Coordinates | | Member | | Area, A | Inertia, I | Length, L | α | β | $k = AE/L$ | $B = EI/L$ |
	X	Y									
1	0	0	1	1–2	12	600	240	1	0	1450	72,500
2	240	0	2	2–3	8	300	144	0	1	1611.1	60,416.7
3	240	144									
$E = 29,000$											

Define the following joint displacement matrices:

$$\mathbf{U}_1 = \begin{Bmatrix} u_1 \\ v_1 \\ \theta_1 \end{Bmatrix} \quad \mathbf{U}_2 = \begin{Bmatrix} u_2 \\ v_2 \\ \theta_2 \end{Bmatrix} \quad \mathbf{U}_3 = \begin{Bmatrix} u_3 \\ v_3 \\ \theta_3 \end{Bmatrix}$$

Member 1:

$$\begin{Bmatrix} \mathbf{F}_{11} \\ \mathbf{F}_{12} \end{Bmatrix} = \begin{bmatrix} \mathbf{K}_{11} & \mathbf{K}_{12} \\ \mathbf{K}_{12}^T & \mathbf{K}_{13} \end{bmatrix} \begin{Bmatrix} \mathbf{U}_1 \\ \mathbf{U}_2 \end{Bmatrix} + \begin{Bmatrix} \mathbf{F}_{f11} \\ \mathbf{F}_{f12} \end{Bmatrix} \quad \mathbf{F}_{11} = \begin{Bmatrix} F_{1x1} \\ F_{1y1} \\ M_{11} \end{Bmatrix} \quad \mathbf{F}_{12} = \begin{Bmatrix} F_{1x2} \\ F_{1y2} \\ M_{12} \end{Bmatrix}$$

$$\mathbf{K}_{11} = \begin{bmatrix} 1,450 & 0 & 0 \\ 0 & 15.1 & 1,812.5 \\ 0 & 1,812.5 & 290,000 \end{bmatrix} \quad \mathbf{K}_{12} = \begin{bmatrix} -1,450 & 0 & 0 \\ 0 & -15.1 & 1,812.5 \\ 0 & -1,812.5 & 145,000 \end{bmatrix}$$

$$\mathbf{K}_{13} = \begin{bmatrix} 1,450 & 0 & 0 \\ 0 & 15.1 & -1,812.5 \\ 0 & -1,812.5 & 290,000 \end{bmatrix} \quad \mathbf{F}_{f11} = \begin{Bmatrix} 0 \\ 30 \\ 1,200 \end{Bmatrix} \quad \mathbf{F}_{f12} = \begin{Bmatrix} 0 \\ 30 \\ -1,200 \end{Bmatrix}$$

Example 17.2 (continued)

Member 2:

$$\left\{ \begin{matrix} \mathbf{F}_{22} \\ \mathbf{F}_{23} \end{matrix} \right\} = \begin{bmatrix} \mathbf{K}_{21} & \mathbf{K}_{22} \\ \mathbf{K}_{22}^T & \mathbf{K}_{23} \end{bmatrix} \left\{ \begin{matrix} \mathbf{U}_2 \\ \mathbf{U}_3 \end{matrix} \right\} + \left\{ \begin{matrix} \mathbf{F}_{f22} \\ \mathbf{F}_{f23} \end{matrix} \right\} \qquad \mathbf{F}_{22} = \left\{ \begin{matrix} F_{2x2} \\ F_{2y2} \\ M_{22} \end{matrix} \right\} \qquad \mathbf{F}_{23} = \left\{ \begin{matrix} F_{2x3} \\ F_{2y3} \\ M_{23} \end{matrix} \right\}$$

$$\mathbf{K}_{21} = \begin{bmatrix} 34.96 & 0 & -2,517.4 \\ 0 & 1,611.1 & 0 \\ -2,517.4 & 0 & 241,667 \end{bmatrix} \qquad \mathbf{K}_{22} = \begin{bmatrix} -34.96 & 0 & -2,517.4 \\ 0 & -1,611.1 & 0 \\ 2,517.4 & 0 & 120,833 \end{bmatrix}$$

$$\mathbf{K}_{23} = \begin{bmatrix} 34.96 & 0 & 2,517.4 \\ 0 & 1,611.1 & 0 \\ 2,517.4 & 0 & 241,667 \end{bmatrix} \qquad \mathbf{F}_{f22} = \left\{ \begin{matrix} 5 \\ 0 \\ -180 \end{matrix} \right\} \qquad \mathbf{F}_{f23} = \left\{ \begin{matrix} 5 \\ 0 \\ 180 \end{matrix} \right\}$$

STEP 2 Write the equilibrium equations at each of the three joints. These equations can be assembled into the final matrix for the entire structure with those equations being obtained by the summation of the appropriate 3×3 matrices, $\mathbf{K}_{ei}$ established in step 1.

Define the following applied joint load matrices:

$$\mathbf{P}_1 = \left\{ \begin{matrix} R_{1x} \\ R_{1y} \\ R_{1m} \end{matrix} \right\} \qquad \mathbf{P}_2 = \left\{ \begin{matrix} 0 \\ 0 \\ 0 \end{matrix} \right\} \qquad \mathbf{P}_3 = \left\{ \begin{matrix} R_{3x} \\ R_{3y} \\ R_{3m} \end{matrix} \right\}$$

The equilibrium equations at the three joints become

Joint 1: $\mathbf{P}_1 = \mathbf{F}_{11} = \mathbf{K}_{11}\mathbf{U}_1 + \mathbf{K}_{12}\mathbf{U}_2 + \mathbf{F}_{f11}$

Joint 2: $\mathbf{P}_2 = \mathbf{F}_{12} + \mathbf{F}_{22} = \mathbf{K}_{12}^T\mathbf{U}_1 + \mathbf{K}_{13}\mathbf{U}_2 + \mathbf{F}_{f12} + \mathbf{K}_{21}\mathbf{U}_2$
$\qquad\qquad + \mathbf{K}_{22}\mathbf{U}_3 + \mathbf{F}_{f22}$

Joint 3: $\mathbf{P}_3 = \mathbf{F}_{23} = \mathbf{K}_{22}^T\mathbf{U}_2 + \mathbf{K}_{23}\mathbf{U}_3 + \mathbf{F}_{f23}$

Collecting these equations together gives the final set of equations for the whole structure. Using the matrix definitions already established, the equations become

$$\mathbf{P} = \mathbf{KU} + \mathbf{F}_f$$

or

$$\left\{ \begin{matrix} \mathbf{P}_1 \\ \mathbf{P}_2 \\ \mathbf{P}_3 \end{matrix} \right\} = \begin{bmatrix} \mathbf{K}_{11} & \mathbf{K}_{12} & \mathbf{0} \\ \mathbf{K}_{12}^T & \mathbf{K}_{13} + \mathbf{K}_{21} & \mathbf{K}_{22} \\ \mathbf{0} & \mathbf{K}_{22}^T & \mathbf{K}_{23} \end{bmatrix} \left\{ \begin{matrix} \mathbf{U}_1 \\ \mathbf{U}_2 \\ \mathbf{U}_3 \end{matrix} \right\} + \left\{ \begin{matrix} \mathbf{F}_{f11} \\ \mathbf{F}_{f12} + \mathbf{F}_{f22} \\ \mathbf{F}_{f23} \end{matrix} \right\}$$

Example 17.2 (continued)

and the full numerical values of the terms in the $\mathbf{K}$ and $\mathbf{F}_f$ matrices using kip and inch units are

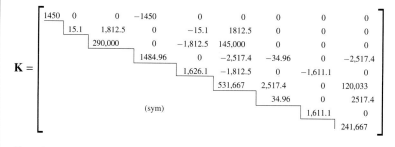

$$
\mathbf{K} =
\begin{bmatrix}
1450 & 0 & 0 & -1450 & 0 & 0 & 0 & 0 & 0 \\
 & 15.1 & 1{,}812.5 & 0 & -15.1 & 1812.5 & 0 & 0 & 0 \\
 & & 290{,}000 & 0 & -1{,}812.5 & 145{,}000 & 0 & 0 & 0 \\
 & & & 1484.96 & 0 & -2{,}517.4 & -34.96 & 0 & -2{,}517.4 \\
 & & & & 1{,}626.1 & -1{,}812.5 & 0 & -1{,}611.1 & 0 \\
 & & & & & 531{,}667 & 2{,}517.4 & 0 & 120{,}033 \\
 & & \text{(sym)} & & & & 34.96 & 0 & 2517.4 \\
 & & & & & & & 1{,}611.1 & 0 \\
 & & & & & & & & 241{,}667
\end{bmatrix}
$$

STEP 3   Separate out the six equations associated with the reaction components and the three equations associated with the unknown joint displacements and rotation. Using the conditions that $u_1 = v_1 = \theta_1 = u_3 = v_3 = \theta_3 = 0$, the first and last three columns of the $\mathbf{K}$ matrix are multiplied by zero, so those columns can be dropped from the equations. Following the notation of Eq. (16.9) and incorporating the fixed end force matrices $\mathbf{F}_{fr}$ and $\mathbf{F}_{fa}$ for the fixed end forces at the reaction joints and at the joint with the unknown displacements gives $\mathbf{U}_u = \mathbf{U}_2$ and

$$\mathbf{R} = \mathbf{K}_{ru}\mathbf{U}_u + \mathbf{F}_{fr} \qquad \text{and} \qquad \mathbf{P}_a = \mathbf{K}_{uu}\mathbf{U}_u + \mathbf{F}_{fa}$$

$$\mathbf{R} = \begin{Bmatrix} \mathbf{P}_1 \\ \mathbf{P}_3 \end{Bmatrix} \quad \mathbf{F}_{fr} = \begin{Bmatrix} \mathbf{F}_{f11} \\ \mathbf{F}_{f23} \end{Bmatrix} \quad \mathbf{P}_a = \mathbf{P}_2 \quad \mathbf{F}_{fa} = \mathbf{F}_{f12} + \mathbf{F}_{f22} = \begin{Bmatrix} 5 \\ 30 \\ -1380 \end{Bmatrix}$$

$$
\mathbf{K}_{ru} =
\begin{bmatrix}
-1450 & 0 & 0 \\
0 & -15.1 & 1{,}812.5 \\
0 & -1{,}812.5 & 145{,}000 \\
-34.69 & 0 & 2{,}517.4 \\
0 & -1{,}611.1 & 0 \\
-2{,}517.4 & 0 & 120{,}833
\end{bmatrix}
$$

$$
\mathbf{K}_{uu} =
\begin{bmatrix}
1{,}484.96 & 0 & -2{,}517.4 \\
0 & 1{,}626.1 & -1{,}812.5 \\
-2{,}517.4 & -1{,}812.5 & 531{,}667
\end{bmatrix}
$$

STEP 4   Solve the equation $\mathbf{P}_a = \mathbf{K}_{uu}\mathbf{U}_u + \mathbf{F}_{fa}$ for the displacements, $\mathbf{U}_u$ and backsubstitute for the reactions from Eq. (16.9) with the addition for the $\mathbf{F}_{fr}$ matrices. The member end forces and moments are obtained

---

Example 17.2 (continued)

from the expression $\mathbf{F}_e = \mathbf{K}_e\mathbf{U}_e + \mathbf{F}_{fe}$ which is referenced to the global coordinate system with the matrices being defined in step 1.

$$\mathbf{U}_u = \left\{ \begin{array}{c} 0.0009492129 \\ -0.015609153 \\ 0.0025468910 \end{array} \right\}$$

With the displacements known the reactions and member end forces can be calculated. The final moment diagram for the structure is shown below with the axial forces in the members shown in parentheses.

$$\mathbf{R} = \mathbf{K}_{ru}\mathbf{U}_u + \mathbf{F}_{fr} = \left\{ \begin{array}{c} -1.376 \\ 4.852 \\ 397.591 \\ 6.378 \\ 25.148 \\ 305.359 \end{array} \right\} + \left\{ \begin{array}{c} 0 \\ 30 \\ 1200 \\ 5 \\ 0 \\ 180 \end{array} \right\} = \left\{ \begin{array}{c} -1.376 \\ 34.852 \\ 1597.591 \\ 11.378 \\ 25.148 \\ 485.359 \end{array} \right\}$$

$$\left\{ \begin{array}{c} F_{x1} \\ F_{y1} \\ M_1 \\ F_{x2} \\ F_{y2} \\ M_2 \end{array} \right\} = \left\{ \begin{array}{c} -1.376 \\ 4.852 \\ 397.591 \\ 1.376 \\ -4.852 \\ 766.890 \end{array} \right\} + \left\{ \begin{array}{c} 0 \\ 30 \\ 1200 \\ 0 \\ 30 \\ -1200 \end{array} \right\} = \left\{ \begin{array}{c} -1.376 \\ 34.852 \\ 1597.591 \\ 1.376 \\ 25.148 \\ -433.110 \end{array} \right\}$$

$$\left\{ \begin{array}{c} F_{x2} \\ F_{y2} \\ M_2 \\ F_{x3} \\ F_{y3} \\ M_3 \end{array} \right\} = \left\{ \begin{array}{c} -6.378 \\ -25.148 \\ 613.110 \\ 6.378 \\ 25.148 \\ 305.359 \end{array} \right\} + \left\{ \begin{array}{c} 5 \\ 0 \\ -180 \\ 5 \\ 0 \\ 180 \end{array} \right\} = \left\{ \begin{array}{c} -1.378 \\ -25.148 \\ 433.110 \\ 11.378 \\ 25.148 \\ 485.389 \end{array} \right\}$$

25.148^k

485.359$^{k\text{-in.}}$ 11.378^k

(25.148^k)

333.88$^{k\text{-in.}}$

34.852^k

(1.376^k)

1.376^k

784.650$^{k\text{-in.}}$ 433.110$^{k\text{-in.}}$

1597.591$^{k\text{-in.}}$

is 9×9. The six reaction restraints yield zero values of the two displacements and rotation at joints 1 and 3. Because of these zero values, the terms in the first and last three columns of the stiffness matrix, $\mathbf{K}$, are multiplied by zero, so that they do not contribute to the nine equilibrium equations. These six columns are dropped from the matrix, reducing the number of columns to three.

The equations at the first and last joints, which are the first and last three rows of the $\mathbf{K}$ and the $\mathbf{F}_f$ matrices, are the equilibrium equations that will be used to calculate the reactions of the frame. They are extracted from these matrices and identified with the notation $\mathbf{K}_{ru}$ and $\mathbf{F}_{fr}$. The remaining three equations are for joint 2, where the two displacements and rotation are unknown and appear in the $\mathbf{U}_u$ matrix.

The solution of the three equations in step 4 provides a complete set of joint displacements and rotations for the frame. The reactions and member end forces and moments can be calculated by direct substitution into the stiffness equations established for reactions in step 3, and member end forces and moments from the equations established in step 1. The member end forces and moments, $\mathbf{F}_e$, are expressed in the global coordinate system but can be expressed in the local coordinate system if they are transformed by the relation in Eq. (17.9a), $\mathbf{f}_e = \mathbf{TF}_e$. The final moment diagram is shown for the frame. The moment diagram for each member is plotted following the techniques presented in Chapters 11 and 12. The reaction forces for the frame are also shown on the moment diagram, and member axial forces are shown in parentheses. This frame is indeterminate to the third degree, but this condition did not enter into the solution process.

The solution of plane frames problems is lengthy and not well suited to hand computation methods. The procedure illustrated in Chapter 16 for trusses and in this chapter for frames is a series of operations that can easily be programmed on a computer. The extended matrix manipulations and the solution of simultaneous equations are operations that are quickly and efficiently executed by computer programs.

17.4   Summary and Limitations

The stiffness matrices for the analysis of plane frame structures are developed in the local member coordinate system. The displacements and rotations of the ends of the member undergoing bending action are related to the member end moments and shears by the slope-deflection equations presented in Chapter 11. The axial force behavior of the member includes the addition of a distributed applied axial load per unit length and axial tension due to member end displacement is identical to the action of truss members in the local coordinate system given by Eq. (16.18). The use of equations in this form to solve problems is similar to solving problems by the slope-deflection method.

The local member relations are transformed to the global coordinate system relations with the definition of suitable transformation matrices. In this form the individual member stiffness relations can be assembled into the equilibrium equation for a general plane frame with members having arbitrary orientations. The effects of axial deformation are included in the formulation, which can be programmed for solution on a computer.

As indicated in Chapter 16, the capability of computer programs to produce analyses of very large and very complicated structures underscores the need to understand the limitations on both individual member and overall structural behavior. It is imperative that the assumptions and limitations listed below are understood by the structural engineer when computer analyses are evaluated.

- Because of the use of simple bending theory principles, the relations between member end forces and moments and member end displacements and rotations are subject to all of the same limitations and assumptions that are listed at the end of Chapter 5 and in Section 6.2 as well as those associated with the slope-deflection method of analysis presented in Chapter 11.

- All material deformations must be continuous, without any cracks or gaps.

- All structural deformations must be continuous, without any violation of the conditions of constraint at points of support or at points (joints) of attachment.

- All axial deformations are assumed to be small, and axial forces have no significant effect on the bending behavior of the members of the structure.

- The distribution and orientation of all loads acting on a structure are not changed by the deformations of the structure.

- The orientation of all reaction components at supports is unchanged by the action of applied loads or any movement of the support.

- Any movements that occur at supports are assumed to be small and do not affect the initial geometry of the structure.

- The principle of superposition of deformations with all of its limitations is required for the analysis of symmetric structures under general loading if the analysis is carried out on the symmetric and antisymmetric forms of the structure.

- All members are assumed to be prismatic so that simple deformation behavior from mechanics of materials is valid.

- The structure as well as the individual members of the structure are stable under the application of the loads.
- The centroidal axes of all members connected at each joint intersect in a common point.

The analysis of plane frames using matrix methods provides the structural engineer with a valuable tool, but one that should always be used responsibly and with caution. Some questions that might be asked are:

- How can nonprismatic members be analyzed by the methods presented in this chapter without modifying the stiffness matrices?
- What kinds of concerns should be addressed as the number of joints and members in a structure becomes very large?
- How can a computer program that does plane frame analysis be validated?
- If an input error to a computer program causes a failure of a structure or some part of it, who is responsible?
- If a computer program produces an incorrect analysis, even though all input to the program is correct, and a failure results, who is responsible?

PROBLEMS

17.1 Using the stiffness matrix method, obtain the joint displacements and rotations for the plane frame shown.

Member 1–2: $A = 10$ in^2, $I = 450$ in^4
Member 2–3: $A = 8$ in^2, $I = 300$ in^4
Both members: $E = 29,000$ ksi

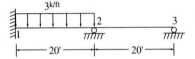

17.2 For the plane frame structure, obtain the final values of the displacements and rotation at joint 2 and all of the reaction forces and moments using the stiffness matrix method of analysis. For both members, A, E, I, and L are identical.

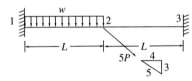

17.3 For the plane frame structure, obtain the final values of the displacements and rotation at joint 2 and all of the reaction forces and moments using the stiffness matrix method of analysis. Both members: $A = 220$ cm^2, $I = 2400 \times 10^6$ mm^4, $E = 200$ GPa.

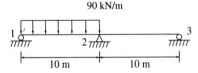

17.4 Repeat the solution of Problem 17.1 with the addition of a vertical support settlement of 0.02 ft at joint 2.

17.5 Repeat the solution of Problem 17.3 with the addition of a vertical support settlement of 6 mm at joint 1.

17.6 Repeat Problem 17.1 for the loading shown.

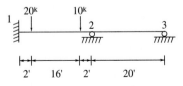

17.7 Repeat Problem 17.3 for the loading shown.

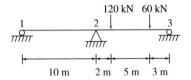

17.8 The two-member frame shown is to be analyzed by the stiffness matrix method.

(a) Obtain the stiffness matrix for the two members in the global coordinate system.

(b) Assemble the complete set of equations for solution in matrix form.

(c) Identify the equations that define reaction components.

(d) Identify the equations of solution for the unknown joint displacements and rotations and solve.

(e) Obtain the reactions and the member end forces and moments in the global coordinated system.

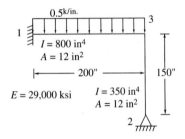

17.9 The two-member frame shown is to be analyzed by the stiffness matrix method.

(a) Obtain the stiffness matrix for the two members in the global coordinate system.

(b) Assemble the complete set of equations for solution in matrix form.

(c) Identify the equations that define reaction components.

(d) Identify the equations of solution for the unknown joint displacements and rotations and solve.

(e) Obtain the reactions and the member end forces and moments in the global coordinate system.

Both members: $E = 30$ GPa
Member 1–3: $A = 1300$ cm^2; $I = 3000 \times 10^6$ mm^4
Member 2–3: $A = 1500$ cm^2; $I = 3300 \times 10^6$ mm^4

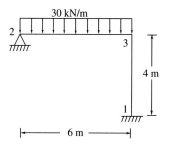

17.10 Use the stiffness matrix method to obtain the final displacements and rotation of joint 3 of the frame shown. Also compute the member end forces and moments. $AL^2/I = 1500$ for both members.

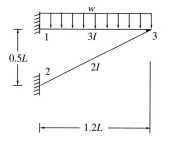

17.11 Use the stiffness matrix method to analyze the structure shown. Note that members 2–3 and 4–2 are truss members and that member 1–2 is a plane frame member. As a result, the only nonzero displacements in this structure are the horizontal, vertical, and rotational displacements at 2.

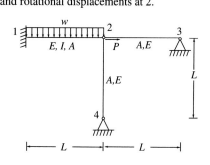

COMPUTER PROBLEMS

The following problems should be solved using a computer program that does plane frame analysis by the stiffness matrix method.

C17.1   Analyze the frame of Example 10.5.

C17.2   Analyze the frame of Fig. 10.17.

C17.3   Analyze the frame of Example 11.3. Take AL^2/I of all members to be 1200.

C17.4   Analyze the frame of Example 11.6. Take AL^2/I of all members to be 1200.

C17.5   Analyze the frame of Fig. 14.7. Take the ratio of the moment of inertia of beams to moment of inertia of columns as 4, AL^2/I of all members as 1500, and the ratio of L/H as 2.

C17.6   Analyze the frame of Fig. 14.8. Take the ratio of the moment of inertia of beams to moment of inertia of columns as 4, AL^2/I of all members as 1500, and the ratio of L/H as 2.

C17.7   Analyze the tapered beam of Problem 15.5 by introducing one or more joints at equal spacings between 1 and 2.

Proceed as indicated in parts (a), (b), and (c). Compare the computer solution with the exact solution. Take AL^2/I of each segment to be 600 in. (a), 400 in. (b) and 200 in. (c).

(a)   Divide the beam into two equal-length segments and use for the moment of inertia of the segments the inertia at the ¼ and ¾ points, respectively, of the original member.

(b)   Divide the beam into four equal-length segments and use for the moment of inertia of the segments the inertia at the ⅛, ⅜, ⅝, and ⅞ points, respectively, of the original member.

(c)   Divide the beam into eight equal-length segments and use for the moment of inertia of the segments the inertia at the ¹⁄₁₆, ³⁄₁₆, ⁵⁄₁₆, . . ., ¹⁵⁄₁₆ points, respectively, of the original member.

C17.8   Repeat Problem C17.7 for the beam of Problem 15.9.

C17.9   Repeat Problem C17.7 for the beam of Problem 15.11.

Appendix: Determinants, Matrices, and Linear Equations

The material presented in this appendix is for the convenience of the student user. Terminologies and techniques used in several of the chapters, particularly Chapters 16 and 17, are reviewed here. Interested readers who want more in-depth treatments or derivations of techniques are referred to books on numerical analysis, such as References 13, 16, 19, 20, 34, and 37.

A.1   Determinants

A square array of numbers can have a unique magnitude assigned to it which is called its *determinant*. The value of determinants can be used in the calculation of the solution of a set of linear simultaneous equations or show that a solution does not exist. The determinant of an $n \times n$ square array of numbers is defined symbolically as

$$\det A = \left| A \right| = \left| a_{ij} \right| = D_n \tag{A.1}$$

in which the subscript i represents the row and j the column of the individual terms, a_{ij}, of the square array. The order of the determinant, n, is defined as the number of rows (and columns) of the array. A third-order determinant is, by Eq. (A.1),

$$\det A = \left| A \right| = \begin{vmatrix} a_{11} & a_{12} & a_{13} \\ a_{21} & a_{22} & a_{23} \\ a_{31} & a_{32} & a_{33} \end{vmatrix} = D_3 \tag{A.2}$$

The evaluation of the determinant is defined in a very specific way mathematically, but it can be obtained simply from a set of defined operations. When the order of A is 1, the determinant, D_1, is obviously just a_{11}, and the sign of D_1 is also the sign of a_{11}. For D_2, the following calculation can be made:

$$\det A = \begin{vmatrix} a_{11} & a_{12} \\ a_{21} & a_{22} \end{vmatrix} = D_2 = a_{11}a_{22} - a_{21}a_{12} \tag{A.3}$$

For higher-order determinants, it is convenient to make some futher definitions. The minor M_{ij} of any term a_{ij} in the square array A is defined as the determinant of the square array that is obtained when the ith row and jth column of the original array has been deleted. Thus, for the third-order array A,

$$\begin{matrix} a_{11} & a_{12} & a_{13} \\ a_{21} & a_{22} & a_{23} \\ a_{31} & a_{32} & a_{33} \end{matrix}$$

the minor of a_{32} is

$$M_{32} = \begin{vmatrix} a_{11} & a_{13} \\ a_{21} & a_{23} \end{vmatrix} = a_{11}a_{23} - a_{21}a_{13}$$

For an order 2 array such as the one shown above in Eq. (A.3), the minors are:

Term	Minor	Cofactor
a_{11}	a_{22}	a_{22}
a_{12}	a_{21}	$-a_{21}$
a_{21}	a_{12}	$-a_{12}$
a_{22}	a_{11}	a_{11}

The cofactor A_{ij} of any term a_{ij} is defined by the expression

$$A_{ij} = (-1)^{(i+j)}M_{ij} \tag{A.4}$$

The cofactor of a_{32} in the earlier example is

$$a_{32} = (-1)^{(3+2)}M_{32} = (-1)(a_{11}a_{23} - a_{21}a_{13})$$

and the cofactors of the order 2 array of Eq. (A.3) are presented in the table above.

The evaluation of the determinant of the square array A can now be defined as the sum of the product of all terms in any row or column of the array with their corresponding cofactors. Thus, by definition, for a square A of order n, the expansion by column j gives

$$\det A = |A| = |a_{ij}| = \sum_{i=1}^{n} a_{ij}A_{ij} = D_n \tag{A.5a}$$

where $j = 1$ or $j = 2 \cdots$ or $j = n$. Also, the expansion by row i gives

$$\det A = \left| A \right| = \left| a_{ij} \right| = \sum_{j=1}^{n} a_{ij} A_{ij} = D_n \qquad (A.5b)$$

where $i = 1$ or $i = 2 \cdots$ or $i = n$.

With these definitions it is easy to verify that Eq. (A.3) gives the correct value of D_2. For D_3 using the first row ($i = 1$) in Eq. (A.5b) gives

$$D_3 = \begin{vmatrix} a_{11} & a_{12} & a_{13} \\ a_{21} & a_{22} & a_{23} \\ a_{31} & a_{32} & a_{33} \end{vmatrix} = a_{11}A_{11} + a_{12}A_{12} + a_{13}A_{13}$$

$$= a_{11}\begin{vmatrix} a_{22} & a_{23} \\ a_{32} & a_{33} \end{vmatrix} + a_{12}(-1)\begin{vmatrix} a_{21} & a_{23} \\ a_{31} & a_{33} \end{vmatrix} + a_{13}\begin{vmatrix} a_{21} & a_{22} \\ a_{31} & a_{32} \end{vmatrix}$$

$$= a_{11}(a_{22}a_{33} - a_{32}a_{23}) - a_{12}(a_{21}a_{33} - a_{31}a_{23}) + a_{13}(a_{21}a_{32} - a_{31}a_{22})$$

It is tedious but straightforward to show that the same result is obtained if the second or third row or the first, second, or third column of A had been used for the expansion. The expansion defined in Eqs. (A.5) is known as the *Laplace expansion*.

The definitions established in Eqs. (A.5) lead to several important observations (OBS):

OBS 1. If all the terms of any row or column of A are multiplied by a constant k, the value of the determinant of A is also multiplied by k.

OBS 2. If all the terms of any row or column are zero, the determinant is zero.

OBS 3. The interchange of any two rows or any two columns of the array only changes the sign of the determinant of the array.

OBS 4. If any two rows or any two columns of an array are identical, the determinant is zero.

OBS 5. If any row (or column) is a constant multiple of any other row (or column) in the array, the determinant is zero.

OBS 6. If a constant multiple of any row (or column) is added to any other row (or column), the determinant of the array is unchanged.

OBS 7. If any row (or column) is a linear combination of two or more rows (or columns) of the array, the determinant of the array is zero.

The sixth of these observations is very important because it is the basis of an alternative means of evaluation of the determinant of an array. This technique is known as *pivotal condensation.* For large arrays it is the most computationally effective means of obtaining the determinant of an array.

Consider the nth-order array A for which the determinant D_n is to be computed.

$$D_n = \begin{vmatrix} a_{11} & a_{12} & a_{13} & \cdots & a_{1n} \\ a_{21} & a_{22} & a_{23} & \cdots & a_{2n} \\ a_{31} & a_{32} & a_{33} & \cdots & a_{3n} \\ \vdots & & & & \\ a_{n1} & a_{n2} & a_{n3} & \cdots & a_{nn} \end{vmatrix}$$

Factor a_{11} out of the first row to obtain the new expression for D_n (by virtue of OBS 1):

$$D_n = a_{11} \begin{vmatrix} 1 & a_{12}/a_{11} & a_{13}/a_{11} & \cdots & a_{1n}/a_{11} \\ a_{21} & a_{22} & a_{23} & \cdots & a_{2n} \\ a_{31} & a_{32} & a_{33} & \cdots & a_{3n} \\ \vdots & & & & \\ a_{n1} & a_{n2} & a_{n3} & \cdots & a_{nn} \end{vmatrix}$$

Successively multiply the first row by a_{21} and subtract it from the second row, by a_{31} and subtract it from the third row, and so on, until the first row has been multiplied by a_{n1} and subtracted from the last row. By virtue of OBS 6, the value of D_n is not changed by any one and hence all of these operations. The new array obtained from this set of operations and its determinant is

$$D_n = a_{11} \begin{vmatrix} 1 & a_{12}/a_{11} & a_{13}/a_{11} & \cdots & a_{1n}/a_{11} \\ 0 & a_{22} - a_{21}a_{12}/a_{11} & a_{23} - a_{21}a_{13}/a_{11} & \cdots & a_{2n} - a_{21}a_{1n}/a_{11} \\ 0 & a_{32} - a_{31}a_{12}/a_{11} & a_{33} - a_{31}a_{13}/a_{11} & \cdots & a_{3n} - a_{31}a_{1n}/a_{11} \\ \vdots & & & & \\ 0 & a_{n2} - a_{n1}a_{12}/a_{11} & a_{n3} - a_{n1}a_{13}/a_{11} & \cdots & a_{nn} - a_{n1}a_{1n}/a_{11} \end{vmatrix}$$

If D_n is evaluated by using Eq. (A.5a) and expanding the determinant with the first column ($j = 1$), the result is

$$D_n = a_{11} \begin{vmatrix} b_{22} & b_{23} & \cdots & b_{2n} \\ b_{32} & b_{33} & \cdots & b_{3n} \\ \vdots & & & \\ b_{n2} & b_{n3} & \cdots & b_{nn} \end{vmatrix}$$

where

$$b_{ij} = a_{ij} - \frac{a_{i1}a_{1j}}{a_{11}} \quad \begin{array}{l} i = 2, 3, \ldots, n \\ j = 2, 3, \ldots, n \end{array}$$

The B array is of order $n - 1$ since it is the cofactor of the a_{11} of the modified A array. Note that the other cofactors of the modified A array, A_{i1} ($i = 2 \cdots n$), are nonzero but the column terms a_{i1} ($i = 2 \cdots n$) are zero, hence by Eq. (A.5a) D_n is correctly defined in terms of a_{11} and the determinant of the B array. If this process is repeated by factoring b_{22} out of the first row of B, then successively multiplying the modified first row by the first term of each succeeding r ($r = 2, 3, \ldots, n - 1$) row in B and subtracting it from that rth row, the result becomes

$$D_n = a_{11}b_{22} \begin{vmatrix} 1 & b_{23}/b_{22} & \cdots & b_{2n}/b_{22} \\ 0 & c_{33} & \cdots & c_{3n} \\ \vdots & & & \\ 0 & c_{n3} & \cdots & c_{nn} \end{vmatrix}$$

or

$$D_n = a_{11}b_{22} \begin{vmatrix} c_{33} & \cdots & c_{3n} \\ \vdots & & \\ c_{n3} & \cdots & c_{nn} \end{vmatrix}$$

where

$$c_{ij} = b_{ij} - \frac{b_{i2}b_{2j}}{b_{22}} \quad \begin{array}{l} i = 3, 4, \ldots, n \\ j = 3, 4, \ldots, n \end{array}$$

Now the array C is of order $n - 2$. Continued computation in this way will eventually reduce the original A array of order n to the product of $a_{11}b_{22}c_{33} \cdots$ and an array that is 1×1 (the last remaining modified term). The process just described is known as pivotal condensation.

The use of these two techniques for computing the value of a determinant is shown in Example A.1 for a 4×4 array.

Example A.1.   Compute the determinant of the array shown by (a) Laplace expansion and (b) pivotal condensation.

$$|A| = \begin{vmatrix} 2 & 2 & 4 & -2 \\ 1 & 3 & 2 & 1 \\ 3 & 1 & 3 & 1 \\ 1 & 3 & 4 & 2 \end{vmatrix}$$

(a) *Laplace expansion* (use the first row in each determinant expansion)

$$D_4 = |A| = 2\begin{vmatrix} 3 & 2 & 1 \\ 1 & 3 & 1 \\ 3 & 4 & 2 \end{vmatrix} - 2\begin{vmatrix} 1 & 2 & 1 \\ 3 & 3 & 1 \\ 1 & 4 & 2 \end{vmatrix}$$
$$+ 4\begin{vmatrix} 1 & 3 & 1 \\ 3 & 1 & 1 \\ 1 & 3 & 2 \end{vmatrix} - (-2)\begin{vmatrix} 1 & 3 & 2 \\ 3 & 1 & 3 \\ 1 & 3 & 4 \end{vmatrix}$$

699

$$= 2\left[3\begin{vmatrix} 3 & 1 \\ 4 & 2 \end{vmatrix} - 2\begin{vmatrix} 1 & 1 \\ 3 & 2 \end{vmatrix} + 1 \cdot \begin{vmatrix} 1 & 3 \\ 3 & 4 \end{vmatrix}\right]$$

$$- 2\left[1 \cdot \begin{vmatrix} 3 & 1 \\ 4 & 2 \end{vmatrix} - 2\begin{vmatrix} 3 & 1 \\ 1 & 2 \end{vmatrix} + 1 \cdot \begin{vmatrix} 3 & 1 \\ 1 & 4 \end{vmatrix}\right]$$

$$+ 4\left[1 \cdot \begin{vmatrix} 1 & 1 \\ 3 & 2 \end{vmatrix} - 3\begin{vmatrix} 3 & 1 \\ 1 & 2 \end{vmatrix} + 1 \cdot \begin{vmatrix} 3 & 1 \\ 1 & 3 \end{vmatrix}\right]$$

$$+ 2\left[1 \cdot \begin{vmatrix} 1 & 3 \\ 3 & 4 \end{vmatrix} - 3\begin{vmatrix} 3 & 3 \\ 1 & 4 \end{vmatrix} + 2\begin{vmatrix} 3 & 1 \\ 1 & 3 \end{vmatrix}\right]$$

$$= 2[3(6 - 4) - 2(2 - 3) + 1(4 - 9)]$$
$$- 2[1(6 - 4) - 2(6 - 1) + 1(12 - 3)]$$
$$+ 4[1(2 - 3) - 3(6 - 1) + 1(9 - 1)]$$
$$+ 2[1(4 - 9) - 3(12 - 3) + 2(9 - 1)]$$
$$D_4 = 2[3 \cdot 2 - 2(-1) + 1(-5)] - 2[1 \cdot 2 - 2 \cdot 5 + 1 \cdot 9]$$
$$+ 4[1(-1) - 3 \cdot 5 + 1.8] + 2[1(-5) - 3 \cdot 9 + 2 \cdot 8]$$
$$= 2 \cdot 3 - 2 \cdot 1 + 4(-8) + 2(-16) = -60$$

(b) *Pivotal condensation* (operations indicated row-wise using letters to designate all terms in a row).

$$D_4 = \begin{vmatrix} 2 & 2 & 4 & -2 \\ 1 & 3 & 2 & 1 \\ 3 & 1 & 3 & 1 \\ 1 & 3 & 4 & 2 \end{vmatrix} \begin{matrix} \text{row } a \\ \text{row } b \\ \text{row } c \\ \text{row } d \end{matrix}$$

Step 1:

$$D_4 = 2\begin{vmatrix} 1 & 1 & 2 & -1 \\ 0 & 2 & 0 & 2 \\ 0 & -2 & -3 & 4 \\ 0 & 2 & 2 & 3 \end{vmatrix} \begin{matrix} a_1 = a/2 \\ b_1 = b - 1 \cdot a_1 \\ c_1 = c - 3 \cdot a_1 \\ d_1 = d - 1 \cdot a_1 \end{matrix}$$

Step 2:

$$D_4 = 2 \cdot 2\begin{vmatrix} 1 & 1 & 2 & 1 \\ 0 & 1 & 0 & 1 \\ 0 & 0 & -3 & 6 \\ 0 & 0 & 2 & 1 \end{vmatrix} \begin{matrix} a_2 = a_1 \\ b_2 = b_1/2 \\ c_2 = c_1 - (-2)b_2 \\ d_2 = d_1 - 2b_2 \end{matrix}$$

Step 3:

$$D_4 = 2 \cdot 2(-3) \begin{vmatrix} 1 & 1 & 2 & -1 \\ 0 & 1 & 0 & 1 \\ 0 & 0 & 1 & -2 \\ 0 & 0 & 0 & 5 \end{vmatrix} \begin{array}{l} a_3 = a_1 \\ b_3 = b_2 \\ c_3 = c_2/(-3) \\ d_3 = d_2 - 2c_3 \end{array}$$

Step 4:

$$D_4 = 2 \cdot 2(-3) \cdot 5 \begin{vmatrix} 1 & 1 & 2 & -1 \\ 0 & 1 & 0 & 1 \\ 0 & 0 & 1 & -2 \\ 0 & 0 & 0 & 1 \end{vmatrix} \begin{array}{l} a_4 = a_1 \\ b_4 = b_2 \\ c_4 = c_3 \\ d_4 = d_3/5 \end{array}$$

Step 5:

$$D_4 = 2 \cdot 2(-3) \cdot 5 = -60$$

Note that the determinant of the remaining 4×4 array in step 4 is 1 by using the simple Laplace expansion, Eq. (A.5a), and the first column ($j = 1$).

The Laplace expansion and the pivotal condensation techniques for evaluating the determinant, D_n, have been presented. A count of multiplications and divisions (not addition and subtractions) can be made for each method assuming that all terms in the array are nonzero. In making this count the division of the a_{11} term by a_{11}, the b_{22} term by b_{22}, and so on, in the pivotal condensation technique are not counted because the division is not necessary in the process. The following results are obtained:

Technique/Order of Array, n	2	3	4	5	n
Laplace expansion	2	9	40	205	$n!\left(1 + \sum_{r=2}^{n-1} \dfrac{1}{r!}\right)$
Pivotal condensation	3	10	23	44	$\dfrac{n^3 + 2n - 3}{3}$

Clearly, the pivotal condensation technique is superior for all D_n when n is greater than 3.

A.2 Solution of Linear Equations Using Determinants _____

There are several techniques for solving a system of linear equations. Their solution through the use of determinants is not the most computationally efficient, but there are some aspects to the solution of linear equations that are brought out in this process that are worthy of discussion. For systems of up to three equations and three unknowns, the solution by determinants is computationally practical. Computationally efficient solutions of systems of four or more linear equations are presented in Section A.5.

The system of linear equations

$$\sum_{j=1}^{n} a_{ij}x_j = c_i \quad i = 1, 2, \ldots, n \tag{A.6}$$

or

$$
\begin{array}{ccccccccc}
a_{11}x_1 & + & a_{12}x_2 & + & a_{13}x_3 & + & \cdots & + & a_{1n}x_n & = & c_1 \\
a_{21}x_1 & + & a_{22}x_2 & + & a_{23}x_3 & + & \cdots & + & a_{2n}x_n & = & c_2 \\
a_{31}x_1 & + & a_{32}x_2 & + & a_{33}x_3 & + & \cdots & + & a_{3n}x_n & = & c_3 \\
& & \vdots & & & & & & & & \\
a_{n1}x_1 & + & a_{n2}x_2 & + & a_{n3}x_3 & + & \cdots & + & a_{nn}x_n & = & c_n
\end{array}
$$

can be solved for the x_j through use of the definition

$$x_j = \frac{D_{nj}}{D_n} \quad j = 1, 2, \ldots, n \tag{A.7}$$

where D_n is the determinant of the array A and D_{nj} is the determinant of the array that is obtained when the jth column of A is replaced with a column made up of the c_i. Thus D_{n2} is given by

$$D_{n2} = \begin{vmatrix} a_{11} & c_1 & a_{13} & \cdots & a_{1n} \\ a_{21} & c_2 & a_{23} & \cdots & a_{2n} \\ \vdots & & & & \\ a_{n1} & c_n & a_{n3} & \cdots & a_{nn} \end{vmatrix} \tag{A.8}$$

The solution of simultaneous equations in this manner is known as *Cramer's rule*.

The solution of n equations requires the evaluation of $n + 1$ determinants and n divisions. Using the Laplace expansion this requires a total of

$$(n + 1)! \left(1 + \sum_{r=2}^{n-1} \frac{1}{r!}\right) + n$$

operations, while a total of

$$\frac{(n + 1)(n^3 + 2n - 3) + 3n}{3}$$

operations is needed with pivotal condensation. Note that these calculations assume that each new determinant D_{nj} must be evaluated completely. In fact, some computations made in obtaining D_n and the other D_{nj} do not need to be repeated, and with careful organization the number of computations is less than those given above. However, if no account is made for this, the number of operations to solve two, three, or four equations in a like number of unknowns is 8, 39, and 20 for the Laplace expansion and 11, 43, and 119 for pivotal condensation.

The solution of a set of linear equations is defined by Eq. (A.7) and the solution is unique since each of the $(n + 1)$ determinants represents a different combination of the $(n + 1)$ columns made up of the a_{ij} and c_i for the set of equations. The solution process fails if the determinant D_n is zero. In this case, no solution exists since the x_j in Eq. (A.7) cannot be obtained due to the division by D_n. If all the D_{nj} are zero, all the corresponding x_j are indeterminate because of the $0/0$ division, but a solution to the equations in terms of an unknown constant can be obtained by a process similar to the one illustrated in the next section. However, no unique solution exists under any circumstances if D_n is zero. In this case, the set of equations is said to be singular.

A.3   Homogeneous Linear Equations

There are occasions when it is necessary to solve a set of homogeneous equations, that is, a set of equations like Eq. (A.6), but where all c_i are zero. As can be seen in Eqs. (A.7) and (A.8), the value of all D_{nj} and hence all x_j will be zero since one column of all D_{nj}, the column on the right-hand side, c_i, is all zeros. Indeed, this is the correct and only solution for the homogeneous equations if D_n is nonzero. If D_n is zero, Eq. (A.7) gives indeterminate values for the x_j because of the ratio $0/0$ but does not preclude the possibility of a nonzero solution for the x_j. In what follows, a technique is illustrated which produces a solution to the homogeneous equations under the condition that D_n is zero.

Take the equations defined in Eq. (A.6) but in which all the c_i are zero and rewrite them in the form

$$\sum_{j=1}^{n-1} a_{ij}x_j = -a_{in}x_n \qquad i = 1, 2, \ldots, n \qquad (A.9)$$

One of the n equations in Eq. (A.6) must be some linear combination of two or more of the remaining $(n - 1)$ equations by virtue of OBS 7 and the fact that D_n is zero. The equations may be reordered such that the dependent equation is nth or last. Thus, any solution of the preceding $(n - 1)$ equations must also be a solution of the nth equation and it may be put aside.

Assume that D_{n-1} is not zero. If x_n is taken as some undetermined constant, C, the $(n - 1)$ equations left can be solved and a solution for the n homogeneous equations becomes

$$X_j = C\frac{D_{n-1,j}}{D_{n-1}} \qquad j = 1, 2, \ldots, n - 1$$

$$X_n = C \qquad\qquad\qquad\qquad (A.10)$$

where $D_{n-1,j}$ is the determinant of the order $(n-1)$ array composed of the $(n-1)$ columns of D_{n-1} but with the jth column replaced by $-a_{in}$. The solution of the n equations is given by Eq. (A.10) and is known in terms of the constant C.

If D_{n-1} is also zero, two of the n equations must be some linear combination of two or more of the remaining $n-2$ equations by OBS 7. In this case the set of equations can be reordered such that the last two equations are the ones that are a linear combination of some or all of the first $n-2$. In this case a solution can be found by proceeding as above, using the first $n-2$ equations but with the last two columns of the original A array being transferred to the right-hand side. The solution in this case (assuming that D_{n-2} is nonzero) can be calculated in terms of two unknown constants, one corresponding to each of the unknowns x_n and x_{n-1}. In this case the solution will be

$$x_j = \frac{C_1 D^n_{n-2,j} + C_2 D^{n-1}_{n-2,j}}{D_{n-2}}$$

$$x_{n-1} = C_2 \qquad\qquad\qquad (A.11)$$

$$x_n = C_1$$

where the superscript in the $D_{n-2,j}$ indicates that the jth column of $D_{n-2,j}$ is replaced with $-a_{in}$ when the superscript is n and by $-a_{in-1}$ when it is $n-1$.

A.4 Matrix Operations

Definitions

A *matrix* is defined as a rectangular array of terms that obeys a prescribed set of operational laws. The following notation for matrices is used:

Square or rectangular matrix: **A** or [A]
Column matrix: **a** or {a}

The size of a matrix is defined by the number of its rows and columns. An $m \times n$ matrix has m rows and n columns. If $m = n$ the matrix is called *square*. If $n = 1$ (one column), the matrix is called a *column matrix* or *column vector*. If $m = 1$, the matrix is called a *row matrix*. The elements of a matrix are identified by two subscripts. In the matrix **A**, a_{ij} is the term in the ith row of **A** and jth column.

$$Example: \quad \overset{\text{2nd row}}{} \begin{bmatrix} 1 & 2 & \overset{\text{3rd column}}{3} & 4 \\ 5 & 6 & 7 & 8 \\ 9 & 10 & 11 & 12 \end{bmatrix} = \mathbf{A}$$

$$a_{23} = 7$$

Symmetric Matrix.   A *symmetric matrix* is a square matrix that has the property that $a_{ij} = a_{ji}$; that is, the term that appears in the ith row and jth column is equal to the term that appears in the jth row and the ith column for all rows and columns.

$$\textit{Example:} \quad \begin{bmatrix} 1 & 2 & 3 & 4 \\ 2 & 5 & 6 & 7 \\ 3 & 6 & 8 & 9 \\ 4 & 7 & 9 & 10 \end{bmatrix} = \mathbf{A} \quad \textit{Note:} \quad \begin{matrix} a_{12} = a_{21} \\ a_{13} = a_{31} \\ a_{14} = a_{41} \\ a_{23} = a_{32} \\ a_{24} = a_{42} \\ a_{34} = a_{43} \end{matrix}$$

Diagonal Term.   A *diagonal term* is a term that appears in the ith row and ith column of a square matrix. In the example above, 1–5–8–10 are diagonal terms.

Diagonal Matrix.   A *diagonal matrix* is a square symmetric matrix that has nonzero terms in the diagonal positions only.

$$\textit{Example:} \quad \begin{bmatrix} 1 & 0 & 0 & 0 \\ 0 & 2 & 0 & 0 \\ 0 & 0 & 0 & 0 \\ 0 & 0 & 0 & 4 \end{bmatrix}$$

Null Matrix.   A *null matrix* is a matrix that has all terms equal to zero. The null matrix usually is denoted as

$$[0] \text{ or } \mathbf{0}$$

Identity Matrix.   An *identity matrix* is a diagonal matrix where every diagonal term is 1. The identity matrix is always given the notation

$$[I] \text{ or } \mathbf{I}$$

Matrix Operations

Addition.   Matrices are added term by term. All matrices must be $m \times n$.

$$\mathbf{A} + \mathbf{B} = \mathbf{C}$$

$$c_{ij} = a_{ij} + b_{ij} \quad \begin{aligned} i &= 1, 2, \ldots, m \\ j &= 1, 2, \ldots, n \end{aligned} \tag{A.12}$$

Subtraction.   Matrices are subtracted term by term. Again all matrices must be $m \times n$.

$$\mathbf{A} - \mathbf{B} = \mathbf{C}$$

$$c_{ij} = a_{ij} - b_{ij} \quad \begin{aligned} i &= 1, 2, \ldots, m \\ j &= 1, 2, \ldots, n \end{aligned} \tag{A.13}$$

705

Multiplication.   Matrices are multiplied by summing the products of the terms of the rows of the premultiplying matrix [A in Eq. (A.14)] with the columns of the postmultiplying matrix [B in Eq. (A.14)] . Taking A to be an $m \times n$ matrix and B a $n \times p$ matrix, the product C matrix will be $m \times p$ and the product is defined as

$$AB = C$$

$$c_{ij} = \sum_{k=1}^{n} a_{ik}b_{kj} \quad \begin{matrix} i = 1, 2, \ldots, m \\ j = 1, 2, \ldots, p \end{matrix} \qquad (A.14)$$

The product is not defined unless the matrices in the product are *conformable;* that is, the number of columns of the premultiplying matrix (A) is equal to the number of rows of the postmultiplying matrix (B). This means that BA is not even defined unless $m = p$. A geometric arrangement of the product of two matrices is shown below to emphasize the row-by-column nature of multiplication.

$$
\begin{bmatrix} a_{11} & a_{12} & a_{13} \\ a_{21} & a_{22} & a_{23} \\ a_{31} & a_{32} & a_{33} \end{bmatrix} \qquad \begin{bmatrix} & & & \\ & & c_{23} & \\ & & & \end{bmatrix}
$$

$$c_{23} = a_{21}b_{13} + a_{22}b_{23} + a_{23}b_{33} \qquad \begin{bmatrix} b_{11} & b_{12} & b_{13} & b_{14} \\ b_{21} & b_{22} & b_{23} & b_{24} \\ b_{31} & b_{32} & b_{33} & b_{34} \end{bmatrix}$$

Note that even if A and B are square matrices of order n, the products AB and BA are not equal, that is,

$$AB = C \neq BA$$

Inverse Matrix.   The matrix B is defined as the *inverse* of the matrix A if the conditions

$$AB = BA = I$$

where I is the identity matrix, are satisfied. The notation

$$B = A^{-1}$$

is used. Thus

$$AB = A^{-1}A = I \qquad (A.15)$$

Singular Matrix.   A square matrix is *singular* if its determinant is zero. Hence A is singular if

$$|A| = 0$$

Matrix Transpose.   If the terms of the matrix **B** are defined as

$$b_{ji} = a_{ij} \quad \begin{matrix} i = 1, 2, \ldots, m \\ j = 1, 2, \ldots, n \end{matrix} \tag{A.16}$$

where the a_{ij} are the terms of the matrix **A** which is an $m \times n$ matrix, then **B** is defined as the *transpose* of **A**, that is,

$$\mathbf{B} = \mathbf{A}^T$$

$$\mathbf{A}^T \text{ is } n \times m$$

Note that the transpose of a symmetric matrix is equal to the original matrix. If **A** is symmetric, then

$$\mathbf{A}^T = \mathbf{A} \tag{A.17}$$

The transpose of a product is given as the product of the transposes in reverse order. Thus

$$(\mathbf{AB})^T = \mathbf{B}^T\mathbf{A}^T \tag{A.18}$$

Also, the transpose of a sum is the sum of the transposes. Thus

$$(\mathbf{A} + \mathbf{B})^T = \mathbf{B}^T + \mathbf{A}^T = \mathbf{A}^T + \mathbf{B}^T \tag{A.19}$$

Determinant of a Product.   The *determinant* of a product of two square matrices is equal to the product of the determinants.

$$|\mathbf{AB}| = |\mathbf{A}| \cdot |\mathbf{B}| \tag{A.20}$$

Differentiation and Integration of Matrices.   If the terms of an $n \times m$ matrix **A** or **B** are functions of x, the following definitions hold:

$$\frac{d}{dx}\mathbf{A} = \mathbf{B} \quad b_{ij} = \frac{da_{ij}}{dx} \quad \begin{matrix} i = 1, 2, \ldots, n \\ j = 1, 2, \ldots, m \end{matrix} \tag{A.21}$$

$$\int \mathbf{A}\, dx = \mathbf{B} \quad b_{ij} = \int a_{ij}\, dx \quad \begin{matrix} i = 1, 2, \ldots, n \\ j = 1, 2, \ldots, m \end{matrix} \tag{A.22}$$

Finally,

$$\frac{d}{dx}(\mathbf{AB}) = \left[\frac{d}{dx}\mathbf{A}\right]\mathbf{B} + \mathbf{A}\left[\frac{d}{dx}\mathbf{B}\right] \tag{A.23}$$

Positive Definite Matrix.   A square matrix, **A**, is considered to be *positive definite* if its quadratic form, Q, given by

$$Q = \frac{1}{2}\mathbf{x}^T\mathbf{A}\mathbf{x} \tag{A.24}$$

707

is positive for all column vectors $\mathbf{x}$ and is zero only if $\mathbf{x}$ is the null matrix. If $\mathbf{A}$ is also symmetric, $\mathbf{A}$ is positive definite if all the determinants that follow are greater than zero.

$$|a_{11}| > 0 \qquad \begin{vmatrix} a_{11} & a_{12} \\ a_{21} & a_{22} \end{vmatrix} > 0 \qquad \begin{vmatrix} a_{11} & a_{12} & a_{13} \\ a_{21} & a_{22} & a_{23} \\ a_{31} & a_{32} & a_{33} \end{vmatrix} > 0$$

$$\vdots$$

$$\begin{vmatrix} a_{11} & a_{12} & \cdots & a_{1r} \\ a_{21} & a_{22} & \cdots & a_{2r} \\ & & \vdots & \\ a_{r1} & a_{r2} & \cdots & a_{rr} \end{vmatrix} > 0$$

$$\vdots$$

$$|\mathbf{A}| > 0$$

The latter definition is extremely useful in structural engineering problems.

Band Width.   If in a square matrix $\mathbf{A}$, all nonzero terms are positioned within r columns on either side of the diagonal term in every row, the matrix is said to be banded with bandwidth $2r + 1$.

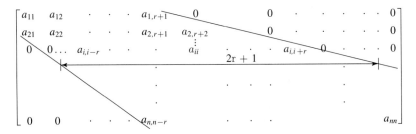

For symmetric matrices a half-bandwidth of $r + 1$ is also frequently used.

Partitioning of Matrices.   In the manipulation of matrices it sometimes is convenient to partition a matrix into two or more submatrices. The partitioning operation is one of subdividing a matrix into rectangular arrays. The division of the $m \times n$ matrix $\mathbf{A}$ into two, three, or four partitioned matrices is illustrated.

$$\left[\begin{array}{c:c} \mathbf{A}_1 \\ m \times r & \mathbf{A}_2 \\ & m \times (n-r) \end{array} \right] \quad \text{or} \quad \left[\begin{array}{c} \mathbf{A}_1 \\ r \times n \\ \hdashline \mathbf{A}_2 \\ (m-r) \times n \end{array} \right] \qquad \text{two submatrices}$$

$$\left[\begin{array}{c:c:c} \mathbf{A}_1 & \mathbf{A}_2 & \mathbf{A}_3 \\ m \times r & m \times p & m \times (n-r-p) \end{array}\right]$$

or

$$\left[\begin{array}{c} \mathbf{A}_1 \\ r \times n \\ \hdashline \mathbf{A}_2 \\ p \times n \\ \hdashline \mathbf{A}_3 \\ (m-r-p) \times n \end{array}\right]$$

three submatrices

$$\left[\begin{array}{c:c} \mathbf{A}_1 & \mathbf{A}_2 \\ r \times p & r \times (n-p) \\ \hdashline \mathbf{A}_3 & \mathbf{A}_4 \\ (m-r) \times p & (m-r) \times (n-p) \end{array}\right]$$

four submatrices

If the matrices in a matrix equation involving multiplication are partitioned, the size of the submatrices must be compatible with the operations in the equation. The partitioning of the matrices $\mathbf{A}$, $\mathbf{B}$, $\mathbf{X}$, $\mathbf{Y}$, and $\mathbf{C}$ in the matrix equation

$$\mathbf{AX} + \mathbf{BY} = \mathbf{C}$$

into the submatrices $\mathbf{A}_{11}$, $\mathbf{A}_{12}$, $\mathbf{A}_{21}$, $\mathbf{A}_{22}$, $\mathbf{B}_{11}$, $\mathbf{B}_{12}$, $\mathbf{B}_{21}$, $\mathbf{B}_{22}$, $\mathbf{X}_1$, $\mathbf{X}_2$, $\mathbf{Y}_1$, $\mathbf{Y}_2$, $\mathbf{C}_1$, and $\mathbf{C}_2$, respectively, is possible as long as the multiplications indicated below are all properly defined.

$$\begin{bmatrix} \mathbf{A}_{11} & \mathbf{A}_{12} \\ \mathbf{A}_{21} & \mathbf{A}_{22} \end{bmatrix} \begin{Bmatrix} \mathbf{X}_1 \\ \mathbf{X}_2 \end{Bmatrix} + \begin{bmatrix} \mathbf{B}_{11} & \mathbf{B}_{12} \\ \mathbf{B}_{21} & \mathbf{B}_{22} \end{bmatrix} \begin{Bmatrix} \mathbf{Y}_1 \\ \mathbf{Y}_2 \end{Bmatrix} = \begin{Bmatrix} \mathbf{C}_1 \\ \mathbf{C}_2 \end{Bmatrix}$$

or, upon expansion,

$$\mathbf{A}_{11}\mathbf{X}_1 + \mathbf{A}_{12}\mathbf{X}_2 + \mathbf{B}_{11}\mathbf{Y}_1 + \mathbf{B}_{12}\mathbf{Y}_2 = \mathbf{C}_1$$
$$\mathbf{A}_{21}\mathbf{X}_1 + \mathbf{A}_{22}\mathbf{X}_2 + \mathbf{B}_{21}\mathbf{Y}_1 + \mathbf{B}_{22}\mathbf{Y}_2 = \mathbf{C}_2$$

which requires that the number of rows in $\mathbf{A}_{11}$, $\mathbf{A}_{12}$, $\mathbf{B}_{11}$, $\mathbf{B}_{12}$, and $\mathbf{C}_1$ be equal, the number of rows in $\mathbf{A}_{21}$, $\mathbf{A}_{22}$, $\mathbf{B}_{21}$, $\mathbf{B}_{22}$, and $\mathbf{C}_2$ be equal, the number of columns of $\mathbf{A}_{11}$ and $\mathbf{A}_{21}$ equal the number of rows of $\mathbf{X}_1$, the number of columns of $\mathbf{A}_{12}$ and $\mathbf{A}_{22}$ equal the number of rows of $\mathbf{X}_2$, the number of columns of $\mathbf{B}_{11}$ and $\mathbf{B}_{21}$ equal the number of rows of $\mathbf{Y}_1$, and finally, the number of columns of $\mathbf{B}_{12}$ and $\mathbf{B}_{22}$ equal the number of rows of $\mathbf{Y}_2$.

A.5 Computations with Matrices: Solution of Linear Equations

General Solution of Simultaneous Equations: Gauss Reduction

The Gauss reduction scheme is a method of successively reducing by elimination the number of equations to be solved until only one equation remains. That equation is solved and a back-substitution process establishes the solution of the remaining equations. For the system of equations that is identical to those defined in Eq. (A.6),

$$\mathbf{AX} = \mathbf{C}$$

the matrix $\mathbf{A}$ can be transformed into the matrix $\mathbf{L}$, which is upper triangular (i.e., $\mathbf{L}$ is a matrix in which all terms below the diagonal are zero). Thus

$$\mathbf{A} = \begin{bmatrix} a_{11} & a_{12} & \cdots & a_{1n} \\ a_{21} & a_{22} & \cdots & a_{2n} \\ & & \vdots & \\ a_{n1} & a_{n2} & \cdots & a_{nn} \end{bmatrix} \longrightarrow \begin{bmatrix} l_{11} & g_{12} & g_{13} & \cdots & g_{1n} \\ 0 & l_{22} & g_{23} & \cdots & g_{2n} \\ 0 & 0 & l_{33} & \cdots & g_{3n} \\ & & & \vdots & \\ 0 & 0 & . & \cdots & l_{nn} \end{bmatrix}$$

This same transformation operates on the vector $\mathbf{C}$, which is transformed into $\mathbf{X}^*$: that is,

$$C = \begin{Bmatrix} c_1 \\ c_2 \\ \vdots \\ c_n \end{Bmatrix} \longrightarrow \begin{Bmatrix} x_1^* \\ x_2^* \\ \vdots \\ x_n^* \end{Bmatrix} = \mathbf{X}^*$$

These transformations are defined by means of the computations given in the expressions

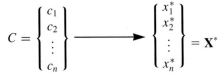

$$
\left.
\begin{aligned}
l_{ii} &= a_{ii} - \sum_{k=1}^{i-1} g_{ik}g_{ki} \\
g_{ij} &= a_{ij} - \sum_{k=1}^{i-1} g_{ik}g_{kj} \quad i \geq j \\
g_{ij} &= \frac{a_{ij} - \sum_{k=1}^{i-1} g_{ik}g_{kj}}{l_{ii}} \quad i < j
\end{aligned}
\;\left.
\begin{aligned}
\\
\\
\end{aligned}
\right\} j = i+1, i+2, \ldots, n
\right\} i = 1, 2, \ldots, n
\tag{A.25}
$$

Note that it is implied here that $l_{11} = a_{11}$, $g_{1k} = a_{1k}/a_{11}$, and $g_{k1} = a_{k1}$, $k = 2, 3, \ldots, n$.

It can be shown that Eq. (A.25) is a formalization of the pivotal condensation technique discussed earlier for evaluation of determinants. Note that the l_{ii} are left in the diagonal position in the **L** matrix, and the size of the **L** matrix is the same as the original **A** matrix. The transformation of **C** to **X*** is given by

$$x_i^* = \frac{c_i - \sum_{k=1}^{i-1} g_{ik} x_k^*}{l_{ii}} \qquad i = 1, 2, \ldots, n \qquad \text{(A.26)}$$

where it is implied that $x_i^* = c_1/a_{11}$. Note that the calculations are identical to the third of Eq. (A.25) if the c_i are considered as a column of the **A** array.

Once the reduction is complete, the backsubstitution process can be initiated. This process transforms **X*** into the solution to the set of equations, that is,

$$\mathbf{X}^* = \begin{Bmatrix} x_1^* \\ x_2^* \\ \cdot \\ \cdot \\ \cdot \\ x_n^* \end{Bmatrix} \longrightarrow \begin{Bmatrix} x_1 \\ x_2 \\ \cdot \\ \cdot \\ \cdot \\ x_n \end{Bmatrix} = \mathbf{X}$$

where

$$x_i = x_i^* - \sum_{k=i+1}^{n} g_{ik} x_k \qquad \text{(A.27)}$$

The x_i are calculated in reverse order with $x_n = x_n^*$. An important by-product of this reduction process is that the value of the determinant of **A** can easily be computed. It can be shown that

$$\left| \mathbf{A} \right| = l_{11} \, l_{22} \, l_{23} \cdots l_{nn} \qquad \text{(A.28)}$$

An operations count shows that the solution of a set of n equations in n unknowns for a single right-hand side **C** requires a total of $(n^3 + 6n^2 + 2n)/3$ multiplications and divisions.

Example A.2. Consider the **AX** = **C** problem, where

$$\mathbf{A} = \begin{bmatrix} 2 & 2 & 4 & -2 \\ 1 & 3 & 2 & 1 \\ 3 & 1 & 3 & 1 \\ 1 & 3 & 4 & 2 \end{bmatrix} \quad \text{and} \quad \mathbf{C} = \begin{Bmatrix} 10 \\ 17 \\ 18 \\ 27 \end{Bmatrix}$$

711

The reduction process of **A** and **C** to produce **L** and $\mathbf{X}^*$ is summarized step by step below.

x_1	x_2	x_3	x_4	c	
2	2	4	-2	10	row a
1	3	2	1	17	row b
3	1	3	1	18	row c
1	3	4	2	27	row d

Step 1 ($i = 1$)

x_1	x_2	x_3	x_4	c	
②	1	2	-1	5	$a_1 = a/2$
0	2	0	2	12	$b_1 = b - 1 \cdot a_1$
0	-2	-3	4	3	$c_1 = c - 3a_1$
0	2	2	3	22	$d_1 = d - 1 \cdot a_1$

Step 2 ($i = 2$)

x_1	x_2	x_3	x_4	c	
②	1	2	-1	5	$a_2 = a_1$
0	②	0	1	6	$b_2 = b_1/2$
0	0	-3	6	15	$c_2 = c_1 - (-2)b_2$
0	0	2	1	10	$d_2 = d_1 - 2b_2$

Step 3 ($i = 3$)

x_1	x_2	x_3	x_4	c	
②	1	2	-1	5	$a_3 = a_1$
0	②	0	1	6	$b_3 = b_2$
0	0	⊖③	-2	-5	$c_3 = c_2/(-3)$
0	0	0	5	20	$d_3 = d_2 - 2c_3$

Step 4 ($i = 4$)

x_1	x_2	x_3	x_4	c	
②	1	2	-1	5	$a_4 = a_1$
0	②	0	1	6	$b_4 = b_2$
0	0	⊖③	-2	-5	$c_4 = c_3$
0	0	0	⑤	4	$d_4 = d_3/5$

$x_4 = 4$

$x_3 = -5 - (-2)(4) = 3$

$x_2 = 6 - (0)(3) - (1)(4) = 2$

$x_1 = 5 - (1)(2) - (2)(3) - (-1)(4) = 1$

The determinant of **A** is given by

$$|\mathbf{A}| = (2)(2)(-3)(5) = -60$$

The circled l_{ii} terms are given that designation to indicate that the division of the diagonal term by its circled value gives the value 1 that is required as part of the reduction process for the terms below it. However, there is no need to execute this division in the sequences of operations, and the saving of the values of the l_{ii} provides the terms required to compute the determinant of **A**.

Solution of Symmetric Simultaneous Equations: Gauss Reduction

For symmetric matrices the number of operations in the Gauss reduction process can be reduced by nearly one-half since the g_{ij} will be symmetric in the process. In this case the reduction is defined by the expressions

$$
\left.
\begin{array}{l}
l_{ii} = a_{ii} - \displaystyle\sum_{k=1}^{i-1} \frac{h_{ki}^2}{l_{kk}} \\[3mm]
h_{ij} = a_{ij} - \displaystyle\sum_{k=1}^{i-1} \frac{h_{ki}h_{kj}}{l_{kk}} \\[3mm]
d_i = c_i - \displaystyle\sum_{k=1}^{i-1} h_{ki}x_k^* \\[3mm]
x_i^* = \dfrac{d_i}{l_{ii}}
\end{array}
\;\middle|\;
\begin{array}{l}
j = i+1, i+2, \ldots, n
\end{array}
\;\middle|\;
\begin{array}{l}
i = 1, 2, \ldots, n
\end{array}
\right\}
\tag{A.29}
$$

The back-substitution process is identical to the one described previously in Eq. (A.27). For the symmetric matrix problem the number of operations is reduced to $(n^3 + 12n^2 + 11n - 6)/6$, which for large n is about half the number of operations for the unsymmetrical **A**.

Example A.3

x_1	x_2	x_3	x_4	c	
1	2	3	4	1	row a
2	6	8	16	0	row b
3	8	12	30	6	row c
4	16	30	160	10	row d

713

Step 1 ($i = 1$)

x_1	x_2	x_3	x_4	c	
①	2	3	4	1	$a_1 = a/1$
	2	2	8	-2	$b_1 = b - 2 \cdot a_1$
		3	18	3	$c_1 = c - 3 \cdot a_1$
			144	6	$d_1 = d - 4 \cdot a_1$

Step 2 ($i = 2$)

x_1	x_2	x_3	x_4	c	
①	2	3	4	1	$a_2 = a_1$
	②	1	4	-1	$b_2 = b_1/2$
		1	10	5	$c_2 = c_1 - 2 \cdot b_2$
			112	14	$d_2 = d_1 - 8 \cdot b_2$

Step 3 ($i = 3$)

x_1	x_2	x_3	x_4	c	
①	2	3	4	1	$a_3 = a_1$
	②	1	4	-1	$b_3 = b_2$
		①	10	5	$c_3 = c_2/1$
			12	-36	$d_3 = d_2 - 10 \cdot c_3$

Step 4 ($i = 4$)

x_1	x_2	x_3	x_4	c	
①	2	3	4	1	$a_4 = a_1$
	②	1	4	-1	$b_4 = b_2$
		①	10	5	$c_4 = c_3$
			⑫	-3	$d_4 = d_3/12$

Back substitution

$x_4 = -3$

$x_3 = 5 - (10)(-3) = 35$

$x_2 = -1 - (4)(-3) - (1)(35) = -24$

$x_1 = 1 - (4)(-3) - (3)(35) - (2)(-24) = -44$

Note that $\left| \mathbf{A} \right| = (1)(2)(1)(12) = 24$. Again the circled l_{ii} terms carry the same meaning as they do in the reduction process described in Example A.2 of the unsymmetric simultaneous equation solution.

References

1. AMERICAN ASSOCIATION OF STATE HIGHWAY AND TRANSPORTATION OFFICIALS, *Standard Specifications for Highway Bridges,* 14th ed., AASHTO, Washington, D.C., 1989.

2. AMERICAN CONCRETE INSTITUTE, *Building Code and Commentary,* ACI 318–92/318R–92, ACI, Detroit, Mich., 1992.

3. AMERICAN INSTITUTE OF STEEL CONSTRUCTION, *Manual of Steel Construction, Load and Resistance Factor Design,* AISC, Chicago, 1986.

4. AMERICAN SOCIETY OF CIVIL ENGINEERS, *Minimum Design Loads for Buildings and Other Structures, ASCE 7–93,* ASCE, New York, 1993.

5. BATHE, K., *Finite Element Procedures in Engineering Analysis,* Prentice Hall, Englewood Cliffs, N.J., 1982.

6. BAZANT, Z. P., and CEDOLIN, L., *Stability of Structures,* Oxford University Press, New York, 1991.

7. BEER, F. P., and JOHNSTON, E. R., *Mechanics of Materials,* 2nd ed., McGraw-Hill, New York, 1992.

8. BLEICH, F., *Buckling Strength of Metal Structures,* McGraw-Hill, New York, 1952.

9. BORESI, A. P., SCHMIDT, R. J., and SIDEBOTTOM, O. M., *Advanced Mechanics of Materials,* 5th ed., John Wiley, New York, 1993.

10. CLOUGH, R. W., and PENZIEN, J., *Dynamics of Structures,* 2nd ed., McGraw-Hill, New York, 1993.

11. COOK, R. D., MALKUS, D. S., and PLESHA, M. E., *Concepts and Applications of Finite Element Analysis,* 3rd ed., John Wiley, New York, 1989.

12. CRANDALL, S. H., *Engineering Analysis: A Survey of Numerical Procedures,* McGraw-Hill, New York, 1956.

13. CURTIS, G., *Applied Numerical Analysis,* 2nd ed., Addison-Wesley, Reading, Mass.

14. FUNG, Y. C., *Foundations of Solid Mechanics,* Prentice Hall, Englewood Cliffs, N.J., 1965.

15. GALAMBOS, T., *Structural Members and Frames,* Prentice Hall, Englewood Cliffs, N.J., 1968.

16. GOLUB, G. H., and VANLOAN, C. F., *Matrix Computations,* 2nd ed., Johns Hopkins University Press, Baltimore, 19.

17. HIBBELER, R. C., *Mechanics of Materials,* Macmillan, New York, 1991.

18. HIBBELER, R. C., *Structural Analysis,* 3rd ed., Prentice Hall, Englewood Cliffs, N.J., 1995.

19. HILDEBRAND, F. B., *Introduction to Numerical Analysis,* McGraw-Hill, New York, 1956.

20. HILDEBRAND, F. B., *Methods of Applied Mathematics,* Prentice Hall, Englewood Cliffs, N.J., 1952.

21. INTERNATIONAL CONFERENCE OF BUILDING OFFICIALS, *Uniform Building Code,* ICBO, Whittier, Calif., 1991.

References

22. KINNEY, J. S., *Indeterminate Structural Analysis,* Addison-Wesley, Reading, Mass., 1957.

23. LAIBLE, J. P., *Structural Analysis,* Holt, Rinehart and Winston, New York, 1985.

24. MATHESON, J. A. L., *Hyperstatic Structures: An Introduction to the Theory of Statically Indeterminate Structures,* Vol. 1, Academic Press, New York, 1959.

25. MAUGH, L. C., *Statically Indeterminate Structures,* John Wiley, New York, 1946.

26. MUVDI, B. B., and MCNABB, J. W., *Engineering Mechanics of Materials,* Macmillan, New York, 1980.

27. NAKAI, H., and YOO, C. H., *Analysis and Design of Curved Steel Bridges,* McGraw-Hill, New York, 1988.

28. NERVI, P. L., *The Works of Pier Luigi Nervi,* F.A. Praeger, New York, 1951.

29. NERVI, P. L., *Buildings, Projects, Structures,* F.A. Praeger, New York, 1963.

30. NORRIS, C. H., WILBUR, J. B., and UTKU, S., *Elementary Structural Analysis,* 4th ed., McGraw-Hill, New York, 1991.

31. OLEINIK, J. C., and HEINS, C. P., Diaphragms for curved box bridges, *Proceedings of the American Society of Civil Engineers,* Vol. 101, No. ST-10, pp. 2161–2178, Oct. 1975.

32. PARCEL, J. I., and MOORMAN, R. B. B., *Analysis of Statically Indeterminate Structures,* John Wiley, New York, 1955.

33. PAZ, M., *Structural Dynamics Theory and Computation,* 3rd ed., Van Nostrand Reinhold, New York, 1991.

34. PRESS, W. H., FLANNERY, B. P., TEUKOLSKY, S. A., and VETTERING, W. T., *Numerical Recipes,* Cambridge University Press, New York, 1986.

35. RAMBERG, W., and OSGOOD, W. R., *Description of Stress–Strain Curves by Three Parameters,* Technical Note 902, National Advisory Committee on Aeronautics, Washington, D.C., 1943.

36. STORRER, W. A., *The Architecture of Frank Lloyd Wright,* 2nd ed., MIT Press, Cambridge, Mass., 1978.

37. STRANG, G., *Linear Algebra and Its Applications,* 2nd ed., Academic Press, New York, 1980.

38. TARANATH, B. S., *Structural Analysis and Design of Tall Buildings,* McGraw-Hill, New York, 1988.

39. TIMOSHENKO, S. P., and GERE, J. M., *Theory of Elastic Stability,* 2nd ed., McGraw-Hill, New York, 1961.

40. TIMOSHENKO, S. P., and GOODIER, J. N., *Theory of Elasticity,* 3rd ed., McGraw-Hill, New York, 1970.

41. TORROJA, E. M., *The Structures of Edwardo Torroja,* F.W. Dodge Corp., New York, 1958.

42. VLASOV, V. Z., *Thin-Walled Elastic Beams,* National Science Foundation, Washington, D.C., 1961.

43. WANG, C. K., *Statically Indeterminate Structures,* McGraw-Hill, New York, 1953.

44. WEST, H. R., *Fundamentals of Structural Analysis,* John Wiley, New York, 1992.

Answers to Selected Problems

CHAPTER 2

2.1　a) Stable, Indeterminate; b) Unstable; c) Stable, Indeterminate; d) Stable, Determinate; e) Unstable;
f) Stable, Determinate; g) Stable, Indeterminate; h) Stable, Determinate; i) Unstable; j) Unstable

2.4　$H_1 = 0$; $V_1 = 30^k \uparrow$; $V_2 = 100^k \uparrow$; $V_4 = 30^k \uparrow$;

2.7　$H_A = 20 \text{Kn} \leftarrow$; $V_A = 11.7 \text{kN} \uparrow$; $V_B = 63.3 \text{kN} \uparrow$; $V_C = 45 \text{kN} \uparrow$;

2.10　$H_1 = 4^k \leftarrow$; $V_1 = 2.5^k \uparrow$; $H_4 = 0$; $V_4 = 7.5^k \uparrow$;

2.13　$H_A = 135 \text{kN} \rightarrow$; $V_A = 0$; $H_B = 165 \text{kN} \leftarrow$; $V_B = 50 \text{kN} \uparrow$;

2.15　$H_A = 0$; $V_A = 80 \text{kN} \uparrow$; $V_B = 320 \text{kN} \uparrow$; $V_E = 106.7 \text{kN} \uparrow$; $V_F = 26.7 \text{kN} \downarrow$

2.19　$V_A = 0$; $V_B = 40^k \uparrow$; $H_B = 0$; $V_C = 0$; $V_D = 20^k \downarrow$;

2.21　$H_1 = 42^k \leftarrow$; $V_1 = 14.4^k \downarrow$; $H_5 = 12^k \leftarrow$; $V_5 = 14.4^k \uparrow$;

CHAPTER 3

3.2　$L_0 L_1 = -120 \text{kN}$; $L_1 L_0 = 0$; $U_o U_1 = 0$; $U_1 U_2 = 120 \text{kN}$; $L_0 L_0 = 0$; $L_1 = U_1 = -80 \text{kN}$; $L_2 U_2 = 80 \text{kN}$;
$L_0 L_1 = 144.2 \text{kN}$; $L_1 L_2 = -144.2 \text{kN}$;

3.6　$L_0 L_1 = 15^k$; $L_1 L_2 = 15^k$; $L_2 L_3 = 5^k$; $L_3 L_4 = 5^k$; $L_0 U_1 = -21.21^k$; $U_1 U_2 = -7.91^k$; $U_2 U_3 = -7.91^k$;
$U_2 L_3 - 7.07^k$; $L_1 U_1 = 20^k$; $L_2 U_2 = 5^k$; $L_3 U_3 = 0$;

3.10　$F_a = 11.1^k$; $F_b = 22.2^k$; $F_c = 0$; $F_d = 17.8^k$; $F_e = -26.7^k$;

3.14　$F_a = 46.25 \text{kN}$; $F_b = 82.79 \text{kN}$; $F_c = 47.14 \text{kN}$; $F_d = -56.35 \text{kN}$;

3.17　$F_a = 4.11^k$; $F_b = 0$; $F_c = -1.0^k$; $F_d = -4.0^k$;

3.20 $F_a = -64.3\text{kN}$; $F_b = 38.6\text{kN}$; $F_c = -22.5\text{kN}$;

3.23 $F_a = 9.52^k$; $F_b = 13.81^k$; $F_c -47.62^k$; $F_d = 2.25^k$;

3.25 $F_a = 2.4^k$; $F_b = -3.5^k$; $F_c = -4.0^k$; $H_1 = 5.7^k \leftarrow$; $V_1 = 155^k \downarrow$; $H_2 = 2.3^k \leftarrow$; $V_2 = 15^k \uparrow$;

CHAPTER 4

4.2	x	0	7.5	20	32.5	40	feet
	Shear	30	0	−50, 50	0	−30	Kips
	Moment	0	112.5	−200	112.5	0	Kip-ft.

4.6	x	0	2	6	10	12	meters
	Shear	100	100, 60	60, 20	20, −20	−20	kN
	Moment	−480	−280	−40	40	0	kN-m.

4.9	x	0	2	3	5	6.5	7.5	meters
	Shear	0	0,−68.9	−68.9	31.1	31.1,160	160,0	kN
	Moment	−100	−100	−168.9	−206.7	−160	0	kN-m.

4.13	x	0	2	5	11	meters
	Shear	−90	−90	−120	0	kN
	Moment	990	810	480	0	kN-m.

4.15	x	0	5	8	10	feet
	Shear 1-2	0	−10	−16	−20	Kips
	Moment 1-2	195	145	131	95	Kip-ft.
	Shear 2-3	−25	−25, 10	10, 0	——	Kips
	Moment 2-3	95	−30	0	——	Kip-ft.

4.18	x	0	10	20	30	feet
	Shear	3.33	−6.67	−20.7, 16.7	16.7, −3.33	Kips
	Moment	0	0	−166.7	0	Kip-ft.

	x	40	45	50	feet
	Shear	−3.33, −16.7	−16.7, 13.3	13.3	Kips
	Moment	−33.33	−66.67	0	Kip-ft.

4.21	x	0	5	10	15	20	meters
	Shear	100	100, −50	−50, −200	−200, 150	150	kN
	Moment	0	500	250	−750	0	kN-m.

4.24	x	0	5	9	meters
	Force Diagram	32	32	0	kN

4.25 | x | 0 | 15 | 30 | feet
Force Diagram | 45 | 33.75 | 0 | Kips

CHAPTER 5

5.1 $M_{CD} = 136$ k-ft compression on bottom

$M_{DB} = 98.5$ k-ft compression on left

5.4 $M_A = 112.5$ kN-m compression on left

$M_{CB} = 37.5$ kN-m compression on top

$M_{CD} = 37.5$ kN-m compression on right

5.7 $V_{AB} = 15$ Kips

$V_{BC} = 17$ Kips @ B; -17 Kips @ C

$V_{CD} = -15$ Kips

$M_{BA} = M_{BC} = M_{CB} = M_{CD} = 255$ k-ft. compression on bottom

5.9 $M_{BA} = 18.6$ kN-m compression on left

$M_{BC} = 18.6$ kN-m compression on top

$M_{DC} = 48.9$ kN-m compression on bottom

$M_{DE} = 48.9$ kN-m compression on left

$M_F = 48.9$ kN-m compression on right

5.12 $M_{12} = M_{13} = 0$

$M_{43} = 200$ k-ft compression on bottom

$M_{45} = 258.7$ k-ft compression on right

$M_{46} = 458.7$ k-ft compression on bottom

$M_{76} = 258.7$ k-ft compression on top

$M_{78} = 258.7$ k-ft compression on right

5.16 $M_{BA} = 18.6$ kN-m compression on left

$M_{BE} = 18.6$ kN-m compression on top

$M_{CB} = 126.75$ kN-m compression on left

719

$M_{CD} = 126.75$ kN-m compression on top

$M_{DC} = 126.75$ kN-m compression on bottom

$M_{DE} = 126.75$ kN-m compression on left

$M_{EB} = 234.9$ kN-m compression on bottom

$M_{EF} = 234.9$ kN-m compression on left

5.18 $M_{CB} = M_{CD} = 400$ kN-m compression on bottom

$M_{ED} = M_{EF} = 0$

CHAPTER 6

6.1 $\Theta_B = 0.00133$ rad. counterclockwise; $v_C = 0.04.8$ in. ↑

6.4 $\theta_C = \dfrac{500}{EI}$ clockwise; $v_C = \dfrac{300}{EI}$ ↓

6.7 $\theta_{B\ \text{left}} = \dfrac{562.5}{EI}$ clockwise; $v_B = \dfrac{5625}{EI}$ ↓ ; $\Delta\theta_B = \dfrac{1062.5}{EI}$

6.10 $v_C = 8.9$ mm ↓ ; $v_{\text{center}} = 6.0$ mm ↓

6.14 $T_A = \dfrac{160}{EI}$ counterclockwise; $v = \dfrac{800}{EI}$ ↑ ; $\Delta\theta = \dfrac{560}{EI}$

6.17 $v_{\text{max}} = \dfrac{339.4}{EI}$ ↑ at 8.485 ft. from A

6.19 $v_C = 0.0166$ ft. ↓

6.22 $u = 0.0037$ ft. →

6.25 $u = \dfrac{250}{AE}$ →

CHAPTER 7

7.1 $v_{L1} = 10.64$ mm ↓

7.3 $v_{u3} = 0.03$ ft. ↓ ; $u_{U3} = 0.003$ ft. →

7.6 $v_{L2} = 0.00448$ ft. ↓

7.10 $u_{L8} = 45$ mm→

7.13 $v_{L1} = 0$

7.16 $v_a = \dfrac{272.6}{EI} \downarrow$; $v_c = \dfrac{41.48}{EI} \uparrow$; $\theta_d = \dfrac{11.85}{EI}$ clockwise

7.19 $u = \dfrac{1200.5}{EI} \leftarrow$

7.23 $u = 0.001664$ ft. →

CHAPTER 8

8.1

x	0	3.5	7	9	meters
R_A	1.0	0.5	0	$-2/7$	Force
R_B	0	0.5	1.0	$9/7$	Force
M_B	0	0	0	-2.0	Force-m
$V_{B\,\text{left}}$	0	0.5	1.0, 0	$2/7$	Force
M	0	1.75	0	-1.0	Force-m

8.3

x	0	5	10	13	20	feet
R_A	1.0	0	0	0	0	Force
R_E	0	-0.5	0	0.3	1.0	Force
V_D	0	0.5	0	$-0.3, 0.7$	0	Force
M_D	0	-3.5	0	2.1	0	Force-ft.

8.7 Maximum positive $= 12.75$; Maximum negative $= -21.0$

8.10

x	0	24	32	48	64	80	feet
M_D	-5.33	2.67	5.33	10.7	0	-10.7	Force-ft.
V_{DE}	0.33	-0.17	-0.33	-0.67	0	-0.33	Force
M_C	-10.7	5.33	10.7	5.33	0	-5.33	Force-ft.
V_{BC}	0.33	-0.17	0.67	0.33	0	-0.33	Force

8.12

x	0	12	30	meters
V_{CD}	0.25	$-0.35, 0.65$	-0.25	Force
M_C	-3.75	3.25	-1.25	Force-m

8.17	x	0	100	200	feet
	F_A	0	-1.82	0	Force

8.20	x	0	28	56	63	98	meters
	R_C	0	-0.667	0	0.167	1.0	Force
	F_a	0	1.491	0	-1.863	0	Force
	F_b	0	-2.67	0	0	0	Force

CHAPTER 9

9.1   a) $v = \dfrac{0.1875 \, PL^3}{EI} \downarrow$;   b) $v = \dfrac{0.1939 \, PL^3}{EI} \downarrow$

9.4   Eq. 9.3: $0.0268 \dfrac{PL^3}{EI_0}$;   Eq. 9.2: $0.0289 \dfrac{PL^3}{EI_0} \downarrow$

9.7 $v_1 \times \dfrac{wl^4}{EI_0} \downarrow$ $\theta_1 \times \dfrac{wl^3}{EI_0}$ counterclockwise

Eq. 9.3   0.01972   0.0286

Eq. 9.2   0.02043   0.0289

9.10 $v = \dfrac{183.02}{E} \downarrow$ (Eq. 9.3, $n = 4$)

9.13   a) u $= 0.75 \dfrac{PL}{AE} \rightarrow$; b)   Eq. 9.3   Eq. 9.2   Eq. 9.5   Eq. 9.6

$u \times \dfrac{PL}{AE}$ 0.7039   0.7042   0.7038   0.7039

9.16 $u = 0.4307 \dfrac{pL^2}{AE} \rightarrow$ (Eq. 9.3 $n = 4$)

9.19 $u = 3.418 \in_0 L \rightarrow$

9.22 $\theta_1 = 0.3851 \, \phi_0 L \, c'$ clockwise; $v_1 = 0.3038 \, \phi_0 L^2 \downarrow$ (Eq. 9.3, $n = 4$)

9.25 $\theta = 0.571 \, \phi_0 L \, c'$ clockwise; (Eq. 9.3, $n = 4$); $v = 0.4072 \, \phi_0 L^2$exact

CHAPTER 10

10.1 $H_{UO} = 83^k \uparrow$; $V_{UO} = 44^k \leftarrow$; $H_{LO} = 44^k \rightarrow$; $V_{L2} = 17^k \uparrow$;
$F_{UQU1} = F_{U1U2} = -22.7^k$; $F_{L0L2} = 0$; $F_{L0U0} = 33^k$; $F_{L2U2} = -17^k$;
$F_{M1U1} = -100^k$; $F_{L0M1} = -55^k$; $F_{M1U2} = 28.5^k$; $F_{M1U0} = 83,3^k$

10.4 $u_c = \dfrac{3968}{AE} \rightarrow$

10.7 $F_{df} = 39.4\text{kN}$

10.10 $H = 6.26^k$

10.13 $M_A = 12.02\text{kN-m}$ compression on bottom;
$M_{BA} = M_{BC} = 3.711\text{kN-m}$ compression on top

10.16 $M_A = 0$
$M_{BA} = 7.82\text{k-ft}$ compression on right
$M_{BC} = 7.82\text{k-ft}$ compression on bottom
$M_{CB} = M_{CD} = 9.16\text{k-ft}$ compression on bottom

10.19

x	0	3.33	6.67	10	13.33	13.67	20	feet
Moment	0	2.84	1.73	21.1	18.0	27.5	0	k-ft

10.23 $M_{BA} = M_{CD} = 101.7\text{kN-m}$ compression on left
$M_{BC} = 101.7\text{kN-m}$ compression on top
$M_{CB} = 101.7\text{kN-m}$ compression on bottom

10.25 $M_a = 124.3\text{k-ft}$ compression on right
$M_{ba} = 27.42\text{k-ft}$ compression on left
$M_{bc} = 27.42\text{k-ft}$ compression on top
$M_{cb} = 1.19\text{k-ft}$ compression on bottom
$M_{cd} = 1.19\text{k-ft}$ compression on left

10.30 $M_{BA} = 12\text{kN-m}$ compression on right
$M_{BC} = M_{CB} = 12\text{kN-m}$ compression on bottom
$M_{CD} = 12\text{kN-m}$ compression on left

CHAPTER 11

11.1 $M_{BA} = -M_{BC} = -41.66\text{kN-m}; \; M_{CB} = -M_{CD} = -41.66\text{kN-m}$

11.3 $M_{AB} = 30.68\text{kN-m}; \; M_{BA} = -M_{BC} = -6.14\text{kN-m}$

11.6 $M_{AB} = 80\text{k-ft}; M_{BA} = -20\text{k-ft}; M_{BC} = 12\text{k-ft}; M_{BD} = 8\text{k-ft}; M_{DB} = 4\text{k-ft}$

11.9 $M_{AB} = -163.8\text{k-ft}; M_{BA} = -M_{BC} = 26.85\text{k-ft}$

11.12 $M_{21} = -M_{23} = -347.8\text{kN-m}$

11.16 $M_{AC} = -19.96\text{k-ft}; \quad M_{CA} = -39.93\text{k-ft}; \quad M_{CB} = -26.47\text{k-ft}; \quad M_{CD} = 66.4\text{k-ft}$

11.19 $M_{AB} = -2M/9; \; M_{CB} = M/9$

11.22 $M_{AB} = M_{DC} = 198\text{k-ft}; M_{BA} = M_{CD} = 177\text{k-ft}; M_{BC} = M_{CB} = -177\text{k-ft}$

CHAPTER 12

12.1 $M_{AB} = 16.62\text{k-ft}; M_{BA} = -M_{BC} = -2.77\text{k-ft}$

12.4 $M_{21} = -M_{12} = 63.07\text{kN-m}$

12.7 $M_{21} = -M_{23} = -118\text{k-ft}; M_{32} = -M_{34} = -166\text{k-ft}; M_{43} = -M_{45} = 118\text{k-ft}$

12.9 $M_{BA} = -M_{BC} = -62.5\text{kN-m}; M_{CB} = -12.1\text{kN-m}; M_{CF} = 4.9\text{kN-m}; M_{CD} = 7.2\text{kN-m};$
     $M_{DC} = -M_{DE} = 1.7\text{kN-m}$

12.13 $M_{AB} = -M_{BC} = -68.6\text{k-ft}; M_{CB} = -M_{CD} = 27.4\text{k-ft}$

12.16 $M_{BA} = -M_{BC} = 151.2\text{kN-m}; M_{CB} = -M_{CD} = -172.8\text{kN-m};$

12.20 $M_{AB} = 255.2\text{kN-m}; M_{BA} = -M_{BC} = 94.4\text{kN-m}$

12.23 $M_{12} = -M_{14} = M_{34} = M_{32} = 61.0\text{k-ft}; M_{41} = -M_{43} = -30.5\text{k-ft}; M_{21} = -M_{23} = -152.6\text{k-ft}; M_{24} = M_{42} = 0$

12.26 $M_{12} = -179\text{k-ft}; M_{21} = -M_{23} = -266\text{k-ft}; M_{32} = -M_{34} = -222\text{k-ft}$

12.27 $M_{21} = -M_{24} = -148.5\text{kN-m}; \quad M_{42} = -55.4\text{kN-m}; \quad M_{43} = -338.9\text{kN-m}; \quad M_{54} = -M_{56} = -102.6\text{kN-m};$
     $M_{34} = -344.3\text{kN-m}; M_{45} = 394.2\text{kN-m};$

12.31 $M_{AB} = 612\text{k-ft}; M_{BA} = -M_{BC} = 262\text{k-ft}; M_{CB} = -M_{CD} = -112\text{k-ft}; M_{DC} = 138\text{k-ft}$

CHAPTER 13

13.2 a) $R_A = 0.9375^k \downarrow$; $R_B = 6.88^k \uparrow$; $R_C = 906^k \uparrow$;

b) $R_A = 16.09^k \uparrow$; $R_B = 42.84^k \uparrow$; $R_C = 1.084^k \uparrow$

13.4 a) $x = 28.453$ m $R_A = 9.623$ kN $\downarrow$; b) 437.5 kN $\uparrow$

13.6

x/L	0	1/6	1/3	1/2	2/3	5/6	1	7/6	4/3
M_B/L	0	−0.054	−0.095	−0.111	−0.111	−0.066	0	−0.066	−0.101
R_A	1.0	0.780	0.572	0.389	0.233	0.101	0	−0.066	−0.101

x/L	3/2	5/3	11/6	2
M_B/L	−0.111	−0.095	−0.054	0
R_A	−0.111	−0.095	−0.054	0

13.12 a) 117.8^k Tension; b) 164.9^k Tension

CHAPTER 14

14.2 Girder end moments: 150k-ft both girders

Column end moments: 150k_ft exterior, 300k-ft interior

14.5 Upper story: Girder and column end moments $3P/4$

Lower story: Middle girder end moments $9P/4$

Outside girder end moments $3P/2$

Exterior column end moments (top) $3P/2$

Interior column end moments (top) $3P$

14.8 Upper story girder and column end moments 36k-ft

Lower story girder end moments 144k-ft

Lower story column end moments 108k-ft

14.10 Upper story girder and column end moments $PL/4$

Lower story girder end moments $5PL/4$

Lower story column end moments (top) PL

14.14 Portal Method Member end moments are positive
counterclockwise on end in units of kN-m

1-2: $-325/3$; 3-4: $-350/3$; 5-6: $32/3$; 7-8: $350/3$;

1-3, 2-4: $325/3$; 3-5, 4-6: $25/3$; 5-7, 6-8: $-350/3$

CHAPTER 15

15.1 $u = 0.4 \dfrac{PL}{A_0 E}$

15.4 a) $u = \dfrac{(1 + 2a)}{2} \dfrac{PL}{AE}$; b) $u = \dfrac{1}{2(1 - a)} \dfrac{PL}{AE}$

15.6 $v = 0.02835 \dfrac{PL^3}{EI_0} \downarrow$

15.11 $v_1 = \dfrac{48771}{E} \downarrow$

15.15 $u = \dfrac{PL^3}{9EI}$ $M_{21} = M_{43} = M_{65} = -M_{24} = -M_{64} = PL/3$; $M_{42} = M_{46} = -PL/6$

15.18 Single Bay Building: $u = \dfrac{5PL^3}{36EI} \rightarrow$; $\theta_{\text{Base}} = \dfrac{-7PL^2}{36EI}$; $\theta_{\text{Top}} = \dfrac{-7PL^2}{36EI}$

$M_{24} = M_{42} = M_{46} = M_{64} = -PL/4$; $M_{21} = M_{43} = M_{65} = PL/3$

CHAPTER 16

16.3 $u_3 = 0.01664$ in.; $u_4 = 0.017645$ in.; $v_3 = v_4 = 0$;

$S_{23} = S_{13} = -6.187^k$; $S_{24} = S_{14} = 6.569^k$; $S_{34} = 2.426^k$

16.6 Displacements in mm and forces in kN. $u_1 = -1.333$;

$v_1 = -15.26$; $u_2 = -0.6665$; $v_2 = -2.574$; $u_3 = 0.3893$; $v_3 = -2.532$; $S_{21} = -200$; $S_{31} = 206.2$; $S_{23} = 16.70$;

$S_{42} = -266.6$; $S_{43} = -34.42$; $S_{52} = 74.47$; $S_{53} = 240.6$

16.9 Displacements in mm. $u_1 = -7.333$; $v_1 = -39.26$; $u_2 = -6.6665$; $v = -14.574$; $u_3 = -2.6107$;

$v_3 = -14.532$; $u_4 = -6.00$; All member forces unchanged from 16.6.

CHAPTER 17

17.1 $u_2 = u_3 = 0$; $\theta_2 = 0.003678$ rad.; $\theta_3 = -0.001839$ rad.

17.3 $u_1 = u_3 = 0$; $\theta_1 = -0.005859$ rad.; $\theta_2 = 0.003906$ rad.;
$\theta_3 = -0.001953$ rad.

17.5 $u_1 = u_3 = 0$; $v_1 = -6.0$ mm; $\theta_1 = -0.005109$ rad.;
$\theta_2 = 0.004206$ rad.; $\theta_3 = -0.002103$ rad.

17.10 $u_3 = 0.0007953\dfrac{wL^4}{EI}$; $v_3 = -0.004394\dfrac{wL^4}{EI}$; $\theta_3 = 0.002111\dfrac{wL^3}{EI}$

Index

INDEX

Index

Index